XIX International Symposium in Honour of Academician M. A. Usov: Problems of Geology and Subsurface Development 2015

IOP Conference Series: Earth and Environmental Science
Volume 27

Tomsk, Russia
6 – 10 April 2015

Editors:

G. M. Ivanova

ISBN: 978-1-5108-1632-9
ISSN: 1755-1307

Printed from e-media with permission by:

Curran Associates, Inc.
57 Morehouse Lane
Red Hook, NY 12571

Some format issues inherent in the e-media version may also appear in this print version.

Printed by Curran Associates, Inc. (2016)

For permission requests, please contact the Institute of Physics
at the address below.

Institute of Physics
Dirac House, Temple Back
Bristol BS1 6BE UK

Phone: 44 1 17 929 7481
Fax: 44 1 17 920 0979

techtracking@iop.org

Additional copies of this publication are available from:

Curran Associates, Inc.
57 Morehouse Lane
Red Hook, NY 12571 USA
Phone: 845-758-0400
Fax: 845-758-2634
Email: curran@proceedings.com
Web: www.proceedings.com

Table of contents

Volume 27

XIX International Scientific Symposium in honor of Academician M.A. Usov "Problems of Geology and Subsurface Development"
6–10 April 2015, Tomsk, Russia

Published online: 10 November 2015

Preface

Papers

Mineralogy, Geochemistry and Petrography

Ore Deposits. Current Prospecting and Exploration Methods

Petroleum Geology. Current Exploration and Extraction Methods

Geoecology and Environmental Protection. Geoinformation Systems in Geoecology and Environmental Protection

Latest Researches in Reservoir Engineering

Latest Researchers in Hydrocarbon Refining

Mining. Ore and Non-metallic Minerals Development

Modern Oilfield Equipment and its Upgrading

Latest Technology of Oil-and-Gas Pipelines and Petroleum Storage

Preface

XIX International Scientific Symposium in honor of Academician M.A. Usov "Problems of Geology and Subsurface Development" (for students and young scientists) was organized under the guidance of the Ministry of Education and Science of the Russian Federation and the Russian Foundation for Fundamental Research within the National Research Tomsk Polytechnic University (NR TPU). TPU is one of the oldest technical higher education institutions in Russia, training specialists in the domain of geoscience and enhancing their further research in this area. The Institute of Natural Resources, National Research Tomsk Polytechnic University (INR TPU) was chosen to hold the International Scientific Symposium. In 2015 the Institute of Natural Resources celebrated its 114th anniversary. It was founded by V.A. Obruchev in 1901, first Siberian geologist, member of USSR Academy of Sciences, Hero of Socialist Labor, and first Laureate of the Lenin Prize. He was recognized as a prominent scientist in the area of geology all over the world.

INR is the first institute of geological education and geosciences in Asian Russia. Even today the Siberian Mining and Geological School, established by V.A. Obruchev and M.A. Usov, has retained its significance in the discovery, exploration and development of mineral resources not only in Siberia, the Far East and North-East of Russia, but also in Central Asia.

There are numerous outstanding scientists and engineers among alumni of The Institute of Natural Resources. The institute is proud of such outstanding people as: M.A. Usov, student and first postgraduate of V.A. Obruchev, first professor and academician in Siberia, whose name is associated with the mining industry in Siberia; Academician K.I. Satpaev, founder and first president of the Academy of Sciences of Kazakhstan; Professor N.N. Urvantsev, discoverer of the unique Norilsk ore deposits in the North of East Siberia and Professor M.K. Korovin, who, in the 30s of the 20th century, considered West Siberia deposits to be prospective in hydrocarbons. There are over 35,000 graduates of the institute, 450 of whom became explorers of mineral deposits including one Nobel laureate, 50 laureates of Lenin and State Prizes, more than 250 doctors (DSc) and 1600 associate professors (PhD), 15 academicians and corresponding members of the USSR Academy of Sciences and five Heroes of Socialist Labor.

Within the framework of the XIX International Scientific Symposium, there were 21 panels and workshops in four days. The Symposium was unique in the fact that it embraced all scientific fields of geology and subsurface development, including mineral and hydrocarbon prospecting and exploration methods, space geological research and geoinformation systems, as well as ecological problems of the planet and humanity, rational utilization of mineral resources, land management, natural resource law and economics. Another important issue was technology application in mineral resource exploitation.

The participants were students, post graduates, Master degree students and young scientists from Russia, foreign and CIS countries. The research issues presented at the Symposium highlighted the latest achievements via modern technology and comprehensive interpretation methods; revealed the experimental research results and computer technology in geology and geoecology; provided the analysis of theoretical and experimental research on various geological and environmental protection problems.

The reports embraced the vital issues and latest scientific achievements in stratigraphy, paleontology, tectonics, historical and regional geology, mineralogy, geochemistry, petrography, lithology, metallogeny, hydrogeology and engineering geology, geophysics, petroleum geology, oil and gas field development and processing of hydrocarbon and mineral resources, geoinformation systems (GIS) in geology, geospace research, oilfield equipment upgrading; modern technological achievements in mineral exploration, oil and gas production, transportation and storage, drilling, mining engineering, geoecology, hydrogeoecology, environmental protection engineering, rational utilization of mineral resources, land management, mining and natural resources law, and economical problems of mineral resource sectors in Russia and CIS countries. There was a special panel for those who wanted to present reports in English and / or German.

The presented reports were of scientific importance due to the fact that new concepts and original ideas were suggested by the authors. A number of research topics introduced fundamentally new discoveries. There is also a practical aspect: the findings of the young scientists' research, both fundamental and experimental, could result in DSc and PhD theses. Some authors were granted patents and licenses for their significant discoveries.

Within the framework of the XIX International Scientific Symposium, 1020 reports were presented (including poster presentations), including 262 non-residents from other cities and 52 from foreign countries (22- non-CIS countries and 30-CIS countries, correspondingly).

All in all, 1250 students and young scientists from Russia, CIS countries and foreign countries (52 to be exact) participated in this Symposium. More than 262 submission applications were sent from other cities.

The number of participants was diverse and numerous: non-CIS countries (12 cities), 22 from 16 universities, R&D organizations; CIS countries (15 cities)- 30 from 30 universities, R&D organizations; Russia (48 cities) - 968, including 210 non-residents from 78 universities, R&D organizations; and Tomsk 758 of which 710 were from TPU. It should be mentioned that 54 participants presented their reports in English and German.

Within the 21 different panels and RT 810 reports were presented (including 690 from TPU) of 1020 submitted reports, as well as 210 poster presentations. Nonresidents actively participated in the Symposium – 110 reports.

The status of the participants was: students -694, post-graduates- 190, researchers – 80, engineers – 32, young teachers- 20, and 4 school-students.

The representatives of different organizations were as follows: higher education institutions – 954, Russian Academy of Science – 24, National Academy of Science -7, R&D Institutes – 20, industrial enterprises -11 and schools -4.

Of the 810 presented reports (690 -TPU) there were: 646 students and master degree students (including 608 from TPU), 160 post-graduates and young researchers (including 82 from TPU) and 4 school students. During the 4-day Symposium more than 1900 people attended 21 different panels.

Twelve non-CIS countries presented 22 reports: Germany, Great Britain, Netherlands, Switzerland, Mexico, France, China, Italy, Vietnam, Mongolia, New Zealand and Nigeria.

Seven CIS countries presented 30 reports: National Academies of Sciences and Universities of Belarus, Kazakhstan, Latvia, Armenia, Uzbekistan, Kyrgyzstan and Tajikistan.

Russian participants came from various areas of the country: in the east from Sakhalin, Petropavlovsk- Kamchatsky, Yuzhno-Sakhalinsk, Vladivostok, Blagoveshchensk, Krasnoyarsk, Chita, Irkutsk to Barnaul, Kemerovo, Novokuznetsk, and etc.; in the north from Mirny, Yakutsk, Neryungri, Magadan, Nizhnevartovsk, Khanty-Mansiysk, Nefteugansk to Tyumen, Ulan-Ude, Syktyvkar, and etc.; in the west from Minsk, Kiev, Moscow, St. Petersburg, Yekaterinburg to Samara, Kazan, Ufa, Perm, Novosibirsk;in the south from the cities of the Central Asian republics to Ivano-Frankovsk, Odessa, Novocherkassk, Simferopol, Novorossiysk, Vladikavkaz, Voronezh, Stavropol, Astrakhan, and etc. A great number of young people from Urals, Western and Eastern Siberia took an active part in the Symposium. CIS countries were presented by participants from Uzbekistan (Tashkent), Tajikistan (Dushanbe), Azerbaijan (Baku), Kazakhstan (Almaty, Semipalatinsk, Karaganda, Pavlodar), Belarus (Minsk, Gomel), Armenia (Yerevan, Gyumri), Ukraine (Kiev, Odessa, Ivano-Frankovsk, Dnepropetrovsk, Donetsk, etc.), Kyrgyzstan (Bishkek), Moldova (Chisinau). The students and young scientists from Tomsk representing Tomsk Polytechnic University, Tomsk State University, the Institute of Petroleum Geology and Geophysics (SB RAS) and other organizations and institutions took an active part in the symposium.

The research results of the Symposium can be found in a two-volume edition which is available at http://portal.tpu.ru/science/konf/usovma/eng and /or www.usovma.tpu.ru

The Editorial Boards of this Symposium and *IOP Conference Series: Earth and Environmental Science* consider that the articles would be of great interest for university researchers and young scientists, research and academic institutes, academies of sciences and their branches, engineering and technical staff of ministries and government departments – for those who explore and develop the Earth subsurface.

Many people have contributed in a variety of ways in the preparation of this edition. We would like to thank the administration of the Institute of Physics and Professor P.S. Chubik, Rector of National Research Tomsk Polytechnic University.

Executive Editor
XIX International Symposium
"Problems of Geology and Subsurface Development" - 2015,
PhD in Geology and Mineralogy,
Associate Professor G.M. Ivanova

Peer review statement

All papers published in this volume of *IOP Conference Series: Earth and Environmental Science* have been peer reviewed through processes administered by the proceedings Editors. Reviews were conducted by expert referees to the professional and scientific standards expected of a proceedings journal published by IOP Publishing.

Hydrothermal alteration mapping of Siberian gold-ore fields based on satellite spectroscopy data

Yu S Ananyev[1,2], A A Maskov[1] and R N Abramova[1]

[1] National Research Tomsk Polytechnic University, 30 Lenin Ave., Tomsk, 634050, Russia

E-mail: [2] AnanyevYS@ignd.tpu.ru

Abstract. The mapping of the hydrothermal alterations in Urjahskoe and Fedorov-Kedrov gold-ore fields was conducted by applying channel relationship method (band ratio) based on ASTER spectral-zonal satellite image data. It was determined that the calculated mineral indices in ore-bearing structures are zonal. Outer ore-bearing structures revealed increased ferric mineral index values, while inner – high epidote- chlorite- calcite and muscovite- siderite mineral index values. Detected regularities could be used in identifying potential gold-ore bearing areas within identical fields based on remote sensing survey data.

1. Introduction

Spectral methods in remote sensing of geological surveying, mineral deposit exploration activities became applicable in the 70s of the last century due to the introduction of multispectral Earth surface photographs. Today, there are numerous operating spectro- radiometers, including not only satellite (ASTER, MODIS, Hyperion, Landsat ALI and others), but also aeronautic (SFASI, MASTER, SEBASS, AVIRIS, Hymap) ones, as well as a significant data archive of space systems.

Based on the technical features of the spectral channel configuration in the radiometer and data accessibility, ASTER data in geological surveying is more optimal in application.

The system characteristics include the following number of channels: 3- visible and near infrared (VNIK); 6- shortwave infrared (SWIR) and 5- thermal infrared (TIR). Spatial resolution is 15m, 30m and 90m, respectively.

The research target is Urjakhskoe ore field, located within Sulban deep-fault distribution area in Baikal-Patom upland and Fedorov-Kedrov ore field within Orton-Baliksinore district in southern Kuznetsk-Alatau.

North N-W trending Sulban deep fault within Urjahskoe gold-ore field divides the host rocks into two blocks: eastern (significantly metaterrigenic and metacarbonate) and western (predominately metavolcangenic) [1]. Intrusive rocks as plagiogranite dykes can be found only in the western block of this ore field. Host ore rocks metamorphosed in environments of green schist and amphibolite facies [2], while in affected Sulban deep fault zones these rocks transformed into dynamoschists and subjected to metasomatic alteration resulting in beresite-listvenite profile. In these metasomatite zones veined and stringer-porphyry mineralization have been detected. Quartz-carbonate and carbonate veins and veinlets have been revealed in the axial Sulban fault region. Productive gold-sulphide mineralization is confined to pre-selvages in these veins. Stringer-porphyry mineralization is followed by beresitization and carbonizing of host rocks and this, in its turn, is confined to dynamoschist development areas within the Sulban fault. Sulban deep fault is the major ruptured ore field zone. In the north field area it is a single joint branching southward and forming lense-boudin blocks. Here, it is trending S-W, while in the south area - N-E.

Fedorov-Kedrov gold-ore field is located in the south of Orton-Baliksinsk gold-ore area, southern Kuznetsk-Alatau. Structurally, this area is confined to the joint area of deep-medium and vari-oriented structural complexes: N-E joint zone of mesozonal structure and N-W zone of Kuznetsk-Alatau epizonal deep fault. Host, predominately volcangenic-carbonate, rocks are cross-cut by a dense network of meridional mafic dykes. Ore-bearing Fedorov-Kedrov zone, enclosing Fedorov gold-ore deposit, is located in the S-E edge of the North-East joint being structurally, lenticular and cross-grained in apparent thickness of more than 2km. Several linear areas of intensive N-E trending dynamometamorphism were identified within this area, followed by metasomatic propylitic-listvenite-beresite alterations. Gold mineralization involves structural-morphological veins (Fedorov deposit) and stringer-porphyry mineralization (Kedrov, Zolotoe, Anomalnoe ore occurrences) [3].

2. Research methods

Mapping of hydrothermal alterations associated with ore mineralization, including ASTER data, involves band ratio methods, Spectral Angle Mapper, its modifications and constrained energy minimization. Such an approach was applied to identify potential areas of gold-ore, gold-silver, copper-porphyry, copper- molybdenum-porphyry targets [4 – 7]. It should be noted that mapping hydrothermal alterations and identifying potential areas are conducted in conditions of rather favorable exposure.

Investigated area embraces partially forestry and peat zones which restrict the possible application of Spectral Angle Mapper and constrained energy minimization in mapping the hydrothermal alterations. In this case, the basic research method is band ratio.

The designed procedure in processing ASTER satellite images and further interpretation of obtained calculations included:

1) image screen preparation;
2) calculation of spectrum indexes;
3) interpretation of obtained material.

Image screen preparation involves the following: projecting ASTER data to a rectangular cartographic system, atmospheric correction, noise masking, spatial resolution to 30m, recalculation of radiometeric resolution of TIR to 8bit/pixel.

Calculation of band ratio is based on ArcGIS. Thirty different indices were calculated for each area and further evaluated in terms of informativeness (table 1).The interpreted information involved the data of geological structure and areal minerageny.

Table 1. Information relationships to map hydrothermal alterations based on ASTER data.

Index	Calculation formula	Reference source
RBD6 (muscovite+siderite+illite+ smectite)	(band 5+ band 7)/band 6	[8]
RBD8 (epidote+chlorite+calcite)	(band 7+ band 9)/band 8	[8]
Quartz I	(band11/band 10)/(band 11/band 12)	[9]
"Hematite" (Gossan)	band 4/band 2	[10]
"Ferric" oxides	band 4/band 3	[11]

3. Research results and discussion

The most informative indices within Urjahskoe gold-ore field are quartz, muscovite-siderite, epidote-chlorite-calcite and ferric.

Increased quartz index value distinctly shows silicification along the developing Sulban deep fault zones revealing a regional ruptured hanging wall block structure. Muscovite- kaolinite index values are also predominately high within Sulban fault zone. However, it should be noted that the increased index values are correlated to the location of metaterrigenic rocks in the eastern ore field block, while in the western metavolcanogenic block the index value is lower. The conjugation domain of high muscovite- siderite index values indicates beresitization areas.

High epidote-chlorite-calcite index values are typical only for the western metavolcanogenic ore field block. In developing Sulban fault zones the high values of this mineral index are observed sporadically in areas of low quartz and muscovite-siderite index values. Accordingly, it is supposed that epidote-chlorite-calcite index values could reveal areas of developing propylitization.

"Hematite" and "Ferric" indices have exactly the same behavior. These indices are mininum in the influence zone of the Sulban fault and progressively increasing outward from the fault. Particularly, "Hematite" index is revealed within the metavolcanogenic environment of the western ore field, while "Ferric" index- in the eastern area within shale metaterrigenic environment.

Based on the obtained consistent spatial attitude values of the above-described mineral indices according to ASTER data, a model of metasomatic zoning in Urjahskoe gold-ore field was plotted (figure 1).

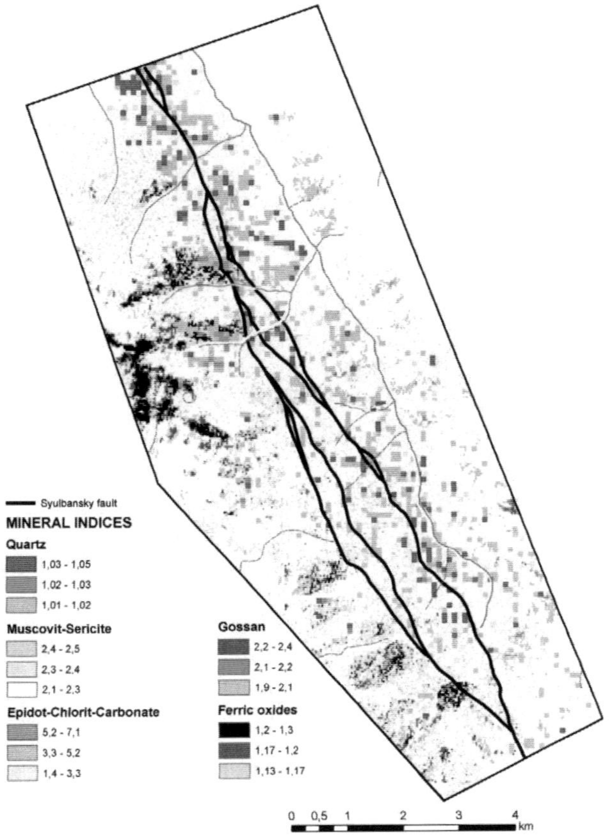

Figure 1. Model of metasomatic zoning in Urjahskoe gold-ore field.

The spectral model of metasomatic zoning in Urjahskoe gold-ore field revealed the following: significant inner quartz index zone; first intermediate quartz- muscovite- siderite index zone; second intermediate epidote-chlorite-calcite index zone within the metavolcanogenic thickness; outer "ferric": in metavolcanogenic thickness – ferric oxide, while in metaterrigenic thickness- hematite index zones.

The most informative indices were those of epidote-chlorite-calcite and ferric oxide within Fedorov-Kedrov gold-ore field. In this case, quartz and muscovite- siderite indices only supplement the above-described ones and, in some cases, accentuate the inner structure of potential ore-bearing zones.

High "Ferric" oxide index values embrace north-western, south-western and southern potential ore-bearing joints in Fedorov-Kedrov gold-ore field disaggregating its lenticular structure. Conversely, increased epidote-chlorite-calcite index values reveal the inner part of these lenses. In some cases, the increased muscovite- siderite index values reveal the inner part of local joints in potential ore-bearing zones. However, the quartz index within this area does not demonstrate any consistency to gold mineralization. This could be explained by development of crust weathering in the investigated area. Based on the obtained consistent spatial attitude values of the above-described mineral indices, according to ASTER data, a model of metasomatic zoning in Fedorov-Kedrov gold-ore field was plotted (figure 2).

Figure 2. Model of metasomatic zoning in Fedorov-Kedrov gold-ore field.

The spectral model of metasomatic zoning in Fedorov-Kedrov gold-ore field revealed the following: inner muscovite- siderite; intermediate epidote-chlorite-calcite and outer ferric oxide index values.

Obtained consistent metasomatic structure areolas in Urjahskoe and Fedorov-Kedrov gold-ore fields closely correlate to the classical metasomatism concept of gold-ore bearing areas, where the inner zones are silicificated, while intermediate zones embrace metasomatites of beresite-listvenite thickness, which, in its turn, are replaced progressively outwards by propylites and propylitized rocks. Ultimately, outer ferric zones correlate to frontal basification zones.

4. Conclusions

1) Muscovite- siderite, epidote-chlorite-calcite, hematite (gossan) and ferric oxide mineral indices, calculated from ASTER spectral-zonal satellite image data, consistently reflect characteristic gold-ore fields.

2) Designed spectral models distinctly showed several zones: inner with hematite and ferric oxide, intermediate composed of increased epidote-chlorite-calcite and inner with muscovite- siderite mineral indices. Quartz mineral index does not always reflect unambiguous results. This is associated with the ASTER spatial resolution TIR channels (90 meters), landscape environments and specific feature of silicification.
3) This consistency could be used in collaboration with other minerogenic data as potential criteria for gold-ore bearing areas.

Acknowledgement
The investigations were supported by National Research Tomsk Polytechnic University. The research was performed with the financial support from Government Project «Nauka», 114.2014

References

[1] Kucherenko I, Gavrilov R 2012 Strukturno-dinamicheskie rezhimy obrazovanija zoloto-sul'fidno-kvarcevoj mineralizacii v Sjul'banskoj zolotorudnoj zone (bassejn srednego techenija r. Vitim) Urjahskoe rudnoe pole *TPU Izvestiya* **320** 19–27 (in Russian)
[2] Zorin Yu, Mazukabzov A, Donskaya T, Presnyakov S and Sergeev S 2008 Silurijskij vozrast glavnyh skladchatyh deformacij rifejskih otlozhenij Bajkalo-Patomskoj zony *RAS Proceedings* **423** 228–33 (in Russian)
[3] Konishev V 2014 Methodology in Au content evaluation in ore occurrences from significant Orton-Baliksin gold area, Kuznetsk-Alatau *Ores and metals* **2** 44–64
[4] Di Tommaso I and Rubinstein N 2007 Hydrothermal alteration mapping using ASTER data in the Infiernillo porphyry deposit, Argentina *Ore Geology Reviews* **2** 275–90
[5] Zhang X, Pazner M and Duke N 2007 Lithologic and mineral information extraction for gold exploration using ASTER data in the south Chocolate Mountains (California) *ISPRS Journal of Photogrammetry & Remote Sensing* **62** 271–82
[6] Gabr S, Ghulam A and Kusky T 2010 Detecting areas of high-potential gold mineralization using ASTER data *Ore Geology Reviews* **38** 59–69
[7] Pour A and Hashim M 2011 Identification of hydrothermal alteration minerals for exploring of porphyry copper deposit using ASTER data, SE Iran *Journal of Asian Earth Sciences* **42** 1309–23
[8] Crowley J K, Brickey D W and Rowan L C 1989Airborne imaging spectrometer data of the Ruby Mountains, Montana: mineral discrimination using relative absorption band-depth images *Remote Sensing of Environmen*t **29** 121–34
[9] Ninomiya Y 2003 Advanced remote lithologic mapping in ophiolite zone with ASTER multispectral thermal infrared data *Proc. IEEE 2003 Int. Geoscience and Remote Sensing Symposium (IGARSS'03) (21–25 July 2003)* 3 1561–63
[10] Volesky J C, Stern R J and Johnson P R 2003 Geological control of massive sulfide mineralization in the Neoproterozoic Wadi Bidah shear zone, southwestern Saudi Arabia, inferences from orbital remote sensing and field studies *Precambrian Research* **123** 235–47
[11] Hewson R D, Cudahy T J, Mizuhiko S, Ueda K and Mauger A J 2005 Seamless geological map generation using ASTER in the Broken Hill–Curnamona province of Australia *Remote Sensing of Environmen*t **99** 159–72

Petrogenetic characteristics of mafic-ultramafic massifs in Nizhne-Derbinsk complex (East Sayan Mountains)

T Cherkasova[1,3], A Chernishov[1,4], Yu Goltsova[1,5], T Timkin[1,6] and R Abramova[2]

[1] Department of Mineral Exploration, Institute of Natural Resources, National Research Tomsk Polytechnic University, 30 Lenin Ave., Tomsk, 634050, Russia
[2] Department of Foreign Language, Institute of Natural Resources, National Research Tomsk Polytechnic University, 30 Lenin Ave., Tomsk, 634050, Russia

E-mail: [3] tomasstik@yandex.ru, [4] aich53@yandex.ru, [5] y.goltsova@mail.ru, [6] timkin@tpu.ru

Abstract: The article describes the results of petrographic, petrochemical, petrofabric, mineralogical and geochemical studies of the major rock groups potentially Cu, Ni, Pt ore-bearing mafic-ultramafic massifs in the Nizhne- Derbinsk complex (Eastern Sayan Mountains). Based on the data interpretation the investigated massifs can be classified as peridotite-pyroxenite-gabbronorite formation of geosynclinal regime in Altai-Sayan folding area. Significant massif deformation occurred during the final post-consolidation formation stage. The petrographic features of gabbro and petrofabric patterns of the rock-forming minerals in the Burlakski and Nizhne-Derbinsk massifs indicated the fact that massifs were involved in the accretion-collisional development stage of the Central Asian folding belt during the final formation stages the Nizhne-Derbinsk complex.

1. Introduction

Most ultramafic rock bodies are various in forms, petrographic and chemical composition, while only a few massifs are enriched with commercially ore [1]. The physical rock form and geological position of massifs excludes their classification into ore-bearing and ore-free. In this case, the mineragenic type of the massifs and their potential mineralization could be evaluated only by petrological and geochemical criteria. Therefore, the study of the natural geological features embraces the formation morphology and distribution of not only the deposits and occurrences but also genetically related mineral deposits (Cu, Ni, Cr, Pt) in the crust [2-4]. Mafic-ultramafic complexes play an important role in the reconstruction stages of the Earth's history [5].

2. Methodology

The investigation of the rock composition was based on the following methods: petrographic analysis (National Research Tomsk Polytechnic University and National Research Tomsk State University), silicate analysis / solution chemistry (Institute of Geochemistry and Mineralogy SB RAS-Irkutsk) and microprobe analysis ICP-MS (Rb, Sr, U, Zr, Ta, Nb, Hf, REE) (Institute of Geology and Mineralogy SB RAS-Novosibirsk). The integrating analysis results revealed the basic petrographic types with a detailed composition description of the major rock-forming minerals (olivine, pyroxene, plagioclase).

To reconstruct the thermodynamic and tectonic formation conditions of studied massifs, the petrofabric analysis of the main rock-forming minerals was conducted.

3. Regional setting

The research target was the massif area of Nizhne-Derbinsk mafic-ultramafic complex, which is located in the N-W of Eastern Sayan, left bank of Krasnoyarsk water reservoir, interstream areas of Sisim, Kizhart, Derbina and Tubil rivers. The massifs are confined to the Eastern-Sayan and East-Kuzbass fault zones intruding Early Proterozoic and Late Riphean Derbinsk and Urmansk suites. Nizhne-Derbinsk intrusive complex merges mafic-ultramafic massifs as a lateral zone extending 40km. and 100km. southward of Krasnoyarsk.

The mineragenic type and metallogeny of the Nizhne- Derbinsk mafic-ultramafic massifs complex (Derbinsk area – East Sayan), identified by S.S. Serduk, V.A. Kirilenko, G.R. Lomaeva, V.E. Babushkin, A.V. Tarasov et al. [6] as highly potential explorable commercial Cu, Ni and Cr concentrations, still remains a controversial issue. T.Y. Kornev, A.P. Romanov consider these massifs complex as Late Archaean ophiolites of intrusive magmatism, having emerged in the Kuzeevsk greenstone belt [7]. A.E. Izokh, R.A. Shelepaev and others [8] consider these rocks to be derivatives of gabbro-monzodiorite magmatism within the Altai-Sayan fold area (ASFA). S.S. Serdkom, A.I. Zverev do not exclude such a fact as the simultaneous occurrence of both Late Riphean and Ordovician gabbro in the Nizhne-Derbinsk complex [6]. The geological setting, genesis and metallogeny of above-described massifs are still under discussion.

4. Research results

In accordance with the results of previous studies, several differentiated ultramafic-mafic massifs have been defined within the Nizhne- Derbinsk complex: Ashtatski, Azertakski, Nizhne-Derbinsk, Pravoderbinski, Burlakski, Medvezhi and Tubilski, Verhne-Tubilski. To investigate the rock material composition two centrally located massif complexes – Burlakski and Nizhne-Derbinsk – were taken into consideration. Burlakski massif could be considered a reference complex massif as practically all magmatic rock types can be found here, comparable to other massifs; whereas only ultramafites are found in Nizhne- Derbinsk massif.

Ultramafites embrace wehrlites, websterites (aluminite) and clinopyroxenites with subordiante apodunite and apoharzburgite serpentinite. Websterites, clinopyroxenites and wehrlites reveal gradual rock alteration interchange from one type to another. The above-described ultramafites are mainly composed of clinopyroxene, including a small content of olivine and orthopyroxene and still smaller content of hornblende, and, it is this composition ratio that indicates their difference. The rocks often reveal cumulate structure where the cumulus phase is composed of tabular and sub-isometric clinopyroxene grains and, rarely, of orthopyroxene ones; the interstices of which are xenomorphic olivine and hornblende grains. The formation of banded ultramafites is obviously due to the fractional melt crystallization in the steady-state chamber.

Hornblende pyroxenite and hornblendites (hornblende being the primary igneous mineral) in the upper horizon indicate a rather high (over 3%) water saturated magmatic melt formed during the late-magmatic crystallization stage. Wehrlites and pyroxenite are usually fresh. Secondary changes are associated with the replacement of pyroxene by tremolite, actinolite or uralite, or the formation of pseudomorphs. Olivine is replaced by cellular lizardite. Under intrusion conditions dunites and harzburgites are unstable and completely serpentinized, which, in its turn, is obvious in the auto-metasomatic processes. Olivine is usually replaced by platy and sectorial lizardite aggregates, while orthopyroxene is usually replaced by tabular bastite grains.

The Burlakski massif gabbro mainly includes unchanged leucocratic gabbronorites and partially olivine gabbro, which are characterized by an ophitic texture and, rarely, gabbroic or trachytoid texture. Gabbroids are subjected to extensive plastic deformation, observed in the curved elongated prismatic plagioclase crystals, recrystallization on the periphery of their crystals and, sometimes, disintegration within separate sub-blocks. Trachytoid rocks demonstrate the plane surface and direction of rock flowage. Structural study of Burlakski and Nizhne- Derbinsk massifs rocks shows the evidence of the final stages of active dynamic massif environment formation yielding subsequent structural elements synformal with the rim rocks. It should be noted that Burlakski massif gabbro are more susceptible to the rock flowage, which could be observed in the gabbro distinct directive texture. This, in its turn, mineralogical-petrographic features is consistent with the plastically deformed gabbronorites.

Olivines in Burlakski and Nizhne- Derbinsk massif wehrlite are compositionally similar to high-magnesia chrysolites (Fa = 14.3...15.3%). NiO content in olivines of Burlakski massif increases as iron content increases. A similar trend is observed for olivines in Kingashski massif ultramafites, located eastward of the Kansky greenstone belt. Significant differences were observed in the olivines of the studied Ospinski massif intrusions and metamorphic ultramafic rocks (ophiolite complex SE of East Sayan). The latter has significant low iron content under consistent high NiO content. Such an element behavior in olivines is typical for magmatic rocks formed as a result of crystallization differentiation. A distinctive feature of Nizhne- Derbinsk complex olivines is their insignificant deformation which is only characteristic of olivines from alpinotype ultrabasites.

Petrostructural study of olivine in Burlakski and Nizhne- Derbinsk massif wehrlites revealed that the crystallization of the magmatic melt apparently occurred in steady-state conditions forming semi-isotropic fabric of olivine crystal optical axes due to its crystal decantation providing insignificant influence of the laminar flow in the subhorizontal plane. Further superimposed plastic deformation, which wehrlites have been subjected to, presumably occurring within active tectonic environment during massif consolidation. Plastic deformation of olivine at this stage obviously occurred at decreasing temperatures and increasing deformation rate with a heterogeneous intracrystalline slippage, where the temperature shift is from high {0kl}[100] to low (100)[001] [9]. Olivine shows petrofabric patterns as a result of primary magmatic processes including metamorphic processes in the final massif formation stages. In this case, the number of crystals increases and they respond to the dynamic load under the influence of the external stress field.

The most abundant and predominate mineral rock in Nizhne- Derbinsk complex is clinopyroxene. A specific feature of this mineral rock is that it forms euhedral grains comparable to the xenomorphic minerals, especially olivine. In wehrlites, websterites and clinopyroxenites its chemical composition corresponds to augite and diopside. Enstatite component (En = Mg / (Mg + Fe + Ca) · 100%) of clinopyroxene ranges from 56% in ultramafites and up to 41% in gabbroids. Ferrosilite component (Fs = Fe / (Mg + Fe + Ca) · 100%) varies from 5 to 18%. Wollastonite component (Wo = Ca / (Mg + Fe + Ca) · 100%) ranges from 29 to 46%. Orthopyroxene has wide chemical composition variation: in ultramafites it corresponds to bronzite with iron content of 17 to 21% and in gabbronorites corresponds to hypersthene (Fs = 32 ... 33%), which is consistent with the magmatic matter differentiation.

Orthopyroxene in gabbronorites differs from that in ultramafites by the increased content of Ti, Mn, Fe, Na and decreased content of Al, Cr, Mg. According to the optical (symmetric extinction angle) and chemical properties, plagioclase could be aligned to labrador № 55…57. Fine plagioclase grains are fresh, however, their composition is more acidic comparable to megaphenocrysts related to andesine № 45...47. Plagioclase grains are sometimes slightly sericitized and pelitized. Brownish-yellow iron hydroxides, greenish-yellow chlorite flakes and fine carbonate grains are observed in the cracks. Plagioclase forms not only large banded units in gabbronorites forming porphyric structure but also fine grains. The size of porphyritic megaphenocrysts is 4...8 mm., sometimes up to 15 mm. Plagioclase forms on liquidus after olivine and pyroxene, thus it has a more acidic composition corresponding to labrador-andesite, whereas plagioclase crystallizes with olivine being more mafic in composition (anorthite-bytownite) in troctolitic massifs.

Hornblende shows obvious allotriomorphism in relation to units of olivine and clinopyroxene. Its chemical composition corresponds to edenitic hornblende which is characterized by increased Mg and alkalinity. In Nizhne- Derbinsk massif monomineral rocks, composed of magmatic hornblende, top the layered intrusion cross-section. The results of petrographic studies showed Burlakski and

Nizhnederbinsk massifs are related to layered mafic-ultramafic intrusions of intra-geosynclinal folded areas.

5. Discussion

The major geological feature revealing the restitic nature of ophiolite complexes (relict fragments of ancient oceanic crust) is their tectonic contacts with hosting rocks of banded and lenticular-banded structure with predominant dunite-harzburgite rock association. A.E. Izokh [10] identified intrusive contacts of Burlakski massif with hosting rocks in the Fadeev stream (left tributary of Derbin River). Here, taxitic irregularly grained gabbroids with large porphyritic plagioclase segregations (including small hornfel xenoliths) outcrop. Nizhne- Derbinsk massif- Urmansk shale suite contact was defined in a channel within the Bezimyanni stream, where these shales are hornfelsed [10]. According to the chemical composition of Nizhne- Derbinskif massif rocks are divided into three groups: ultramafic (dunites, wehrlites), sub-ultramafic (clinopyroxenites, websterites) and mafic (gabbronorites, gabbro) [11].

Obtained rare earth element (REE) distribution spectra, standardized to chondrites for Burlakski and Nizhne- Derbinsk massif rocks, has the same REE spectra. Depleted content of light REE is typical under conditions of heavy REE" planar" distribution and most rocks are characteristic of a weak Eu minimum. The multi-element spectra, standardized to primitive mantle rocks of Nizhne-Derbinsk complex, have pronounced typomorphic features typical for magma melting out of depleted supra-subduction mantle. In this case, they are enriched by ion lithophile elements (Rb, Ba, U, K) and Sr, as well as, depleted highly- charged elements (Nb, Zr, Hf) [12]. The similar distribution pattern of REE and rare elements in Burlakski and Nizhne-Derbinsk massifs are indicators of a single formation source for both ultramafic and mafic sequences.

6. Conclusions

Based on the integrated research of the Nizhne-Derbinsk massif composition it was possible to evaluate its mineragenic type-peridotite-pyroxenite-gabbronorite. The cumulative structures in Burlakski and Nizhne-Derbinsk massif ultramafites and ophitic in gabbroids indicate the fact that these massifs formed in mesoabyssal conditions at relatively shallow depths. Mineral plastic deformation showed the severe plastic deformation of massif rocks during the post-magmatic stage. Petrochemical characteristics of rocks reflect the evolution path of magmatic melt from ultramafic to gabbronorites.

According to the geological, mineralogical, petrographic and petrogeochemical data, Burlakski and Nizhne-Derbinsk massif rocks are derivates of one and the same magma forming system, the parent melt of which was high-magnesia basaltic magma resulting from partially significant melting of the suprasubduction mantle under the influence of metasomatized water. Geographically, Nizhne-Derbinsk complex is confined to the magmatic associations of the Altai-Sayan folding area, which, in its turn, is located within the Central Asian folding belt. The Burlakski massif is related to Ordovician period (~ 490 million years) [11].

Another important fact is that during the post consolidation formation stage of the Nizhne-Derbinsk complex, these rock massifs were involved in Early Ordovician accretion-collisional stage of the Altai-Sayan folding area (490-475 mln. years). Based on the research results the Nizhne-Derbinski complex massifs could be associated with both massive copper-nickel ores, confined to deeper horizons and unexposed at the ore zone erosion, and isolated dotted PGE mineralization. The latter fact was identified in the investigated mafic-ultramafic Idzhinsk massif complex (Western Sayan) via scanning electron microscope [13]. By applying the electron microscopy the mineral phases in PGE mineralization could be observed in the thin sections at 5 mill micron [14].

Acknowledgments

The research was supported by National Research Tomsk Polytechnic University, project No. 114-2014

References

[1] Li C, Ripley EM and Naldrett AJ 2003 Compositional variations of olivine and sulfur isotopes in the Noril'sk and Talnakh intrusions, Siberia: Implications for ore forming processes in dynamic magma conduits. *Economic Geology* **98** 69–86

[2] Maier WD, Gomwe T, Barnes S-J, Li C and Theart H 2004 Platinum-Group Elements in the Uitkomst Complex, South Africa. *Economic Geology* **99** 499–516

[3] Taranovic V, Ripley EM, Li C and Rossell D 2015 Petrogenesis of the Ni-Cu-PGE sulfide-bearing Tamarack Intrusive Complex, Midcontinent Rift System, Minnesota. *Lithos* **212–215** 16–31

[4] Mao Y-J, Qin K-J, Li C, Xue, S-C and Ripley EM 2014 Petrogenesis and ore genesis of the Permian Huangshanxi sulfide ore-bearing mafic-ultramafic intrusion in the Central Asian Orogenic Belt, western China. *Lithos* **200–201** 111–25

[5] Pearse J A 2008 Geochemical fingerprinting of oceanic basalts with application to ophiolite classification and the search for Archean oceanic crust *Lithos* **100** 14–48

[6] *Geology and prospects of sulfide Pt-Cu-Ni mineralization of Eastern part of the Altai-Sayan folded area* ed S S Serdyak, V A Kirilenko, G L Lomaeva, B E Babushkin, A V Tarasov and A I Zverev 2010 (Krasnoyrsk: Gorod) p 184

[7] Kornev TY, Romanov A P, Ekhanin AG, Knyazev VN and Sharifulin SK 2008 Platinum greenstone belts of Eastern Sayan and the Yenisei Ridge *Platinum of Russia* **5** 358–80

[8] Izokh A E, Shelepaev R A, Lavrenchuk A V, Borodina E V, Yegorova V V, Vasjukova E A and Gladkochub D P 2005 Cambro-Ordovician Variety of ultramafic-mafic associations of the Central Asian fold belt as a reflection of the interaction of the plume and lithospheric mantle. Geodynamic Evolution of the Lithosphere of the Central Asian Mobile Belt (from Ocean to Continent) *Proceedings of scientific conference on the Programme of Basic Research* **1** 106–108

[9] Nicolas A 1976 *Crystalline plasticity and solid state flow in metamorphic rocks* ed Poirier J P (New York: Wiley-Interscience) p 444

[10] Izokh A E, Kargopolov S A, Shelepaev R A, Travin V A and Yegorova V V 2001 Mafic magmatism of Cambro-Ordovician stage of the Altai-Sayan folded area and the connection of metamorphism of high temperatures and low pressures *Actual issues of geology and minerageny of Southern Siberia: The materials of scientific conference (31 October – 2 November 2001 Elan village, Kemerovo Oblast, Novosibirsk, Lavrentyev Institute of Hydrodynamics, Siberian Branch of the Russian Academy of Sciences)* pp 68–72

[11] Cherkasova T and Chernyshov A 2009 Petrochemical characteristics of the stratified mafic-ultramafic massifs of the Nizhne-Derbinski complex (NW of the Eastern Sayan Mountains) *Tomsk State University Journal* **324** 390–94

[12] *Interpretation of geochemical data* 2001 ed E V Sklyrova (Intermed Engeneering) p 288

[13] Cherkasova T and Korotchenko T 2014 Noble-metal mineralization in olivine clinopyroxenite within Idgimski gabbro-peridotite-pyroxenite complex (West Sayan Mountains) *IOP Conf. Ser.: Earth Environ. Sci.* **21** 012008

[14] Cherkasova T, Timkin T and Savinova O 2015 First native silica findings in bismuth from garnet skarns of Ribny Log – 2 gold ore target in Topolninsk ore deposit (Gorny Altai) *IOP Conf. Ser.: Earth Environ. Sci.* **24** 012033

Rear polymineral zone of near-veined metasomatic aureole in mesothermal Zun-Holba gold deposit (Eastern Sayan)

T Cherkasova[1,3], I Kucherenko[1,4] and R Abramova[2]

[1] Department of Mineral Exploration, Institute of Natural Resources, National Research Tomsk Polytechnic University, Lenin Ave., Tomsk, 634050, Russia
[2] Department of Foreign Language, Institute of Natural Resources, National Research Tomsk Polytechnic University, Lenin Ave., Tomsk, 634050, Russia

E-mail: [3] tomasstik@yandex.ru, [4] kivr@tpu.ru

Abstract: Unique data of the mineralogical and petrochemical zoning of near- veined metasomatic aureole of mesothermal Zun-Holba gold deposit are presented and discussed. It was established that mineralogical and petrochemical zoning order is based on Korzhinsky theory describing the differential component mobility. However, the internal polymineral zone structure of metasomatic column in Zun-Holba deposit does not comply with Korzhinsky concept describing the mono-mineral composition of axial (ore-bearing quartz veins) and binary-mineral rear (quartz-sericite) zones. Mineral zoning complication is governed by component diffusion (from fractured fluid to pores) and pulsation mode of metalliferous fluid input into the mineralization area.

1. Introduction

Since the 50's of the twentieth century, most experts accepted the non-alternative Korzhinsky theory of metasomatic processes, particularly, metasomatic zoning [1]. However, there still exists a misunderstanding in assessed mineral composition of metasomatic rocks in the axial zone (quartz vein) and adjacent rear zone involving zonal near-veined (wallrock) metasomatic aureole. It is a well-known fact that metasomatic zoning theory embraces Korzhinsky concept of differential component mobility determined by thermodynamic and physico-chemical regimes in fractured porous rock-fluid mineralization systems. The bulk of rock-forming components include most mobile (H_2O, CO_2), highly mobile in all conditions (S, Cl, Na, K), mobile in certain conditions (O_2, Si, Mg, Ca, Fe) and inert (Al, P, Ti).

According to this theory, in near-fractured metasomatism, where invading external metalliferous fluids into cracks intensify rock alteration, sequential transformation of components from inert into mobile state accompanied by mineral dissolution in parent rocks occurs. According to Korzhinsky, when stagnant conditions merge into a mobile state and pore solutions with filtrating fluids from cracks form hydraulically connected system, the components diffuse into the fractured solutions and are removed. The remaining minerals in the columns decrease to the extent of monomineral axial zone. Zonal metasomatic columns are formed, where in each frontier mineral zone there is one mineral less comparable to less altered rocks in adjacent closer frontal zone

However, more and more attention is focused on the fact in natural, for example, gold-bearing beresite columns with medium-low temperature (380...250°C), potassium-sulfur-carbon dioxide metasomatism of binary-mineral (quartz-sericite) compositions adjacent to the axial rear (beresite)

zone does not occur [2]. The contradiction of natural columns to their theoretical model would have required a thorough investigation of the structure of both beresites and near-fractured columns, as well as other acid metasomatites – greisen, argillizated rocks, argillizites, propylites; however, if there exists such a contradiction then its reasons should be studied and metasomatic zoning theory should be fully defined. In this case, there is still a shortage of mineral zoning descriptions of natural hydrothermal metasomatic columns. The theoretical model of metasomatic zoning with monomineral axial (quartz) zone characterized as natural ones has been described in different publications [3,4]. Based on the chemical analysis data [3,4], it was found that in the adjacent to axial "quartz-sericite (muscovite) zone" the columns formed from carbonate-free and sulfide-free granodiorites. As CO_2 and recomposed S (up to tenths of w. %) are evident in this zone, this indicates the fact that in addition to quartz and sericite minerals, there are two other mineral classes – carbonates and sulfides. Beresite according to its mineral-chemical pyrite-quartz-sericite-carbonate composition is often associated with metasomatites, but due to the fact that it is not found in near-ore altered quartz-sericite and quartz as described in the metasomatic zoning theory [5], this fact in itself is often neglected. The above-mentioned fact is based on the general concept that the polymineral structure of metasomatic columns adjacent to quartz-vein zone is naturally-formed which could be governed by either underdevelopment process or complex evolution of thermodynamic and physico-chemical fluid regimes, filtrating in wallrock pore spaces [6-8] which, in its turn, is a characteristic stagnant regime of pore-filling solutions. Therefore, the transportation of components in wallrocks is possible not by fluid flow, but by their concentration diffusion. To explain the polymineral metasomatite composition in a column rear zone based only on the above-mentioned facts is rather ambiguous and, in this case, requires further investigation.

Excluding the ambiguity discussed issue and defining some aspects of the existing metasomatic zoning theory is impossible within the walls of either a classroom or laboratory. This could be explained by a detailed study of natural near-ore metasomatic columns in hydrothermal deposits of metallic mineral resources, accumulation of experimental data on the mineral zoning structure of hydrothermally altered rock aureoles in wallrock space and research facts proving the formation conditions of rear and axial metasomatic column zone mineral composition. It should be noted that the mineral zoning sequence, i.e. the succession of mineral zones in columns, could be described by the differential mobility of components. This paper involves the detailed investigation of near-veined metasomatic aureole in Zun-Holba mesothermal gold deposits.

2. Results and discussion

The deposit is located in the S-E Eastern Sayan in the watershed of the riverheads of Urik and Kitoy rivers. This deposit is included in the Urik-Kitoy gold-ore zone controlled by Okino-Kitoy system of northwest trending deep faults, separating Gargan ridge of Archean Siberian platform basement from Proterozoic and Paleozoic margins. The deposit embraces in the south- gneiss-granite Gargan ridge; in the north – Middle Paleozoic Ambartogolsk massif granitoids (400 – 420m.a.) [9] and Holbinsk (Sumsunursk) complex; and host ore bodies of Riphean-Vendian volcanogenic-sedimentary formation in Ilchirsk suite, all of which are interlaying within Holbinsk fault, found in deep faulted Okino-Kitoy system. This formation of up to 400m. includes alternating layers and bands of sandstones, siltstones, limestones, mafic and acid effusive rocks, quartzite, black shales. These rocks have been subjected to intensive folding deformation –this multi-folding process is combined with intensive hydrothermally altered rocks throughout the thickness, including granites adjacent to volcanic-sedimentary formation of Ambartogolsk massif, hosting gold-bearing veins. Subvertical orientation of stringer-porhyry ore deposits is analogous to oriented tectonic joints in subvertical deep fault zones. Near- veined metasomatic column in Ambartogolski massif plagiogranites includes the following mineral zones (table 1).

Table 1. Mineral zoning of wallrock metasomatic aureoles in mesothermal Zun-Holba deposit.

Mineral zones	Rock mineral composition (italicized minerals fading out to outer rear zones)
Frontal	*quartz+ sericite+ calcite+ leucoxene+ rutile+ ilmenite+ magnetite+ albite+ chlorite+ zoisite+ epidote+* biotite (parent rock)
Epidote-chlorite	*quartz+ sericite+ calcite+ leucoxene+ rutile+ ilmenite+ magnetite+ albite+ chlorite+ zoisite-epidote*
Chlorite	*quartz+ sericite+ calcite+ leucoxene+ rutile+ ilmenite+ albite+ chlorite*
Rear (beresite)	*quartz+ sericite+ calcite+ leucoxene+ rutile+ sulfides+gold+silver*
Axial (quartz vein)	*quartz+ calcite+ sericite+ sulfides+gold+silver*

The frontal zone with a thickness of several tens meters embraces massive coarse-grained (grain size of up to several mm) gray plagiogranite composed of randomly oriented narrow tabular plagioclase crystals (andesite up to 60 vol. %) intertwining with gray and dark gray quartz (up to 25 vol.%) incorporated into isolated large oval grains of nested lenticular aggregates and brownish biotite grains and flakes (up to 30 vol.%) with impurities of xenomorphic orthoclase grains, single microcrystals of hypersthene, sphene, zircon, apatite. Rocks preserve a hypautomorphic structure – weakly hydrothermal alteration with new growth volume of up to 10 %. This alteration is reflected in the formation of thin albite rims on the plagioclase crystal periphery corroded by ultimate quartz, in uneven, usually weakly sericite replaced plagioclase crystals incorporated with calcite, variolated zoisite segregation and mature epidote. Brownish biotite forms large flakes (up to 3 mm) and accumulation of small pale green chlorite- substituted scales often incorporated with epidote. Biotite cleavage results in leucoxene-rutile accumulation and lenticular quartz segregation.

In the several meter epidote-chlorite zone there are no biotite and pyroxene – they are dissolved on the inner frontal zone boundary. Rock alteration intensity has increased due to increasing neocystallisation up to 25...30 vol. %, basically, epidote, chlorite, sericite. Because of abundant green minerals the rocks have a green tint. Hypautomorphic structure groups with rounded elongated clastic grains.

Short-columnar (up 1 mm) plagioclase crystals include heavy, usually not massive and, rarely, impregnated sericite and/or and / or variolated zoisite segregation (up to 30 vol.%), partially epidote-substituted in microcracks. Although polysynthetic plagioclase twinning is veiled, it is sometimes visible. On the periphery some plagioclase crystals are rimmed by pure albite bands, indicating plagioclase deoxidation. In sericite-substituted plagioclase generating silica excess is crystallized into thin fine-grained quartz aggregates within the crystal rims, whereas, the boundaries between quartz aggregates and plagioclase crystals are not distinct, but can be visible only under a microscope.

Quartz is large (up to 2…3 mm) and suboval segregations, but the number of fine-grained aggregate intergrowths, being nests, lenses and veins, increased. There is little pale green chlorite, which form as isolated rare flakes up to tenths of a millimeter or as a cluster of up to 3 mm in diameter. Sericite-substituted chlorite associated with leucoxene (rutile), quartz, rare magnetite and ilmenite grains are found along the cleavage. There is an insignificant impurity of xenomorphic calcite grains can be observed in the rocks. In the chlorite zone within the first meters there is no epidote-zoisite, but the rock still has a pale green color because of chlorite. There is a massive structure, the coarse crystalline texture predominately includes rounded elongated clastics with hypautomorphic relics.

Plagioclase of up to 55...60 vol.% is practically substituted by sericite including impurities of xenomorphic quartz and calcite grains or without. Under conditions of complete sericite aggregate-substituted plagioclase, these clusters preserve their former crystal form. There are rare preserved xenomorphic orthoclase grains, which are poorly sericite- substituted and pelitized.

Quartz (up to 35 vol. %.) with coarse, sometimes suboval grains, can be found in fine-grained aggregates. Coarse grains have smooth, linear or even winding contour; the aggregates develop quartz-like structure with toothed bay-like grain intergrowths. Green sericite-substituted chlorite flakes (up to 2…3 vol.%, size to tenths millimeter) are found along the cleavage, where fine-flaked muscovite incorporated with fine xenomorphic grains (hundredths of mm) of calcite (up to about 4…5 vol.%), quartz, ilmenite, leucoxene and rutile.

Well-developed metasomatites in the quartz-vein exocontacts are greenish-gray fine-grained massive porphyritic rock with rounded elongated clastic mass. Alternating composed sericite and quartz, including proportionally 50 * 50 vol.% with impurities of fine xenomorphic grains of calcite (up to 5 vol.%) and comparably sized (to tenths of a mm) crystals of pyrite (up to 2 vol.%) and leucoxene (rutile).

Porphyritic-like rocks (up to 3 mm) are a result of oval dark gray quartz grains comparable to those in plagiogranites immersed in fine flaked sericite, quartz-sericite mass. The grains have a smooth non-corroded contour in contrast to fine (up to tenths mm) quartz grains, having jagged, crow contours and, intergranular gradation forming clusters of subisometric, lenticular, vein-lenticular sericite mass.

Substituted plagioclase crystals by pure sericite clusters preserve their shape. Sericite clusters having substituted pigment minerals – chlorite, epidote and further biotite are contaminated with impurities of leucoxene and rutile. Pyrite crystals are usually rimmed by quartz grains, including flame-shaped ones.

According to the mineral composition these rocks are classified as beresite. Quartz-sericite rear zone rimmed axial monocomponent gold-ore with carbonates, sulfides, gold-bearing quartz veins in the metasomatic halos of deposits were not found. Formation of calcite, pyrite and other sulfides, abundant potassium mica – sericite in rear (beresite) zone of near -veined metasomatic columns formed in carbonate-free, sulfide-free plagiogranites indicate the fact of potassium, carbon dioxide, sulfur, metals input (diffusion) [10] from fractured metal-bearing fluids into porous fluids of wallrocks, that is potassium- sulfur- carbon dioxide metasomatism type.

3. Conclusion

Described metasomatic zoning was developing in accordance to the phenomenon of differential component mobility – highly intensive rock alteration, where one mineral dissolved on the inner boundary of each of the forming zones sequentially– biotite, epidote, chlorite, albite, while reacting mobile components diffused into fracture solutions and/or incorporated into the forming minerals.

Two-component and mono-component composition of column inner zones is not developed due to the invading counter diffusion with component fluids, from the fractured fluids in pores, sericite and their further incorporation into rock masses of sericite, calcite and sulphides. Counter-diffusion of components complicates the mineral composition of the inner metasomatic column zones. It is also involved in the polymineral composition formation inner mineral metasomatic column zones the proof of which is the pulsating regime of hydrothermal systems [11], as a result of which, there exists a continuous spatial alignment of mineral associations and complexes – derivatives of following fluids significantly different from the dissolved solid compositions. Korzhinsky metasomatic zoning theory was developed for continuous outflowing fluid from the generation source throughout the process. The above-described metasomatic columns embrace two metasomatite formations – beresite in inner zones and propylitic – in peripheral zones, which could be found in many gold deposits [13].

Acknowledgments

Research was supported by National Research Tomsk Polytechnic University, project No. 114-2014

References

[1] Korzhinskii D S 1982 *Theory of Metasomatic Zoning* 104
[2] Borodaevsky N I and Cher S D 1967 The near-ore changes in gold deposits. *Papers of Central Geological Research Institute for Nonferrous and Precious Metals* **76** 113–26
[3] Zharikov V A, Rusinov V L, Marakushev A A Zaraiskii G P et al 1998 *Metasomatism and metasomatic rocks* (Scientific World Publishing House) p 492

[4] Korobeynikov A F, Ananiev Y S, Gusev A I, Voroshilov V G, Nomokonova G G, Pshenichkin A Y and Timkin T W 2013 *Ore-metasomatic and geochemical zoning of gold fields and deposits of fold belts of Siberia* (Tomsk: Publishing House TPU) p 458

[5] Rusinov V L, Borisovsky S E, Rusinova O V, Kryazhev S G, Shchegolkov Yu V and Alysheva E I (2008) Wall-rock metasomatism of carbonaceous terrigenous rocks in the Lena gold district *Geology of Ore Deposits* **50** 3–46

[6] Barsukov V L and Borisov M V 1987 Modeling of geochemical phenomena related to self-organization of a structure of the hydrothermal waters flow *Geochemistry International* **1** 87–101

[7] Korenbaum S A 1988 Moving and separation of substance in hydrothermal halos *Bulletin of Moscow Society of Naturalists* **63** 94–104

[8] Korotaev M Y 1990 Zoning of heterogeneous hydrothermal systems *Proceedings of the USSR Academy of Sciences Geological series* **7** 133–45

[9] Mironov A G, Roschektaev P A, Zhmodik S M, Kulikov A A and Karmanov N S *Zun-Kholbinskoe deposi*t URL: http://www.geokniga.org/books/4678

[10] Kucherenko I V 2010 Hydrodynamics of the fracture-pore fluid-rock interactions and mass transfer mechanism in the adjacent-fracture hydrothermal metasomatic processes *Prospect and protection of mineral resource*s **11** 37–43

[11] Baksheev I A 2001 Genesis of metasomatic rocks and mineralized veins at the berezovskoe deposit, central urals: Evidence from fluid inclusions and stable isotopes *Geochemistry International* **39** S129–S144

[12] Kucherenko I V and Zhang Yuxuan 2015 Metallogenic problems of hydrothermal gold deposit formation: facts and arguments *IOP Conf. Ser.: Earth Environ. Sci.* **24** 012024

[13] Zharikov V A 1966 Some regularities of metasomatic processes *Metasomatic alteration of wall rocks and their role in ore formation* 123–25

Post – sedimentation influence on filtration capacity reservoir rock properties (Pur-Tazov oil\gas-bearing area)

E Isaeva[1,3], N Stolbova[1,4] and T Dolgaya[2]

[1] Department of Geology and Mineral Exploration, Institute of Natural Resources, National Research Tomsk Polytechnic University, 30 Lenin Ave., Tomsk, 634050, Russia
[2] Department of Foreign Languages, Institute of Natural Resources, National Research Tomsk Polytechnic University, 30 Lenin Ave., Tomsk, 634050, Russia

E-mail: [3] elena_isayeva.89@mail.ru, [4] dtf2@yandex.ru

Abstract. The processes of the second mineral formation (kaolinite, carbonates and micas) were identified during the post-sedimentation transformation studies in oil\gas deposits. Besides, quartz regeneration, solid product destructive formation processes and hydrocarbon oxidation processes were -determined. Correlation analysis of the mineralogy and petrophysics data revealed the post-sedimentation influence factors on the reservoir properties of deposits. It should be noted that the second kaolinite composition increase results in water saturation and density decrease, porosity and, especially, permeability increase. Quartz regeneration and second mica formation deteriorate the reservoir properties or poorly influence them. The hydrocarbon decay and oxidation products, as well as secondary carbonate seal the void space, replace the soluble rock debris and sharply deteriorate the reservoir properties of oil andgas deposits.

1. Introduction

Nowadays, oil and gas–bearing forecast provides necessary knowledge on hydrocarbon accumulation mechanisms, concepts of sedimentary rocks formation and transformation, especially organic geochemistry, post-sedimentary reservoir and cap rocks alterations for enhanced oil recovery and deposit identification [1-4]. To understand the processes related to the formation of reservoir properties in hydrocarbon accumulation zones we should determine and specify the geophysical techniques data with the help of mineralogy-petrography and other analyses [5].

The mineralogy-petrographic research results were based on transparent, polished (thin) sections (302), laboratory data of open porosity, permeability, water saturation, rock density identification, as well as X-ray analysis. The research area includes Jurassic-Cretaceous sediments from 8 deep stripped wells (Vankor-11, North-Vankor-1, West-Lodochnaya-1, Ichemminsk-1, Yachindinsk-1, North-Tukolandsk-1, Tukolando-Vadinsk-320) drilled in Purtazov oiland gas-bearing area (N-E Western Siberia). The following suites were stripped: Tanamsk K_2tn, Salpadayakhinsk K_2sl, Nasonovsk K_2ns, Dorozhkovsk K_2dr, Dolgansk K_2dl, Yakovlevsk K_1jak, Malokhetsk K_1mch, Sukhodudinsk K_1sd, Nizhnekhetskaya K_1nch, Yanovstansk J_3-K_1 jan, Sigovsk J_3sg, Tochinsk J_{2-3}tch, Malyshevsk J_2ml, Leontyevsk J_2ln and Vymsk J_2vm.

The Jurassic-Cretaceous sediments are confined to the Northern part of the Western Siberia Platform being subjected to differentiated uplifting [6]. According to V.A. Krinin data the investigated area is inclined towards the western border of Southern Hudoseysk rift zone, which is characterized as a linear tectonic structure via gravimetric and seismic data [7]. This indicates the fact of existing dislocation processes (permeable zones) within the investigated area and possible abyssal fluid penetration into Jurassic-Cretaceous sediments.

So, these sediments were subjected to the following alterations: diagenetic, katagenetic, metagenetic during sedimentation and evolution in the process of sedimentary basin subsidence (phased epigenesis), as well as dislocation-metasomatic [2] transformations associated to fluid migration and oil and gas-bearing formation processes during the areal inversion development. It is rather difficult to identify the mineral associations of phased and superposed epigenesis because of their property convergence [8].

During diagenesis and katagenesis the rocks were subjected to compaction, unstable feldspar debris disintegrated and underwent argillization. Similar, but more considerable sequence could have been observed during dislocation-metasomatic processes – intensive dissolving of terrigenous components and sandstone cement resulting in secondary void space formation. Besides, carbon dioxide metasomatosis stimulates secondary minerals formation: mica, clayey and flinty carbonates and other minerals [2, 9]. As a result, diagenetic and katagenetic intergranular structure is destroyed; intergranular channels are formed in- situ complex contacts. Moreover, under favorable conditions the zones of secondary highly-porous rocks with integrated-void spaces are formed.

2. Results and discussion

To identify the carbon dioxide metasomatosis influence on the oil and gas-bearing deposits, the relationships between mineralogical-petrographical and petrophysical parameters have been studied. Based on statistical analysis results in the most common influence relationships of the secondary minerogenesis and hydrocarbon oxidation product formation on reservoir properties were determined (table 1).

Table 1. Open porosity and permeability mean values (during post-sedimentation).

Secondary processes	Porosity, %	Permeability, n*10^{-3} мкм^2
Kaolinite formation		
Weak or absent	12.47	9.58
Moderate	15.19	7.96
Intensive	17.83	65.44
Carbonate formation		
Weak or absent	14.58	21.71
Moderate	15.44	17.97
Intensive	16.66	10.92
Quartz regeneration		
Weak or absent	15.74	16.06
Moderate	14.40	20.57
Mica formation		
Weak or absent	15.30	14.89
Moderate	14.88	23.52
Intensive	13.16	–

As a result of the carbon dioxide influence on the rocks, petrogenetic components were evacuated and pores in debris and cement were formed and cement to the extent of complete cement dissolution. Such rocks are considered to be more permeable. These evacuated components redeposit and form compacted cementation zones.

Muscovite and feldspar replacement of kaolinite occur with little evacuation. During stage processes, three-layer clay minerals are formed; high zone of secondary kaolinite is developed close to carbon dioxide deposits [2]. Either cations or significant silica (which is redeposited in the form of regenerated quartz rims) portions are evacuated during this process.

Statistic data identified that during secondary kaolinite formation porosity and, especially, rock permeability improve (table 1). Usually, cement microporosity is formed (figure 1).

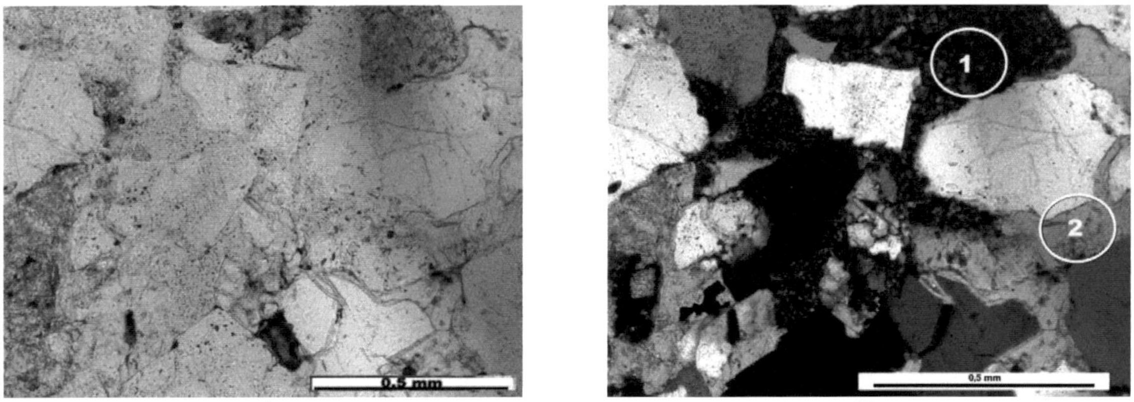

A B

Figure 1. Microporosity in kaolinite cement (1). Secondary carbonate formation (2). North-Tukolandskaya well -1, depth 2525. A – nicols ||, B – nicols ×. Lens magnification 10.

Besides, kaolinite is a hydrophobic mineral and has the lowest absorption capacity, i.e., the mineral does not absorb oil in the productive layers, but enhances recovery. During formation testing these kaolinite properties decrease due to the influence of alkaline -based clayey drill mud, which, in its turn results in hydrophobic filming of crystals, sealing of cracks and rock cavities and further recovery constraints; and, eventually, mud filtrate with an oil film. To eliminate this problem relevant drilling technology should be applied, for example, CO_2 injection [1, 10-12].

Table 2 shows the results of statistical data manipulation (coefficient permeability (Kper), water saturation (Sw), rock density and kaolinite content) calculated for Ichemminsk well -1. The following relationships have been identified: amount of secondary kaolinite is directly proportional to rock permeability and, inversely, proportional to water saturation and density of deposits (table 2).

Table 2. Petrophysical properties of clastic sedimentary rocks and newly-formed kaolinite ratio (Ichemminsk well -1).

№	Suites	Rock	Reservoir class	Kper, n 10^{-3} mkm^2	Volume density, g/cm^3	Sw, %	Kaolinite, %
1	Yakovlevsk	Sandstone	2	737.1	1.99	13.3	84
2	Yakovlevsk	Sandstone	3	299.6	1.98	19.8	87
3	Yakovlevsk	Sandstones	3	236.07	1.99	20.9	80
4	Yakovlevsk	Sandstones	4	58.71	2.04	26.8	72
5	Yakovlevsk	Sandstones	5	1.78	2.16	41.7	66
6	Nizhnekhetsk	Sandstones	1	1310,.8	2.13	14.7	31.5
7	Nizhnekhetsk	Sandstones	2	702.42	2.12	16.1	27.7
8	Nizhnekhetsk	Sandstones	3	228.78	2.14	20.4	33.3
9	Nizhnekhetsk	Sandstones	4	59.45	2.21	25.9	35.7
10	Nizhnekhetsk	Sandstones	3	188.22	2.13	19.5	42
11	Nizhnekhetsk	Sandstones	4	39	2.18	34.4	26.7
12	Nizhnekhetsk	Sandstones	5	3.44	2.2	47.1	38
13	Nizhnekhetsk	Sandstones	No	0.13	2.53	89.7	33

Secondary carbonate (one of the end products of carbon dioxide metasomatosis) formation influences clastic rocks, the occurrence of which depends on the pressure gradient decrease and environment pH increase up to slightly alkaline (~8 and higher) conditions. Compacted zones of secondary carbonate might be formed in terrigenous rocks under conditions of repeated reactions. In broader terms, carbonate formation sharply decrease reservoir properties, rock permeability, in particular (table 1).

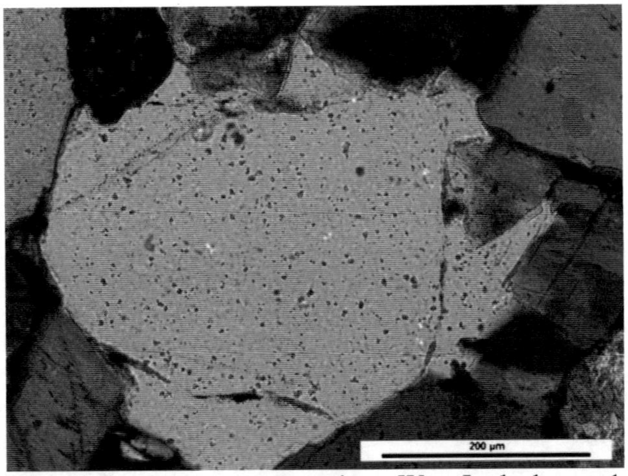

Figure 2. Quartz regeneration. West-Lodochnaya-1 well. Depth 3350. Nicols ×. Lens magnification 20.

Besides the two above-mentioned evacuation zone (kaolinite formation) and substance deposit zones (carbonate formation), there is an intermediate zone of quartz regeneration, as well as insignificant secondary kaolinite formation, excluding carbonates.

Quartz is the most stable to disintegration and mineral dissolution of all sedimentary rocks. Experimentally, quartz regeneration is related to structure dissolution under pressure. Thus, at pH=9.88 (according to Kashik data, 1965) [13]) silica becomes mobile. If there are aggressive compounds (K_2CO_3 and Na_2CO_3),and in the case of pressure increase, quartz can be dissolved even in conditions with lower pH (close to 8-9).

During dissolution and corrosion, wavy debris rims could be observed in quartz. As a result, there is a considerable amount of silica acid in the migrating solution, redepositing as regenerated growth rims on quartz debris (figure2). Due to granulometric properties increase of quartz grains, void space intake occurs, i.e. reservoir properties deteriorate.

Observed rims often reflect debris defects and differ from those of initial quartz grains in purity. As a result of epigenetic rock alterations, a band- zone was identified between the growth rim and quartz debris nucleus as microscopic foreign inclusions of ore-forming environment.

Intensive silica dissolution and its migration into adjacent layers could result in the formation of flinty interlayers and lenses which, in its turn, decrease reservoir properties (table1).

Intensive muscovite and biotite mica (from hydromica) formation could be observed in samples with increased shaliness. According to statistic data, reservoir rock properties are insignificantly decreased during these processes (table 1).

In the studied oil and gas-bearing deposits solid products of oxidation and degradation in hydrocarbons could be found in cracks, intergranular spaces and along argillizied feldspars, the amount of which is almost 15-20% in terrigenous deposits.

Statistical data manipulation identified that hydrocarbon oxidized products formed in the systems defined by B.A. Lebedev classification "H-H", "R-H" (rock hydrocarbon), "W-H" (water

IOP Conf. Series: Earth and Environmental Science **27** (2015) 012004 doi:10.1088/1755-1315/27/1/012004

hydrocarbon) [2] deteriorate the reservoir properties filling the pore space and preventing further migration.

Besides rock chemical alterations, dislocation processes are observed. It is rather difficult to evaluate their ambiguous influence on secondary void space formation. Gaseous-interstitial fluids via brittle deformations as migration channels could either provide secondary integrated void space or seal permeable zones by hydrocarbon solid products of oxidation (bitumen) and newly-formed minerals (carbonates, quartz, mica, etc.), resulting in decreasing reservoir volume.

Dislocation processes can particularly be observed on the feldspar and quartz clastics. Brittle and plastic deformations are identified among tectonic deformations. Destruction (shearing) of clastics is characteristic for brittle deformations, resulting in cataclasis of separate fragments, weak fracture zones, interfragmental stripping and induced porosity are determined. Plastic deformation is followed by displacement of quartz clastic extinction, its granulation, regeneration and recrystallization. Under conditions of intensive plastic deformation, oriented element destruction and distinctive cataclastic structures are formed. Sometimes, thin cracks of stripping are filled with hydrocarbon products, rarely – with newly-formed minerals.

Induced porosity with integrated void space, caverns and microporosity are formed under positive dislocation-metasomatic processes. At the same time, reservoir properties improve.

Based on the investigation of Jurassic-Cretaceous sediments Lower Cretaceous Nizhnekhetsk and Yakovlevsk suites are characterized by highly intensive dislocation-metasomatic activity and, consequently, probable improved reservoirs. Sukhodudinsk, Malokhetsk and Dolgansk suites are characterized by moderate and intensive rock transformation. These zones are characterized by cataclasis, rock fracturing, increased porosity and permeability and significant secondary mineral formations.

3. Conclusion

Obtained data showed that Jurassic-Cretaceous terrigenous deposits in N-E Western Siberia were subjected to both stage and dislocation-metasomatic processes, which resulted, in one case, in the formation zones of secondary high porosity rocks with integrated void space, or in the second case, the decrease of reservoir volume.

It was identified that as a result of carbon dioxide fluid influence, rock dissolution and secondary kaolinite formation occurred, which, in its turn, is hydrophobic mineral forming cement microporosity. These processes improve the reservoir properties, especially rock permeability. Secondary mineral formations (quartz regeneration, carbonates and mica formation) and solid bitumen formation either insignificantly influence or deteriorate reservoir properties.

Thus, forecasting effective reservoir zones forecast is related to the zones of intensive carbon dioxide metasomatosis developing in the “Water-Rock” system and resulting in integrated void space formation. It is necessary to consider the following: the more active carbon dioxide metasomatosis in the sedimentary basin is, the more pores, heterogeneous cavities in newly-formed void space are. In thus case, it becomes more difficult to identify and detect effective and non-effective reservoirs.

References

[1] Vilkovskaya I Yu, Serdyuk Z Ya, Zubareva L I, Kirilova N V, Stefanenko O N and Bubnova E S 2013 Postsedimentary changes in rocks of the Jurassic-Neocomian strata, their influence on reservoir properties and oil recovery *Materials of VII All-Russia Conference of lithologists “Sedimentary basins, sedimentary and post-sedimentary processes in geological history” Research Institute of Geology and Geophysics* (SNIIGGIMS) **1** 163–67

[2] Lebedev B A 1992 Geochemistry of epigenetic processes in sedimentary basins (JT) 239

[3] Lifshits S H 2008 The role of deep fluids in the formation and genesis of petroleum deposits *M. Geos* 278–81

[4] Perozio G I 1990 *The epigenesis of terrigenous sediments sedimentary rocks of the Jurassic and Cretaceous of the South-Eastern part of the West Siberian lowland* (Moscow: Nedra) 159

[5] Konovalenko S I and Nedolivko N M 1990 The possibility of using mineralogical information to solve applied problems oil Geology *Prirodokompleks Tomskoy oblasti. Vestnik TSU* 34–38

[6] Nikishin A N et al 2010 Tectonic history of the Siberian platform in the Vendian and the Phanerozoic *Vestnik Moscow University, Geology* **1** 3–16

[7] Krinin V A 2011 Tectonics of the basement and evaluation of oil resources the Jurassic and Cretaceous sediments of the North-East of the West Siberian plate within the Krasnoyarsk territory *Gorniye Vedomosty* **9**(88) 16–24

[8] Sergeeva E I 2004 Epigenesis of sedimentary rocks *Proc. the manual.* (SPb.: SPb University) p 152

[9] Sahibgareev R S 1989 *Secondary changes of collectors in the process of formation and destruction of oil deposits* (Leningrad: Nedra) p 260

[10] Predtechenskaya E A 2002 Effect of CO_2-bearing hydrothermal solutions on reservoir properties of deep sandy rocks (on experimental data) *Izvestia TPU* **305**(8) *Geology and exploration of oil and gas fields* 62–70

[11] Frank T H and Burchfield T E 1987 Researched aimed at immiscible CO_2 flooding *Oil and gas Geology* **85** 76–82

[12] Hervey J R and Jakovakis A S 1991 Performance review of miscible CO_2 tertiary project: Rangely Weber Sand Unit, Colorado *SPE Reservoir* **2** 163–68

[13] Kashik S A 1965 Replacement of quartz by calcite in sedimentary rocks *Geochemistry Journal* **2** 180–87

Oriented core application in texture analysis of J_1 formation in Kazan oil-gas condensate field (Tomsk Oblast)

L Krasnoshekova[1,4], D Cherdansteva[1,5], M Yurkova[2] and R Abramova[3]

[1] Department of Geology and Mineral Exploration, Institute of Natural Resources, National Research Tomsk Polytechnic University, 30 Lenin Ave., Tomsk, 634050, Russia
[2] Department of Oil and Gas Industrial Facilities Design Construction, Institute of Natural Resources, National Research Tomsk Polytechnic University, 30 Lenin Ave., Tomsk, 634050, Russia
[3] Department of Foreign Language, Institute of Natural Resources, National Research Tomsk Polytechnic University, 30 Lenin Ave., Tomsk, 634050, Russia

E-mail: [4] krasnl@tpu.ru, [5] dannia@tpu.ru

Abstract. The paper describes the results of the characteristic structure features of oil-bearing rocks via paleomagnetic oriented cores. Volume core model is plotted on the basis of circular panoramic images. In applying scanned panoramic core it is possible to determine the azimuth of terrigenous material transportation and its course and to describe its sedimentation environment in details.

1. Introduction

Kazan oil/gas-condensate field is located in east central Western Siberian lowlands within Tomsk Oblast. The geological profile embraces argillo-arenaceous sediments of Mesozoic-Cenozoic sedimentary mantle, superimposing the metamorphosed rocks of Paleozoic folded basement. Layer J_1 of Vasugan suite is a major hydrocarbon-bearing formation, extending horizontally and vertically and tapped in all drilled wells within Kazan field structure.

2. Research methods

Research target involves core samples of layer J_1 oil-bearing rocks from Kazan field wells. Sampling description embraced a core column of 130m. Paleomagnetically, there are 25 oriented core samples from two field wells. Core panoramic scanning was conducted on the basis of present sample data. Such core sweeping provides the possible tracing of changing texture features in volume, including orientation and layer incidence angle, distribution patterns and relations, quantitative ratio and other characteristic features of sedimentary formations.

3. Results and discussion

Layer J_1 sediments include medium-fine grained sandstones and argillo-arenaceous rocks. Previous studies in lithological and mineralogical-petrographical rock features have been described [1]. Based

IOP Conf. Series: Earth and Environmental Science **27** (2015) 012005 doi:10.1088/1755-1315/27/1/012005

on the classification of textures and bedding types of core samples from investigated intervals, the following texture and rock stratification types were determined (table 1, figure 1).

Facies dismembering involves the macrofauna classification of Jurassic sediments [2]. Analysis of identified texture-structure features and rock granulometry, which, in its turn, include fauna remains and clastics revealed the sedimentation environment as basin type. It was determined that sedimentation occurred in shallow-water basins under conditions of flowing shallow open basin water (SBW) or under conditions of offshore (open) basins (OB).

The first (SBW) are characterized by intensive matter accumulations, transported from outboard shore to its rear area by alongshore currents. The rocks formed in such environments are typically well-sorted and rounded clastics, low-inclined cross-bedding and regionally banded structures.

The second (OB) can be found in outboard basin shores. The environment here is predominately coastal-marine, associated with lower lowland rivers. This macrofacies embraces clay, aleurolite and sandstone formations of marine sediments Depending on the marine fauna, which can be found in the sediments, these sediments are transported to this or that facies. The rocks formed under such conditions possess horizontal bedding, and are well-sorted.

Table 1. Structure and bedding types in J_1 sandstone layer, well X, Kazan field.

Structure	Bedding
Unstratified: homogeneous (massive), nodular and lumpy	Lenticular (lenticular-wavy), linguloid, weakly-strongly disrupted
Indistinct bedding: placic with rare inclusions and laminae	Wavy (cross) non-parallel, weakly-strongly disrupted
Disrupted creeping and roiling	–

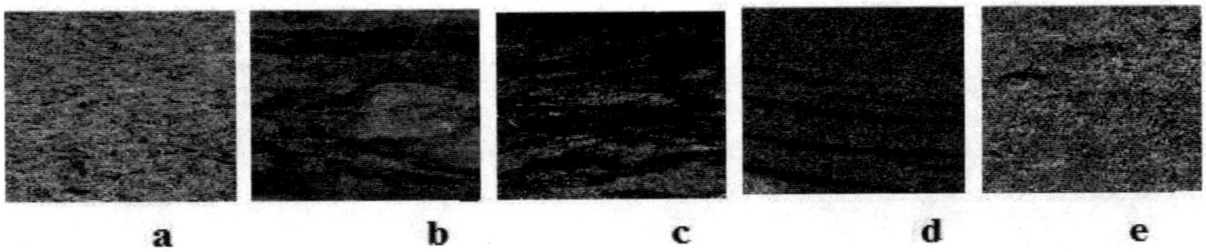

Figure 1. Examples of structure and bedding types of rock layer J_1

a) Fragment of homogeneous massive structure of solid fractures shells.

b) Fragment of placic fine-grained alliterate-argillite sandstones.

c) Fragment of lenticular bedding, governed by the argillite in sandstones as laminae and lenses of first millimeters to 3 cm thickness and strong irregular gradual boundaries.

d) Fragment oriented to locations of weakly disrupted aleurolite-argillite sandstones.

e) Fragment of indistinct structure with rare inclusions.

Applying circular scan made it possible to plot a volume model of the investigated core samples in order to determine and define the rock structure features. Preliminarily, via paleomagnetic method, the core was north – oriented, and in this case, the viscous magnetization component, i.e. the angle between the present north line and sediment formation was scanned on the magnetometer, determining the sample orientation to Jurassic period. Plotting volume core scan is less time-consuming and resource -consuming comparable to 3D-scanning of inner well bore (figure 2).

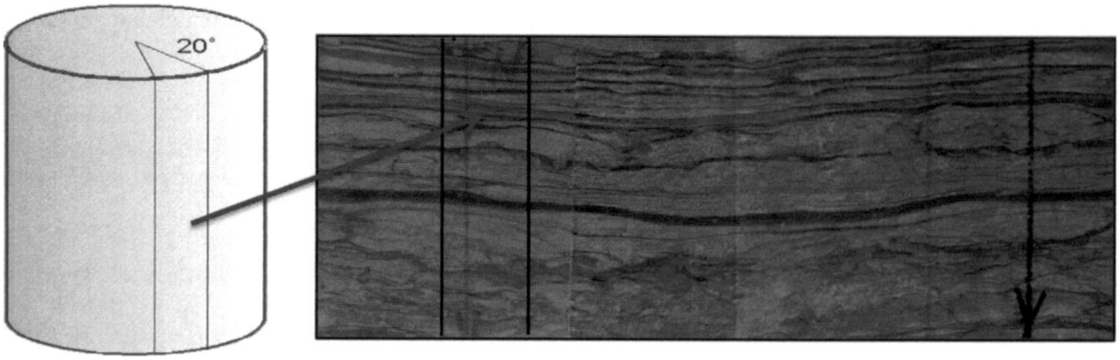

Figure 2. Core scan profile.

Based on produced core samples the structure types were classified and bedding types were also defined. In general, a log of similar structure types was plotted northward in depth (figure 3).

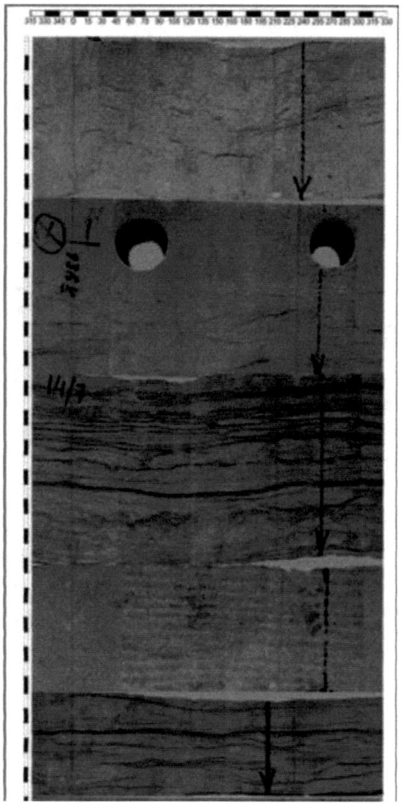

Figure 3. Core scan fragment at 360°, northward- blue arrows (Kazan oil field, layer J_1, well X).

Based on paleomagnetically-oriented core samples, the azimuth of matter transportation and course was determined, and respectively, sedimentation environment was described in details. In this case, the circular scan of core sample was adjusted so as north corresponded to zero degree (zero mark). Further, the maximum and minimum layer mark was established and the corresponding angles were determined by a scale.

For example, in figure 4, the maximum and minimum curvature point in discussed sample layer is located on the azimuth of 87° and 270^0,which, in its turn, indicates that the direction of matter transportation and course is W-S-W to E-N-E. The low angle wavy cross bedding reveals existing unstable environments in which matter transportation and further accumulation occurs. This is characteristic of coastal-marine environments. Obtained results are completely verified by facies distribution data and microstructural – morphological investigations of core samples from oil-producing reservoirs [3].

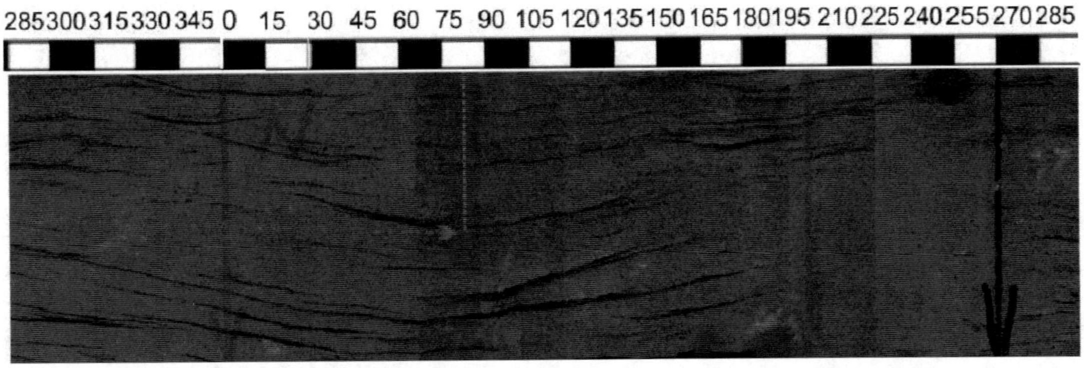

Figure 4. Azimuth line determination of matter transportation and lamina formation based on sample X (Kazan oil-gas condensate field, layer J_1).

4. Conclusion

It should be noted that recent paleomagnetically-oriented coring has been applied in investigating the sedimentary rocks in hydrocarbon-bearing deposits. The produced circular (volume) core scan not only reveals the spatial image of bedding and structure features, but also presents details of the rock fabric and, respectively, accurate data of sediment formation environments.

References

[1] Krasnoshchekova L et al 2015 Litho-mineralogical characteristics and facies formation conditions of oil-reservoir rocks J_1 and J_2 in Kazan oilfield (Tomsk Oblast) *IOP Conf. Ser.: Earth Environ. Sci.* ***24*** 012018

[2] Alekseev V 2007 *Atlas fatsiy yurskih terrigennyih otlozheniy (uglenosnyie tolschi Severnoy Evrazii)* (Ekaterinburg: Izd-vo UGGU) p 209

[3] Krasnoshchekova L and Cherdantseva D et al 2014 Microstructual analysis of quartz grains in Vasugan suite sandstones of layer U1-21 in Kazanskoe deposit *IOP Conf. Ser.: Earth Environ. Sci.* **21** 012007

Mineral-petrochemical wallrock alteration of rocks in Bericul gold-ore deposit (Kuznetsk Alatau)

I Kucherenko[1,3], Zhang Yuxuan[1] and R Abramova[2]

[1] Department of Geology and Mineral Exploration, Institute of Natural resources, National Research Tomsk Polytechnic University, 30 Lenin Ave., Tomsk, 634050, Russia
[2] Department of Foreign Language, Institute of Natural Resources, National Research Tomsk Polytechnic University, 30 Lenin Ave., Tomsk, 634050, Russia

E-mail: [3] kucherenko.o@sibmail.com

Abstract. The distribution of mineral associations in near-veined zonal propylite-beresite metasomatic columns of mesothermal Bericul gold-ore deposit was analyzed. However, the polymineral composition in the inner (axial and adjacent with it rear) zones is inconsistent to the existing metasomatic column theoretical model. According to Korzhinskii metasomatic zoning theory, implied monomineral (quartz) and binary-mineral (quartz, sericite) compositions are characteristic of axial and rear zones, respectively. In common with above-mentioned facts, the zoning formation of differential component mobility is influenced by two additional factors: counter diffusion of components from fractured fluids into pores and diffusion mechanism of mass transfer it's from pores fluids into fractured of rock-fluid systems.

1. Introduction

Near-veined zonal metasomatic aureolas involve both beresite and propylite metasomatite formations in mesothermal gold-ore deposits. The inner zones of aureolas are composed of beresites and associated rocks, while outer zones of aureolas are propylites.

There is misunderstanding in defining the mineral composition of the inner zones - axial (ore-bearing quartz veins) and adjacent more outer zone. According to Korzhinskii metasomatic zoning theory [3] and corresponding theoretical model of metasomatic column, the rear zone is composed of quartz-sericite and is termed as beresite. Taking into consideration the previous 50-year observation data [2], the rear zone of naturally-occurring near-veined metasomatic columns are composed of mineral enriched rocks – quartz, sericite, carbonates, pyrite (sulphides). This rock is termed as beresite too. It should be mentioned that this misunderstanding still exists [4, 7, 8].

The investigation of naturally-occurring metasomatic columns could prove how the theory of near-fractured (near-veined) metasomatic zoning responds to the natural process itself, as well as the reasons of above-mentioned misunderstanding.

The research involved the investigation of the wallrock alterations in the mesothermal quartz-veined Bericul gold-ore deposit. Information on the geological structure of this deposit has been imported from [5].

Bericul gold-ore deposit is located in the northern area of Kuznetsk Alatau, 70km. southward from Trans-Siberian railway station Tjazhin (figure 1).

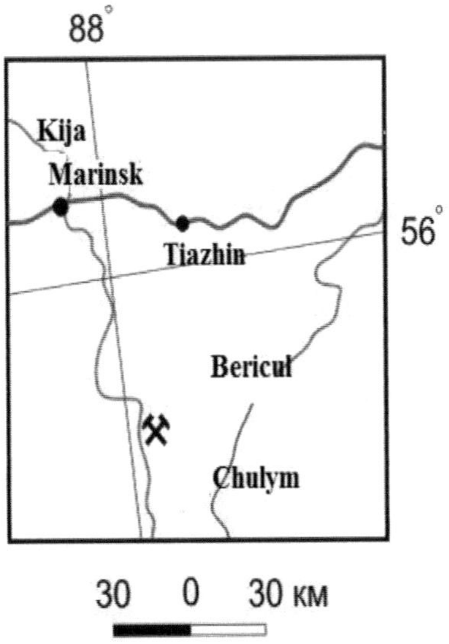

Figure 1. Geographical location of Bericul gold-ore deposit.

According to geology, there are several development stages [1]: oceanic (R_3–Є_1), peri-continental (R_3–Є_1), island-arc (V–Є_2^1), colliding (Є_2^2–S), rifting intraplate continental (D_{1-2}–K). This deposit developed during the colliding stage (Є_2^2–S) and it is 474 Ma.
Bericul deposit parallel with other deposits (Komsomolsk, Kommunar, Central, Fedorovsk and others) is controlled by Kuznetsk Alatau deep-seated faults.

2. Geological structure of Bericul deposit

Bericul deposit is located in the south-western flank of the Bericul syncline. The syncline fold is composed of Belokamen suite mramorized limestones (Є_1) and Bericul suite andesite-basalt volcanites (Є_2). In the eastern deposit area there is intermediate-mafic intrusive stock which formed the western margin of Dudetsk granitoid pluton in Early Palaeozoic Martayginsk complex. The stock rocks (intrusives) graded slowly to andesite-basalts and basalts in volcanogenic thickness. This factor indicates that gabbroids and dioritoids were formed as a result of magmatic replacement. The contacts between limestone thicknesses including volcanites and the eastern stock boundaries dip towards one another, resulting in preserved volcanites gradually converging N-W, while in depth, as a wedge increasing in width towards S-E. The volcanites in the southern area of the deposit host stock rocks identical to the eastern stock rocks.
Alternating basalt and andesite porphyrites and aphanitic volcanites of similar composition can be found in ore-hosting volcanic thickness. These rocks cut through numerous dykes of acid and mafic rocks.
Basalt porphyrite is composed of the following minerals: augite, labradorite with protobase and hornblende impurities, biotite having porphyritic structure, while there is basic plagioclase in porphyrite segregations. Andesite porphyrites are composed mainly of andesite and hornblende. Dykes of aplite include quartz, potassic feldspar, albite-oligoclase with impurities of apatite, zircon, magnetite and sphene. The pre-ore generation in dolerite dykes involves labradorite, biotite and titanaugite.
The ore bodies are confined to the volcanite thicknesses within quartz-veined ore deposits. There are two types of veins: massive (up to 20 m.) discontinuous shallow-dipping (up to 20°) vein № 4 trending EW was formed in the superimposed bed of the central displacement and submerging southward; steeply-dipping veins (more than 100) underlying vein № 4 with thicknesses ranging from 0.3 to 3 m., submerging NW at an angle of 30...60° and pinching into the contacts of SW limestones with the eastern stock rocks in the NE.

The ores embrace five mineral complexes. The first complex is formed as a result of the two generations of quartz, pyrite and arsenopyrite with impregnated gold, calcite, pyrrhotite, bismuth, bismuthinite emulsion. The second mineral complex is predominately composed of sulphides, zinc, copper, lead, as well as quartz with gold impurities. The third mineral complex includes carbonates intertwined with growths of quartz, arsenide and sulphite-arsenide cobalt and nickel with impurities of gold and other minerals and is located not only in gold-ore veins but also in lateral pre-ore and inter-mineral displacements. The fourth complex embraces predominately calcite and zinc, copper and lead sulphides. The final complex is composed of pure calc spar.

3. Results and discussion

Near-veined alterations in overlying basaltic and andesitic volcanic porphyrites within ore-hosting volcanic thicknesses and in aplite and dolerite dykes were investigated.
Near-veined metasomatic columns [6] with typical mineral zoning formed in all rocks (the following italicized minerals disappear in direction to rear zone).

Frontal zone:	Quartz+ sericite + leucoxene + rutile + magnetite ± pyrite ± calcite + albite ± chlorite± zoisite ± *actinolite* ± *tremolite*
Chlorite zone:	Quartz+ sericite + leucoxene + rutile + magnetite+ pyrite ± calcite ± dolomite + albite ± *chlorite* ± *zoisite* ± *clinozoisite* ± *epidote*
Albite zone:	Quartz+ sericite + leucoxene + rutile + magnetite+ pyrite ± calcite ± dolomite-ankerite ± siderite+ *albite*
Rear zone:	Quartz+ sericite + leucoxene + rutile + magnetite+ pyrite ± calcite ± ankerite ± siderite ± breunnerite

The thickness of the frontal zone extents to hundreds of meters, chlorite zone – tens of meters, albite zone – several meters, while the rear zone (beresite) ranges from 1...1.5 m. The rocks are slightly alternated in the frontal zone. Masses of newly-formed minerals increase towards the rear zone.
The decrease of the number of newly-formed minerals from the frontal to the rear zones, resulting from the dissolution of actinolite-tremolite, epidote (either zoisite or chlorite) and albite within the inner boundaries zones, respectively, is compensated by the formation of new minerals – ankerite, siderite and breunnerite, predominately within the inner zones.
According to the balance calculations, sodium is almost completely disposed (up to 97 wt. %) from the rocks, while silica- only 25wt. % during metasomatism. Beresitization is characterized by the increase of content of potassium-up to 540wt%, carbon dioxide (up to 2000 wt. %) and sulphide sulphur (up to 11705 wt. %). Potassium is captured in sericite, carbon dioxide-in carbonates, sulphur – in sulphides (pyrite).
The described rock hydrothermal alterations during the deposit formation involve factors determining the specific composition of neocrystallisation and the structure (mineral petrochemical zoning order) of near-veined metasomatic columns. The specific mineral composition correlates with the average-low fluid temperature regime during metasomatism. At the same time the specific mineral composition is insignificantly influenced by the wallrock chemical composition as impurities in nonstoichiometric minerals. In this case, the combinations of newly-formed minerals are regularly recurrent in apobasalt, apoandesite, apoaplite and apodolerite metasomatic columns. This dependency is possibly less extensive within the inner zones of the metasomatic columns. The decrease of the number of newly-formed mineral phases from the frontal to rear zones results from the dissolution of actinolite-tremolite, epodite-zoisite, chlorite, albite within the inner zones boundaries, respectively. Differential component mobility could describe this fact. This major factor forms the basis of the metasomatic zoning theory itself [3]. It also provides evidence of the correlation between the described naturally-occurring columns and the Korzhinskii theoretical model. Polymineral composition of the rear zone columns responds to the counter diffusion of components from fractured fluids into pores, which, in its turn, proves the diffusion mechanism of mass transfer into fractured porous, rock-fluid systems [9] and

is based on the results of calculation balance of component migration [6]. However, this is inconsistent with the existing theory of quartz-muscovite and mono-quartz composition. The spatial superposition of mineral assemblages deposited as a result of gradual alternating metalliferous fluids in the existing pulsing regime [5] of hydrothermal ore-bearing rock-fluid systems impedes the mineral-petrochemical zoning of the columns.

4. Conclusion

Based on the Bericul deposit investigation results, gold mesothermal deposits are characteristic of potassium-sulfur-carbon dioxide petrochemical profile of near-veined metasomatism, while in all rocks - polymineral composition within the rear (beresite) zone of near-veined metasomatic columns. The latter mineral-petro-chemical zoning sequence could be correlated to the Kucherenko model. This model reflects the structure of naturally-occurring near-ore metasomatic columns, the mineral petrochemical zoning of which is recurrent in rocks of different age and origin, and in hosting metasomatic gold deposits located in mountain-fold structures in southern Siberia [6]. Relationship of newly-formed minerals in all mineral -petrochemical zones of near-veined metasomatic deposit columns is evident. The mineral composition in inner, albite and rear zones is regularly recurrent; whereas, the composition in the rear zone is determined predominately thermodynamic and physico-chemical regimes of metalliferous fluids.

References

[1] Alabin LV and Kalinin Yu A 1999 *Gold Metallogeny of Kuznetsk Alatau* (Novosibirsk: SB RAS RC UIGGM) p 237
[2] Borodaevski N I and Sher S D 1976 Ob okolorudnyh izmenenijah v mestorozhdenijah zolota *Trudy CNIGRI Vyp.* **76** 113–126
[3] Korzhinskii D S 1982 *Teorija Metasomaticheskoj zonal'nosti* (Moscow: Nauka) p 104 (in Russian)
[4] Korobeinikov A F, Ananev Yu S, Gusev A I, Voroshilov V G et al. 2013 *Rudno-metasomaticheskaja i Geohimicheskaja Zonal'nost' Zolotorudnyh Polej i Mestorozhdenij Skladc atyh pojasov Sibiri* (Tomsk: TPU) p 458 (in Russian)
[5] Kucherenko I V, Gavrilov R Yu, Martinenko V G and Verkhozin A V 2008 Strukturno-dinamicheskaja model' zolotorudnyh mestorozhdenij, obrazovannyh v neslancevom i chernoslancevom substrate. P. 1. Berikul'skoe mestorozhdenie (Kuzneckij Alatau) *TPU Izvestija* **313** 11–26
[6] Kucherenko I V 2014 Prognozno-poiskovyj kompleks dlja mezotermal'nyh mestorozhdenij zolota. Ch. 3. Petrohimicheskij i geohimicheskij kriterii. *TPU Izvestija* **324** 39–45
[7] Rafailovich M S 2010 Netradicionnye mestorozhdenija zolota Kazahstana *Sovremennye problemy geologii i razvedki poleznyh iskopaemyh: Int. Conference Proceedings (Tomsk, 5-8 October 2010)* (TPU Publ.) 368–73
[8] Rusinov V L, Rusinova O V, Krjazhev S G, Shhegolkov Yu V, Alysheva E I and Borisovskii S E 2008 Okolorudnyj metasomatizm terrigennyh uglerodistyh porod v Lenskom zolotorudnom rajone. *Geologija rudnyh mestorozhdenij* **50** 3–46 (in Russian)
[9] Kucherenko I V 2010 Gidrodinamika treshhinno-porovyh porodno-fljuidnyh vzaimodejstvij i mehanizm massoperenosa v processah okolotreshhinnogo gidrotermal'nogo metasomatizma *Razvedka i ohrana nedr* **11** 37–43 (in Russian)

Crystal morphology and thermal EMF of pyrites in the western flank of Sukholozhsky gold ore field (Lenski ore area)

R Gavrilov[1,3], A Pshenichkin[1], M Ponamarenko[1] and R Abramova[2]

[1] Department of Mineral Exploration, Institute of Natural Resources, National Research Tomsk Polytechnic University, 30 Lenin Ave., Tomsk, 634050, Russia
[2] Department of Foreign Language, Institute of Natural Resources, National Research Tomsk Polytechnic University, 30 Lenin Ave., Tomsk, 634050, Russia

E-mail: [3] GavrilovRY@tpu.ru

Abstract. The investigated crystal morphology and thermal EMF of pyrites in the western flank of Sukholozsky ore field showed that the pyrite crystals have cubic habitus with a weakly-developed face {210}. The crystal faces {100} and {210} are covered with multiple irregular-oriented growth laminae. It has been determined that pyrites have such a property as p-type conduction and embrace insignificant thermal EMF variations. The results of the research indicated the fact of upper ore zone erosion in the western flank of Sukholozhsky ore field and its area potential at depth.

1. Introduction

Although gold mining in Lenski ore area has been conducted for more than two centuries, early geologic setting of this gold placer area and its developed socio-economic environment were described by V.A. Obruchev only at the beginning of the 20th century. Further geological prospecting involved a total gold placer exploration and increasing placer resources development. It was only during the latter half of the 20th century that the priority of geological prospecting embraced the exploration of gold orebodies. Regardless the prolonged exploration of gold orebodies [1, 2], their genesis and substance sources still remain the object in issue, as well as the factors and features of gold mineralization. The major task in geological exploration is to determine the upper ore zone erosion. The topomineralogical analysis effectively identifies the mineral formation and distribution in different geological systems, as well as the potential mineralization at depth [3]. The most widespread and extensively studied mineral in gold ore deposits is pyrite [4].

Based on the multi-year investigations of pyrite in gold ore deposits with composite genesis, a series of common factors was established [3, 5].

The crystal morphology of pyrite consistently changes within ore bodies and deposits. For example, pyrite in the top of ore bodies and deposits has predominantly cubic crystal habitus, where the face {210} is either absent or weakly and unevenly developed; while in the mid of ore bodies, especially in "ore chimneys" containing maximum gold grade, the face {210} is habitus in crystals and pyrite crystals are predominantly pyritohedron.

Pyrite is classified as a mineral with p-type impurities. Stoichiometrically-pure pyrite has such a property as p-type conduction. Monad and dyad elements (Ag, Sb, Hg, Pb, Zn, Cu), isomorphous constituent mineral, increase and / or decrease p-type constituent thermal EMF value, while triad and tetrad elements (Ni, Co, As, Ti, W), isomorphous substituent iron or sulphur, change pyrite p-type conduction to n-type conduction. It was also determined that [3] the following factors influence the value and sign of pyrite thermal EMF: temperature differential between cold and hot electrodes (the higher the temperature differential between the electrodes, the higher the p-type and n-type conduction), natural background and artificial radiation of the mineral, crystallographic line (embracing the measurement itself), deposit formation temperature and depth, and, respectively, the pyrite isomorphic volume of the crystal lattice.

The research showed that in the case of thermal EMF pyrites from ores and wallrock metasomatites of gold ore deposits with different genesis, the thermal EMF is responsive typomorphic attribute and consistently changes in time and space from pyrites with electron conduction within early high-temperature mineral associations and root section of orebodies to pyrites with mineral associations and mixed-type conduction from medium-temperature and mid-section of orebodies, as well as pyrites with p-type conduction during the final deposit formation stages and upper section of orebodies. According to the thermal EMF pyrites, it is possible to identify the deposit formation conditions, orebody zone erosion, and, consequently, their area potential at depth and on the flanks [3].

The following research is focused on the investigation of crystal morphology and thermal EMF of pyrites in the western flank of Sukholozsky ore field to determine the mineralization zone erosion.

2. Regional setting

The structure and composition of rocks and ores in Sukholozsky ore field have been described in details in many publications [1, 2, 6, 7]. A brief description of the western flank of Sukholozsky ore field is as follows (figure 1).

The ore field is within the Bodaibo syncline, Kropotkinore ore cluster, where its western flank is stripped by surface mining and core drilling. Sukholozsky anticline is the major structure controlling the ore localization. It is a linear, highly compressed asymmetric fold, elongated to the E-W and southward vergent. The northern and southern anticline flanks dip at the angles of 15…20° and 30…45°, respectively.

The orebodies in the western ore field flank is limited to the sediments of the upper Khomolkho suite subformation and the lower Imnyakh suite subformation, formed of sandstones, aleurolites and phyllites of different grain sizes. The rocks within these suites were exposed to regional metamorphism of chlorite-sericitic subfacies in greenschist facies.

Hydrothermal carbonization reveals carbonate, sulphide and quartz-veined mineralization. Ore mineralization is confined to disjunctive folded dislocation zones, embracing vein-disseminated quartz-sulphide and quartz-veined mineralization.

3. Research methods

The study of pyrite crystal morphology (figure 2) was based on the selected samples [3] from the western flank ore zones of Sukholozsky ore field.

Pyrite thermal EMF identification was conducted on an integrated double-microscope MBS-10, designed to measure the thermal EMF of small solid mineral fractions ranging from 0.2 to 1.0mm (in the Geology of Gold Lab, TPU). A cold solid 20×10×0.5cm plate brass electrode and a hot brass needle-shaped electrode are connected to the digital milliammeter M-800F, automatically measuring either electron (*–n*) or p-type (*+p*) conduction of a mineral.

The measurement was conducted under the conditions of the constant temperature difference of 150 °C between cold and hot electrodes. The calibration testing of the electrode- establishment of temperature difference between cold and hot electrodes and conduction type – was performed by the reference sample-galena, which has only n-type conduction. Check measurements of reference galena were performed at 10-15 sampling intervals. Thermal EMF measurement time of a mineral is

15…20 sec. Every sampling includes 30-50 thermal EMF measurements of grains and / or mineral crystals.

The crystal morphology and thermal EMF were studied in tens of pyrite crystal samples.

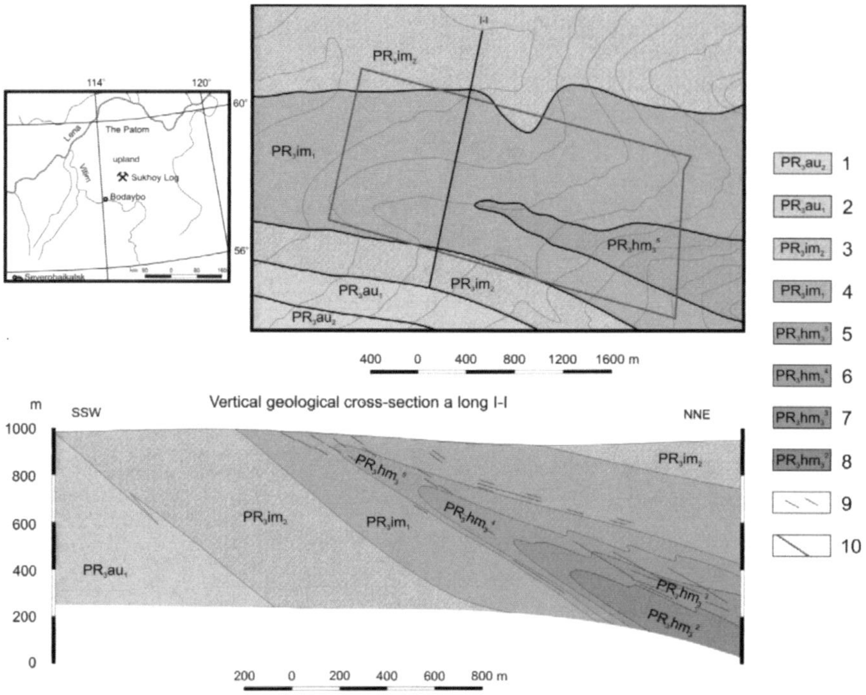

Figure 1. Sketch map of the geological structure of the western flank of Sukholozsky ore field according to [8] and commentaries of the authors.

Legend: *Aunakit suite* (1–2): 1 – middle subformation: quartzitic sandstone interlayers; 2 – lower subformation: alternating carboniferous shale, quartzite and limestone sandstone and limestone; *Imnyakh suite* (3–4): 3 – upper subformation: white and / or creamy, slightly greenish limestone, limestone sandstone layers; 4 –lower formation: rhythmical alternating shale, limestone sandstone and limestone layers; *Khomolkho suite* (5–8): 5 – 5th horizon: phyllitic shale, carbon interlayers of aleurolites and quartzite-mica sandstones; 6 – 4th horizon: quartzite-mica aleurolites; 7 – 3rd horizon: rhythmical alternating carbon shale, aleurolites and quartzite-mica sandstones; 8 – 2nd horizon: high-carbon shale; 9- small ore-hosting faults; 10 – licenced area contour.

Figure 2. Inclusions of pyrite crystal in phyllite.

4. Research results

Pyrites in the western flank of Sukholozsky ore field are cubic{100} or cubic with weakly-developed pyritohedron faces {100}+{210}, ranging from 1.o to 4.0cm. Random crystals have a regular shape, but more often, they are flattened or weakly pseudorhombohedral. The faces {100} are covered with intricated sheet growth laminae. Faces {210} are weakly developed and are also covered with growth laminae. The growth laminae on crystal faces are very thin and do not have regular outlines relative to crystal edges. Crystal faces often show irregular imprints of hosting rocks (argillites, phyllites) and quartz. Internal crystal structures are not always compact due to entrapping hosting rocks and weak conchoidal fracture.

The investigated thermal EMF showed that pyrites have only p-type conduction with thermal EMF value scatter from 27 to 83mV and near mean value ranging from 58.4…67.0 mV (table. 1). Thermal EMF value involves a regular distribution in every investigated crystal (V=12.2…17.3 %).

Table 1. Statistic parameters of pyrite thermal EMF distribution in the western flank of Sukholozsky ore field.

No.	Sample No.	Number of measurements	Thermal EMF, mV			Standard error (S)	Coefficient of variation (V), %
			min	*max*	$x_{cp.}$		
1	3-4	47	33	80	59.6	9.7	16.3
2	3-5	38	40	78	66.9	8.7	12.2
3	3-6	42	27	83	67.0	12.5	18.7
4	3-7	50	35	77	58.4	11.2	17.3
5	3-8	39	31	75	62.5	10.1	15.3

5. Conclusions

Investigated pyrite crystals have only cubic habitus which is characteristic of the pyrites located in the top of ore bodies. The p-type conduction of pyrites indicated the fact of upper ore zone in the western flank of Sukholozsky ore field [3]. The research results do not contradict the stated conclusions of previous investigated gold ore areas [9, 10].

Therefore, the analysis of the crystal morphology and thermal EMF of pyrites revealed insignificant (upper ore) gold mineralization zone erosion. Another important factor is the potential gold resources in the central and eastern areas of Sukholozsky ore field within Khomolkho suite horizons and stratigraphically located vertically below.

Acknowledgements

The investigations were supported by Tomsk Polytechnic University, project № 114-2014.
The research was performed with the financial support from Government Project «Nauka», 1.1488.2015.

References

[1] Sher S D 1966 Okolorudnye izmeneniya, soputstvuyushchie zolotokvartsevym zhilam v Lenskom zolotonosnom rayone. *Metasomaticheskie izmeneniya bokovykh porod i ikh rol v rudoobrazovanii*. Ed. N I Nakovnik (Moscow: Nedra) pp 282–291 (in Russian)

[2] Burjak V A and Hmelevskaja N M 1997 *Sukhoj Log – odno iz krupnejshikh zolotorudnykh mestorozhdenij mira (genezis, zakonomernosti razmeshchenija orudenenija, kriterii prognozirovanija)* (Vladivostok: Dalnauka) p 156 (in Russian)

[3] Korobejnikov A F, Narseev V A, Pshenichkin A Y and Revjakin P S 1993 *Pirity zolotorudnykh mestorozhdenij (svojstva, zonalnost, prakticheskoe znachenie)* (Moscow: CNIGRI) p 213 (in Russian)

[4] Koglin N, Frimmel H E, Minter W E L and Brätz H 2010 Trace-element characteristics of different pyrite types in Mesoarchaean to Palaeoproterozoic placer deposits *Mineralium Deposita* **45** 259–280
[5] Pshenichkin A Y, Oskina Y A, Savinova O V and Dolgaya T F 2014 Gold in Pyrite within Altai-Sayan Folded Belt Gold Deposits *IOP Conf. Ser.: Earth Environ. Sci.* **21** 1–4
[6] Large R R, Danyushevsky L V, Chang Z, Maslennikov V V and Robert F 2007 Multistage sedimentary and metamorphic origin of pyrite and gold in the Giant Sukhoi Log deposit, Lena Gold Province, Russia *Economic Geology* **102** 1233–67
[7] Gavrilov A M and Kryazhev S G 2008 Mineralogo-geokhimicheskie osobennosti rud mestorozhdeniya Sukhoy Log *Razvedka i okhrana nedr* **8** 3–16
[8] Gavrilov R Y 2014 Geohimicheskaja zonalnost rudovmeshchajushchego oreola mezotermalnogo zolotorudnogo mestorozhdenija Zapadnoe (Lenskij rudnyj rajon) *Geologija i mineralno-syrevye resursy Sibiri* **17** 59–67
[9] Timkin T, Lavrov D, Askanakova O and Korotchenko T 2014 *IOP Conf. Ser.: Earth Environ. Sci.* **21** 1–6
[10] Cherkasova T Y, Timkin T V and Savinova O V 2015 First native silica findings in bismuth from garnet skarns of Ribny Log-2 gold ore target in Topolninsk ore deposit (Gorny Altai) *IOP Conf .Ser.: Earth Environ. Sci.* **24** 1–5
[11] Voroshilov V G, Savinova O V, Ananjev Y S and Abramova R N 2014 Anomaly Geochemical Fields in Siberian Hydrothermal Gold Deposits *IOP Conf. Ser.: Earth Environ. Sci.* **21** 1–6

Exploration and local forecast of gold-ore deposits based on typomorphic properties of pyrite

A Ya Pshenichkin[1,3], Yu S Ananyev[1], A I Bushmano[1] and R N Abramova[2]

[1] Department of Mineral Exploration, Institute of Natural Resources, National Research Tomsk Polytechnic University, 30 Lenin Ave., Tomsk, 634050, Russia
[2] Department of Foreign Languages, Institute of Natural Resources, National Research Tomsk Polytechnic University, 30 Lenin Ave., Tomsk, 634050, Russia

E-mail: [3] paya@tpu.ru

Abstract. The article describes the data in exploration and local forecast of gold-ore deposits based on typomorphic pyrite properties. The pyrite properties: crystal shape, impurity-elements and thermal EMF change in relation to the deposit formation conditions are consistent with the mineralogical and geochemical zoning of ore bodies and deposits. In this case, it is possible to evaluate the ore zone erosion, prospectivity and productivity of the ore bodies at depth and flanks. Mineralogical sampling on pyrite and gold should be conducted on the basis of other methods during exploration and mining.

1. Introduction

Nowadays, effective prospecting-evaluation survey is impossible without the development and implementation of mineralogical and geochemical methods in forecasting, exploration and evaluation of mineral deposits. The conducted integrated study of pyrites in genetically different gold-ore deposits established the fact that pyrite is one of the most abundant and informative mineral in a deposit where the typomorphic properties change in relation to deposit formation conditions [4, 5, 8]. The information encoded in a mineral furthers the deciphering of the pyrite crystallization conditions and the deposit genesis, in general [1, 2, 6, 7].

2. Pyrite crystallomorphology

Pyrite crystal shape (figure 1) is one of the contrastive and sharp typomorphic features of pyrites in different mineragenic gold-ore deposits [3, 4, 7, 8]. The unique morphological features of pyrite crystals are specific for every deposit individually.

The pyrites in gold-quartz deposits predominately have a cubic crystal habit. Although simple crystals – {100}, {210} and {111} are rarely found in hyperthermal quartz-gold ore associations, they are more often found in combined forms (for example, Maslov, Podlunntinovsk and Konstantinovsk deposits). In medium-temperature gold-sulphide associations the pyrite crystals are hexanedron (in Balakhchinsk, Kuzhnetsk, Znamenitsk and Medvezhe deposits), while in medium-low temperature quartz-gold-polymetallic sulphide associations are hexanedron and / or pentagonal-dodecahedral habit (in Lisogorsk and Saralinsk deposits). Pyrites in gold-skarn deposits have octahedral crystal habit (in

Kalistrovsk and Tardansk deposits). According to the strike and down-dip direction of ore-bodies (trending to pinch-out), pyrite morphology changes regularly, revealing the alteration of pyrite crystal habit from pentagonal-dodecahedral to cubic (Saralinsk ore field, figure 2) or from cubic to cubic pentagonal-dodecahedral (Balakhchinsk, Lisogorsk, Medvezhe and Konstantinovsk deposits). The most significant morphological variations and the least combinative stability of pyrite crystals have been observed in ore chimneys, predominately revealing a pentagonal-dodecahedral habit (figure 3, table 1).

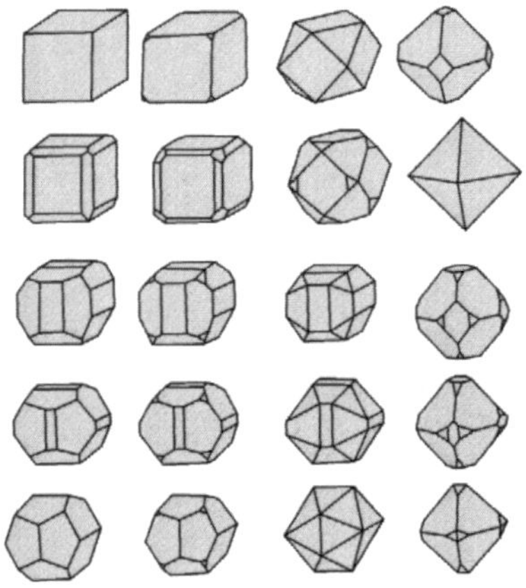

Figure 1. Basic morphological types of pyrite crystals in pyrite gold-ore deposits.

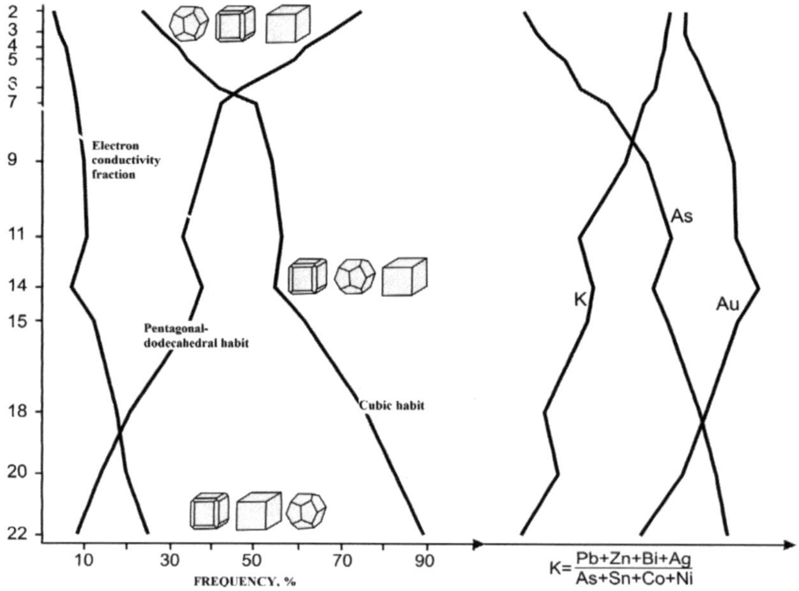

Figure 2. Content alteration of element impurites, crystallomorphology and thermal EMF in pyrites within gold veins of Kaskad- Saralinsk gold-ore field.

3. Research results and discussion

The pyrites in Kaskad and Andreevsk gold deposits (Saralinsk gold-ore field) were investigated in detail (more than 0.9km. down-dip and about 3km. along the strike). The spatial zoning of crystal morphological alteration was defined, which, in its turn, reflected the regular pyrite crystal habit changes from pentagonal-dodecahedral to cubic in the directions to strike, down-dip and across the strike (figure 2). It was determined that pyrites with pentagonal-dodecahedral habit is predominate in significantly rich ore chimneys (figure 3) in crystal types and morphologically diverse; whereas, in pinching-out areas, along the strike and down-dip direction, the combination of pyrite crystals decreases to a minimum. Based on the obtained information on the pyrite crystallomorphological regularities, the ore zone erosion and its prospectivity more than 100 ore veins were evaluated and defined. The ore zone erosion in the ore field increases westward and south-westward (figure 4).

Chemical composition is another specific typomorphic feature of pyrites in gold-ore deposits (table 1). Comparable to other genetically different pyrites, pyrites in gold-ore deposits exhibit significant static parameter scattering of impurity-element distribution ($\overline{X}$, S^2, V). This is particularly relevant for element complexes in gold-ore deposits, where significant concentrations of Au, Ag, Cu, Pb and Zn were accumulating constantly in pyrites of gold-ore deposits.

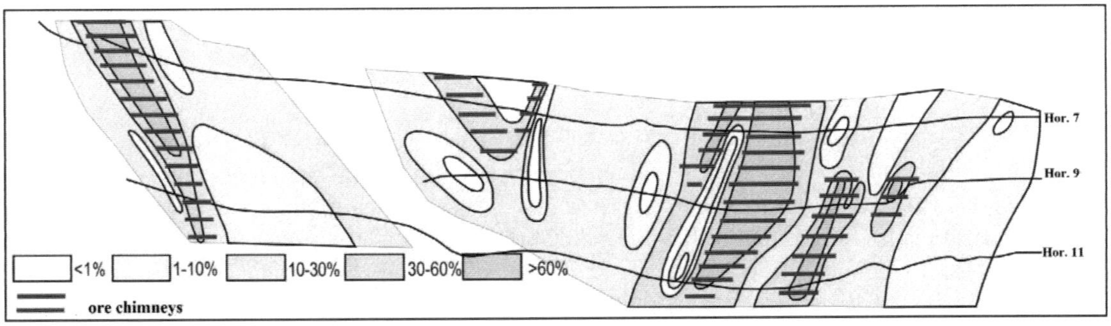

Figure 3. Distribution of pentagonal-dodecahedral pyrite crystals (%) in bedded veins (Kaskad-Saralinsk ore field).

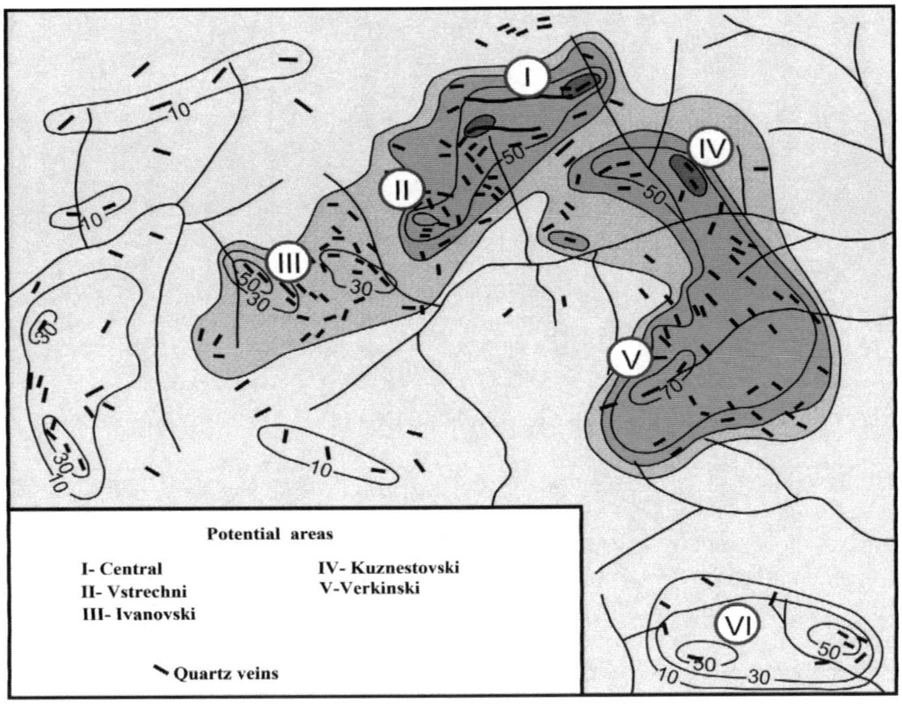

Figure 4. Distribution of pentagonal-dodecahedral pyrite crystals within Saralinsk ore field.

Gold-quartz formation exhibits 1.5-2-fold pyrite enrichment by Au, Ag, As, Pb, Bi, Sn, Cd, and Ba comparable to the pyrites in gold -skarn deposits. There is a higher concentration accumulation of Mo, Zn and rarely, Ti, Co in the latter. However, in this case, pyrites in high-temperature gold-quartz associations are enriched by Mo, Ti, Cr, Co, Ni, V, As and Bi (for example, in Kommunarovsk and Tardansk deposits); pyrites in medium-temperature gold-quartz-sulphide associations are enriched by Au, Cu, Pb and Zn (in Balakhchinsk and Medvezhe deposits); while pyrites in medium-low-temperature gold-quartz- polymetallic sulphide associations are enriched by Hg, Ba, Ag, Sb, and As (in Lisogorsk and Saralinsk deposits).

Pyrites in gold-ore deposits are rarely enriched by gold in comparison to other genetically different pyrite deposits [2, 4, 5]. However, a rather high gold concentration is observed in pyrites from stringer-porphyry and porphyry ores in black shale formations, where these concentrations are up to 50gr/t and more (figure 5). The gold concentration level in the pyrites of above-mentioned deposits increases from diagenetic pyrite to metamorphogenic and ore-forming pyrites. The latter is 100-1000 times (and even more) enriched by gold comparable to diagenetic pyrites.

The zonal (wavelike, the amplitude of which is 200-240 m) distribution of impurity elements in pyrites at deep-seated ore bodies was determined. It was observed that the Ba, Hg, Ag and Sb concentrations (sometimes As) are predominant in the pyrites in the upper part of ore bodies and supraore haloes; pyrites in the mid part of ore bodies are dramatically enriched by Au, Cu, Pb, Zn, while pyrites in the bottom of ore bodies accumulate Ni, Co, Ti, and Cr (sometimes As, Cu) and the remaining elements are sharply depleted [5, 6, 10]. In this case, the chemical composition of pyrite could be used to evaluate the ore zone erosion of orebodies (figures 2, 5).

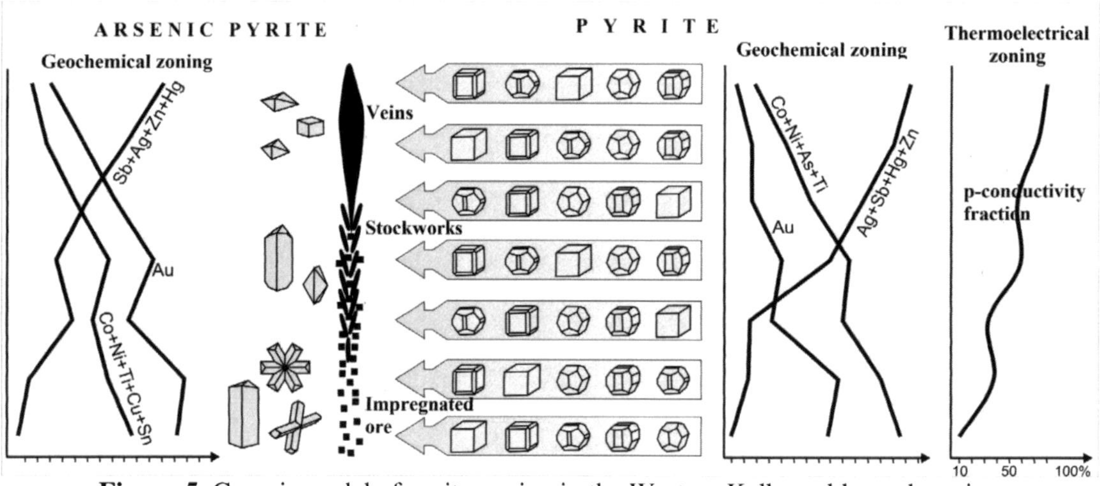

Figure 5. Generic model of pyrite zoning in the Western Kalba gold-ore deposits.

The most informative physical property for pyrites [4, 9] is thermal EMF (figure 5, table 1). Pyrites in orebodies and near-ore mesomatites of gold-ore deposits, Altai-Sayan folding area, exhibit thermal EMF parameters, governed by the characteristic features of pyrite formation and having mixed conductivity, predominately, electronic (Spasskoe, Kaliostovskoe, Konstantinovsk and Tardansk deposits), or p-type (Saralinsk, Balakhchinsk, Lisogorsk and Medvezhe) or p- and n- pyrite types in equal numbers (Znamenitinsk, Kuznestsk, Javarskoe, Maslovskoe and Podlunni deposits). Pyrite can be rarely observed with electronic conductivity (Tarchinsk) [4, 9].

It has been determined that in most deposits there is a regular exchange of pyrite conductivity type from electronic to electronic-p-type and / or p-type in relation to the transition from the early high-temperature mineral associations to the final deposit formation stages (figure 5). Thermal EFM of pyrites changes areally: from the upper part of orebodies to the bottom part where the exchange of pyrite conductivity is from p-type to electronic-p-type and/or electronic. Moreover, pyrites of productive mineral associations and highly – gold enriched areas of orebodies exhibit mixed conductivity type (table 1).

Table 1. Evaluation criteria of ore zone erosion of orebodies in gold-ore deposits and their prospectivity at depth based on pyrite typomorphic features.

Pyrite location	Crystallomorphology	Impurity-elements	Thermal EMF
Supraore pyrite	Morphologically simple crystals (2-3 -type), often cubic habit; crystal face Грани {210}, {111}, {321} weakly developed	Hg, Ba, Sb and other light mobile elements; increased concentration of Ag, Pb, Cu and As (comparable to metacrystals from ore-free zones); content of gold: Au $n{\cdot}10^{-5}$–$n{\cdot}10^{-4}$ %	P-type conductivity (90 % and more)
Pyrite from upper part of orebodies	Crystals – 2-5 types (rarely more); morphological types- cubic and/or pentagonal-dodecahedral habits; often no combined crystal faces {321}	Increased concentration of Hg, Ba, Ag andSb, rarely As; no and/or trace content of Co, Ni, Ti and Cr; Au content ranges from $n{\cdot}10^{-4}$ to $n{\cdot}10^{-3}$, however, a sequence higher in near-ore pyrite	Predominate p-type conductivity (75–90%)
Pyrite from mid part of orebodies	Maximum number of morphological crystal types (30 and more); sharp combination instability;' irregular developed crystal faces; number of crystal habits is maximum, predominately developed cubic and/or pentagonal-dodecahedral	Significant concentration of Pb, Zn, Cu, As and other elements and their distribution parameter ($\bar{X}$, S^2, V); Au content – $n{\cdot}10^{-3}$ and higher	Electronic-p-type conductivity
Pyrite from bottom part of orebodies	Number of morphological crystal types 2-5: cubic and/or pentagonal-dodecahedral habits; no combination of crystal faces {321}and other with complex indexes	Minimum content of Au ($n{\cdot}10^{-5}$–$n{\cdot}10^{-4}$ %), absence of Hg, Ba, Sb; increased concentration of Co, Ni, Cr, Ti, V, rarely, As	Predominate electronic conductivity (80–90%)

Note: Pyrites from near-ore rocks (according to ore zone erosion of orebodies) are allied to ore pyrites in relation to their typomorphic features. The only difference very simple crystal morphology and low concentration of impurity elements, the content of which increases towards to the orebody.

4. Conclusion

The detected changes in the typomorphic properties of pyrites are consistent with the mineralogical and geochemical zoning of orebodies and deposits. This fact makes it possible to evaluate the ore zone erosion level, potential mineralization and its extension depth and identify blind ore bodies. In this case, pyrites could be used as a mineralogical criterion in prospect evaluation survey exploration to mine gold.

Acknowledgements

The investigations were supported by National Research Tomsk Polytechnic University, project №: 114-2014. The research was performed with the financial support from Government project "Nauka", №. 1.1488.2015

References

[1] Ananyev Yu S 1989 Tipomorfizm pirita, kak priznak orudenenija v odnom iz zolotorudnyh mestorozhdenij chernoslancevyh tolshhah Kalby *Geology and mineral resources in Western Siberia platform* 174–175
[2] Andreev Yu S 1992 *Pyrite in gold-bearing deposits* (Moscow: Nauka) p 144
[3] Evzikova N Z 1984 *Prospecting crystallomorphology* (Moscow: Nauka) p 143
[4] Korobeinikov A F, Narseev V A, Pshenichkin A Ja, Revjakin P S and Arifulov Ch H 1993 *Pirity zolotorudnyh mestorozhdenij (svojstva, zonal'nost', prakticheskoe primenenie* (Moscow: CRGEInstitute of Ferrous and Nobel Metals) p 213
[5] Korobeinikov A F and Pshenichkin A Ja 1985 Specific typomorphic features of pyrite in gold-ore deposits *Geochemistry* **1** 93−104
[6] Lukyanova E V 1991 Tipomorfnye osobennosti pirita i arsenopirita odnogo izolotorudnyh mestorozhdenij Kazahstana *Rudonosnye metasomaticheskie formacii Urala* 66–68
[7] Prokhorov V G 1970 *Pyrite (geochemistry, mineralogy, economic and industrial applications)* (Krasnoyask: CSRI of Geology, Geophysics & Mineral Resources) p 168
[8] Pshenichkin A Ja 1989 Kristallomorfologija pirita i ejo ispol'zovanie v praktike poiskovo-razvedochnyh rabot na zoloto *Geology and geophysics* **11** 65−75
[9] Pshenichkin A Ja 1977 Termojelektricheskie svojstva piritov nekotoryh zolotorudnyh mestorozhdenij razlichnyh tipov *TPU Izvestya* **247** 74–78
[10] Pshenichkin A, Timkin T, and Oscina Yu 2015 Noble metals in rocks and ores of Maysko-Lebed ore field (Mountain Shoriya) *IOP Conf. Ser.: Earth Environ. Sci.* **24** 012025

Lithologic-facies and paleogeographic features of Mid-Upper Jurassic oil-gas bearing sediments in Nurolsk depression (Western Siberia)

I Rychkova[1,3], M Shaminova[1], V Sterzhanova[1] and A Baranova[2]

[1] Institute of Natural Resources, National Research Tomsk Polytechnic University, 30 Lenin Ave., Tomsk, 634050, Russia
[2] Department of Foreign Language, Institute of Natural Resources, National Research Tomsk Polytechnic University, 30 Lenin Ave., Tomsk, 634050, Russia

E-mail:[3] irina.rychkova@mail.ru

Abstract. Bathonian-Callovian-Oxfordian sedimentation environment reconstruction in SE Nurolsk depression, Western Siberia has been described. Paleogeographic and litho-geochemical features of sediments, numerous plant remains and ichnofossils indicated the fact that this territory during the Naunaksk suite formation was the transition in-situ sedimentogenesis. Based on the integrated research data, the potential litho-facies were identified in the Mid-Upper sediments within Nurolsk depression, Western Siberia.

1. Introduction

The relevant research target involves Bathonian-Callovian-Oxfordian sediments in the Naunaksk suite, comparable to other petroleum-bearing mid-Jurassic sediments in SE Nurolsk depression, West Siberia. The research area is very composite for areal layering and correlation due to the lenticular lithological bodies being governed by transition sedimentogenesis [5]. Although intensive referencing and publications have been conducted on the subject of the Naunaksk suite composition and structure, there still exists unresolved issues concerning the specific features of this suite, the influence of sedimentation and post-sedimentation factors on reservoir properties (porosity and permeability).So far, no consideration has been given to the geographical relation of the Naunaksk suite to the marine analogue-Vasugan suite [4]. There are disputable points in determining the formation conditions of above-mentioned suites.

According to the structural-facies zoning of West Siberian Callovian and Upper Jurassic, only the southernmost Nurolsk depression areas are within the transition sedimentogenesis zone: from Vasugan to Naunaksk suites. Based on the contemporary concepts of paleogeographic environment, Vasugan suite in SE Nurolsk depression was formed in Bathonian-Callovian-Oxfordian period, and this territory was the location of early marine sedimentogenesis. However, the lithological and paleontological features in the S-E Nurolsk depression reflect the transition sedimentogenesis conditions during the formation of the Naunaksk suite itself.

2. Research data

The basic research material was the deposits in the exposed wells: Maiskaya (№ 228, 379, 393, 568, 569, 572, 573, 698, 704), Western Maiskaya (№ 10), Central Maiskaya (№ 715), Eastern Maiskaya (№ 1) and Southern Maiskaya (№ 5, 309, 400, 410) areas in the S-E Nurolsk depression. All in all, 16 wells were investigated, six of which were located in Naunaksk suite. Besides, paleobotanic remains and ichnofossils were found in six wells (Maiskaya area- wells № 572, 573, 393; Southern – Maiskaya - wells № 309, 400, 410) (figure 1).

Biostratigraphic, litho-petrographic, geochemical, fluorescence microscopy, X-ray fluorescence methods were applied in the paleogeographic reconstruction and effective prediction of natural petroleum reservoirs via core analysis, and further geophysical logging interpretation.

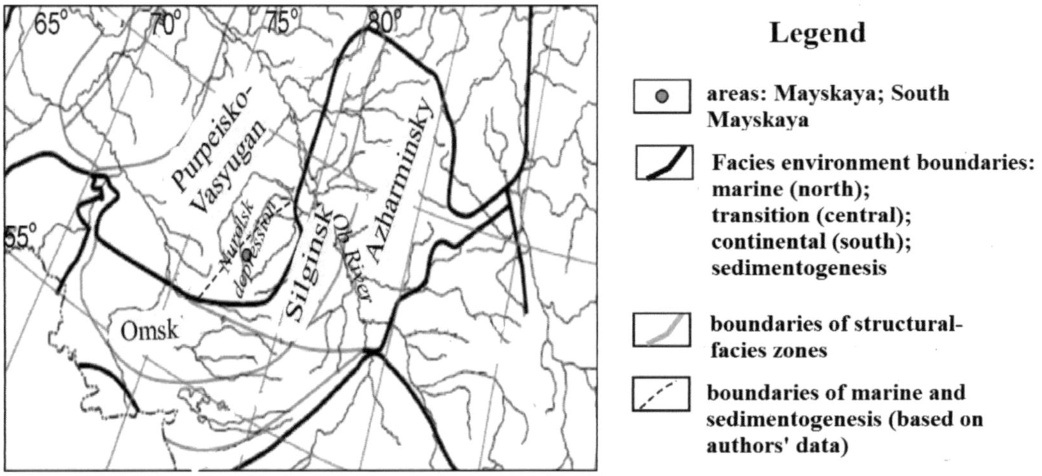

Figure 1. Structural-facies zoning of West Siberian Callovian and Upper Jurassic [3].

3. Research results

Based on the integrated investigation, the observed sediments were described and the following litho-facies groups were identified: silt-mudstones, aleurolites and sandstones. Bathonian-Callovian-Oxfordian rocks embrace the Naunaksk suite and are confined to the intervals with specific well logging-based characteristics, as well as are controlled by orthostratigraphic fauna.

Silt-mudstones are gray-black with laminated structure, distinct interrupted dark-gray aleurolite laminae. The rocks are enriched by carbonaceous wash-outs. There are fern imprints: *Coniopteris simplex, C. depensis, C. vialovae, C. latilobus, Raphaelia diamensis*; seed fern imprints: *Czekanowskia* sp., *Nilssonia urmanica;* pine imprints: *Podozamites eishwaldii*. Large-sized carbonized equisetaceous plant remains, the trunk width of which is more than the core diameter (95.0 mm), could be observed in the sediments.

The bulk mass of silt-mudstones is composed of hydromicas with impurities of fine-grained aleuritic fragments. Clastics include poorly-rounded quartz and feldspar grains. X-ray structure analysis results showed corroded quartz grains and feldspar partially substituted by quartz and kaolinite. These rocks embrace insignificant amount of bituminous matter including predominately resinous epibitumen (brownish luminescence), irregularly distributed throughout the rock mass. Resin-asphaltine bitumen (dark-brownish luminescence) is confined to micro-fractures, oriented by sublateral bedding (figure 2).

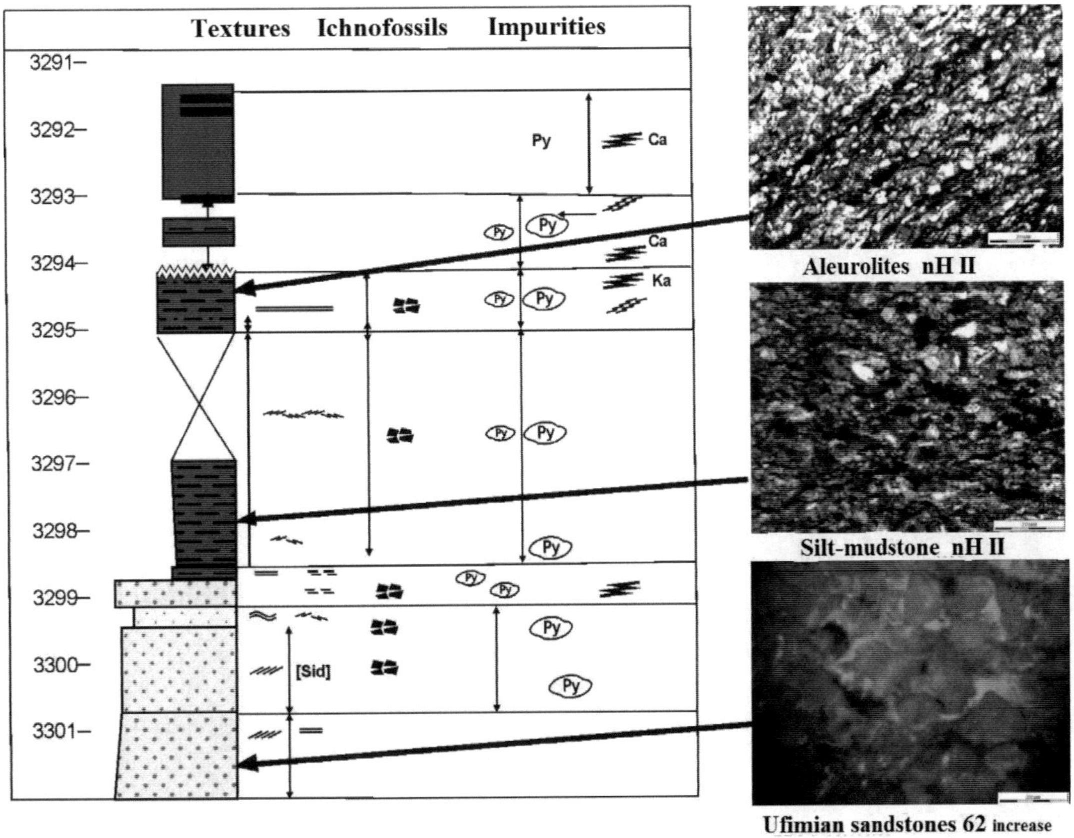

Figure 2. Litho-facies groups, well № 309 (Southern Maiskaya).

Gray aleurolites with rare carbonaceous-fauna debris. There are pyrite concretions of up to 1 cm. The rocks are coarse-grained aleurolite micro-structures. Clastics, predominately, quartz and feldspar, comprise of up to 90% of the thin section. There are single zircon grains. Quartz grains are corroded and often recomposed, K-feldspar is substituted by argillaceous and siliceous matter. Cementing matter- siliceous-hydromica; cement texture- contact-film. Bitumen of mixed genesis was revealed in ultraviolet fluorescent microscope. The bulk mass is impregnated by resinous bitumen (brownish luminescence), greasy resinous bitumen (yellowish luminescence) and resin- asphaltine (dark-brownish luminescence) which is confined to micro-fractures.

Light-gray sandstones, subhorizontal, wavy-cross bedding. Bedding reveals distinct siderite wash-outs and carbonaceous-plant detritus and coal seams (up to 10.0cm.).There are pyrite concretions.

The rocks are fine-medium grained, micro-structure, clastics is 75-90% of the thin section. Clastic composition- quartz (33%), feldspar (47%), rock clastics (20%), which according to Shavnov classification are feldspar greywacke. Post-diagenesis alteration shows corroded and weakly regenerated quartz clastics, most K-feldspar grains are substituted by quartz-kaolinite aggregates. Plagioclase is often substituted by albite (according to X-ray structure analysis results). Cement composition -siliceous-argillaceous; texture- contact-film, rarely, porous. Clay mineral composition includes hydromica and kaolinite, plastically deformed muscovite flakes. Ultraviolet fluorescent microscope revealed sandstone horizons enriched by migrated greasy resinous bitumen (light yellow brownish luminescence) and greasy bitumen (light yellowish luminescence) forming cement bitumen texture, which, in its turn, could be an indicator of hydrocarbon reservoirs.

IOP Conf. Series: Earth and Environmental Science **27** (2015) 012009 doi:10.1088/1755-1315/27/1/012009

Sandstone bedding is disrupted by vertical plant bioturbation (figure 3 – rb), horizontal and vertical ichnofossils (*Skolithos* (sk), *Palacophycus, Chondrites*). The animals that leave such traces are feebly moving eaters of suspended matter, which indicates a relatively high wave energy flow (coastal bars) [1]. These sandstones also include plant root system remains, oriented cross-bedding of pyritized branches. Leaves of Cherkanowsky sperms and pine Podozamites are located along major axis parallel to each other. This indicates the location of basin coastline with mangroves within shallow areas.

Figure 3. Southern Maiskaya area- well № 400.

Sometimes tubercular bedding, "ripple" type is found in these rocks, which, in its turn, is a non-unifrom combination of two or even three litho-types in Naunaksk suite sandstones.

Complex lithological rock composition in Naunaksk suite of Centeral Nurolsk depression is conditioned by alternating sedimentation environment facies within transition conditions.

Rozhkov genetic diagrams "asymmetry- excess" plotted to the granolmetric analysis results show that sedimentation of sandstones occurring in conditions of outgoing waves in shallow waters, intense littoral currents, oncoming sea (coastal-marine facies), continental beach microfacies of large lowland rivers (field V, figure 4).

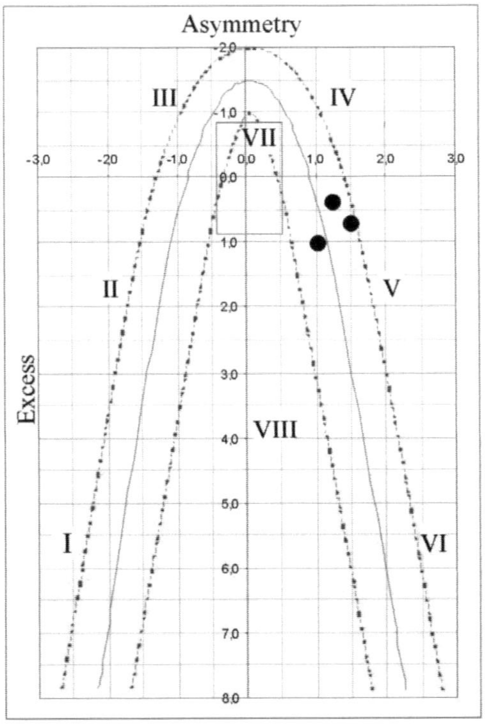

Figure 4. Rozhkov genetic diagram "asymmetry-excess".

Chemical analysis recalculation results and specification of lithochemical models [2] support the coastal-marine formation environment of Naunaksk suite and intensive post-diagenetic rock alteration (table 1).

Table 1. Calculation of lithochemical models.

Model	Characteristic features
TM=TiO_2/Al_2O_3=0,14	Humid climate
HM=Na/Al=0,02	Significant plagioclase alteration
FM=Fe_2O_3/SiO_2	No igneous rocks
B, Cl, Na, Rb, Ca, Al/Ti=3,8	Low paleosalinity of basin
KM=K_2O/Al_2O_3=0,11	Coast-marine sedimentation environment

4. Conclusion

Lithochemical and paleontological analysis data showed that Bathonian-Callovian-Oxfordian Bathonian-Callovian-Oxfordian Maiskaya and Southern Maiskaya areas involved sedimentogenesis transition with active ingression. The marine sedimentogenesis boundaries could have been shifted towards N-W comparable to the plotted map – Callovian and Upper Jurassic structural-facies zoning [3]. Litho-facies sediment alterations in the cross-section is governed by frequent changing sedimentation environment facies.

Litho-facies groups of Naunaksk suite sandstones enriched by greasy epibitumen were identified in the S-E Nurolsk depression, which, in its turn, could be potential indicators of hydrocarbon reservoirs.

References

[1] Vakulenko L G and Jan P A 2001 Jurskie ihnofacii Zapadno-Sibirskoj plity i ih znachenie dlja rekonstrukcii obstanovok osadkonakoplenija *Novosti paleontologii i stratigrafii* **4** 83–93

[2] Panova E G and Shishlov S B 2013 *Strukturno-geneticheskij i geohimicheskij analiz osadochnyh formacij* (St. Petersburg) p 152

[3] *Reshenija 6-go Mezhvedomstvennogo stratigraficheskogo soveshhanija po rassmotreniju i prinjatiju utochnennyh stratigraficheskih shem mezozojskih otlozhenij Zapadnoj Sibiri* 2004 (Novosibirsk: CNIGRI) p 113

[4] Shurigin B N, Nikitenko B L, Devjanov V P et al 2000 *Stratigrafija neftegazonosnyh bassejnov Sibiri.Jurskaja sistema* (Novosibirsk: SB RAS Pub. House) p 480

[5] Shaminova M, Rychkova I, Sterzhanova U and Dolgaya T. Lithologo-facial, geochemical and sequence-stratigraphic sedimentation in Naunak suite (south-east Western Siberia) 2014 *IOP Conf. Series: Earth Environ. Sci.* **21** 012001

Estimating gold-ore mineralization potential within Topolninsk ore field (Gorny Altai)

T Timkin[1,3], V Voroshilov[1], O Askanakova[1], T Cherkasova[1], A Chernyshov[2] and T Korotchenko[1]

[1] National Research Tomsk Polytechnic University, 30 Lenin Ave., Tomsk, 634050, Russia
[2] National Research Tomsk State University, 36 Lenin Ave., Tomsk, 634050, Russia

E-mail: [3] timkin@tpu.ru

Abstract. Based on the results of ore and near-ore metasomatite composition analysis, the factors and indicators of gold-ore mineralization potential were proposed. Integration of the obtained data made it possible to outline magmatic, structural, and lithological factors, as well as direct and indirect indicators of gold-ore mineralization. Applying multidimensional analysis inherent to geochemical data, the spatial structure was investigated, as well as the potential mineralization was identified. Based on the developed and newly-identified mineralization, small (up to medium-sized) mineable gold-ore deposits in skarns characterized by complex geological setting was identified.

1. Introduction

Topolninsk ore field is situated on the right bank of the rivers Anui and Karama between Topolnoye and Stepnoye villages in Soloneshensky region of the Altai Territory (figure 1).

Structurally, this ore field is concentrated within the Anui-Chuya structural-formation zone. It is confined to NE and NW fault systems. The first is characterized by the large plastic Sarasino-Insk fault classified as left-lateral fault. This fault is perfectly traced on geophysical logs and is associated with dykes and intrusions of Topolninsk and Ust-Belovo complexes. The width of this zone is about several dozens of kilometers, while its extension is more than 500 km. Heterochronous intrusive formations, as well as associated gold-ore bodies are identified at the zones where the fault is intersected by NW and near N-S fractures [1].

Contact metamorphic and hydrothermal metasomatic rocks of different composition and occurrences locally extent throughout the ore field located at the contact of Topolninsk granitoid complex (D_2) and carbonate, carbonate-terrigenous sequences (D_2). According to the data, skarns, propylites, and beresites are identified within the ore field [2], where two types of ore have been identified: gold-skarn ore and gold-quartz ore. The content of sulfides varies between 1-5%, with pyrite, arsenic pyrite, magnetic pyrite, and copper pyrite being well developed. Sphalerite, molybdenite, erubescite, chalcosine, galena, fahlite, tellurides and Bi, Pb, Ag sulfotellurides are rarer [3, 4].

The purpose of the present study is to reveal basic localization features of gold-ore mineralization within Topolninsk ore field.

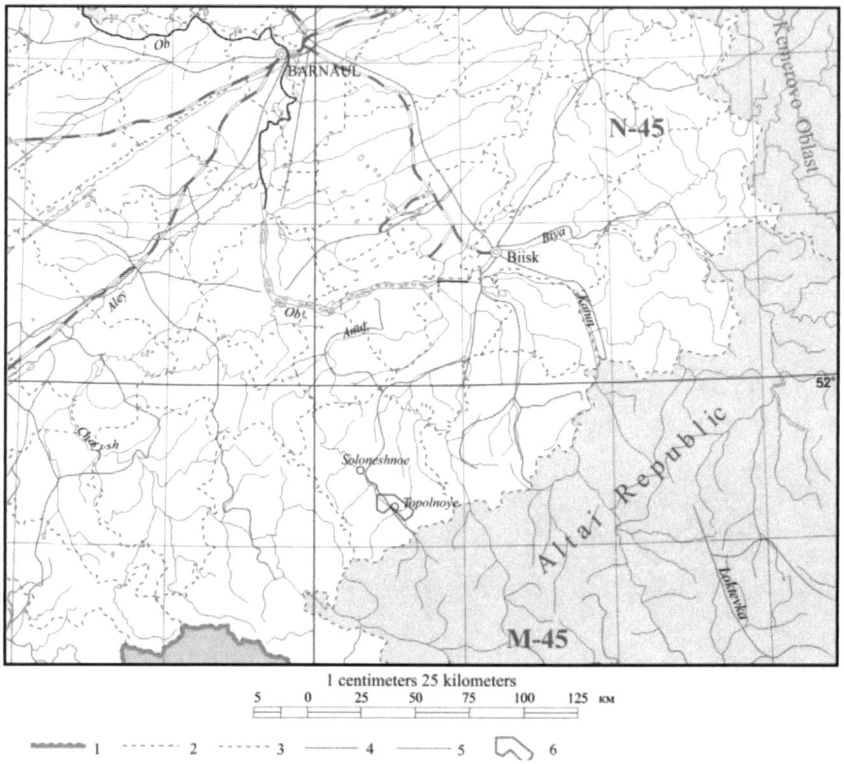

Figure 1. Location map: 1 – state boundaries; 2 – boundaries of RF territorial entities; 3 – boundaries of administrative districts; 4 – roads; 5 – hydrographic network; 6 – exploration area contours.

2. Research technique

The present study is based on the data collected during the research work aimed at investigating the composition of ore and metasomatic formations within Topolninsk ore field over the period of 2012–2014. The work was carried out by Geology and Mineral Prospecting Department, Tomsk Polytechnic University on request of JSC "Gorno-Altaiskaya Ekspeditsiya", Biysk [2-4].

The conducted research involved both office studies and field investigations. The purpose of the field investigations was the rock sampling analysis in correlation with recorded information of trench channel samples, core samples and surface samples. Besides the traditional methods applied in petrography and mineragraphy, the following types of analysis were used as a part of the office studies: inductively coupled plasma optical emission spectrometry (ICP-OES), inductively coupled plasma mass spectrometry (ISP-MS), X-Ray Diffraction Analysis (XRD) of ore metals and metasomatites, scanning electron microscope (SEM), electron microprobe analysis (EMPA) of ore metals, analysis of gas-fluid inclusions within minerals by thermobarogeochemistry methods (TBG) [3].

The processing of geochemical mapping data was carried out on the basis of standard statistical methods of multidimensional analysis [5]. Geometrization was performed by ArcGis, Surfer in accordance with generally accepted methods and techniques [6-9].

3. Research results

The combination of magmatic, geo-structural, lithologic-and-stratigraphic, geochemical, and metasomatic factors, as well as physicochemical properties of ore-forming solutions considerably influenced the gold-ore mineralization localization. Based on the available data on Topolninsk ore field, it is possible to outline the factors and indicators of gold-ore mineralization listed in table 1.

Table 1. The contribution of factors and indicators to localizing gold-ore mineralization within Topolninsk ore field.

Factors and indicators	Gold-skarn mineralization	Gold-quartz mineralization
Magmatic factors:		
gold-ore mineralization is confined to the contact zone of Topolninsk complex intrusive;	+++	++
gold-ore mineralization is confined to dykes and stocks of rhyolite-dacite-andesite Kuyagansk complex.	+	++
Structural factors:		
a) ore localization control is due to NW, near N-S, and NE faults;	+++	+++
b) zones of increased skarn fracturing;	+++	–
c) folded limestone and siltstone beds, delamination in fold curves.	+++	–
Lithological factors:		
Carbonates and aluminosilicate rocks contacts	+++	–
Direct mineralization indicators:		
mineralogical gold haloes	+++	+++
geochemical gold haloes	+++	+++
Indirect mineralization indicators		
Metasomatites a) skarns	+++	–
b) propylites	+++	++
c) steeply dipping linear zones of beresite metasomatites.	++	+++
Concentric-zonal anomalies with accumulations of Cu, Ag, Mo, W, Sb, Bi, As in nuclear zone concentration and Co, Ni, Cr, V in frontal enrichment zone.	+++	+++
Ore mineralization represented by sulfides (pyrite, arsenic pyrite, magnetic pyrite, copper pyrite, erubescite, sphalerite, etc.), tellurides and sulfotellurides of lead, bismuth, silver, oxides and hydroxides of these minerals	+++	+++
Geophysical indicators:		
a) low electric resistivity;	+	–
б) high electric resistivity;	–	+
в) positive anomalies in contact zone with intrusives	+++	–

Note: +++ - strong defined; ++ - moderately defined; + - poorly-defined; – non-defined

Magmatic factors of gold-ore mineralization involve the fact that gold-bearing skarns are confined to endo- and exocontact massifs of Topolninsk complex. The strong positional connection of gold-quartz mineralization is associated with silicic and intermediate dykes of the Middle Devonian Topolninsk and Kuyagansk complexes.

Structural factors are different for gold-ore mineralization types. For gold-skarn mineralization, tectonic traps scattered within the skarns are important structural factor: folded limestone and siltstone beds, delamination in fold curves [4]. Skarn ore deposits are confined to the intersection of the above-mentioned structures and zones of increased fracturing, NW, near N-S, and NE faults. The basic structural factor for gold-quartz mineralization is the intersection of NW, near N-S, and NE faults.

Lithological factors are due to the fact that ore deposition environment had a significant influence on different mineralization types. Lithological contrasts in carbonates and aluminosilicate rocks, i.e. favorable conditions for skarn formation, were of great importance for gold-skarn mineralization. Gold-quartz vein mineralization is basically identified within igneous rocks, and to a smaller degree, in clastic sedimentary rocks. This is explained by physico-mechanical characteristic features of gold-quartz vein mineralization. Despite silicification, persistent quartz veins do not form in skarns which are easily subjected to the increased fracturing of various orientations.

The *direct prospecting indicators* are as follows:
1) mineralogical gold haloes;
2) primary and secondary geochemical gold haloes.

The following *indirect indicators* are the most significant: altered areas, haloes of minerals associated with gold-ore mineralization, geochemical haloes of mineralization trace-elements, geophysical anomalies.

The ore bodies are characterized by definite geological position in ore-metasomatic structures. The strong positional connection of gold-skarn mineralization is associated with skarns of different composition located within Topolninsk ore field. The propylitic alteration is developed in the pinching-out zone of skarn bodies and cross-cutting skarns. This type of alteration was identified within the well-known ore occurrences. The temperature zoning was identified in propylites, changing in distance from skarn bodies, i.e. from medium-temperature epidote-actinolitic facies (305–250° C) to low-temperature epidote-chloritic facies (225–165° C).

Propylite formation results in thin calcite veinlets (up to 1.5 cm) which in most cases cross skarned and propylitized rocks.

Quartz-sericitic metasomatites (beresite type) commonly occur in gold-quartz type; in skarns, beresitization is in the form of silicification, carbonization, and chloritization. The ore mineralization represented by sulfides (pyrite, arsenic pyrite, magnetic pyrite, copper pyrite, erubescite, sphalerite, etc.), tellurides and sulfotellurides of lead, bismuth, silver, gold nuggets is associated with beresites. The beresites associated with gold-quartz mineralization are confined to the same porous structures that control skarn and propylites localization, where propylites are defined as an important prospecting indicator.

The gold nuggets are directly associated with a wide range of ore minerals such as sulfides, tellurides, and sulphosalts. Their mineralogical haloes are an important indirect prospecting indicator of the above-studied mineralization type. Under supergene conditions, these minerals alter to oxides and hydroxides which are also considered as a prospecting indicator. In addition, the typomorphic characteristics of vein minerals which are associated with gold in quartz and calcite are also very informative. The thermometric characteristics and salinity of fluid inclusions in these minerals allow more accurate identification of the mineralization type and zoning.

Geochemical characteristics are of great importance in hydrothermal mineralization prospecting. Gold in ore bodies exposed by prospecting wells №№ 46, 32 correlates with Ag, As, Bi, Co, Cu, Cu, Mn, Mo, Ni, Sn, W, Zn (table 2).

Table 2. Correlation coefficients for Au and trace-elements in ore bodies.

Ore hole	Elements													
	Ag	As	Bi	Co	Cr	Cu	Mn	Mo	Ni	Pb	Sn	V	W	Zn
46	0.49	0.44	0.37	0.56	0.34	0.38	0.55	0.31	0.72	-0.37	0.64	0.57	0.52	0.64
32	0.74	0.32	0.13	0.68	0.26	0.36	0.70	0.56	0.40	0.34	0.71	0.21	0.77	0.48

Note: significant correlations are in red ($r_{crit.}$ = 0.35).

In the secondary geochemical field, Au is also associated with a wide range of trace-elements. The characteristic feature of gold-ore occurrence is concentric zonal abnormal geochemical fields (AGF) with accumulations of Au, Cu, Ag, As, Bi, (W, Mo) in the nuclear zone concentration, and Ni, Cr, V – in frontal enrichment zone [1]. Based on the factor analysis [2], the elements accumulating in the central zones of AGF are classified into three associations: W, Mo, Mn – As, Ag – Au, Cu, Bi, which are confined to both discovered and newly-identified gold-ore mineralization (figure 2).

As correlation of elements is the synchronism of their behavior in geological processes, the identified associations can be interpreted in the following way.

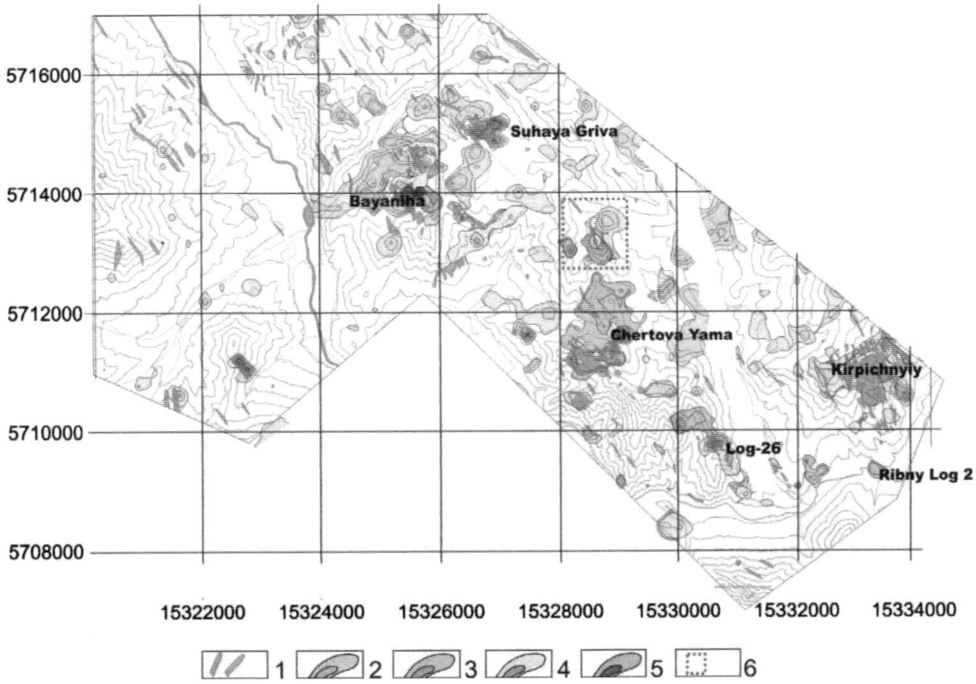

Figure 2. Location of ore manifestation within abnormal geochemical fields of Topolninsk area (secondary lithochemical haloes); 1 – gold anomalies; 2-5 – geochemical associations identified by factor analysis: 2 – W, Mo, Mn; 3 – As, Ag; 4 – Au, Cu, Bi; 5 – Cr, Ni, V; 6 – potential ore manifestation.

The first association of granitophile elements W and Mo with Mn (carbonate deposits) is due to the impact of ganitoid intrusions on the clastic-carbonate sequence, indicating skarnization.

The ore association of As and Ag can be confined to sulfide mineralization developed on the flanks of ore bodies and in the above-ore zone, with the role of gold being insignificant. In this respect, it is essential to consider the area located 2 km northward from ore manifestation Chertova Yama, where the mentioned association is obvious and there is a halo of run-of-mine gold that can be within the above-ore zone (figure 2).

The association Au, Cu, Bi is most strongly associated with developed ore deposits. The correlation of gold with Cu and Bi (also peculiar feature of Sinukhinsk gold-skarn mineralization) indicates that they are synchronously formed during the stage of gold accumulation accompanied by sulfides, bismuth tellurides, and sulphosalts.

The accumulation of Ni, Cr, V in the zones of frontal enrichment is a characteristic feature of all types of hydrothermal deposits and can be applied for the delineation of AGF confined to different geological bodies. This can be used for ranking geochemical fields of ore objects.

The geophysical indicators of localization of different mineralization types have their own peculiarities. Unlike other gold-skarn fields of the investigated area, ore-generating intrusions are not identified in gravitational field due to the absence of massive mafic volcanic rocks in the host thickness.

The weakly magnetic granitoids of Topolninsk intrusion, as well as medium-magnetic diorites of Karaminsk massif in endo- and exocontact zones of which magnetic differences are prominent, are identified on the magnetic field map. In addition, the intensity (up to 670 gamma) of anomalies over

Karaminsk intrusion are displaced to its eastern contact, i.e. skarn zone of ore manifestation Ribny Log 2. The negative anomaly is identified above Topolninsk massif, which could be explained by the predominance of nonmagnetic rocks throughout the massif. Three positive anomalies with the intensity of up to 50 gamma comparable to the negative ones of 100-500 gamma are identified along the contacts of Topolninsk massif. Such values of magnetic susceptibility are characteristic features of thermally altered rocks that have undergone decompaction with decreasing magnetization intensity.

Identification of ore mineralizationin electric fields is conditioned by the association of gold with high-conductivity sulfides. In general, Topolninsk ore field is located in the periphery of extreme positive anomaly of natural electric field.

4. Conclusion

Thus, based on the revealed factors and indicators essential for estimating gold-ore mineralization potential and with due regard to the characteristic features of geo-structural location within Topolninsk ore field including both developed and newly-identified mineralization, it is possible to forecast small (up to medium-sized) mineable gold-ore deposits in skarns characterized by complex geological setting.

Acknowledgement

The investigations were supported by National Research Tomsk Polytechnic University, project №: ВИУ_ИПР_114_2014.

References

[1] Gusev A I, Gusev N I, Tabakaeva E M, Dzagoev E A, and Kukoeva M A 2013 *Petrologiya i rudonosnost' magmo-rudno-metasomaticheskikh sistem Soloneshenskogo rudnogo rayona Altaya* (Biysk: AGAO) p 200 (in Russian)

[2] Timkin T V, Lavrov D S, Askanakova O Y and Korotchenko T V 2014. Structure-geochemical zoning of Topolninsk gold-ore field (Gorny Altai) *IOP Conf. Ser.: Earth Environ. Sci.* **21 (1)** 1–6

[3] Logvinenko O V, Timkin T V 2014 Material composition and sequence of mineral ore manifestation Log-26 Topolninskogo gold field (Gorny Altai*) Tomsk State University Journal* **383** 212–20

[4] Cherkasova T Y, Timkin TV and Savinova O V 2015. First native silica findings in bismuth from garnet skarns of Ribny Log-2 gold ore target in Topolninsk ore deposit (Gorny Altai) *IOP Conf. Ser.: Earth Environ. Sci.* **24** 1–5

[5] Reis A P, Sousa A J and Cardoso Fonseca E 2003 Application of geostatistical methods in gold geochemical anomalies identification (Montemor-o-Novo, Portugal) J. *Geochem. Explor.* **77** 45–63

[6] Voroshilov V G 2009 Anomalous structures of geochemical fields of hydrothermal gold deposits: formation mechanism. Methods of geometrization, typical models, and forecasting of ore mineralization. *Geology of Ore Deposits* **51 (1)** 1–16

[7] Harraz H Z, Hamdy M M 2015 Zonation of primary haloes of Atud auriferous quartz vein deposit, Central Eastern Desert of Egypt: A potential exploration model targeting for hidden mesothermal gold deposits. *J. Afr. Earth Sci.* **101** 1–18

[8] Cheng Zh Zh, Xie X J, Yao W Sh, Feng J Zh, Zhang Q, Fang J D 2014 Multi-element geochemical mapping in Southern China. J. *Geochem. Explor.* **139** 182–91

[9] Gavrilov R Y, Rudmin M A 2015 Methodological approaches in estimating anomalous geochemical field structure *IOP Conf. Ser.: Earth Environ. Sci.* **24** 1–6

Geophysical characteristics and structural composition of clay rocks in the terrigenous complex of the southeastern part of the West Siberian oil and gas bearing province

A Ezhova[1,2] **and N Nedolivko**[1,3]

[1] Department of Geology and Mineral Exploration, Institute of Natural Resources, National Research Tomsk Polytechnic University, 30 Lenin Ave., Tomsk, 634050, Russia

Email: [2] eav@tpu.ru, [3] nedolivkonm@yandex.ru

Abstract. The basic attributes identifying the clay rock composition in the terrigenous complex of the southeastern part of the West-Siberian platform have been determined, based on the correlation of structural composition and geophysical data, including induced potential logging, apparent resistivity, induction logging, radioactive and caliper logging. According to the obtained results it is possible to identify above-mentioned rocks in well logs, even under conditions of limited core samples, their application in well log correlation and back stripping.
Key words: geophysical well logging, the West-Siberian oil and gas bearing province, terrigenous complex, clay rocks, back stripping.

1. Introduction

In the southeastern part of the West Siberian platform the petroleum-bearing formations are composed of clastic, clay carbonaceous and carbonate rock alterations. However, siliceous, aluminous and ferrous rocks have also been revealed. Under the conditions of limited terrigenous profile core samples from Jurassic sedimentary formations and those emerging problems in log analysis interpretation [1], one of the major tasks in the development of an improved profile study of enclosing sedimentary rocks in oil-gas reservoirs is the integration of geophysical and lithological data. In this case one aspect of the above-mentioned problem is to assign rock structural composition to their geophysical field characteristics.

2. Subject and Research Methods

The study data included the following: core samples of clay and carbonaceous rocks and their produced thin sections; results of XD and quantitative spectrum analyses and geophysical well logging data package including induced potential logging (SP log), apparent resistivity logging, inductive electro-magnetic logging, radioactive and caliper logging, describing the drill logs from the south-eastern area of the southeastern part of West Siberian oil and gas bearing province, Tomsk region: Sobolinaya (Sob.), Gerasimovskaya (G), East-Gerasimovskaya EG), Kalinovaya (K), Fobosskaya, Nizhne-Tabaganskaya (NT), South-Tabaganskaya (ST) fields.

3. Results

The clay rocks are composed of 0.01 mm particles, where the basic rock-forming clay minerals are kaolinite, hydromica, montmorillonite, involving alternating layers, such as illite-montmorillonite. Chlorite, ferrum and aluminium oxides and hydroxides and silica could also be important components. Two genetic types of clay rocks have been revealed-chemogenic and clastic. Chemogenic clay rocks within the formation are classified as residual products of weathering crust, i.e. structured clay residual rocks, according to B.B. Polinov [9] and V.P. Kazarinov [5]. The cross-sections show that there are several different structured clay residual rocks, the mineralogical composition of which was revealed by XD analysis data (18 samples).

1. Clays are light gray, practically white in color with sparse angular fragments of highly weathered siliceous-ferruginous rocks (figure 1A). The thin sections show (figure 1B) that the basic rock body embraces kaolinite with fine-grained quartz and siderite as fine oolites, sometimes oxidized oolites.

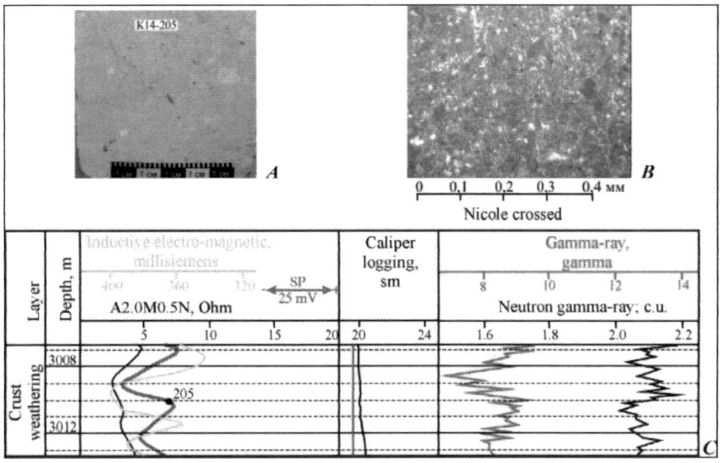

Figure 1. Structure (A), petrographic composition (B) and geophysical characteristics (C) of kaolinite clay rocks of the crust weathering:
A, B, C – Well 14 Kalinovaya; depth=3010.0 m.

According to the XD analysis data the mineral percentage is the following: kaolinite (45-62 %), chlorite (8-10 %), hydromica (8-15 %), montmorillonite (4-5 %), siderite (7-15 %), quartz (10-13 %). The well log shows that clays are defined by positive SP curves, low (up to 5 Ohm) electrical resistance, high (from 350 millisiemens) conductivity, insignificant increase (3-4 cm) of the borehole diameter, increased radioactivity (up to 10 γ) and increased (2.0-2.2 pulse/min.) neutron gamma-ray curve values (figure 1C). Based on the mineralogical composition and the evidence of intensely altered angular fragments and their view location in the profile – in breccia roof, the described clays are neo-formed residual rocks. The latter (according to B.B. Polinov [9]) is a structured clay residual rock horizon formed during the Permo-Triassic period due to the decomposition of loose coarse-grained sediments.

2. Siliceous-clay rocks are white in color with horizontal and vertical cracks (figure 2A).

In the thin sections (figure 2B), as in the previous ones, the basic rock body embraces kaolinite with fine-grained quartz and extra fine hydromica crystalloids. XD analysis data revealed the fact that this composition is as follows: kaolinite (40-45 %), chlorite (5-7 %), hydromica (30-35 %), montmorillonite (up to 5 %), quartz (25-30 %), siderite (often oxidizied) up to 10%. In single samples powder quartz contributed the white color to the rock. The rocks show the following characteristic features (figure 2C): positive SP anomalies, low (about 5 Ohm) electrical resistance, high (from 200 millisiemens) conductivity, insignificant increase (up to 22 cm, if rated value is 19 cm) of the borehole diameter, rather high radioactivity (up to 40γ) and increasingly intensive (1.8-2.0 pulse/min.) neutro gamma-ray curve values. The above-mentioned factors – evidence of powder quartz, mineral composition and location in the profiles, i.e. top of siliceous rocks (spongolites and quartzites) indicate that these clays could be classified as para-residual siliceous sedimentary rocks.

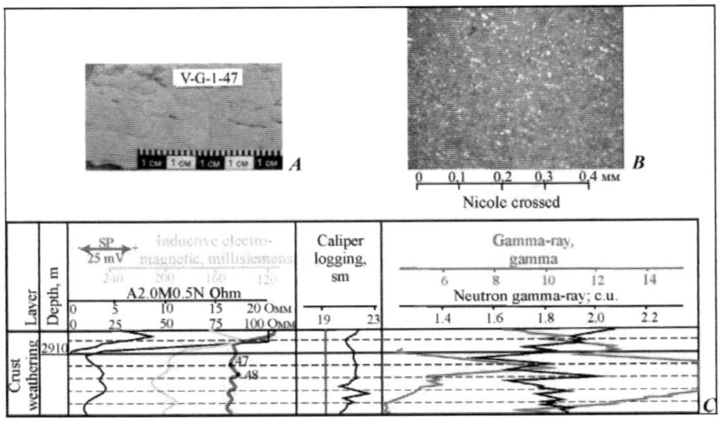

Figure 2. Structure (A), petrographic composition (B) and geophysical characteristics (C) of siliceous clay rocks of the crust weathering:
A, B, C – Well 1 East-Gerasimovskaya; depth=2912.0 m.

3. Chemogenic clay rocks include a limited number of chloritized clays underlying mafic effusive – clay ortho-residual rocks. According to the XD analysis data the mineral percentage is the following: kaolinite (15-20 %), chlorite (25-30 %), hydromica (30-35 %), montmorillonite (up to 5 %), siderite (often oxidizied) up to 10 %, quartz (up to 5%). Geophysical characteristics of lay rocks are similar to the ones described above. Clastic clays are the result of erosion and redeposition of the weathering crust and pre-existing sedimentary rocks. The mineralogical clay material alteration processes in the diagenetic period in the case of free water loss and compaction of sediments. Iron sulphides, carbonates, sulphates, iron oxide and iron hydroxide minerals are formed in clay rocks. During catagenesis clay rocks are affected by increasing temperatures and pressure involving interstital pores, resulting in compacted clays and argillites, plasticity loss and gradual mineral deswelling. Primary clay rock alteration occurs during metagenesis, whereas clay rocks become foliated and further form mud shales and argillaceous slates. According to the degree of composition, clay rocks include the following sequence: clays- compacted clays–argillites–mud shales–argillaceous slates, within which only clays and compacted clays exhibit plasticity and exchange properties. Due to these properties the transformation of clay minerals, alteration of their mineralogical composition and replacement of minerals under amenable conditions occur during the post-sedimentation period. In this respect, the composition of clay rocks in sedimentary thicknesses is rarely correlated to the sedimentation environment [6]. However, particular differences in the mineralogical composition of the diversely genetic clay rocks can be observed within the study area (100 sample results based on XD analysis).

The marine green clays (figure 3A), formed in relatively deep waters in Early Cretaceous and Late Jurassic period [4], are predominately composed of hydromica (35 %), alternating hydromica-montmorillonite sequences (22 %), montmorillonite (10 %), chlorite (18 %) and kaolinite (15 %). Thin-walled pelecypodas, ichthyofauna and foraminifer can also be observed.

The thin sections (figure 3B) show insignificant quartz impurities, fine-flaky hydromica and glauconite inclusions. The well log (figure 3C) shows that clays are defined by low (up to 5 Ohm) electrical resistance, relatively high conductivity (100-120 millisiemens), positve SP anomaly, increased radioactivity (14-16γ), the increasing of the borehole diameter from 24 to 28 cm, if rated value is 19 cm.

Callovian-Oxfordian clays (Nizhne-Vasugan subsuite), formed in shallow marine and coastal environments [8], include hydromica (54 %), chlorite (16 %), alternating hydromica-montmorillonite sequences (16 %), montmorillonite (4 %), and kaolinite (10 %). Fauna remains, ichnofossils (figure 4A) and plant detritus can also be observed. Predominate hydromica with oriented arrangement of flakes showing micro-cleavage, as well as pelitomorphic textured siderite or siderite concretions can be distinctively observed in the thin sections (figure 4B). Geophysical characteristics (figure 4C) of clay rocks are analogs of above-described ones.

Figure 3. Structure (A), petrographic composition (B) and geophysical characteristics (C) of chlorite-hydromica clays of marine genesis:
A, B, C – Well 213 Sobolinaya; depth= 2225.0 m

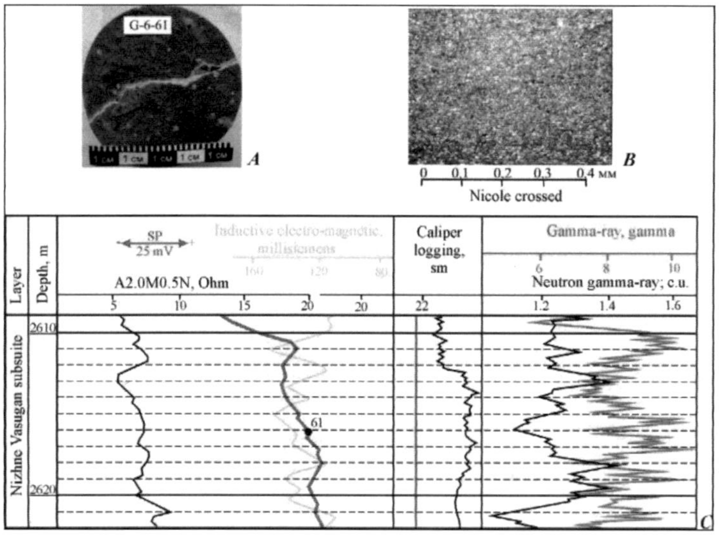

Figure 4. Structure (A), petrographic composition (B) and geophysical characteristics (C) of Nizhne-Vasugan subsuite hydromica clay rocks:
A, B, C – Gerasimovskaya-6; depth= 2616.0 m.

Early and Mid. Jurassic clay rocks, accumulating in laustrine-swamp, deluvial-proluvial-alluvial fresh shallow marine and lagoon environments [3], exhibit variable polymineral composition: include kaolinite (13-44 %), chlorite (12-20 %), hydromica (30-50 %), and alternating hydromica-montmorillonite sequences (4-7 %). Small siderite concretions – up to 10 %, scattered organic matter, ichnofauna and phyllopoda accumulations were also revealed; rocks contain concretions of siderite and pyrite, plant detritus mass, burrow traces and terrigenous impurities. The well log showed these clays are defined by low (5-10 Ohm) electrical resistance, high (120-150 millisiemens) conductivity, increased radioactivity (14-16γ), low (1.2-1.6 pulse/min.) intensity of neutron gamma-ray curve values, the increase of the borehole diameter up to 6-8 cm, positive SP anomaly curves with negative deviation in case of increased terrigenous impurity content.

Togur subsuite brown argillites (figure 5A), apparently formed in fresh marine waters [2], include kaolinite (28-32 %), chlorite (14-18 %), hydromica (30-40 %) and alternating hydromica-montmorillonite sequences (5-25 %). Small siderite concretions (up to 10 %), scattered organic matter, ichnofauna and phyllopoda accumulations were also revealed. The thin sections show finest quartz impurities and extra fine plant detritus amid fine hydromica, siderite matter and brown bituminous substance (figure 5B). The well log (figure 5C) showed increased (15-22 Ohm) electrical resistance (in comparison to the above-described ones), high conductivity (up to 150 millisiemens), high radioactivity (20-22γ), low intensity of neutron gamma-ray curve values (1.4 pulse/min.), the increase of the borehole diameter (more than 40 cm), high DTS values (ΔT – to 335 ms/m).

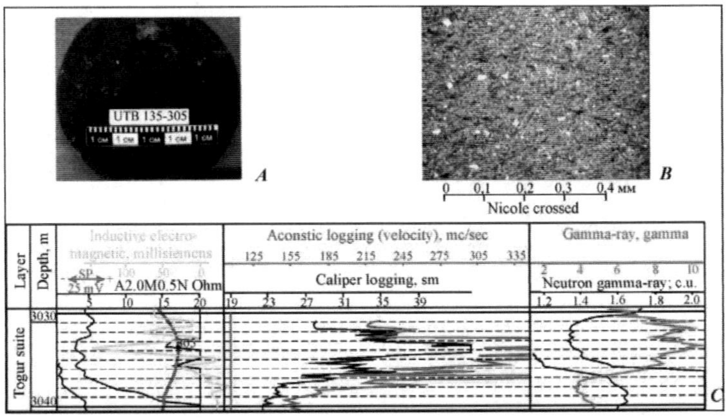

Figure 5. Structure (A), petrographic composition (B) and geophysical characteristics (C) of Togur subsuite bituminous hydromica argillite with pyrites and fine-grained quartz impurities:
A, B, C – Well 135 South-Tabaganskaya; depth=3033.5 m.

Bazhenov suite bituminous argillites (figure 6A), formed in deep-sea conditions [7], have extremely heterogeneous mineralogical composition. Hydromica and chlorite (43-44-72 %), with tributary kaolinite (2-15 %) and alternating hydromica-montmorillonite sequences (9-22 %) and montmorillonite (3-10 %) are predominate in the clay fraction. The thin sections show abundance of bituminous substance, pigmenting the rocks brown. Silicified and calcilated sponge spicula remains, radiolaria and foraminifera were also revealed. Thin weaving micro-fractures with or without bitumen and abundant tiny pyrite and quartz inclusions were also observed (figure 6B).

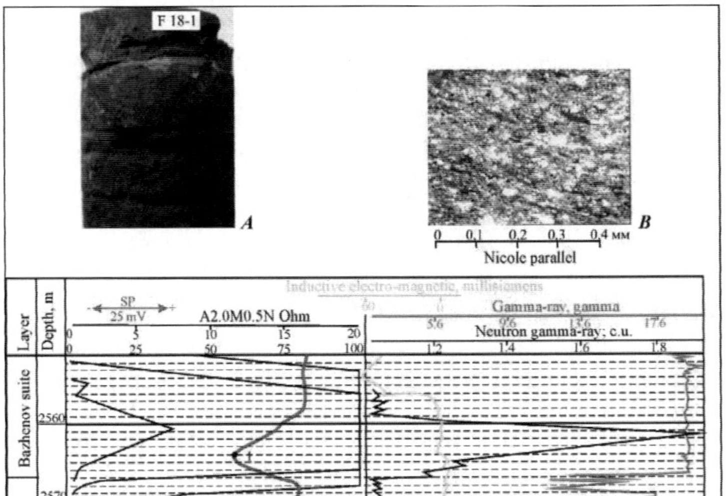

Figure 6. Structure (A), petrographic composition (B) and geophysical characteristics (C) of Bazhenov suite bituminous hydromica argillites with pyrite: A – Well 18 Fobosskaya, depth=2598.0 m; B – Well 18 Nizhe-Tabaganskaya, depth=2564.0 m; C – Well 18 Fobosskaya; depth=2598.0 m.

Bazhenov suite argillites have clear geophysical characteristics: significantly high electrical resistance (100 Ohm) and high radioactivity (more than 40γ), high (2.0-2.2 pulse/min.) neutron gamma-ray curve values if any carbonate interlayers, while, typically, undifferentiated SP curve. However, there are profiles with fractured sections in the argillite itself, indicating negative SP curve deviations of up to 20 mV (figure 6C).

4. Conclusions

The basic characteristic features of clay and carbonaceous rocks and coals in the profiles of studied formation are the following: chemogenic clays are residual products of crust weathering formation – structured clay residual rocks in the profile are neo-formed residual rocks according to petrographic and mineralogical composition (during decomposition of loose coarse-clastic rocks) and para-residual (due to siliceous rocks) and ortho-residual rocks (during weathering of effusives); the mineralogical composition of chemogenic clays depends on the composition of the rocks, showing a significant abundance of kaolinite, chlorite and hydromica content; deswelling component (montmorillonite) is not more than 5 % and differently oxidied quartz and siderite could also be found in the samples; geophysical attributes of clay rocks of crust weathering does not depend on the mineralogical composition, in broader terms: low electrical resistance, significantly high (up to 400 millisiemens) conductivity, positive SP anomalies, insignificant increase of the well diameter, increased raadioactivity values and intensive gamma rays; clastic rocks in investigated formation is related to compacted clays, with kaolinite, chlorite and hydromica content and different alternating hydromica-montmorillonite sequences and dependable of marine montmorillonite-hydromica clays with chlorite; continental: hydromica-kaolinite, hydromicas, often sideritized with abundant flora remains; in geophysical logs homogeneous clays show low electrical resistivity, increased conductivity, high radioactivity, increasing well diameter, positive PS anomaly; terrigeneous impurities initiate negative PS curve deviations, while bituminous matter causes the increase of electrical resistivity and radioactivity; deswelling minerals decrease in agrillites which lessen the plastic property to further fracturing or fracture porosity, which, in its turn, reveals negative deviations on SP curves.

References

[1] Ezhova A V 2012 *Geological interpretation of geophysical data* (Tomsk: Publishing House of Tomsk Polytechnic University) (in Russian)

[2] Ezhova A, Merkulov V, Chekanstev V and Abramova R 2015 Litho-Geophysical Structure of Paleozoic-Mesozoic Contact Zone in North-Ostaninsk Oil Field (Tomsk Oblast) *IOP Conf. Ser.: Earth Environ. Sci.* **24** 012013

[3] Ezhova A, Ostankov A, Panova E and Abramova R 2015 Stratigraphy and correlation of Lower-Middle Jurassic sediments in SE West-Siberian petroleum-bearing province *IOP Conf. Ser.: Earth Environ. Sci.* **24** 012014

[4] Ezhova A V 2008 Refinement of the Geological Model of the Verhnesalat Oil Field (Tomsk District) based on lithofacies investigations of the Middle-Upper Jurassic Productive Bed. *Earth Sciences* **419(2)** 235–238

[5] Kazarinov V P 1958 *Mesozoic and Cainozoic deposits in the Western Siberia* (Moscow: Gostoptehizdat) (in Russian)

[6] Nedolivko N M 2001 Clay minerals of the Jurassic in the eastern part of Nurolsk Basin *Proceedings of the International Scientific and Technical Conference Geological and Mining Education. Oil and Gas Geology* (Tomsk: Publishing House of Tomsk Polytechnic University) 181–184 (in Russian)

[7] Nedolivko N M 2014 Lithogenetic types and Bazhenov suite sediment formation conditions based on Well U-M 413 data (Tomsk Oblast) *Academic Journal of West Siberia* **5**(10) 95–100

[8] Perevertailo T, Nedolivko N and Dolgaya T 2015 Vasyugan horizon structure features within junction zone of Ust-Tym depression and Parabel megaswell (Tomsk Oblast) *IOP Conf. Ser.: Earth Environ. Sci. Scientific and Technical Challenges in the Well Drilling Progress,* **24** 012023 doi:10.1088/1755-1315/24/1/012023

[9] Polyinov B B 1934 *Weathering crust. Processes of the weathering. General facies, forms of weathering and their distribution.* (Leningrad: Publishing house of the USSR Academy of Sciences) (in Russian)

Core acid treatment influence on well reservoir properties in Kazan oil-gas condensate field

A Janishevskii[1] and A Ezhova[1,2]

[1] Department of Mineral Exploration, Institute of Natural Resources, National Research Tomsk Polytechnic University, 30 Lenin Ave., Tomsk, 634050, Russia

E-mail: [2] eav@tpu.ru

Abstract. The research involves investigation of the influence of hydrochloric acid (HCI-12%) and mud acid (mixture: HCl – 10 % and HF – 3 %) treatment on the Upper-Jurassic reservoir properties in Kazan oil-gas condensate field wells. The sample collection included three lots of core cylinders from one and the same depth (all in all 42). Two lots of core cylinders were distributed as following: first lot – reservoir properties were determined, and, then thin sections were cut off from cylinder faces; second lot- core cylinders were exposed to hydrochloric acid treatment, then, after flushing the reservoir properties were determined, and thin sections were prepared. Based on the quantitative petrographic rock analysis, involvin 42 thin sections, the following factors were determined: granulometric mineral composition, cement content, intergranular contacts and pore space structure. According to the comparative analysis of initial samples, the following was determined: content decrease of feldspar, clay and mica fragments, mica, clay and carbonate cement; increase of pore spaces while in the investigated samples- on exposure of rocks to acids effective porosity and permeability value range is ambiguous.

1. Introduction

Bottom-hole formation zone treatment in oil-producing wells is performed to eliminate low production problems. Alternative treatment methods for bottom-hole formation depends on the physico-chemical rock properties and fluid-saturated reservoir.

The research goal is to investigate how hydrochloric acid (HCl – 12 %) and mud acid (mixture of hydrochloric acid HCl – 10 % and fluoric acid HF – 3 %) treatment influences porosity-permeability in reservoir formations. The research tasks involved the multi-analysis of cores before and after rock acid treatment and further comparison of obtained results.

2. Source data

Investigated layers are located within the coal-overlaying thickness of Vasugan suite (U_1^1, U_1^2 b). These layers are composed of fine-grained and medium-grained sandstones with clay and carbonate cement embracing visible traces of authigenic calcite and siderite [2] being to different extents both porous and sometimes oil-saturated.

Samples were taken from three wells 14 and 16 (layer U_1^1) and 673 ($Ю_1^2$ b) in Kazan oil-gas condensate field, located in Parabel region, Tomsk Oblast, Russian Federation. Tectonically, the field is related to local elevations involving the N-W Kalgachsk meso-protusion. Geologically, this field is related to West -Siberian petroleum province.

The sample cores were prepared in the following way: cores from one and the same depth were being cut into three cylinder lots, each of which was 30cm. in diameter and 45cm. in length. The cylinder faces of the first lot were cut off and impregnated with colored resin in a vacuum, and these faces were further made into thin sections. Porosity and permeability in the first cylinders were determined according to the standard method [4, 3]. The second lot of cylinders were inserted into a vessel with 12 % hydrochloric acid solution for 1.5 hours. The third lot of cylinders were exposed to mud acidizing, which is a mixture of hydrochloric acid (HCl)– 10 % and fluoric acid (HF) – 3 %. In this case the samples were exposed to the solution for 5 minutes, where dissolution reaction proceeds violently and, as a result, the solution was discolored to brown. After the samples were flushed, the faces were cut off and then thin sections were made; and porosity and permeability were determined in the cylinders. Therefore, based on the results obtained from 42 cylinders and 42 “colored” thin sections of initial samples (14) after hydrochloric acid and mud acid treatment, the reservoir rock properties were determined. Further, the quantitative petrographic rock analysis in thin sections (before and after acid treatment exposure) was conducted [1, 5]. The above-mentioned results comparable to initial sample data were organized as diagrams, respectively to different colors

3. Research results and discussion

Based on the investigation of the mineral composition, the feldspar content (figure 1) in sandstones after acidizing is significantly different from those in the initial samples. However, the quartz content after acidizing is not different from those in the initial samples. Quartz does not dissolve in hydrochloric acid, while the quartz reaction with fluoric acid (as a constituent of mud acid) proceeds quite slowly.

After the acid treatment a significant change in the rock cement is observed (figure 2). In comparison with clay cement content in the initial samples, the content in above-mentioned samples decreases during the reaction with both hydrochloric acid and mud acid, where a more intensive aggression is closely related to the latter. It is the quick reaction of fluoric acid (HF) (as a constituent of mud acid) which dissolves aluminosilicate, a constituent of clay mineral cement.

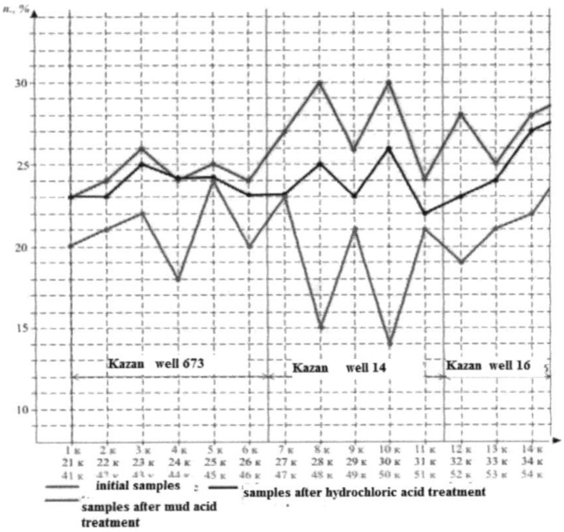

Figure 1. Feldspar content in sandstones according to thin sections.

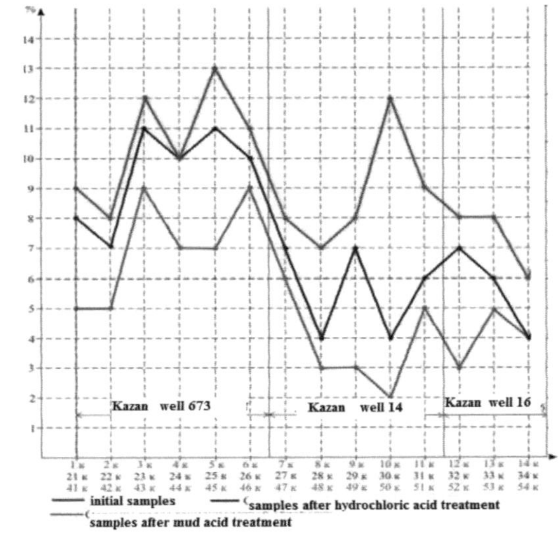

Figure 2. Clay cement content in sandstones according to thin sections.

In rocks with carbonate cement, the cement content decrease is a response to the active hydrochloric acid reaction which dissolves calcite and possibly siderite. On exposure to hydrochloric acid carbonate cement is sometimes preserved as a film around the grains or as filling in small pores. Although mud acid exposure results in the decrease of carbonate cement content, it is less than in the case of hydrochloric acid exposure.

It should be noted that fluoric acid and aluminosilicate and carbonate reaction results in the formation of salts- fluoride silica and fluoride calcite as sediments which are partially removed from the pores spaces during flushing (figure 3). In this case the reaction products are intensively removed during hydrochloric acidizing, pores are decontaminated and the intergranular contacts are clearly defined.

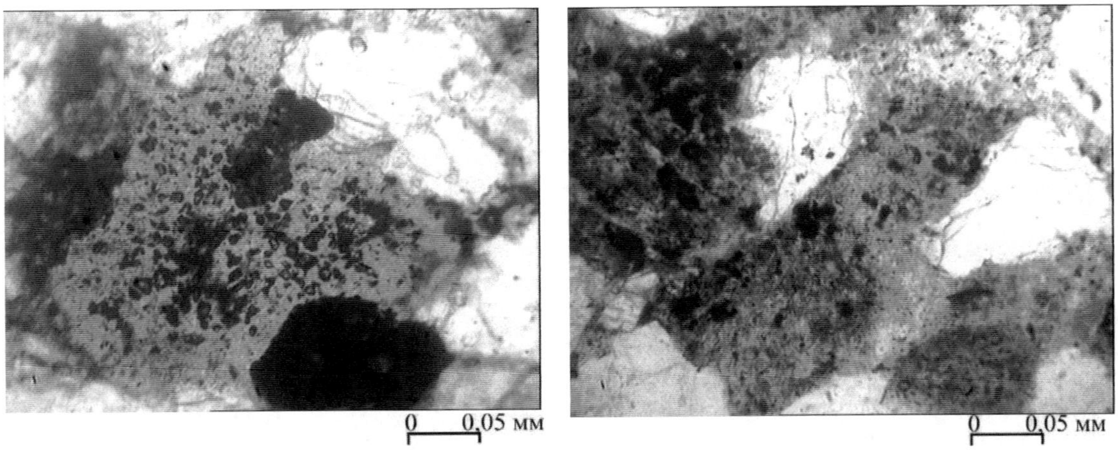

Figure 3. Inclusions of black substance in pores and on grain surfaces (thin sections of samples after mud acid treatment): a) Kazan well 14- thin sections 50к, depth: 2487.0 m. 1 nH; b) Kazan well 16 – thin sections 53к, depth: 2497.8 m.1 nH.

The dissolution of rock components during acidizing resulted in the increase of the pore spaces. The following graphically illustrates the changes of effective porosity and permeability (figures 4, 5).

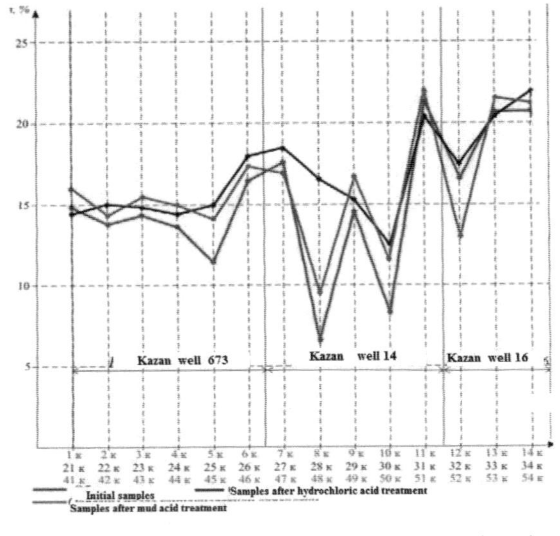

Figure 4. Changes of effective porosity in

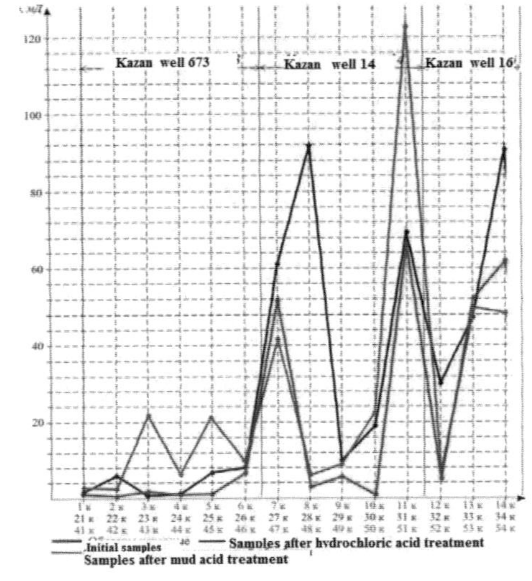

Figure 5. Changes of permeability in samples.

samples.

The increase of effective porosity value is observed, mainly, after mud acid treatment, while the permeability value- after hydrochloric acid treatment. In this case the effective porosity value range is less than the permeability value range. Graphically, it is obvious that in medium-grained sandstones (samples 7k and 12k) the effective porosity value range decreases on exposure to mud acid; while in the samples-7k and 14k- the permeability value range decreases only after mud acid treatment.

During acid treatment the aluminosilcates are dissolved and form salts (fluoride silica and fluoride calcite) in relatively high porosity medium-grained sandstones. These salts precipitate as a black viscous substance in the pore spaces and on grain surfaces, which, in its turn, certainly destroy the reservoir properties.

4. Conclusions

Thus, the above-mentioned research indicated the fact that hydrochloric acid and / or mud acid treatment is highly effective for terrigenous reservoirs, where both clay and carbonate cements are predominate. After mud acid and hydrochloric acid treatment the content of feldspar, clay and mica fragments, carbonate and clay cement decreases in the sandstones, and, consequently, effective porosity and permeability increases.

Based on the samples of medium-grained sandstones with carbonate, as well as argillaceous-silt sandstone and carbonate-argillaceous cement (layers U_1^1 wells 14 and 16 in Kazan field) it was established that hydrochloric acid treatment is more effective in enhancing the reservoir properties.

In relatively high porosity medium-grained sandstones during acid treatment the aluminosilcates are dissolved and form salts (fluoride silica and fluoride calcite), which precipitate as a black viscous substance in the pore spaces and on grain surfaces, and certainly destroy the reservoir properties.

Based on the samples of fine-grained sandstones with carbonate, as well as argillaceous-silt sandstone and carbonate-argillaceous cement (layers U_1^1b well 673 in Kazan field) it was established that both hydrochloric acid treatment and mud acid treatment are effective in enhancing the reservoir properties.

References

[1] Logvinenko N V and Sergeeva E I 1986 *Metody opredelenija osadochnyh porod.* Manual for university students (St. Petersburg: Nedra) p 240

[2] Nedolevko N M 2001. Autigennoe mineraloobrazovanie v jurskih otlozhenijah jugo-vostoka Njurol'skoj vpadiny i ego vlijanie na fil'tracionnye i emkostnye svojstva peschanyh porod. *Petroleum Geology: Proceedings of International Research Conference* (Tomsk: TPU Publishing House) pp 178–81

[3] *Guide for specialist of JSC "Siberian Service Company"* 2009 Tomsk pp 1–17

[4] Khanin A A 1969 *Porody-kollektory nefti i gaza i ih izuchenie* (Moscow: Nedra) p 368

[5] Chernikov O A 1970 *Kompleks metodov kolichestvennogo izuchenija peschano-alevritovyh porod v svjazi s ocenkoj ih kollektorskih svojstv* // Litologicheskie issledovanija porod-kollektorov v svjazi s razvedkoj i razrabotkoj neftjanyh mestorozhdenij (Moscow: Nauka) pp 26–48

Identification of electrofacies on the basis of well logging to determine sedimentation environment of horizon JK_2 in Em-Egovskoe field (Western Siberia)

L Kudryashova[1,2], **V Belozerov**[1,3] **and E Pushtakov**[1]

[1] National Research Tomsk Polytechnic University, 30 Lenin Ave., Tomsk, 634050, Russia

E-mail: [2] KudryashovaLK@tpu.ru, [3] BelozerovVB@hw.tpu.ru

Abstract. Well logging results are one of the ways to study the buried terrigenous rocks genesis. To ensure the most objective characterization of the rock and identification of electrofacies it is necessary to use a complex geological and geophysical survey. The comprehensive investigations of environmental conditions based on well logging have been performed for the horizon JK_2 of Tumenskoe formation in Em-Egovskoe area, Krasnoleninskoe field (Western Siberia). The defined electrofacies were compared with the results of earlier conducted granulometric and mineralogical analyses. The totality of research provided for a conclusion that the investigated sediments of horizon JK_2 had been formed within the destructive tidal delta. Thus, objective facies prediction can only be ensured by analyzing core and well logging data comprehensively.

1. Introduction

Nowadays, lack of core on the one hand but well-developed well logging on the other hand lead to great popularity of genetic interpretation of well logging. Well logging became the main factor in study of the terrigenous rocks genesis. The measurement of deposit geophysical characteristics with the help of well-logging equipment permits obtaining the value of many physical parameters (rocks resistivity, radioactivity, velocity of sound propagation, density, etc.). Moreover, well logging is used to correlate and determine the lithological composition, to define porosity and calculate oil, gas, and water content in reservoirs, as well as to reconstruct depositional environment.

Under the condition of simple structural traps reduction, the particular attention was focused on the composite structured oil and gas traps. Development of these traps is based on knowledge of facies features of terrigenous reservoirs, which, in its turn, makes it possible to predict their location, pinchouts and the change of filtration-capacitive properties of reservoirs [1].

A number of researchers have developed electrometric models of facies based on geometry of spontaneous polarization (PS) curve that reflects definite sequence of graduation change in the process of forming certain facies conditions [1, 2, 3].

As a rule, spontaneous polarization curves are enough to determine sedimentary model of sand formation relying on the core data. However, the more quantity of well logs is used, the more reliable lithological and petrophysical features of rocks are obtained. This decreases the ambiguity and the number of mistakes in facies interpretation [4].

Due to limited amount of core material in deep prospecting, exploratory and producing wells, the most part of parameters determining facies or lithofacies are directly given by well logging data. The term "electrofacies" was introduced by O. Serra in 1970. This term was defined as "the set of log responses, which characterizes a bed and distinguishes it from others" [4].

2. Identification of electrofacies of horizon JK_2 in Em-Egovskoe field

The results of earlier conducted granulometric and mineralogical analyses [5] of horizons $JK_{2\text{-}5}$ in Em-Egovskoe field (Western Siberia) and core description show that the reservoirs are represented by sand and silt varieties of thickness-dependent clay bands. There are fine-grained sandstones, silty with a predominant sand (0.12-0.30 mm) and silt (0.05-0.10 mm) [6].

Granulometric interpretation [6] helped to determine transitional depositional environment of the studied sediments in which the association of tidal plains and deltaic systems are singled out. This association within tidal plains forms tidal channels and estuaries sedimentation environment similar to those in river channel conditions [7].

The detailed analysis was performed only for the horizon JK_2 as Tyumen horizons $JK_{2\text{-}5}$ are characterized by complex geology and maximum set of statistical data that permits making more accurate facies reconstruction.

The top of Abalack formation (horizon JK_1) was used as the marker bed for Jurassic sediments correlation. Low-resistivity cavernous clays of Abalack formation are perfectly recognized and correlated in wells according to the well logging. The horizon JK_1 is represented by marine argillites with calciferous layers and glauconite impurities. Thickness of horizon JK_1 ranges from 23.5 m (well № 1141) to 46.1 m (well № 1841), with average value of 29.4 m.

A special geological feature of Tyumen formation in Em-Egovskoe area lies in the fact that direct qualitative characteristics are not relevant. This does not make possible to identify the reservoirs by traditional way. For example, hole caliper data are usually less informative and poorly correlated with other well logging techniques; microprobe curves often have the increment through the whole well section or do not have the mud cake; the differentiation of spontaneous polarization curve is almost absent. The reasons for the absence of spontaneous polarization curve anomaly are the following: proximity of mud filtrate and formation water mineralization values; high shaliness of the cross-section; ultra-deep flushed zone in reservoir; very high oil saturation of reservoir with hydro-fobization of pore channels surface, etc. Therefore, the facies diagnostics was carried out using the set of well logging techniques, i.e. spontaneous polarization curves (PS), induction logging (IL) and gamma ray logging (GR). These methods were chosen on the basis of well log configuration comparison for the completed wells.

Moreover, it should be noted that the use of well logging technique set permits not only defining more accurately the sedimentation environment but also identifying the bed boundaries in the course of correlation.

About 100 wells of Em-Egovskoe field were interpreted to identify facies environment of the studied territory. Figure 1 presents the correlation of horizon JK_2 in Em-Egovskoe field.

3. Definition of facies sedimentation environment of horizon JK_2 formation

The comparison of granulometric analysis results and identified types of well logging curves resulting from differentiation and correlation shows that the sediments of horizon JK_2 were formed within the estuary (figure 2).

Deltaic sediments refer to the transition facies group. They are formed at the interface of continental and marine sedimentation environments. Both marine and continental deltaic sediments are genetically a single whole. A number of researchers distinguish the set of paleographic zones in the well-studied deltas. These zones are characterized by certain depositional environment [1, 8].

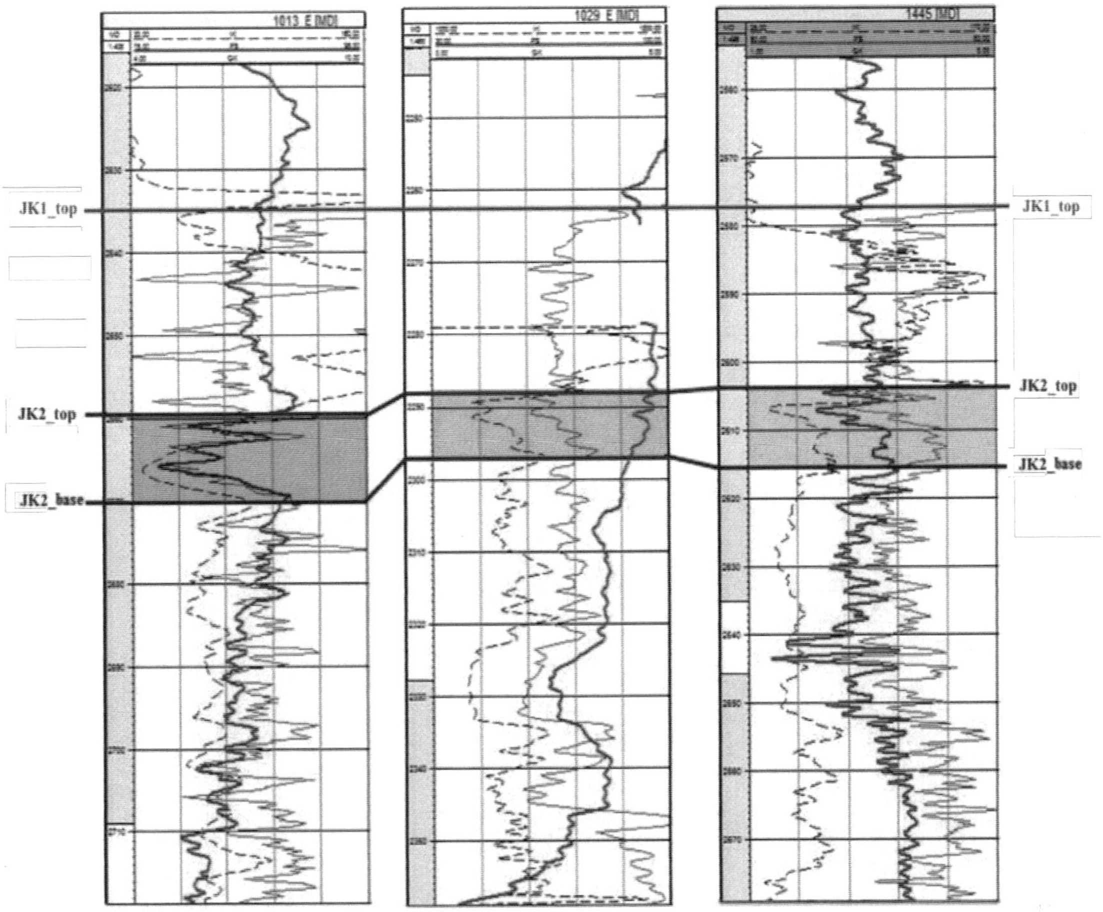

– distributary channel; – distributary mouth bar; – interdistributary bay

Figure 1. Correlation diagram of horizon JK_2 in Em-Egovskoe field through wells № 1013-1029-1445.

The studied deltaic sedimentation environment may be divided in clearly distinguished subenvironments (figure 2) [9].

The distributary channel is a channel with natural stream which transports the part of sediments and river water to the sea.

The distributary mouth bar is a sand bank formed at the interface of the distributary channel and the shoreline. The shallow is formed as a direct consequence of the decrease in flow rate and transport capacity when the flow is outside the channel. The sediments are continuously affected both by the flow itself and the sea waves. The deposits consist of sand and aleurite, there are often thin bands of detritus.

The interdistributary bay is mainly represented by the deposition area of aleurite and clay that are characterized by high rates of deposition [9].

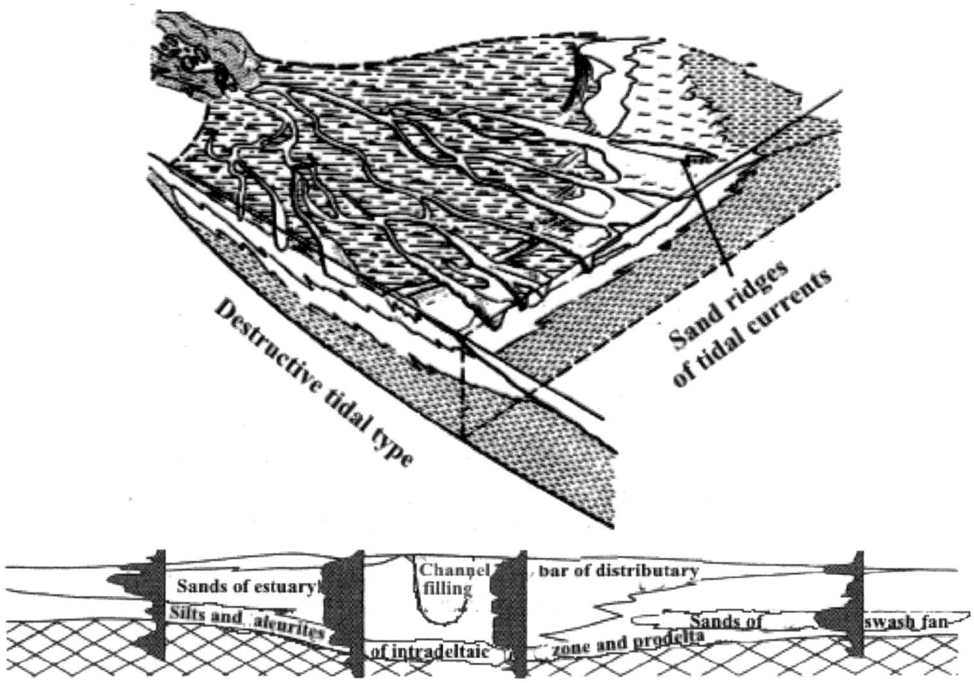

Figure 2. Destructive delta (according to Fisher and Braun) and electric logging profile of deltaic sand bodies [10].

The sand ridge of tidal streams is a continuation of the river channel expanded to the sea.

The channel deposits are represented by coarse-grained and poorly sorted material.

The well logging curves which have been singled out may be compared with the following facies of horizon JK_2 according to the chosen sedimentation model (tidal delta) (figure 3):

- distributary channel;
- distributary mouth bar;
- interdistributary bay.

Facies of distributary channel. PS shows that facies of distributary channel are characterized by "barrel-shaped" and "bell-shaped" forms. These facies are characterized by high amplitudes of PS and low values of GR. Sediments in well № 2548 are represented by light-gray with brownish color, fine-grained, silty, tight sandstone consolidated by clay matter.

Facies of distributary mouth bar. PS shows "funnel-shaped" form. Amplitude of PS is less than one for previous facies. Gamma ray logging curves are characterized by gradual increase in value. Sediments in well № 15r are represented by interlayering of gray, polymistic, fine-grained, dense, tight sandstone and dark-gray, micaceous, tight argillite.

Facies of interdistributary bay are lithologically represented by the sediments that are close to argillaceous sediments. Sediments in well № 505r are represented by thin lamination of light-gray, tight, quartz and clayey siltstone and dark-gray, tight and silty argillites. Thin interlayers of carbonaceous matter and detritus are identified.

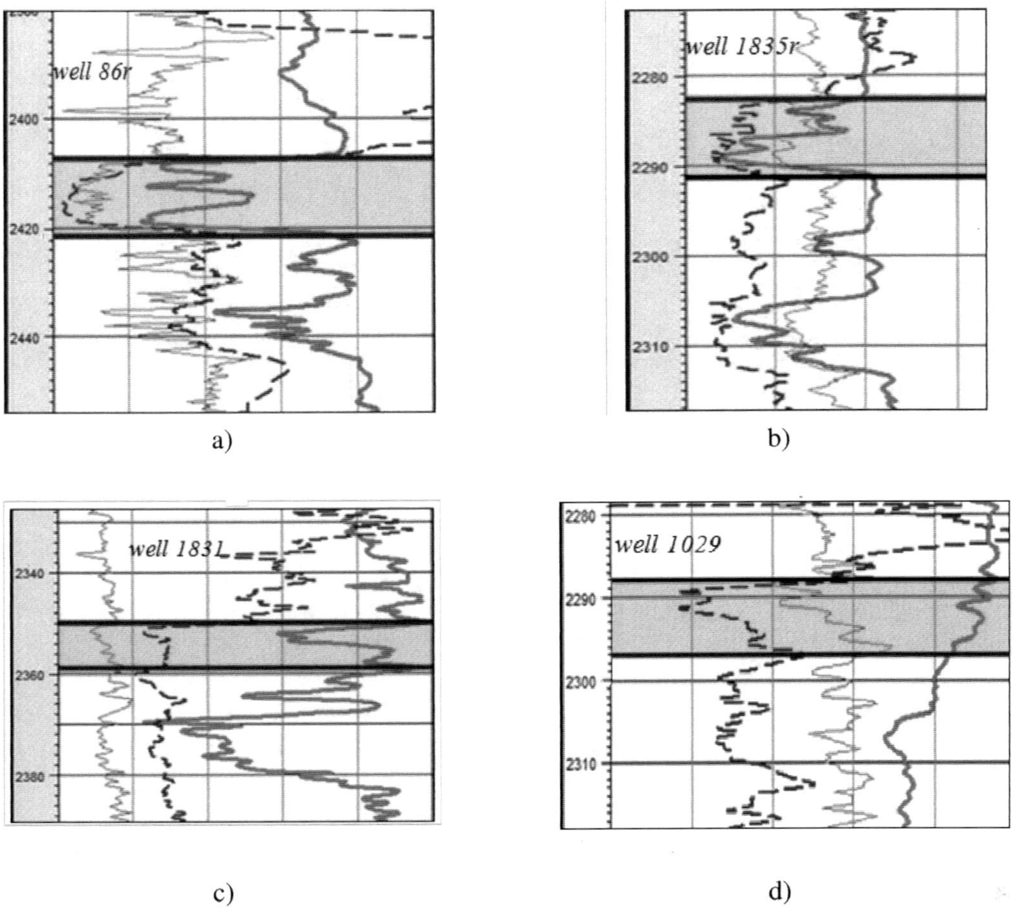

Figure 3. Identified types of electrofacies: a and b) distributary channel, c) distributary mouth bar, d) interdistributary bay.

It should be noted that the determined deltaic depositional environment does not contradict to the accepted lithologic and paleogeographic maps [11].

The horizon JK_2 was formed during Jurassic period – Bathonian stage. At that moment, the studied territory was confined to lowland plain and characterized by the development of the zone containing sand and silt rocks ranging from 30 to 10%. The damp and warm climate is proved by high coal-bearing rock capacity, gray color of rocks, and high content of siderite and significant fraction of kaolinite in horizon sediments [12].

The existing lithofacies schemes of the studied territory predetermine the extensive development of sandstones within the lowland depositional plain, precisely in coastal areas of large water basins during the Early Bathonian. At the second stage of the Bathonian, the transitional facies of these water basins were replaced by marine ones. This contributed to the development of the deltaic depositional environments [12]. The presence of the plain located to the East from Krasnoleninskoe fold which had been covered by the sea also indicates the transitional depositional environments [11].

4. Conclusion

Comparing the results obtained by granulometric, mineralogical and paleogeographic research with the results of facies analysis based on the well logging, the authors draw a conclusion that the sediments of horizon JK_2 were formed by destructive tidal delta. Based on the proposed set of well logging techniques, it is possible to identify electrofacies and compare them with the main facies types of the sedimentation environment: distributary channel, distributary mouth bar and interdistributary bay.

Only comprehensive use of the well logging techniques and the data obtained by core analysis allow identifying the reservoirs even in hard-to-identify sediments which are represented by thin interlayers of sandstone, siltstone, and argillite.

5. References

[1] Muromtsev V S 1984 *Electrometrically geology of sand bodies – lithological traps of oil and gas* (Leningrad: Nedra) p 260

[2] Pettidzhon F, Potter M and Seaver R 1976 *Sands and sandstones* (Moscow: Mir) p 536

[3] Leeder M R 1986 *Sedimentology. Processes and Products* (Moscow: Mir) p 439

[4] Serra O 1985 *Restoration of conditions of sedimentation according to well logging* (Publisher: Schlumberger Limited) p 366

[5] Kudryashova L and Belozerov V 2014 Definition of facies conditions by granulometric analysis on the example of horizons $JK_{2\text{-}5}$ in Em-Egovskoe field (Western Siberia) *IOP Conf. Ser.: Earth Environ. Sci.* **21** 012012

[6] Kudryashova L 2013 Study of lithologic-facies model for enhanced recovery of deposits on the example of Tumenskoe formation in Krasnoleninskoe field *Materials of the I All-Russian Geological youth school "Development of mineral resources base of Siberia from V A Obruchev, M A Usov, N N Urvantsev to our days"* (Tomsk: TPU Press) pp 88-91

[7] Gradzinsky R, Kostetskaya A, Radomski A and Unrug P 1980 *Sedimentology* (Moscow: Nedra) p 640

[8] Broussard M L 1979 *Deltas – models for exploration* (Moscow: Nedra) p 323

[9] Reineck H-E and Singh I B 1981 *Depositional sedimentary environments (with reference to terrigenous clastics)* (Moscow: Nedra) p 439

[10] Reading H G 1990 *Depositional environment and facies* (Moscow: Mir) p 352

[11] Vinogradov A P 1969 *Atlas of lithologic-paleogeographic maps of the USSR* (Moscow: USSR Ministry of Geology Press) p 70

[12] Korzh M V 1978 *The paleogeographic criteria of oil and gas Jurassic in Western Siberia* (Moscow: Nauka) p 136

Specific features of Bazhenov suite sediments in south-eastern Nurolsk sedimentary basin (Tomsk Oblast)

N Nedolivko[1,3], T Perevertailo[1,4], Li Cunyi [1,5] and R Abramova[2,6]

[1] Department of Geology and Mineral Exploration, Institute of Natural Resources, National Research Tomsk Polytechnic University, 30 Lenin Ave., Tomsk, 634050, Russia
[2] Department of Foreign Languages, Institute of Natural Resources, National Research Tomsk Polytechnic University, 30 Lenin Ave., Tomsk, 634050, Russia

Email: [3] nedolivkonm@yandex.ru, [4] ptg@tpu.ru, [5] cunyil1992@yandex.ru, [6] abramova_rn@mail.ru

Abstract. The specific sediment features in Georgiev (J_3kmgr), Bazhenov (J_3vbg) and Kulomzin (K_1bkl) suites, exposed by drilling in the S-E Nurolsk depression (Tomsk Oblast), were defined and described via petrographic, X-ray diffraction and fluorescence-microscopy analysis methods. The classification of agrillites was identified, the structure-texture features, composition, voids and bitumen types and their distribution were determined. It was defined that Bazhenov suite argillites are characteristic of fine-dispersion, high biogenic silica content and scattered organic matter, enriched multi-composite syngenetic bitumen (from light to resin-asphaltine), as well as fractured surface where the migration of light bitumen occurs.

1. Introduction

Bazhenov suite bituminous sediments (Volgian-Early Berriasian) are today a relevant issue in terms of both the formation of source rocks [1, 3] and the potential production of shale gas and shale (light tight) oil [5] within it. The suite, embracing a complex of Domanic source rock formations [8, 9], is characteristic of high lithologic heterogeneity and consists of clay-carbonate-silica rocks, which, in its turn, are organic matter sources of plankton remains with silicic skeletons: radiolarian and diatoms [2]. According to the data [10], siliceous Bazhenov formation «black clay» is regarded as a source rock for more than 85 % of all oil reserves in the Western-Siberian basin. Effective core porosity for argillaceous-siliceous rocks in Bazhenov suite ranges from 3–8 % [4]. Bazhenov suite in studied area overlies Georgiev (Kimmeridgian) suite clays and is overlaid by Kulomzin (Berriasian) suite clays, involving gradation [6].

2. Target formation features and research methods

Investigated Georgiev, Bazhenov and Kulomzin suite sediments were taken from the core depth interval of 2822.3–2795.0 m. in Western-Kvenzersk well № 4 (figure 1), located in similar local elevation in S-E Nurol'sk megadepression within the Central Nurolsk megadepression (Western-Siberian petroleum province, Pudinsk oil-and-gas region, Tomsk Oblast). The cross-section included brownish dark-grey argillites, fine-platy, foliated areas, smooth flinty fracture and containing phosphated and calcitated macrofauna remains. Measured on core, the borders between suites have not been established. In this case, litho-petrographic, X-ray diffraction and fluorescence-microscopy analysis methods were used (18 samples). The thin sections were investigated as following: under the polarization microscope Polam-213 via transmitted-light (nH 1) and polarized light (nH 2); fluorescence-microscopy analysis – fluorescence microscopy Mik-Med II. The research methodology was previously proposed [6, 7].

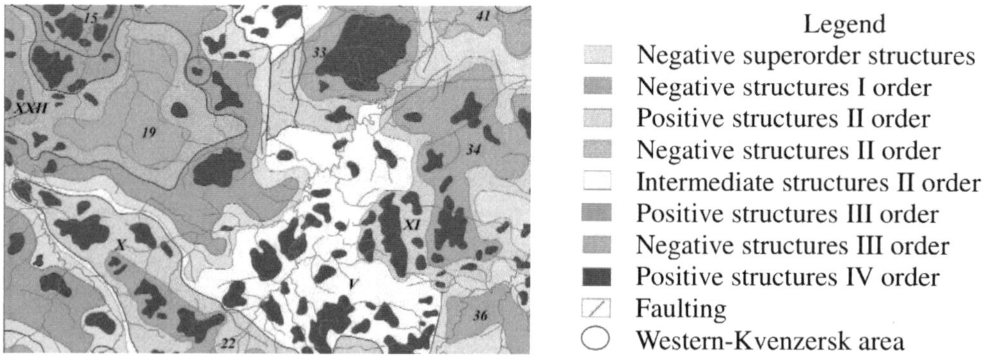

Figure 1. Tectonic map- location of Western-Kvenzersk area, Tomsk Oblast:
V – Chuziksk-Chizhapsko mesoanticline; X – Lavrovsk mesoridge; XI – Pudinsk mesoelevation; XXII – Central-Nurolsk mesodepression; 15 – Festival ridge; 19 – Tamratsk depression; 22 – Lavrovsk ridge; 33 – Luginestsk depression; 34 – Ubileinoe elevation; 36 –South-Pudinsk downfold; 41– South -Sobilin depression.

3. Classification of sediments and lithotype characteristics

According to structure features and composition, five lithotypes were identified.

Lithotype 1 (L-1) – clayey aleurite pyritized agrillites (figure 2) – is abundant in the lower section of studied interval (2822.3–2816.2 m.). Rocks are brownish dark-grey, whereas the color intensifies up-section. The rocks are homogeneous with concealed lamination; contain concretions and pyrite fines, tubular and scallop shell remains, phosphatized onychite layers, as well as isolated poor fossil burrow traces.

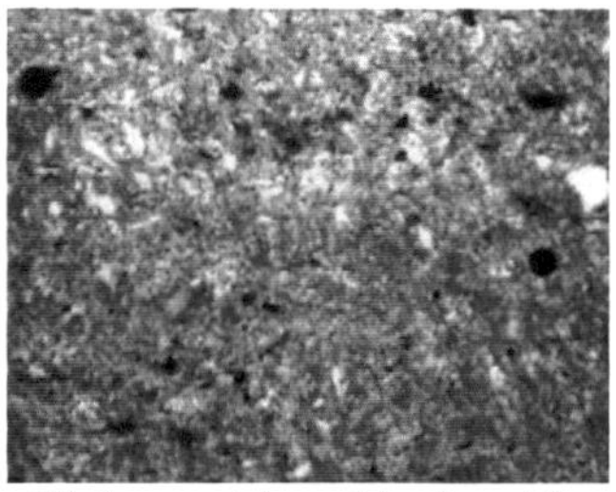

Flaky matrix with aleuritic impurities; Thin section 1; nH 2; depth 2816.5 m.

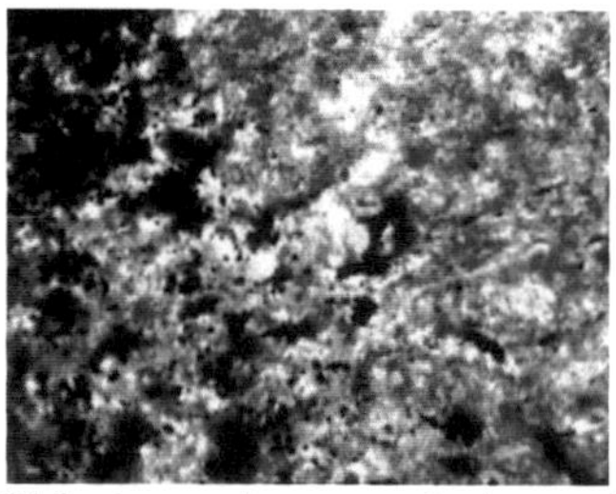

Chlorite and accumulations of loam pyrite; Thin section 1; nH 1; depth 2816.5 m.

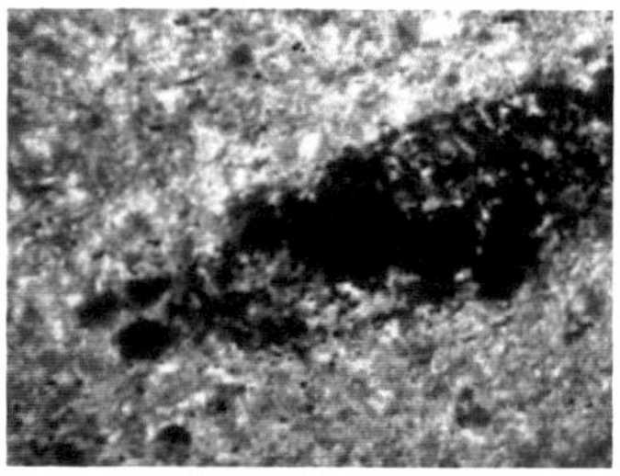

Pyritized burrows; Thin section 1; nH 2; depth 2816.5 m.

0 0,1 0,2 0,3 0,4 мм

Figure 2. Characteristic features of clayey aleurite pyritized agrillites – lithotype 1.

Microscopically, pelitic and lepidoblastic textures were observed in argillites as a result of formed flaky-fibrous argillaceous hydromica (64 %) aggregates. The following was observed throughout the thin section: fine-aleuritic (up to 0.01–0.02 mm.) quartz and biogenic siliceous material (20 %), composing microclusters in the primary rock matrix, and radiolarian shell skeleton remains; irregular distributes loam pyrite (10 %); dispersed tiny (less than 0.01 mm.) rhombohedral and irregular dolomite grains (1 %); isolated chlorite grains. Pyrite being confined to burrows provides shells for radiolaria, belemnite and acline to the extent of complete pseudomorph formation.

Lithotype 2 (L-2) – clayey-siliceous brownish-black bituminous argillites (figure 3) – extends to 2816.2–2814.1 m. These rocks are proportionally composed of pelitic and lepidoblastic clayey (43.2–49.3 %) and microcrystalline and biogenic siliceous (47.6–53.3 %) material which are normally and lenticular distributed and include algae, onychite and belemnite, radiolarian shell and sponge spicule remains. After secondary mineralization uneven occurrences of pyrite (2.1–5.4 %) and calcite trace impurities (2.8 %) are well-defined. Dispersed organic matter, evenly and layerwise distributed, can commonly be found in these rocks and, in this case, the rock matrix has a reddish brown tint, where, thin, relevant to bedding, open cracks are evident.

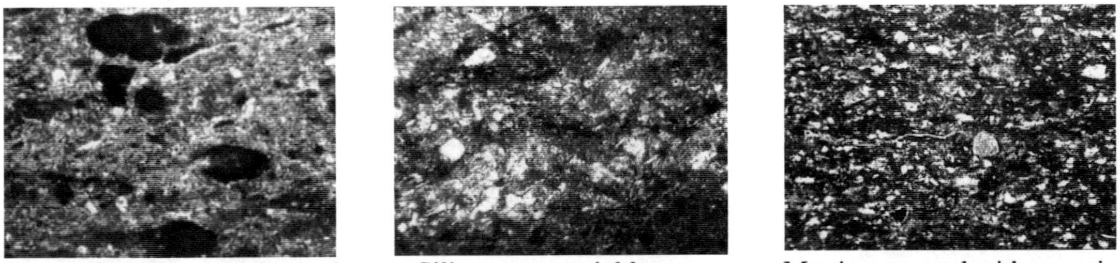

Oreintation of hydromica and radiolarian shells; Thin section 6; nH 2; depth 2813.98 m.

Siliceous material lenses; Thin section 3; nH 2; depth 2815.62 m.

Matrix saturated with organic matter, layer-by-layer cracks; Thin section 5, nH 1; depth 2814.53 m.

0 0,1 0,2 0,3 0,4 мм

Figure 3. Characteristic features of clayey-siliceous argillites – lithotype 2.

Lithotype 3 (L-3) – brownish-black bituminous argillites: clayey (37.1–37.7 %) - *siliceous* (49.2–51.5 %) *carbonated* (figure 4) –overlies vertically at the depth interval of 2814.1–2813.2m., having non-uniform horizontal lenticular and gently dipping foliated bedding and lepidoblastic texture.

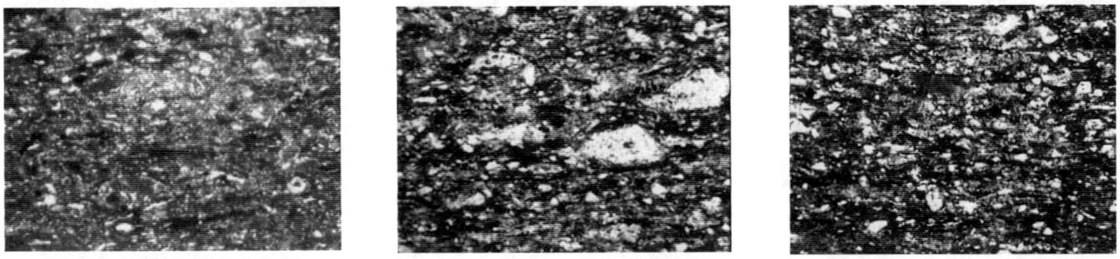

Siliceous material and hydromica; Thin section 8; nH 2; depth 2813.26 m.

Poorly preserved radiolarian; Thin section 8; nH 1; depth 2813.26 m.

Uniformly and lumpy distribution of organic matter; Thin section 7, nH 1; depth 2813.77 m.

0 0,1 0,2 0,3 0,4 мм

Figure 4. Characteristic features of clayey-siliceous argillites – lithotype 3.

Rocks are carbonized (ankerite – up to 0.5 %, calcite – 7.7–10.8 %) weakly pyritized and contain dissolved and recrystallized organogenic remains, often pyrite-calcite replaced: belemnites, onychites, scallop shells, radiolaria, poorly preserved sponge spicules. The organic matter is thin, evenly and layerwise dispersed in the rock matrix, tinting it reddish-brown; and this organic matter is often concentrated as lumps and inclusions. Subvertical and inclined cracks are fractured into large blocks in the top rock, where white calcite can be observed on the sheeting planes. Throughout the layer the characteristic crude oil can be perceived.

Lithotype 4 (L-4) – brownish-black bituminous argillites: clayey (25.1–35.6 %) - *finely-washed high silica* (49.9–64.6 %) (figure 5) – is within the depth interval of 2813.2–2801.9 m. The rocks are composed of weakly dyscrystalline siliceous material and flaky (rarely, fibrous) hydromica aggregates, including thin (up to 0.02 mm.) siliceous bands, concretions and pyrite fines (4.7–10.3 %) and calcite (0–9.6 %).

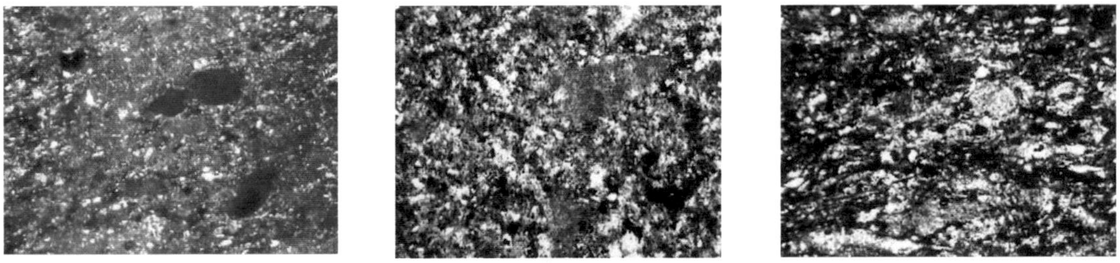

Siliceous material, hydromica flakes, calcite fines; Thin section 10; nH 2; depth 2810.8 m.

Organic matter in radiolarian shells; Thin section 13; nH 1; depth 2804.87 m.

Layerwise organic matter concentration, pyrite fines; Thin section 16; nH 1; depth 2801.96 m.

0 0,1 0,2 0,3 0,4 мм

Figure 5. Characteristic features of clayey-siliceous argillites – lithotype 4.

Macrofauna includes fragments of variform onychite belemites, conical tubular shells, extending to 4 mm. and valva fragments; while microfauna includes desma, dissolved and/or intensely recrystallized siliceous radiolarian skeletons, their pyritized fragments with cellular texture. The rocks are organically saturated, whereas, this organic matter is correlated to either clay substance or siliceous rock constituent. The cavities in radiolarian shells are filled with this organic matter, which, in its turn, could be both dispersed and concentrated forming lumps, often, of bedding orientation. The dense rocks exhibit greasy surface and strong crude oil odor. The cores from above-described wells are split-off layered sections.

Lithotype 5 (L-5) – dark-grey siliceous argillites (38.1–44.9 %) - *clay* (41.6–50.7 %), *homogenous with pelitic texture* (figure 6), poorly – preserves onychites, radiolaria and plant tissue fragments within the depth interval of 2801.9–2795 m.

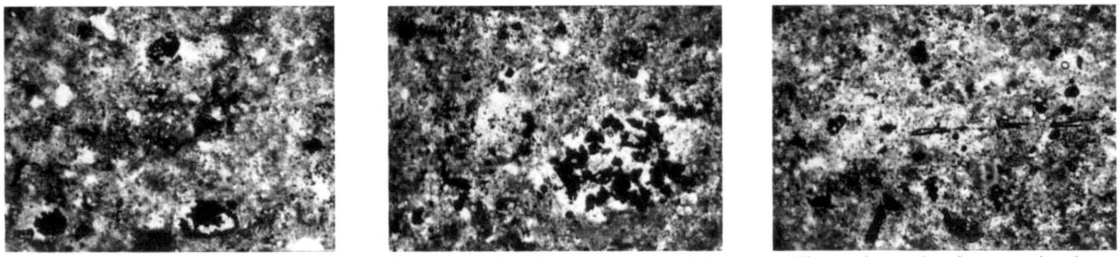

Pyritized radiolaria; Thin section 17; nH 1; depth 2800.4 m.

Pyrite and siliceous material; Thin section 18; nH 1; depth 2797.4 m.

Fine plant detritus, spicula; Thin section; nH 1; depth 2797.4 m.

0 0,1 0,2 0,3 0,4 мм

Figure 6. Characteristic features of clayey-siliceous argillites – lithotype 5.

The pyritizied (8.9–9.1 %) and weakly carbonized (2.3–4.4 %) rocks are composed of non-uniform decrystallized siliceous and clayey pelitic material, containing reddish brown dispersed organic matter, terrigenous quartz (1 %), and remains of fragmented macrofauna , sliced-abraded plant organic detritus. In decrystallization areas of siliceous material to fine-grained chalcedony quartz aggregates, an increasing pyrite content, as thin fines and small irregular grain concentrations, as well as layerwise pyritization have been observed. Comparable to lithotypes 2–4, the organic material content in rocks of lithotype 5 decrease.

4. Lithotypes in cross-section

Clayey-siliceous bituminous argillites (lithotypes 2, 3, and 4) distinguished in the mid-section of the total thickness 14.3 m., comparable to underlying (lithotype 1) and overlying (lithotype 2) agrillites, exhibit more homogeneous micro-texture, smaller component particle size, enriched siliceous material (quantitatively, verging to approximate or exceeding the clay constituent content itself), sometimes enriched calcite and pyrite (figure 7). According to the above-described specific texture and composition, Bazhenov suite is characterized by these argillites. Rocks of lithotype 1 are related to Georgiev suite, while clayey-siliceous argillites (lithotype 5) contribute to the lower section of Kulomzin suite.

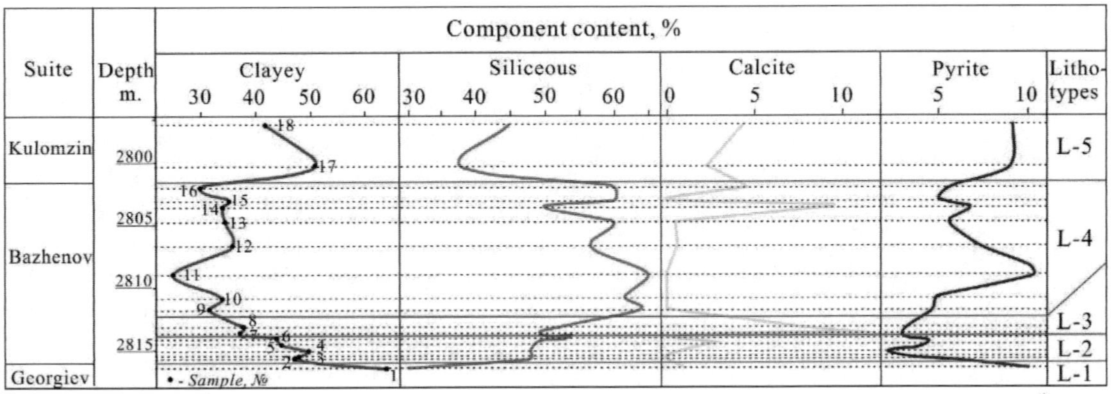

Figure 7. Argillite component composition in Georgiev, Bazhenov and Kulomzin suites in Western-Kvenzersk (well № 4), according to X-ray diffraction analysis data.

5. Voids and bitumen

There are layer-by-layer, sub-parallel and cross-cutting bedding fractures of 0.01–0.02 mm. in the argillites. They are exclusively confined to the lithotypes distinguishing Bazhenov suite. Their formation could be associated with pressure relief, which, in its turn, is conditioned by oil-gas generation within the Bazhenov argillite thicknesses. The factual evidence of this is based on the fluorescence-microscopy analysis results which determined the presence of a wide bitumen spectrum: from light (bluish fluorescence) to resin-asphaltine (dark brown fluorescence), having dotty, uniform, scattered, spotty, and fractured distribution. The Bazhenov argillites are marked by a high content and presence of syngenetic (in rocks) and epigenetic (in cracks) composite bitumen, where resinous (light-yellow fluorescence) bitumen is predominate. The argillaceous suite rocks also contain composite bitumen; however, highly light mobile (blue and light yellow fluorescence) and greasy (yellow fluorescence) epi-bitumen are rather predominate here, having a scattered, dotty and fracture-type distribution. Redistribution of bitumen occurs not only within the Bazhenov formation (where rather light bitumen is located in the cracks), but also via the bitumen migration path: bitumen migrates from Bazhenov suite argillites to underlying and overlying rocks in Georgiev and Kulomzin suites.

6. Discussion

According to the widely-accepted concept, the accumulation of Bazhenov suite rocks is associated with the conditions of Late Jurassic-Early Cretaceous intensive sea basin transgression. Whereas, Georgiev suite sediments deposited under the conditions of low oxygen saturated near-bottom waters (traces of sediment-dwelling organisms); Bazhenov suite sediments deposited under the intensive restoration conditions of depositional environments and stagnant water basins, the indicators of which are the following two facts: finely-dispersed rock and a high content of the indicator mineral-pyrite. Clayey silt was enriched by biogenic and chemical siliceous and carbonate matter and dispersed organic matter as well, and resulted in the formation of the bitumen thickness of fine-washed and laminated argillite source rocks with composite syngenetic bitumen composition, which, in its turn, deposited in areas of relatively deep silt depressions. Sediments in the lower Kulomzin suite accumulated in a stable environment with low hydrodynamic activity, and under more or less deep water conditions (terrigenous impurities, plant detritus).

References

[1] Braduchan Yu V, Gurary F G and Zakharov V A 1986 *Bazhenov Horizon in Western Siberia* (Moscow: Nauka) p 216

[2] Gaivoronskii I N, Leonenko G N and Zamakhaev V S 2000 *Kollertori Nefti i Gaza Western Siberia, ikh Vskritie i Oprobovanie* (Moscow) p 364

[3] Gurary F G 1979 Ob uslovijah nakopleniya i neftenosnosti Bazhenov sviti Western Siberia *Proceedings SRGG&MR* **271** 153–60

[4] Korovina T E, Fedorstov I P and Kropotova E P 2001 Osobennosti sostava, fiziko-chimicheskikh svoistv i emkostnikh charakteristik bituminous argillites *Neftjanoe hozaistvo* **9** 22–25

[5] Lobusev A V, Lobusev M A, Vertievist Yu A and Kulik L S 2011 Bazhenov suite – dopolnitel'ni istochnik uglevodorodnogo sirja v Siberia *Territoriya neftegaz* **3** 28–31

[6] Nedolivko N M 2014 Litogenetic tipi i usloviya obrazovanaya Bazhenov suite, po rezultatam bureniya skvazhina U-M 413 (Tomsk Oblast) *Conference Proceedings. Western Siberia Academic J.* **10**(5) 95–99

[7] Perevertailo T, Nedolivko N and Dolgaya T 2015 Vasyugan horizon structure features within junction zone of Ust-Tym depression and Parabel megaswell (Tomsk Oblast) *IOP Conf. Ser.: Earth Environ.Sci. Scientific and Technical Challenges in the Well Drilling Progress*, **24**, IOP Publishing, UK, 24 (2015) 012023 URL: doi:10.1088/1755-1315/24/1/012023

[8] Poljakova I D, Krol L A, Perozio G N and Predtechenskaya E A 2002 Bazhenov suite: Litho-geochemical classification and sedimentation model *Geology and geophysics* **43** 3 225–36

[9] Predtechenskaya E A, Krol L A, Gurary F G, et al. 2006 On the genesis of carbons within Bazhenov suite of central and south-eastern regions of Western Siberian plate *Litosfera* **4** 131–48

[10] Zakharov V A 2006 Formation conditions of Volga-Berriasian high-carbon Bazhenov suite in Western Siberia based on paleoecology data. *Evolution of biosphere and biodiversity* 552–68

IOP Conf. Series: Earth and Environmental Science **27** (2015) 012015 doi:10.1088/1755-1315/27/1/012015

Terrestrial heat flow and its role in petroleum geology

E N Osipova[1,3], I V Ivanov[1], V A Smirnov[1] and R N Abramova[2]

[1] Department of Mineral Exploration, Institute of Natural Resources, National Research Tomsk Polytechnic University, 30 Lenin Ave., Tomsk, 634050, Russia
[2] Department of Foreign Language, Institute of Natural Resources, National Research Tomsk Polytechnic University, 30 Lenin Ave., Tomsk, 634050, Russia

E-mail: [3] osipovaen@list.ru

Abstract This paper describes an overview of mineral resources exploration survey by geothermal method based on the studies of terrestrial heat flow, anomalies varying in strike, in depth, physical behavior and in time. Applying geothermometry in oil-gas deposit exploration, based on paleotemperature modeling of sedimentation sequences was illustrated.

1. Introduction

Developed producing fields are gradually being depleted, therefore, the application of highly effective methods in exploring new and new hydrocarbon reservoirs is becoming a topical problem today. Analyzing the numerous hydrocarbon exploration methods, the geothermal method is distinguished based on the study of heat flow emerging from the interior earth to the surface.

The term “geothermal survey” was adopted at the Russian Science and Technology Conference in 1972 (Lvov), where the discussed issues involved effective development of hydrocarbon reservoirs for further exploration drilling. The role of geothermics in studying the Earth energy state is in determining the major theoretical geological problem-evolution of our planet.

Based on today's accumulated actual data, the formation behavior of global, regional and local geotemperature fields have been determined. The theoretical concept in evaluating the formation of geological temperature fields involves the physico-mathematical modeling of heat and mass transfer processes in rocks [1].

2. Heat flow and its role in the formation of interior Earth geothermal fields

Heat flow (q) – heat flows through isothermal surface in unit of time, which, in its turn, depends on the heat nature and source intensity, as well as the heat transfer through rocks exhibiting thermal conductivity through radiation and convection. The sources of heat flow are those processes in Earth interior, as well as solar thermal energy.

Solar activity variations (daily, seasonal, annually) result in cyclic air temperature variations. The longer cycle period, the deeper the thermal exposure. So, daily temperature variations are evident up to 1.5m and are associated with solar heat transfer due to molecular rock thermal conductivity and air convection, water vapors, infiltrated precipitation and underground waters. Seasonal and annual solar activity results in the temperature variations at the depth of 20-40m., where heat transfer is due to molecular thermal conductivity, as well as groundwater flow. The neutral layer is located deeper than 40m, where the temperature is practically stable and is on an average of 3.7 °C higher than the annual average air temperature. Below the neutral layer the temperature is higher than the average temperature up to 3 °C at every 100m. Century climatic changes are reflected in the temperature variations at relatively deep depths. For example, cold and warm spells in the Quaternary period influenced the Earth geotemperature regime at the depth of 3-4km.[2]. As hydrocarbon deposits accumulate lower the neutral layer, then the influence of cyclic solar activity is usually not considered.

Heat flow anomalies are distinguished by four attributes in relation to:

the strike- global, regional, local;

the depth- heat source depth: abyssal (mantle), intermediate (crustal), surface (first lithosphere kilometers);

physical nature- alterations of boundary conditions in time and space and environmental thermal physical properties, thermal conductivity features in media of either heterogeneous thermal conductivity or flowing in a solid body, location pattern of thermal sources and different energy-producing processes;

cycle time- stationary, non-stationary.

Deep (abyssal) thermal processes include radiogenic heat as a result of radiogenic isotope element decay (uranium, thorium, potassium) within rocks. Heat also exerts different processes within the Earth interior. For example, exogenous and endogenous chemical reactions, melting, tide deformation and others. The thermal energy of these sources is significantly higher than the energy in tectonic, seismic and hydrothermal processes [2].

Heat flow from the "solid' Earth interior continuously passes into and disperses in the atmosphere. Heat flow density depends on the thermal-physical properties of the geological environment and is closely associated with the regional tectonic structures, as well as volcanic and hydrothermal activities. The distribution of temperature and its sources in the interior and Earth thermal history are based on the geothermometry information of temperature (t) at different depths, geothermal gradients (G) and flow density (q).

Earth heat flow studies involve information on planetary energy balance, energy of geologic-tectonic processes, thermodynamic conditions within the interior, which, in its turn, could be applied in the exploration and development of mineral resources, as well as exploration of geothermal resources – one of the most potential source of future energy.

3. Application of geothermometry in solving petroleum geology problem-tasks

Investigating the distribution of geothermic temperatures in rocks with different thermal-physical properties is a rather difficult task, however, it is solvable. Modeling provides the solution for both direct thermal conductivity problems (solution in differential equations and single-valued conditions is determined by a temperature field) and reverse problems (to given temperature field the boundary conditions are determined, for example, surface heat exchange coefficient).

Significant geothermal field distortion emerges if the rocks of a disturbing body is by 5-10 times different in thermal conductivity comparable to country rocks. This difference is rare in platform regions [3]. Drilling also influences this temperature balance. Temperature recovery time to initial values is ten-fold higher than the drilling time and depends on drilling method.

The processing of factual basin data revealed that, in many cases, geothermal anomalies are confined to oil-gas deposits. Temperatures in these deposits sometimes exceed background temperatures to 10 °C or more. Abnormal heat flows showing hydrocarbons could be the result of organic matter catagenesis and hydrocarbon migration. Emerging from greater depths where the temperature is higher, hydrocarbons provide and distribute the heat while heating the rock mass. According to the thermal exposure the oil composition could be indirectly determined – lighter the oil, more intensive the thermal flow [1].

The technique for the exploration of oil-gas-bearing structures by thermo-tomography was already applied in the 20th century; however, it was not perceived as an effective tool in exploration which could have saved million and million dollars [4]. This method includes plotting a 3D model of heat flow and temperature distribution providing possible cross-sections of a geothermal field at any depth, determining favorable level for hydrocarbon formation, as well as predicting oil-gas field distribution deposit and depth of hydrocarbon accumulations. According to Khutorski this thermo-tomographic model for estimating hydrocarbon potential could be plotted for any territory, and is economically profitable, which, in its turn, shows the advantage of this method.

Other economically effective methods are backstripping and paleotemperature modeling, being based on the geothermal criteria, involve the following: mapping of intensive hydrocarbon generation sources, distribution of relative density of oil resources in oil-gas complex reservoirs, and territorial zoning of potential fields [5]. This method is based on the modeling of heat transmission process in layered sedimentary formation by applying formation temperature measurements, which involved formation temperature measurements and calculated temperatures to vitrinite reflectance.

It is a well-known fact that a sedimentary formation consists of stratigraphic sequences h_i, of which each sequence has given thermal conductivity (λ_i), thermal diffusivity (a_i), heat radiation of radioactive sources (f_i) rocks in sedimentary cross-section and sedimentation time t_i (figure 1).

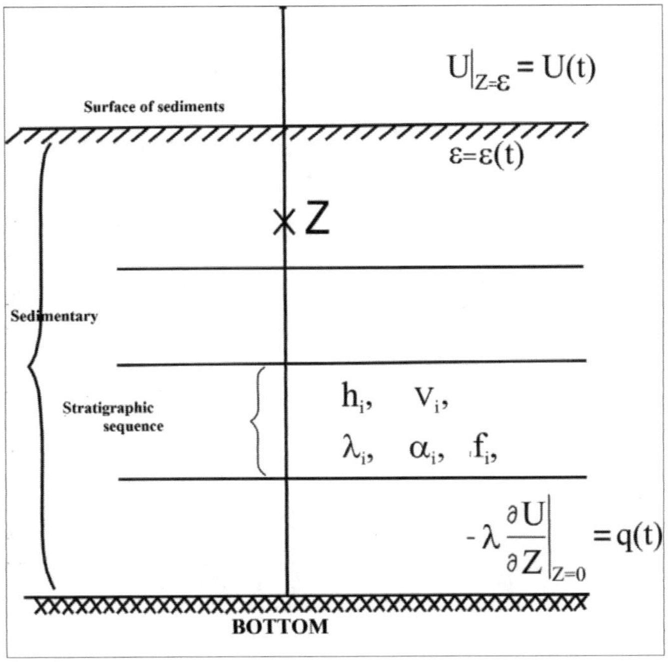

Figure 1. Illustrated scheme of layered sedimentary cross-section in paleotemperature modeling [6]: $\varepsilon = \varepsilon(t)$ – upper boundary of sedimentary thickness; t – sedimentation time; U – temperature; q – hear flow; Z – calculated temperature point; h_i – thickness; v_i – velocity of sedimentation; λ_i – thermal conductivity; a_i – thermal conductivity; f_i – heat radiation density of radioactive sources.

Heat flow from the bottom is determined by geothermic inverse solution. The parameters in modeling paleothermal processes to one well cross-section are depicted in table 1. The values are calculated to 38 deep well cross-sections, drilled in the S-E of Western Siberian platform. This made it possible to identify the distribution of heat flow density within Nurolsk megadepression and its structure contours [5].

Table 1. Example of sedimentary cross-section parameterization, exposed deep well (Vodorasdelnaya 1), determining the parameters of sedimentation and thermo-physical model.

Suite, formation* (stratigraphy)	Thick-ness m*	Age, mil. years	Accumu-lation time mil. years	Density, gr/cm^3	Thermal conducti-vity, W/m·°	T° conducti-vity, m^2/sec.	Heat radiation, W/m^3
Quarternary, *Q*	19	0–1.64	1.64	2.02	1.27	6,5e-007	1.1e-006
Pliocene, N_2	–	1.64–4.71	3.07	–	–	–	–
Miocene, N_1	30	4.71–240.	19.29	2.07	1.31	6,5e-007	1.1e-006
Nekrasov, *nk* $\mathrm{P}g_3$	168	24.0–32.2	8.3	.	1.35	7e-007	1.2e-006
Cheganian, *g* $\mathrm{P}g_{3\text{-}2}$	50	32.2–41.7	9.4	2.09	1.35	7e-007	1.2e-006
Lulinvorean, *ll* $\mathrm{P}g_2$	99	41.7–54.8	131	2.09	1.35	7e-007	1.2e-006
Talich, *tl* $\mathrm{P}g_1$	–	54.8–61.7	6.9	–	–	–	–
Ganlinsk, *gn* $\mathrm{P}g_1\text{-}K_2$	138	61.7–73.2	11.5	2.11	1.37	7e-007	1.25e-006
Slavgorod, *sl* K_2	48	73.2–86.5	13.3	2.11	1.37	7e-007	1.25e-006
Ipatov, *ip* K_2	140	86.5–89.8	3.3	2.18	1.4	7e-007	1.25e-006
Kuznetsov, *kz* K_2	17	89.8–91.6	1.8	2.18	1.43	8e-007	1.25e-006
Pokursk, *pk* $K_{1\text{-}2}$	1057	91.6–114.1	22.5	2.26	1.49	8e-007	1.25e-006
Alymian, a_2K_1	–	114.1–116.3	2.2	–	–	–	–
Alymian, a_1K_1	–	116.3–120.2	3.9	–	–	–	–
Kialinsk, *kls* K_1	459	120.2–132.4	12.2	2.39	1.6	8e-007	1.25e-006
Tarsian, *tr* K_1	93	132.4–136.1	3.7	2.44	1.62	8e-007	1.25e-006
Kulomzian, *klm* K_1	268	136.1–.	9.7	2.44	1.64	8e-007	1.25e-006
Bazhenov, *bg* J_3	26	145.8–151.21	5.4	2.42	1.62	8e-007	1.3e-006
Georgian, *gr* J_3	12	151.2–156.6	5.4	2.42	1.62	8e-007	1.3e.-006
Vasugan, *vs* J_3	62	156.6–162.9	6.3	2.42	1.6	8e-007	1.3e-006
Tumen, *tm* $J_{1\text{-}2}$	313	162.9...	45.1	2.46	1.64	8e-007	1.3e-006

Solving direct geothermic problem-tasks with known value q the temperature U can be calculated in any point of the sedimentary thickness Z within given geological time t.

A geothermal calculation example of one well cross-section is depicted in table 2. Based on the calculated values of 39 deep well cross-sections, maps of geothermal distribution within key geological periods were plotted and intensive intensive hydrocarbon generation sources within Nurolsk megadepression and its structure contours were identified [7, 8].

In early publications, identified intensive hydrocarbon generation sources by geothermal method were described. Comparing the temperatures and time of these sources to Cretaceous and Lower Jurassic oil-gas complex distribution, relative density of generated oil resource distribution was mapped and potential territorial zoning was conducted [9].

Table 2. Thermal history of Bazhenov suite in the sedimentary cross-section of Vodorasdelnaya 1 well.

Suite, formations		Geological age: Index	Geological age: m.a.	Surface heat flow and geothermal in suite center (0C) in paleocross-section (m.a): 0 O	1.64 N_2	4.71 N_1	24.0 Pg_3nk	32.3 $Pg_{3-2}hq$	37.4	41.7 Pg_2ll	54.8 Pg_1tl	61.7 Pg_1-K_2gn	73.2 K_2sl	86.5 K_2ip	89.8 K_2kz	91.6 $K_{1-2}pk$	114.1 K_1a_2
Quarternary		Q	0-1.64	0,3													
Pliocene		N_2	1.64-4.71	0,7	2,6	0,045											
Miocene		N_1	4.71-24.0	1,1	3	4,6	0,045										
Nekrasov		Pg_3h+r	24.0-32.3	4,6	6,7	8,3	7,6	0,045	0,045								
Cheganian		Pg_3r+Pg_2p	32.3-41.7	8,5	10,6	12,2	11,5	10,4	19,9	0,045							
Lulinvorean		Pg_2b+l+i	41.7-54.8	10,9	13,3	14,7	14,1	13	21,6	22	0,045						
Talich		Pg_1t+m	54.8-61.7	12,6	14,8	16,6	15,9	14,8	23,4	23,8	20,5	0,045					
Gankinsk		Pg_1d-K_2m	61.7-73.2	15	17,4	19	18,4	17,3	26	26,2	23	22,1	0,045				
Berezov	Slavgorod	K_2km+st	73,2-86,5	18,4	20,9	22,4	21,6	20,6	29,1	29,5	26,2	25,4	19,9	0,045			
	Ipatov	$K_2st+k+t$	86,5-89,8	21,4	24,1	25,6	24,9	23,8	32,3	32,7	29,4	28,6	23,1	22,2	0,045		
Kuznetsov		K_2t	89.8-91.6	24,1	26,8	28,3	27,7	26,5	35,1	35,4	32,1	31,3	25,9	24,7	20,1	0,045	
Pokrusk		K_2s-K_1al	91.6-114.1	40,9	43,9	45,2	44,7	46,8	52,3	52,6	49,3	48,5	43	41,8	37,3	35,8	0,045
Alymian	Koshaisk member	K_1a	114.1-116.3	57,2	60,6	62	61,3	60,4	69	69,2	65,9	65,1	59,7	58,4	53,9	51,3	21,2
	Layer A	K_1a	116.3-120.2	57,4	60,6	62	61,3	60,4	69	69,2	65,9	65,1	59,7	58,4	53,9	51,3	21,2
Kialinsk		K_1br+q	120.2-132.4	64	67,3	68,8	67,9	67	75,8	75,9	72,6	71,7	66,4	65,1	60,7	58,9	27,9
Tarsian		K_1v	132.4-136.1	71,9	75,2	76,8	75,9	74,9	83,6	83,8	80,5	79,6	74,3	73,1	68,6	65,9	35,8
Kulomzian		K_1v+b	136.1-145.8	76,9	80,2	81,8	81	79,9	88,8	88,9	85,6	84,7	79,4	78,1	73,7	70,9	40,8
Bazhenov		$J_3tt(v)$	145.8-151.2	81	84,4	85,9	85,2	84,1	92,9	93	89,7	88,8	83,5	82,2	77,7	75,1	45
Georgian		J_3km	151.2-156.6	81,6	84,9	86,4	85,7	84,6	93,4	93,5	90,2	89,3	83,9	82,6	78,2	75,6	45,5
Vasugan		J_3 o	156.6-162.9	82,6	86	87,5	86,7	85,7	94,4	94,5	91,3	90,4	85	83,7	79,3	76,6	46,5
Tumen		J_{1-2}	162.9-208.0	87,8	91	92,6	91,8	90,8	99,8	99,7	96,5	95,5	90,3	88,9	84,6	81,8	51,7
Crustal weathering		T-P	208.0-213.0														
Basement		PZ															
Thickness(cross-section),m.			3026	3003	2984	2983	2953	2785	2756,8	2735	2636	2635	2497	2449	2309	2292	1235

4. Conclusion

Deep well drilling is a rather expensive process in exploring the earth interior and results in the destruction of the equilibrium balance of generated formations, which, in its turn, demands significant time for regeneration. Geothermal method is a resource-efficient technique conserving the ecological equilibrium. Potential oil-gas areas could be identified and explored by analyzing the geological structure of this or that area and modeling those heat flow processes within the earth interior. This integrated method makes it possible to determine the density of oil resources in the reservoirs of different oil-gas complexes.

References

[1] A P Kurchikov and B P Stavitskii 1987 *Geotermija neftenosnyh oblastej Zapadnoj Sibiri* 134

[2] Geofizicheskie metody issledovanija zemnoj kory Chapter 5 Termorazvedka http://www.astronet.ru/db/msg/1173309/page47.html

[3] Khutorskoy M D 1982 *Teplovoj potok v oblastjah strukturno-geologicheskih neodnorodnostej* (Moscow: Nauka)

[4] Nauka i tehnologii Rossii: Rossijskie uchenye nashli deshevyj sposob poiska nefti http://strf.ru/material.aspx?CatalogId=21731&d_no=43623#.VXA2a9LtlBc

[5] Isaev V I, Lobova G A and Osipova E N 2014 The oil and gas contents of the Lower Jurassic and Achimovka reservoirs of the Nyurol`ka megadepression *Russian Geology and Geophysics* **55** 1418–28

[6] Gulenok R Yu, Isaev V I, Kosygin V Yu, Lobova G A and Starostenko V I 2011 Estimation of the Oil-and-Gas Potential of Sedimentary Depression in the Far East and West Siberia Based on Gravimetry and Geothermy Data *Russian Journal of Pacific Geology* **5**(4) 273–87

[7] Osipova E N, Prakoyo F S and Kudryashova L K 2014 Petroleum potential of the Neocomian deposit of Nyurolsky megadepression *IOP Conf. Ser.: Earth Environ. Sci.* **21** 0120011

[8] Lobova G, Osipova E, Isaev V and Terre D 2014 Petroleum potential of Lower-Jurassic deposits in Nurolsk megadepression. Scientific and Technical Challenges in the Well Drilling *IOP Conf. Ser.: Earth Environ. Sci.* **21** 012001

[9] Lobova G A, Popov S A and Fomin A N 2013 Lokalizacija prognoznyh resursov nefti jursko-melovyh neftegazonosnyh kompleksov Ust'-Tymskoj megavpadiny *Neftjanoe hozjajstvo* **2** 36–40

Application of geologic-mathematical 3D modeling for complex structure deposits by the example of Lower-Cretaceous period depositions in Western Ust – Balykh oil field (Khanty-Mansiysk Autonomous District)

T Perevertailo[1,3], N Nedolivko[1,4], O Prisyazhnyuk[1,5] and T Dolgaya[2,6]

[1] Department of Geology and Mineral Exploration, Institute of Natural Resources, National Research Tomsk Polytechnic University, 30 Lenin Ave., Tomsk, 634050, Russia
[2] Department of Foreign Languages, Institute of Natural Resources, National Research Tomsk Polytechnic University, 30 Lenin Ave., Tomsk, 634050, Russia

E-mail: [3] ptg@tpu.ru, [4] nedolivkonm@yandex.ru, [5] pon0792@mail.ru, [6] dtf2@yandex.ru

Abstract. The complex structure of the Lower-Cretaceous formation by the example of the reservoir BC_{10}^{1} in Western Ust – Balykh Oil Field (Khanty-Mansiysk Autonomous District) has been studied. Reservoir range relationships have been identified. 3D geologic-mathematical modeling technique considering the heterogeneity and variability of a natural reservoir structure has been suggested. To improve the deposit geological structure integrity methods of mathematical statistics were applied, which, in its turn, made it possible to obtain equal probability models with similar input data and to consider the formation conditions of reservoir rocks and cap rocks.

1. Introduction

Today, the major hydrocarbon resources in West Siberia petroleum – bearing province are related to Jurassic formations. However, due to the increased production and structure depletion of the area, more and more attention is being paid to the Neocomian formations, which, in its turn, is associated with the future reserves growth. Here, the lithologically sealed oil accumulations (clinoforms) are characterized by a complex geological structure, as well as a great variability of reservoir filtration capacity properties.

Nowadays, well drilling justification, oil-gas field development plans, hydrocarbon reserves estimation, cost-effectiveness analysis of geological and technological activities are performed in 3D modeling. In this respect, 3D geological modeling should include the heterogeneity and variability of the reservoir structure both laterally and vertically. It is known that the reliability of the simulated geological unit depends on the exploration certainty. When the information is insufficient, the stochastic approach in 3D modeling is used. It is one of the ways to increase the integrity of the geological structure for oil and gas deposits [2].

2. Unit characteristics and research methods

The targets are Lower-Cretaceous formations of Sortymsk series studied by the example of Western Ust – Balykh oil field located in the area of Surgut and Nefteyugansk (Khanty-Mansiysk Autonomous District). The major reservoir is BC_{10}^{1} layer identified in BC_{10} horizon of Sortymsk series. The stratigraphic layering profile is based on the following documents: "Regional Stratigraphic Sectional Plan of Mesozoic formations in West Siberian Plain" adopted by the Vth Tyumen Agency Stratigraphic Meeting, May18, 1990 and approved by the International Commission on Stratigraphy, January 30, 1991 and "Resolution of the Agency Stratigraphic Meeting, for consideration and approval of Mesozoic depositions stratigraphic schemes in West Siberia territory" approved by ISC on April 2004.

Tectonically, Western Ust – Balykh oil field is divided into two parts: the eastern part is from Surgut arch and is considered to be the Ist order structure in the range of Ust – Balykh Mamontov swell; the western part is found in Tundrinsk basin stretching northwards. Malobalykskaya mega saddle is located on the south border of the area; Salymsk megaswell is on the west.

The characteristic structure for BC_{10} layers is the clinoform development of formations which is isochronous isolated system of sand bodies with variable argillization. These positionally overlap and prolongate each other towards the regressing sea forming regressive sand sheet [3].

Clinoform structure of the studied formations is confirmed by seismic data which identified that the field clinoforms have submeridional orientation, i.e. they can only be traced during the correlation of profiles from the south-east to the north-west. BC_{10}^{1} layer is related to Pokachevsk clinoform of Early Cretaceous Sortymsk series.

To identify the geological structure features and reservoir range relationships correlation of BC_{10}^{1} layer has been conducted, potential oil-bearing formations were mapped, as well as average porosity distribution maps.

When correlating well profiles the datum mark was the bottom of Cheuskinsk clay sequence being extended to thickness of 30-45 m and has specific geophysical characteristic, i.e. is well-identified by geophysical well logging (GIS) by high curve values of gamma logging (GL) and low values of neutron logging (NL). The layer thickness varies from 14 m to 61 m and naturally increases from the north-west to south-east. Lithologically, BC_{10}^{1} layer composition is heterogeneous and characterized by sandstone layering from fine-grained to coarse–grained with clayey and aleurite-clayey layers (figure 1).

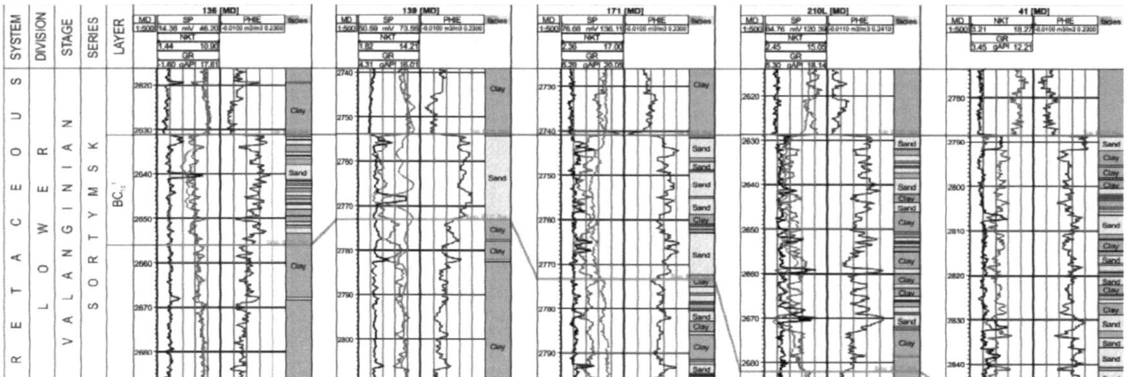

Figure 1. Correlation scheme from NW to SE along well lines 136-139-171-210L-41.

The reflection seismic horizon NBC_{10} is confined to the bottom of Cheuskinsk clays. Taking into account seismic data quality and sufficient seismic density profiles, the reflection horizon surface NBC_{1} was chosen, including drilling data. It should be noted that the wells are located irregularly, so correlation accuracy and reliability within the area would be non-identical.

Generally, Western Ust – Balykh uplift is an isometric structure extruding on the terraced slope, plunging from N-E to S-W.

As a result of studied areal geological and geophysical activities, pinching-out lithological borders of reservoirs have been identified- three deposits in BC_{10}^{1} layer have been determined.

GIS interpretation of the reservoirs and their properties identification was conducted. Porosity coefficient was calculated to the GL curve as in some wells there was no SP curve or brine mud drilling (polymer drilling mud) was performed. The interpretation was made by the dual difference parameter which is equal to the difference between current and minimum GL values. The reservoirs and impermeable rocks were identified by boundary values.

Based on GIS interpretation data general strata maps, efficiency strata maps, oil-bearing strata maps and porosity mean values maps were plotted. Well log maps were used as trend surfaces both for lithology and petrophysical properties modeling [6].

The following relationships were determined: sandstones extent as submeridian strike bands, confined by the lithological layer replacement zones. Within the area reservoir thickness changes from complete pinching-out zone to 56.5 m. Maximum oil-bearing strata thickness is 50.26 m. The deposits are lithologically sealed. Maximum oil column thickness is up to 61m. Absolute OWC Vertical Depth is at 2471m according to GIS data, sampling, perforation and reservoir development data during the first operation months.

Based on porosity mean values maps, the maximum values are identified in the deposit dome, which is 18.6 % and decreases to the pinching lines up to the boundary values. In general, the porosity values of three deposits are distributed uniformly and characterized by high capacity properties.

The model structure carcass was plotted on the basis of the roof formation surface (reflection horizon NBC_{10} related to drilling data) and the bottom formation surface BC_{10}^{1} calculated by the strata maps. All the wells, where GIS data had determined the reservoir roof and bottom locations, were applied.

To calculate 3D grid, the optimal horizontal cell size equals 150×150 m was chosen, including the minimum distance between the wells. Vertical partitioning into layers, based on the geological structure peculiarities, was performed according to the formation roof. The optimal layer thickness was determined by the sampling distribution analysis of the facies thickness. Data presentation as histograms, functions, cross-plots, pie-charts being in compliance with factual log and seismic data analysis were applied for the distribution of properties in the borehole space. The histogram in figure 2 shows the optimal value (0.5 m) for vertical resolution.

The results of log data smoothing in the grid are given in the histogram (figure 3). It shows the comparability of re-scaled cells and GIS initial curves.

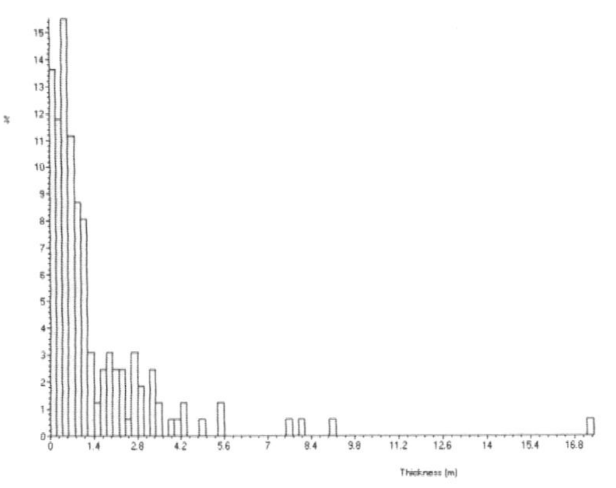

Figure 2. Histogram of sandstone formation layer distribution.

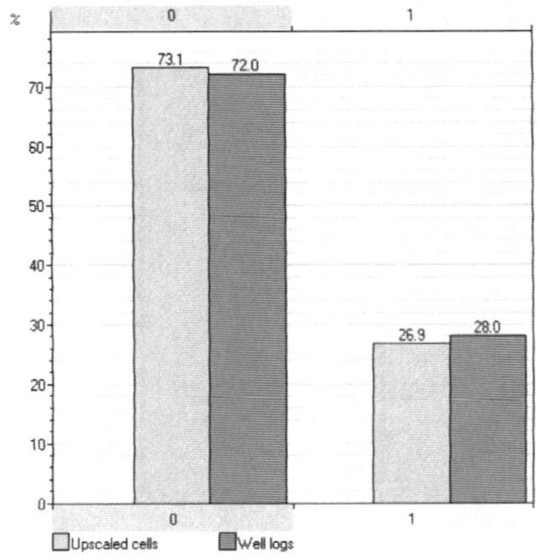

Figure 3. Re-scaled log diagram histogram (green color) and initial log diagram (red color).

IOP Conf. Series: Earth and Environmental Science **27** (2015) 012016 doi:10.1088/1755-1315/27/1/012016

Interpolation of properties in the borehole space is the basic and most complex stage in 3D geological modeling. In this research the mathematical statistics techniques have been applied to obtain more realistic geological models, particularly, in the well zones with low density; to obtain several equal probability models with the same input data set and to consider the reservoir rock and cap rock formation conditions. Variogram analysis was used to identify spatial correlation of geological data. It is based on the correlation bond between two values of a parameter in the adjacent points of space during the properties distribution in the borehole space. The distance increases and variation range also increases [1]. 3D modeling includes plotting the facial model, firstly, and further petrophysical modeling.

The initial parameters for variogram plotting is preset interactively via the Variogram window to the principle direction and perpendicular to the principle direction. The principle direction parameters for sandstones have the following values:

- band width – 1200 m (based on drilling – out grid);
- search radius – 4800 m (should be 0.7-1 of whole field measurement);
- tolerance angle – 32° (orienting pairs of points in chosen direction);
- number of lags– 10 (value is chosen due to the obtained histogram and lag distance delay);
- lag tolerance – 50 % (subsequent and preceding lag pairs in the interval);
- lag distance – 480 m (search radius ratio to the number of lags).

The obtained experimental variogram has the following parameters:

- Major direction– 300 °;
- Minor direction– 210 °;
- Major range – 680 m;
- Minor range – 500 m;
- Nugget – close to 0.

Nugget value close to 0 identifies good functional bond and experimental variogram data can be used for 3D modeling.

Furthermore, the experimental variogram approximated was by a smooth curve. The variogram spherical model characterized as smooth, uniform dispersion increase between the data to specific maximum was applied [4].

Firstly, the curves describing the vertical distribution of sandstone and clay facies due to the model layers were evaluated and edited by the Data Analysis. Secondly, these curves were used to control the vertical facies in the modeling distribution including the variograms.

Lithological modeling was performed by Sequential Indicator Simulation technique based on the experimental variograms and trends data. To approximate to the real geology unit the effective strata map as a horizontal trend including reservoir pinching lines was applied. As a result of obtained plots, the facial model shown in figure 4 was obtained.

For petrophysical modeling the following initial data were used: GIS interpretation results, 3D lithological model; geological relationships variability of petrophysical properties in the area: reservoir pinching polygons; porosity mean values calculated to wells data; petrophysical dependence equations.

To model the reservoir petrophysical properties semideterministic technique called Kriging providing the minimal variance of error was applied.

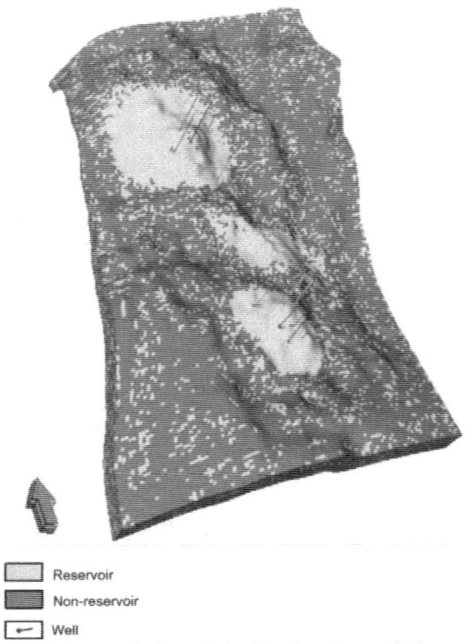

Figure 4. SD lithological model.

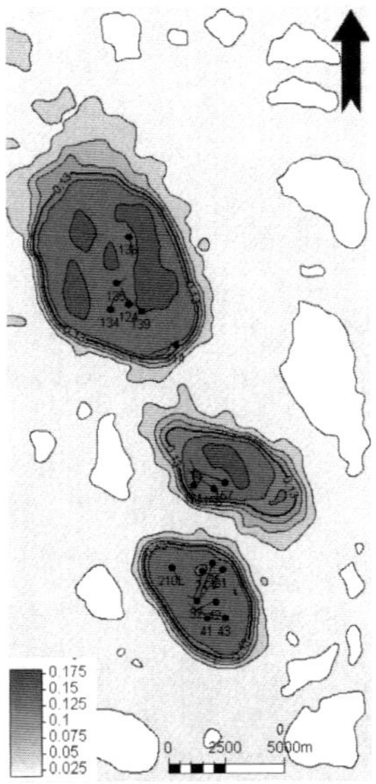

Figure 5. Porosity mean values map obtained from 3D model.

Reliability evaluation of geologo-mathematical model was performed on the basis of correlation values of well data and the obtained 3D geologo-mathematical model data (figure 5).

3.Conclusion.

Formation of BC_{10}^{1} horizon sediments is related to shallow shelf environment, where the lens-convex asymmetrical bodies are formed. These are characterized by sand, impermeable clayey and aleurite-clayey rock layerings.

The total thickness increase is from the North-West to the South-East.

Three lithologically sealed oil accumulations have been identified within the field.

Reservoirs are characterized by high reservoir properties; according to A. A. Khanin classification, they are III^{d} class.

3D geological model is based on geostatistics techniques.

The experimental variogram that precisely corresponds to the spherical model has been calculated for the property distribution relationship in the borehole space.

The obtained 3D geological model corresponds to the well data and represents the complex production horizon BC_{10} structure.

References

[1] Dubrule O 2003 *Geostatistics for seismic data integration in earth models: SEG Distinguished* (USA: Society of Exploration Geophysicists and European Association of Geoscientists and Engineers) p 283

[2] Putilov I and Potekhin D 2015 Creating technological polivariation 3D modeling with quality control of the models realization for raising reliability *Oil Industry* **1** 15–17

[3] Ivanova L, Lapina L, Lebedev M and Yanevits R 2012 Conceptual facies model of clinoform type sedimentary basin *Oil Industry* **10** 7–11

[4] Zakirov S, Zakirov E and Indrupskiy I 2006 New concepts in 3D geological and hydrodynamic modelling *Oil Industry* **1** 34–41

[5] Kovalevskiy E 2007 Fuzzy geological model. Petroleum Geostatistics (Portugal: Cascais)

[6] PerevertailoT and Zakharova A 2011 *3D geological models generation for oil and gas fields by the software Petrel Schlumberger* (Tomsk: TPU) p 93

Investments in oil field development by the example of Tomsk oblast

Ye I Shmidt[1,3], G F Il'ina[1,4] and I A Matveenko[2,5]

[1] Department of Geology and Mineral Prospecting, Natural Resources Institute, National Research Tomsk Polytechnic University, 30 Lenin Ave., Tomsk, 634050, Russia
[2] Department of Foreign Languages, Natural Resources Institute, National Research Tomsk Polytechnic University, 30 Lenin Ave., Tomsk, 634050, Russia

E-mail: [3] lena_shmidt@bk.ru, [4] ilgf@sibmail.com, [5] mia2046@yandex.ru

Abstract. The article describes the geologic structure of the formation located not far from Strezhevoy Tomsk Oblast. The formation has been poorly studied by seismic methods. The reserves categories C_1 and C_2 as well as hydrocarbon potential are presented. 4 exploratory and 39 production wells are designed to be drilled depending on geologic knowledge and formation conditions. The article deals with the investment plan including development, oil export expenditures and implementing cost calculation.

1. Introduction

The field license area of 1192 km^2 is located 460 km from Tomsk to the North-West, 300 km from Strezhevoy to the South-East. The nearest settlement is Kedrovy, which is located 90 km South-East from the field and is the major oil-and-gas producing industry in the southern territory. There are no highways or railways in this area. The territory of license area is considered of little value for agriculture and forestry. The seismic grid interval of CDPM is: in total – 0.77 km/km^2, in 1986-99 – 0.01 km/km^2. The field is poorly studied by seismic CDP methods [5].

The geologic structure of the formation is complex; the oil deposit is confined by terrigenous Upper-Jurassic sandstone formations J_1^2 and J_1^3. The major oil-and-gas formation consists of horizons J_1 and J_2. In terms of reserves, it is referred to small and is of up to 5 mln. tons. It has been developed since 2009, in 2012 the production wells were first drilled. The project drilling depth is at 2961 m. In total, 4 exploratory and 39 production wells are designed to be drilled [3].

2. Hydrocarbon potential and reserves

Oil formation is confined to sandstone formations J_1^2 and J_1^3 horizon J_1 of Vasyugan suite. Formation J_1^2 is a hanging type controlled by OWC from West to South at minus 2345 m, from East – lithologically capillary sealed at minus 2300 m to minus 2320 m. Formation J_1^3 is developed throughout the elevation area, but is significantly shaled out in the East forming a lithological seal, where oil accumulation is controlled by the OWC from the West to the South at minus 2355 m. From the East the boundary of the oil field is a lithological seal revealed within the hypsometric limits from minus 2320 m to minus 2300 m. Oil deposit is a blanket-like accumulation with an overlying seal.

According to the State Reserves Register the parameters of the prospected deposits: gross pay/net pay thickness of producing reservoir J_1^2 – 3.0/2.6 m, effective porosity – 15%, hydrocarbon saturation – 0.59, oil recovery efficiency – 0.3.

Oil reserves (prospective/recoverable) of category C_1 is 2168/542 000 tons and C_2 – 635/159 000 tons [2].

Gross pay/net pay thickness of producing reservoir J_1^3 – 5.6/4.4 m, effective porosity – 16%, hydrocarbon saturation – 0.67, oil recovery efficiency – 0.25.

Oil reserves (prospective/recoverable) of category C_1 is 1113/223 000 tons and C_2 – 2124/425 000 tons.

3. Forecasting development parameters

To forecast the development parameters the ultimate reserve categories C_1 and C_2 were used. It was assumed that the confirmation factor of category C_2 – 50 %.

The following geologic information was used for forecasting: layout of net formation thickness distribution (indirectly reflecting changeability of reservoir properties) and layout of oil-saturated formation thickness distribution [4].

The layout of net formation thickness distribution of J_1^2 and J_1^3 formations shows the improvement of filtration characteristics in the crest.

Production well pattern is based on the layout of oil-saturated formation thickness distribution of J_1^2 and J_1^3 formations.

In terms of exploration and reserve statement the field development is planned in the stages:

I stage – production test of well №1 after its recommissioning, according to well test operation plan (1 year), recommission of exploration well and its test implementation (2 years). Within this period the appraisal drilling of 4 wells is performed to delineate the field and identify other formations. Construction of development and pipeline facilities begins at this stage;

II stage – drilling six production wells and field development, simultaneously, grid-type seismic-geophysical operations are performed to specify the structural field plan and field junction zones. Construction of development and pipeline facilities is completed. Artificial lift is implemented (electric centrifugal pump (ECP) and beam-balanced pump (BBP) with the capacity from 10 m^3/day to 60 m^3/day). The relationship of the well number equipped with ECP and BBP – 40/60. The period lasts 2 years. Then – design of field development plan;

III stage – implementation of the field development plan.

In a period of 15 years to explore the geological structure of the field, it is planned to make the seismic grid denser. The initial cost of geologic exploration for additional field exploration is about 7 mln. roubles (100 lineal km. common depth point seismic 2D) (figure 1) [2].

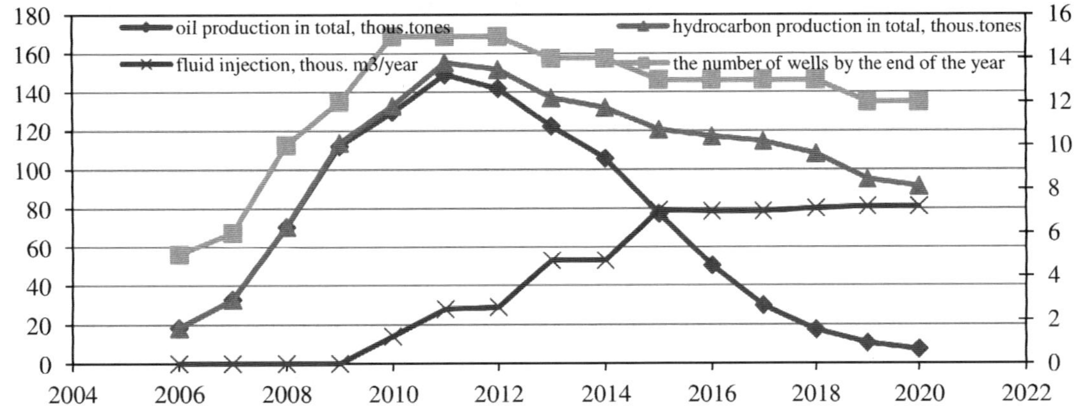

Figure 1. The chart of studied field development.

The formation is planned to be drilled in production well pattern of 700*700 m, boundary net-pay thickness of well placement – 2 m, in total, 10 production wells are designed to be drilled, 3 of them – injection ones. In the course of field development special attention should be paid to the dynamics of gas-oil factor change, as there can be a gas cap.

At the first stage of field development the recovery rate from the reservoir of recoverable resources will not exceed 3 % (of total), the forecasting annual oil production – 18400-33000 tons, cumulative oil – 51400 tons.

At the second stage of field development six production wells are implemented, in two of them hydraulic fracturing treatment is performed. The recovery rate from the reservoir amounts about 10 % of all recoverable resources, annual recovery from 40 to 112 000 tons, water cutting increases to 1.5 %, oil production at the second stage – 182400 tons.

According to the field development plan, at the third stage of field development, 4 more wells are drilled, in 3 of them hydraulic fracturing treatment is performed. The annual production achieves 149.0 000 tons, water cuttings – 92 % by the end of the given period, recovery rate from the recoverable resources decreases from 14.5 % to 1 %. At the given stage 840800 tons of oil are planned to be recovered. The spot contour waterflooding system will be implemented in the field. In total, 3 wells will be converted into injection after oil production.

All in all, over 15 years 1075000 tons of oil will be produced (97.0 % of approved reserves of categories C_1+C_2).

4. Economic analysis and calculations

Commercial efficiency of investments in field development is estimated on the basis of accepted drill footage, production level, well stock conditions.

Calculation of key economic indicators is performed for 15 years of field development. The calculation includes taxes and duties, which are reported and paid by an enterprise for the budgets of different levels in the course of its economic activity (table 1):

1* unified production tax – 340 roubles/t with the coefficient describing the dynamic of the oil world market prices,
2* property tax – 2 %,
3* income tax – 24 %.
4* road fund tax – 1 %,
5* unified social tax –35.6 %.

Commercial evaluation of field development was made under the conditions:

- 30 % export sale,
- 70 % domestic sale.

The sale price in the domestic market is 8200 roubles/t (with VAT), in the international market – 50 USD/barrels.

Export cost for oil transport to the borders is calculated in accordance with the rate equal to 22 USD/t, duty – in accordance with the RF Law "On Custom Tariff "taking into account the average cost rate of Urals crude oil in the world market is included in the calculations [1].

The most critical indicator for calculation of investment projects is oil price. Changes in oil sale price in domestic market and for export, its cuts will result in increase of payback period. Oil price increase (domestic up to 8200 roubles/t and export up to 50 USD/barrels) will allow paying back the investments in field development within 5.1 years.

The present paper considers the reserves confirmation of category C_2 by 50%. The maximum risk in the field development is reserve non-confirmation of category C_2 by more than 50 %. In this case there are changes in dynamics of fluid and oil withdrawal and, in terms of expert appraisal; the payback period for these operations may exceed 8 years, which makes it economically unattractive under the existing norms [3].

IOP Conf. Series: Earth and Environmental Science **27** (2015) 012017
doi:10.1088/1755-1315/27/1/012017

Table 1. Investments in field development of 2010 (thous. USD).

Name, months	Already made	2010											
		1	**2**	**3**	**4**	**5**	**6**	**7**	**8**	**9**	**10**	**11**	**12**
Operations													
License acquisition	3188												
Operation allotting													
Exploration program													
Exploration well drilling project and project evaluation													
Reinterpretation of seismic data													
Well re-suspension and implementation 181, 183 по временной схеме	998												
Preparatory operations for well implementation 187-p (L),190-p (L)		0											
Well re-entry 187-p (L)		101	0										
Well re-entry 190-p(Л)			101	0									
Pre-drilling studies and new pad implementation		288											
Straight-hole drilling 185-p			1143										
Well drilling 187-p				1429									
Well drilling 188-p					1429								
Hydraulic fracturing (5 wells)			214	214	107								
Pre-drilling studies of exploration well 186-p		107			511								
Straight-hole drilling 186-p - exploration							1 096						
Seismic survey 600 l.km2D or 150km2 3D		134	134	134	134	134	134	134					
Drill lease	64	32	32	32	32	32	32	32	32	32	32	32	32
TOTAL COST PER MONTH		686	631	1855	2243	199	1278	181	47	47	47	47	47
TOTAL COST WITH CUMULATIVE TOTAL (DEMAND FOR INVESTMENTS)		4360	5991	7847	10090	10289	11566	11747	11794	11842	11889	11936	11983
TOTAL PRODUCTION PER DAY (m^3)		40	120	255	390	450	450	450	450	450	450	450	450
TOTAL PRODUCTION PER MONTH (m^3)		1200	2400	6900	11100	12900	13500	13500	13500	13500	13500	13500	13500
TOTAL PRODUCTION PER MONTH (barrels)		7536	15072	43332	69708	81012	84780	84780	84780	84780	84780	84780	84780
Oil production of total cumulative per month (barrels)		22608	37680	81 012	150720	231732	316512	401292	486072	570852	655632	740412	825192

The field development is planned to be performed in stages taking into account the geological terms and reserves conditions:

Calculation of commercial efficiency of field development is performed using the system of indicators specified by the current "Methodical recommendations of investment project evaluation and selection for support".

According to the decisions taken in the present paper the development of promising N fields is connected with high investment risk as evidenced by small cash flow (or net present value (NPV)) at the discount factor 40%. Changes in oil prices and possible decrease in tax burden (MET, profit tax relief) may increase the performance indicators of the given operation and make it attractive for investors.

Calculating the indicators sensitivity towards oil prices change with price decline by 28.5% (export up to 35USD/barrels and domestic prices up to 4500 roubles/t) the payback period of capital investments exceeds 10 years. With increase in prices for export oil – up to 55USD/barrels and domestic prices – up to 8000 roubles/t the payback period decreases up to 5.1 years.

5. Conclusion

Thus, the greatest effect can be achieved by cost reduction for construction of new wells by 10 %; the capital exposure decreases by nearly 8 % and, consequently, decreases the payback period of oil field development.

Evaluation of performance factors for oil field development is carried out on the basis of proven oil reserves and suggested operational process design. To forecast the production indicators the ultimate reserves of the categories C_1 and C_2 are used. It is suggested that reserve confirmation of the category C_2 – 50 %, at less confirmation the oil development is non-commercial in the current economic conditions.

The given oil field has a well-developed infrastructure. There are special facilities for oil and gas extraction, different vehicles, cranes, construction equipment, and relocatable buildings for personnel. Oil produced at the given field will be transported in pipelines and then further processed.

References

[1] *The Law of the RF "On custom tariffs"* Informative-legislation portal «Garant» [Electronic resource] http://base.garant.ru/10101366/

[2] Il'ina G F 2012 Issues of oil production and innovative industry development *Gorny Journal. Chernye Metally* I 7–9

[3] Il'ina G F 2013 Innovative development of oil-and-gas production economy sector by the example of Tomsk Oblast *European Applied Sciences* **10-2** 15–18

[4] Il'ina G F 2014 State of resource potential of hydrocarbon reserves, Tomsk Oblast *Vestnik CKR* **1** 40–42

[5] Il'ina G F and Shmidt Ye I 2015 Analysis of development conditions of the unit J_1^1 in Vakha field of Tomsk Oblast. *Proc. the 69-th International Youth Scientific Conference «Oil and Gas – 2015»* (14–16 April 2015) **3** 218

Litho-facies and paleotectonic background of hydrocarbon reservoirs in North Kalinov gas-condensate field (Tomsk Oblast)

T G Ten[1,2], E V Panova[1,3] and R N Abramova[1]

[1] Institute of Natural Resources, National Research Tomsk Polytechnic University, 30 Lenin Ave., Tomsk, 634050, Russia

E-mail: [2] ten@tpu.ru , [3] elenapanova@tpu.ru

Abstract. Based on micro-macroscopic core analysis, geophysical field data, contour mapping and referencing detailed litho-facies and paleotectonic investigation of Upper Jurassic pay thickness in North Kalinov gas condensate field was conducted. Paleotectonic analysis reflected the structure development history and determined the formation and distribution of oil fields.

1. Introduction

In the process of oil and gas exploration geologists and geophysicists often focus their attention only on the present-day prospect areas. However, in complicated developed areas reflecting continuous tectonic restructuring activity, it is essential to find up-dated effective development exploration methods and analyze the following results. Exploration experience indicate the fact that even in established petroleum-bearing areas there are crestal locations deprived of oil-gas deposits, although present-day location structures and other factors determining possible oil accumulation (reservoirs, cap rocks, migration behavior, hydrogeological conditions etc.) do not differ from these productive structures [1].

The research target involves the integrated litho-facies and paleotectonic analysis of the productive sediments in Upper Jurassic formation of North Kalinov field.

2. Geological setting

North Kalinov elevation is located approximately 7 km. S-E of Gerasimovsk structure. Top of Jurassic sediments in North Kalinov elevation is divided into two areas by the deep downfold zone: rather small N-W area and major complex-structure S-E area. N-W area in the vicinity of wells № 30 and 24 to structure contour-2920km. is isolated within a separate sloping local structure of 2.1×1.0 km. and amplitude of 25m.; while the S-E area embraces 12×6 km. A series of divergent tectonic faults penetrate the structure geometry and frequently breaking it into separate blocks. Excluding unaltered strike orientation, the structure geometry in the Bazhenov suite bottom is rather simply. Both areas are bonded by a contour structure of 2475 m., forming north-west trending fold of 3×11 km. (figure 1).

IOP Conf. Series: Earth and Environmental Science **27** (2015) 012018 doi:10.1088/1755-1315/27/1/012018

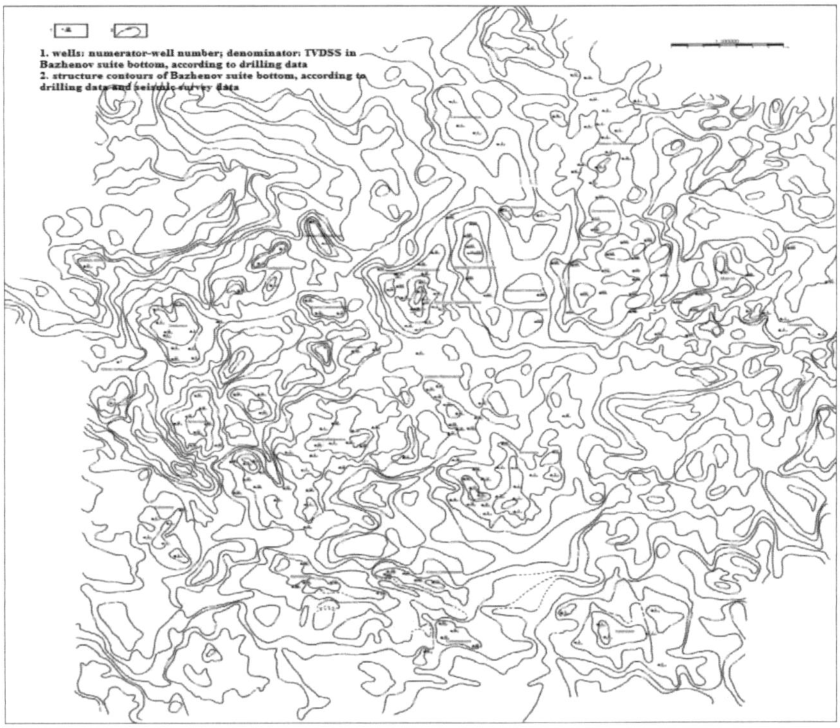

Figure 1. Structural map of junction zones in Nurolsk depression and Pudinbsk mega lithic bank within Bazhenov suite bottom (according to R.V. Belov, 1987).

3. Research data

Well-known geologist-researchers, such as A.A. Bulinnikova, F.G.Gurari, T.I. Gurova, V.P. Kazarinov, A.E. Kontorovich, Yu. K. Mironov, L.G. Markova, I.I. Nesterov, M.Ja. Rudkevich, L.V. Rovnina, S.I. Filina and others have devoted their publications to the issues of paleogeography and sedimentation conditions within Western-Siberian platform.

During a relatively short geological period (Callovian-Oxfordian) S-E Western Siberian platform, as well as Western Siberia, in general, was under the influence of very intensive tectonic activity: Late Callovian -Early Oxfordian regression altered to typical continental conditions of Early Oxfordian and then, at the beginning of Mid- Oxfordian- marine transgression. This fact greatly influenced the composition of investigated formation where one can observe all types of terrigenous rocks-from interstratified conglomerates in the formation bottom to clay and carbonaceous-argillaceous rocks in the formation top [1]. Vertically, rock types have an irregular distribution pattern. The major North Kalinov oil-gas condensate field is associated with sand formations, so-called percarbonate thicknesses. This is the target study discussed in the following paper.

At the end of Mid Oxfordian as a result of transgression the investigated territory gradually became an expanding basin with numerous islands, formerly covered by the sea. Sediments of J_1^2 layer within the North Kalinov area (wells № 25, 31, 34, 35) were formed under rapid current conditions within the boundaries of crest offshore bars (figure 2). The cross-sections showed medium-coarse-grained sandstones, well-sorted clastics. Interbedding is characterized as diverse cross bedding, horizontally interrupted which, in its turn, is conditioned by carbonaceous-micaceous matter -wash. These sediments involve abundant coal lenses and interbeds of conglomerate structure.

IOP Conf. Series: Earth and Environmental Science **27** (2015) 012018 doi:10.1088/1755-1315/27/1/012018

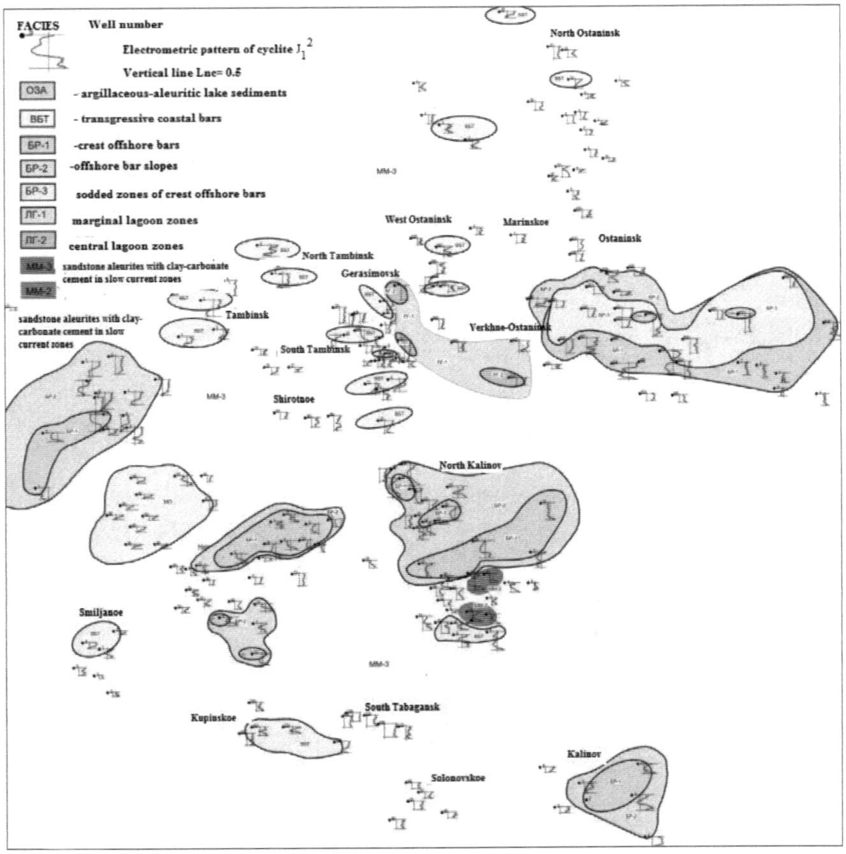

Figure 2. Litho-facies map of J_1^2 layer.

Offshore bar slope facies were determined within the vicinity of wells № 28, 24, 26, 21, 29. The cross-section embraces diverse- grained sandstones with interlayered clay, pebble aleurolite and clay. Interbedding is characterized as diverse cross-bedding and swaley cross bedding. There are also coal, chlorite and glauconite inclusions. Clay interlayers include ichnofossils. Sediments within well № 27 formed under conditions typical for transgressive coastal bars [5].

Layer J_1^2 formed in shallow marine environment at the time of continuous slow transgression. Thefactors -transgression, abundant terrigenous material ingress under conditions of advancing waves resulted in the formation of different accumulative bodies (figure 3).

Facies of offshore bar slopes formed on the crest location limbs. The cross-section is divided into zones with different sequences: bottom zone – alternating fine-grained sandstones and aleurolites; middle zone- predominately, fine-medium grained sandstones; top zone- developed interbedded aleurolite and clay rocks. Interbedding is characterized as diverse cross bedding, one-directional swaley cross bedding, lenticular. Pelecypoda shells could be observed in the core samples (North Kalinov, well № 25). Fauna remains are rare, but could be as detritus.

Sodded zones of crest offshore bars, according to structure geometry, formed within uplifted crest locations (North Kalinov № 21, 35). These facies sediments formed upon the outcropping of crest offshore bars and include aleurolite-clay rocks [2]. The rock structure is lumpy, typical for abundant fauna and root remains.

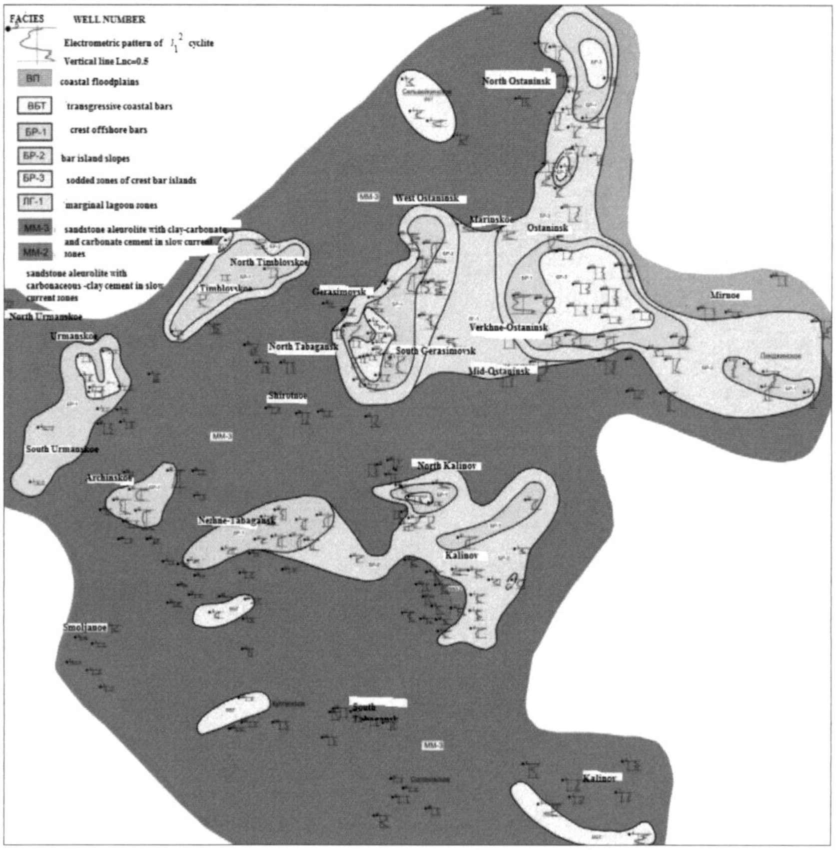

Figure 3. Litho-facies map of layer J_1^1.

Thus, it was determined that sediments in the coal thickness (layers J_1^2 and J_1^1) were formed in shallow marine environment with rapid current [4]. The inmost thickness and maximum sand content is associated with sediments of bar-genesis. The most favorable location in generating traps is the near-crestal zone (wells № 28, 25, 26, 34).

4. Research results and discussion

The target of backstripping is to determine the spatial location of underlying layers to overlying layers. It is necessary to define whether these layers were tilted and contorted in the folds forming oil -gas traps in this or that geological time, i.e. define the relative time of trap formation [3]. Paleotectonic analysis results of individual crest locations are represented (K.A. Mashkovich, 1976) as an isopach triangle, proposed by E.N. Permjakov and Yu. A. Karavashkin. Such a graphical representation provides the possibility to correlate the development history of selected stratigraphical sequences to definite geological periods (figure 4). The paleotectonic analysis markers for investigated territory was Bazhenov suite and coal layers U_1, U_2, U_3 и U_4, the quality and reliability is out of the question. The available graphical results of paleotectonic analysis is the fact that even one sketch visually depicts the different development stages of crest location for various stratigraphical sequences [3].

Examining the development history of above-described structure it was proved that at the end of Mid Callonian sediment thickness increased and crest location began forming in the vicinity of well № 21 (figure 4; III). Tracing the further development of Mid Jurassic sediments to the end of Mid Oxfordian indicated dome displacement north-westward (well № 25) and a distinctive contoured anticline structure (figure 4; IV). At the end of Oxfordian, North Kalinov crest location involved 2 domes within the vicinity of wells № 25 and 24 (figure 4; V).

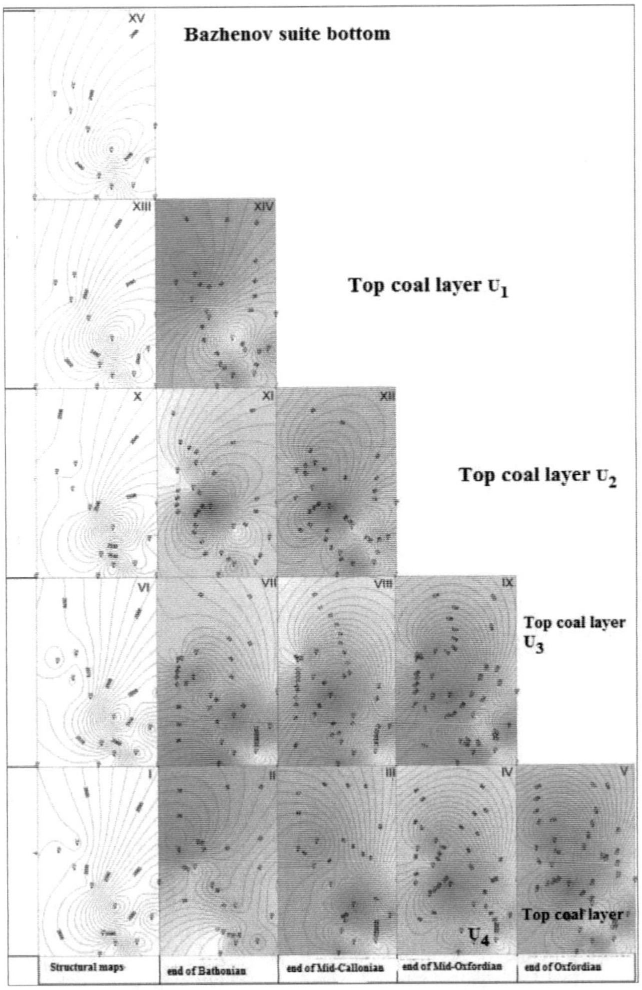

Figure 4. Structural and paleostructural maps of North Kalinov field.

Based on the analysis of maps showing total Mid Jurassic sediment thicknesses at the end of Oxfordian (V) and structural maps plotted according to horizon markers (I, VI, X, XIII, XV) indicates the fact that the reconstruction of the structure geometry itself occurred only after the formation of investigated thickness (based on comparable structural maps to horizon markers).As a consequence of this reconstruction, the hypsometric location of well № 21 is higher, while the location of wells № 24 and 25 are confined to the anticline structure flank.

Geological cross-section of Mid Jurassic sediments in North Kalinov field to line drilling of wells № 30-31-25-21-23 reflects present-day location of anticline structure within the vicinity of well № 21 (figure 5). Hydrocarbon potential of North Kalinov field is associated with the sand layers in Vasugan suite coal thickness (map XIV in isopach triangle). According to well testing, the remaining layers in investigated cross-section are aquiferous. Gas condensate deposit is confined to the W-N crest location flank. The dome within the vicinity of well № 25 has an earlier location, which, in its turn, is conditioned the trap filling. The dome within the vicinity of well № 21 formed a bit later.

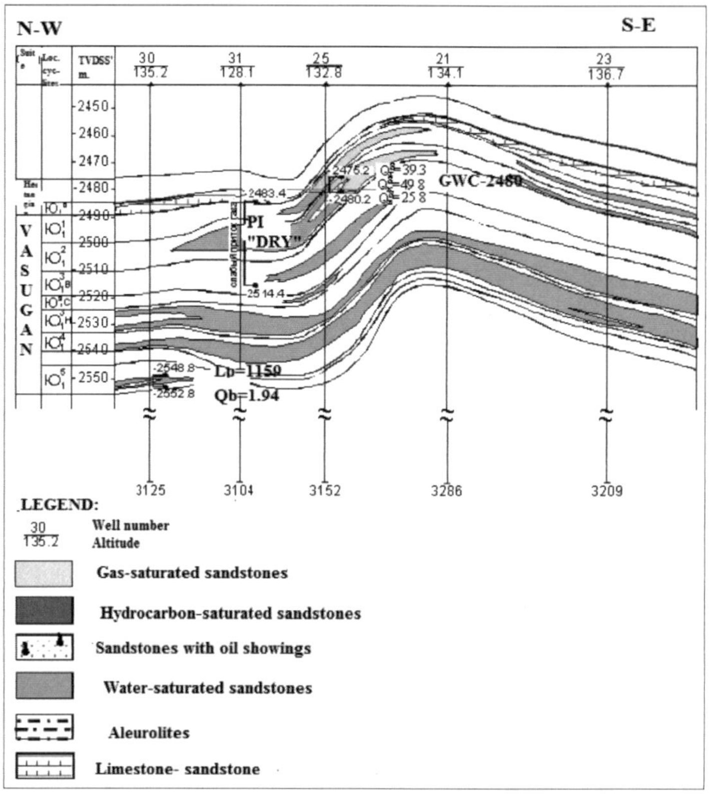

Figure 5. Geological cross-section of Upper Jurassic sediments in North Kalinov field to line drilling 30-31-25-21-23.

5.Conclusion

In exploration a versatile approach in investigating geological features is essential relevant to the changing existing conditions and their development in geological time. Applying an integrated analysis furthers the solving of possible productive structures long before deep prospecting.

References

[1] Belozerov B V, Brilina N A and Danenberg E E 1991 K probleme poiskov lito-stratigraficheskih lovushek v verhnejurskih otlozhenijah jugo-vostoka Zapadnoj Sibiri *Teoreticheskie i regional'nye problemy geologii nefti i gaza* 171–80

[2] Ezhova A V and Ten T G 2013 *Litologija neftegazonosnyh tolshh: Uchebnoe posobie* (Tomsk: TPU Publ. House) p 112

[3] Mashkovich K A 1976 *Metody paleotektonicheskih issledovanij v praktike poiskov nefti i gaza* (Moscow: Nedra) p 221

[4] Muromtsev V S 1984 *Jelektrometricheskaja geologija peschanyh tel – litologicheskih lovushek nefti i gaza* (Moscow: Nedra) p 260

[5] Reading H G 1990 *Sedimentary environments. Processes. Facies and stratigraphy* (UK: Wiley and Blackwell) p 411

Perspectives of oil and gas presence in pre-jurassic sediments on the example of one West Siberian deposit

A I Tseplyaeva[1]

[1] Tyumen State Oil and Gas University, Volodarskogo st. 38, Tyumen, 625000, Russia

Email: [1] ce_anna@mail.ru

Abstract. Conducted geological and production researches were reviewed and methods in determining excessive fracturing for mapping and geological model of the pre-Jurassic complex traps were proposed. The volume of fractured rocks in the calculation of reserves was taken into account The work on mapping of oil and gas bearing potential of the pre-Jurassic complex on the territory of research was performed.

1. Introduction
Current problems of petroleum geology are conditioned by the development of oil- and gas fields with complex geological structure. High drilling cost and, therefore, a limited number of wells require risk reduction at the stage of searching, fieldwork and initial stages of development. Study of complex structure reservoirs in conditions of sediment accumulation, specific to Krasnoleninsk arch, makes it necessary to study and represent the details of geomorphological structure of different trap types sufficiently and accurately, and, simultaneously, to create an initial geological model. Recently, more and more attention is being paid to the regularities related with genesis and oil and gas deposit placement which earlier was neglected by many oil and gas companies [1].

These deposits are related to crystalline sedimentary and metamorphic rocks, on the one hand, and magmatic and effusive rocks or crustal weathering, on the other hand. Despite high prospectivity and potential, the interval of this geological column is still poorly studied, particularly in Western Siberia. There are a number of objective reasons being "constraining" factors in successful prospecting, exploration and development of such deposits.

It is believed that there are no sufficiently up- to- date proved methods and methodical solutions in estimating the physical fields in magmatic rocks, which reflect fractured areas for oil and gas accumulation. To determine the reservoir development zones (fracture, cavern, structural porosity) and their porosity and permeability properties and saturation behavior, there are practically no appropriate study methods and interpretation techniques of geophysical well logging data.

There are still debates on a number of geological aspects related to the oil source, its migration and accumulation, nature of typical to deep depth factors and processes [2, 3].

2. Petroleum potential of the territory
The subject of the study is one field in West Siberia, where oil deposits were discovered related to the crust weathering and Paleozoic, to Jurassic formations, and also to Lower Cretaceous Vikulovsk suite. The oil and gas horizon reaches more than 1000 meters. Geological structure, effective mapping methods for potential oil and gas bearing sites furthers their active involvement in the field

development. There are three structural levels within the profile: lower- undivided folded metamorphic Proterozoic-Paleozoic; medium – intermediate Triassic sequence; higher –Mesozoic-Cenozoic sedimentary mantle. Mesozoic-Cenozoic oil and gas bearing complex embraces UK2-5 productive layers of Tyumen suite and VK1 composed of Lower Cretaceous Vikulovsk suite. The deposit is located within the Uvat-Khanty-Mansi median massif, confined to Krasnoleninsk arch within Baikal folded area, being altered by Hercynian tectogenesis. Investigated area is confined to the structures of deutero-orogenic development stage -superimposed depressions and anticlines within median and resistant massifs, Baikal fold – horst complexes. Oil and gas bearing content has been found in the sediments of Vikulov, Tyumen, and Pre-Jurassic formations [4].

3. Geological and geodynamic analysis

Investigated area is in the initial development stage, stage of increasing petroleum extraction. The project sites under development are related to sediments of Vikulovsk and pre-Jurassic formations. It is the project site Pz, associated with pre-Jurassic sediments (crust weathering and Paleozoic sediments) that is of interest in terms of exploration methods and the mapping of its potential productive areas, in terms of creating a non-trivial approach to reservoir modeling and to estimation of its reserves, as well as development methods

Pre-Jurassic basement includes Paleozoic metamorphosed rocks, mainly crystalline schists (garnet-amphibole, amphibole, amphibole-biotite and biotite-quartz-feldspar, two-mica, quartz-feldspar, chlorite) and amphibolites. There are also local areas composed of gneiss (mica, quartz-feldspar, and rarely, hornblendes). There are no distribution patterns in the cross-section -but only alternating ones. The degree of rock weathering decreases with depth.

3D seismic data interpretation results and drilling data indicated that oil deposit locations are confined to local basement protrusions with deep discontinuous disturbances, which could control blind localized flows and hydrocarbon fluid flows [2]. Commercial oil and gas accumulations are confined to multiple activated system of disjunctive dislocations, formed at the early stages of formation consolidation, whereas recent geological activities still continue to influence this process. Enhanced reservoir areas are associated with areas of maximum fracturing within the activity zone of young or renewed faults, through which deep hydrothermal solutions contribute to the formation of decompacted zones. Carbonization along cracks could be observed.

The research has shown that rock porosity and permeability according to core samples change within a very small range. Porosity coefficient changes from 0% to 15%, while permeability coefficient from 0.001 mD up to several mD, as observed in a number of samples. The cap rocks, reservoirs and disjunctive dislocations interlinking with geodynamic activity constitute favorable conditions for oil and gas formation [1]. It is assumed that the primary factor determining the location of hydrocarbons in low-permeability formations is the secondary formation of their reservoirs, usually fractured, cavern and mixed types [3].

According to the geological, geophysical and field information analysis, the identified oil deposits in the investigated area is the fact that their current geometrization is rather arbitrarily. First of all, this concerns oil-water contact (OWC) because in adjacent wells water-saturated intervals have not been exposed, so OWC is conventional. Consequently, there is no edge water. Based on the results of formation testing and well performance dynamics, sharp oscillatory variationsof well flow rates were registered ranging from a few hundred to zero cubic meters per day.

Above-mentioned characteristics may indicate the relation between the structure and productive intervals, and zonal morphology with excessive tectonic fracturing; while density of hydrocarbon reserves in different deposit areas involves secondary cavities. According to V.I. Popkov [3], characterizes this deposit type as "vein type".

Morphology of pre-Jurassic reservoir complex (including the investigated area) is mainly determined by tectonic fractured areas confined to basement ruptures and protrusions. Therefore, within the framework of set tasks, secondary reservoir development zones and their potential is to be evaluated in order to map the oil and gas deposits in low-permeability rocks, and to identify the most potentially productive areas and zones. Firstly, geophysical, geomorphological and remote-sensing methods are applied and secondly, direct methods are used providing information about the influence of deposits on geophysical and geochemical fields [3]. The data on field fracturing of the geological environment are obtained on the basis of both ground and complex well geophysical investigations, where, seismic acoustic methods are predominate.

Detecting and determining the configuration of fracture zones via tectonic models are used to identify fracturing as a result of complex stress-strain environment, are being made. There is evidence that demonstrate extremely uneven development of rock fracturing in the sedimentary mantle and on the deposit territory forming subvertical zones embracing basement rocks of the whole sedimentary complex. [3-5].

Recently, when mapping fractured areas, special seismic methods are being used (three-component data recording and scattered waves), correlating the intensity of the fracturing activity and one of the seismic parameters (amplitude of duplex waves) [6]. Based on such correlations permeability values in boreholes are determined. Distribution of permeability field reflects the distribution nature of the amplitudes of duplex waves within the investigated areas and deposits, providing a clearer idea about the features of zone distributions with higher permeability.

3D-seismic survey was conducted in this area, thus project potential assessment and oil-and-gas bearing potential were considered, including attribute maps calculated in seismic cube. The attribute map of root-mean-square amplitudes in the slot below the reflector A (roof pre-Jurassic basement) was used to geometrize deposits related to excessive fracture zones. The earlier works conducted in nearby areas proved that this attribute is effective for geometrization of deposits in pre-Jurassic complex. The areas of enhanced reservoirs, related to the areas of excessive fracturing within the zone of activity of young or renewed faults, confined to the most significant basement protrusions, were detected as a result of a comprehensive analysis of 3D-seismic and well test data and of dynamics of well performance (figure 1).

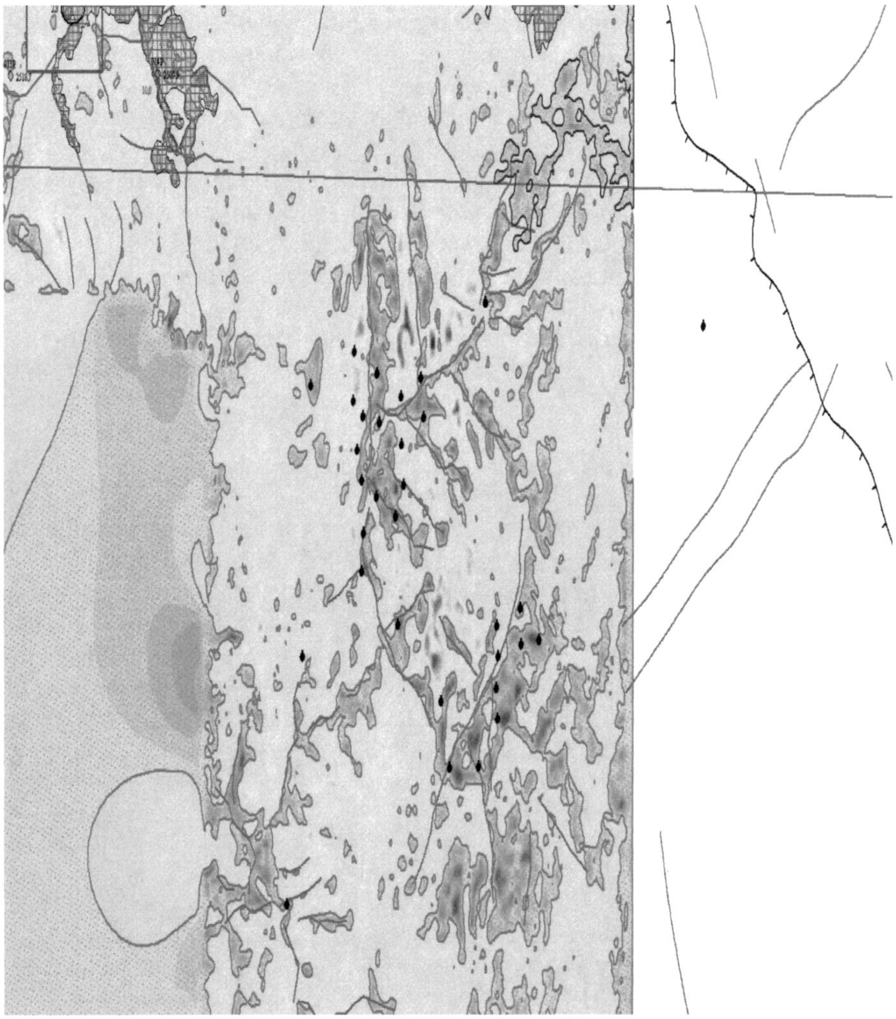

Figure 1. RMS amplitudes attribute map in the slot of 0-70 ms below the reflector A.

4. Conclusions

Evaluation of geological reserves volume in such project sites may be significantly underestimated if excluding the intervals of fractured rocks in well columns in geometrization via volumetric method. There are no officially recognized methods of estimating reserves in such areas with pronounced fractured elements.

It is suggested to include volume of fractured rocks when calculating oil reserves as an alternate solution of the problem. As a result, the volume of geological oil reserves has increased several percent units to the volume of fractured rocks. It is recommended to construct horizontal wells in a path of development of the detected enhanced fracture zones in order to develop such areas effectively. In addition, in this case, an obvious way to involve the highest volume of fractured rock around the well into development is to use hydraulic fracturing, which definitely increases the coverage even if the hit to fracture zones is imprecise. Basing on a comprehensive analysis of 3D-seismic, well test, well log survey and core sample data, areas of enhanced reservoirs, related to the areas of excessive fracturing were detected. Recommendations for well location and further field development were given.

References

[1] Tseplyaeva A I 2014 *Geology and petroleum potential of West Siberian megabasin (Experience, innovation)* (Tyumen: TSOGU) 106–108

[2] Dmitrievsky A N, Shuster V L, Punanova S A et al 2012 Modeling of the geological structure and the mechanisms of formation and location of oil and gas accumulations in pre-Jurassic complex of Western Siberia *Georesources. Geoenergetika. Geopolitics* **2(6)** http://oilgasjournal.ru/vol_6/dmitrievsky.pdf

[3] Popkov V I 2006 *The genesis of hydrocarbon fluids and fields* ed Acad A N Dmitrievsky, B M Valyaev (Moscow: GEOS) 277–285

[4] Akhpatelov E A, Volkov V A, Goncharova V N et al. 2004 *Atlas "Geology and petroleum potential of Khanty-Mansiysk Autonomous Okrug"* (Ekaterinburg: Publishing House "IzdatNaukaServis")

[5] Aarre V et al 2012 Seismic detection of subtle faults and fractures *Oilfield Review Summer* **24** 28–43

[6] Khromova I Y 2008 Migration duplex waves – a method of mapping of tectonic genesis fractured zones *Geology of oil and gas* **3** 37–47

IOP Conf. Series: Earth and Environmental Science **27** (2015) 012020 doi:10.1088/1755-1315/27/1/012020

Stability of the liquid particles separation in the apparatus of oil and gas systems

M V Vasilevsky[1,2], P N Zyatikov[1,3], A Y Burykin[1,4] and V S Deeva[1,5]

[1] Department of geology and oil exploration development, National Research Tomsk Polytechnic University, 30 Lenin Ave., Tomsk, 634050, Russia

E-mail: [2] Vasmix40@mail.ru, [3] zpavel@tpu.ru, [4] BurykinAI@pet.hw.tpu.ru, [5] veradee@tpu.ru

Abstract. The article considers the methods of associated gas purification from liquid particles. The sintering of liquid particles occurs during the separation process and the trapped droplets can be removed as a liquid stream, i.e. there is no need for unloading units. The droplet size depends on the energy input during their fragmentation. The efficiency of drops separation depends on the flow rate and the intensification of droplets coalescence, film formation and liquid flow to the receiver. The dispersion of the liquid particles is the main drawback of the existing purification methods, i.e. lack of sustainability of particle separation. The comparison of the separation system methods and the devices with flow control elements is carried out. The estimation of gas purification efficiency is conducted. It is concluded that the efficiency of associated gas purification gives the possibility to use it in turbine generators, heating furnaces, etc. It significantly reduces the proportion of gas being flared.

1. Introduction

The presence of moisture in gas after separation degrades the quality of gas for in-situ use and further transportation to gas processing plant due to the formation of crystalline hydrates. The same techniques which are used to purify gases from solid particles are applied to purify natural gas from liquid impurities. The separation of drops has advantage over the separation of solid particles. The advantage is that the liquid droplet agglomeration occurs during separation and the trapped droplets can be removed as a liquid stream, i.e. there is no need for unloading units. The droplet size depends on the energy generated during their fragmentation. If the liquid fragmentation is due to the energy of the gas flow then droplets form, with the average size being from the tens of micrometers, and sometimes hundreds of micrometers [1]. In the most cases, the various types of nozzles are used to separate droplets from the gases. They are installed at the gas outlet from devices [2]. High-speed filters are used as mist-liquid trap to protect the compressor from damage, with the first stage being agglomerator, where the particles of all sizes are precipitated but the liquid in the form of large droplets is removed by stream and precipitates in liquid trap [3]. The irregularity of the flow cross-section resistance, the deformation of filter, and the reduction in purification efficiency are the main disadvantages of such a gas-purifier.

In fields and oil treatment units, gas is commonly purified from the impurities (solids, oil droplets, and water) by means of louver separator [4]. The advantages of louver apparatus are as follows: the low hydraulic resistance and ease of layout in gas separators, scrubbers, absorbers, and gas dehydrators. The efficiency of drops separation depends on the flow rate within this separator, i.e. the inertial effects of drops interaction with the apparatus surfaces increase with the flow rate rise up to the maximum value. This leads to the intensification of the droplet coalescence and film formation, as well as it causes liquid to flow to the receiver. The separation of drops is defined by further increase in flow rate, with liquid droplets being entrained with gas flow.

2. Louver separator speed modes

The optimal rate in the cross section of vertical louver separator is defined by formula [1] $\upsilon_G = 0{,}122\sqrt{\frac{\rho_L - \rho_G}{\rho_G}}$. The paper [1] provides information about residual moisture content in the various separators provided by the Kutateladze criterion $K = \frac{\upsilon'_G\sqrt{\rho_G}}{\sqrt[4]{g\sigma(\rho_L - \rho_G)}}$. Here υ'_G is speed of the flow incoming to the nozzle, m/sec; ρ_L, ρ_G are the liquid and gas density respectively, kg/m^3; σ is surface tension at the gas-liquid interface, N/m; g=9.81 m/sec^2. The sharp increase in droplet entrainment begins at K >0.42. At K<0.3 the efficiency decreases, and the required area of flow cross-sections decreases. The area of the nozzle cross-section (m^2) is determined by the formula [4] $F = \frac{Q_G T z p_{at}}{86400 T_0 p \upsilon'_G}$. Here Q_G is gas consumption, m^3/days; T_0 ,T – gas temperature under standard conditions and current, respectively K; p_{at}, p are atmospheric pressure and current respectively; z is gas compressibility factor. The gas density increases with pressure rise, while the gas speed must decrease under the condition K <0.42. In order to endure the required high-speed modes of louver separator, it is necessary to apply the devices which allow regulating cross-section area of the nozzle or use several separators, which makes it possible to switch gas flow.

3. The proposed separator specifications

3.1. The proposed separator with adjustable gas flow

The separator (figure1) consists of a scroll 1 with enclosing surfaces 2, 3, curved channel 4 the input section 5 and the admixture discharge connection 6, purified gas discharge connection 7, regulating blade 8, springs 9, gap 10, liquid seal 11.

The operation principle of the separator is as follows: the gas-liquid mixture passes through the input section 5 in curved channel 4, liquid is discharged through the admixture discharge connection 6 and liquid seal 11. Then the gas-liquid flow passes into the vortex scroll 1, in which the drops are separated in the enclosing surfaces 2, 3 and removed by gas flow through the gap 10 in curved channel 4 in the form of a film. The position of the blade 8 determines the flow rate in the channel 4, i.e. flow rate is influenced by narrowing the area between blade 8 and channel outer wall 4. The magnitude of the stream momentum through the cross section defines the magnitude of the ejected stream from the scroll 1 to the channel 4. The springs 9 are configured so that pressure differential at the end of the blade 8 is enough to for separation carried out in a wide range of gas-liquid mixture flow rates. The stability of process is achieved by increasing the coefficient of the gas ejection, which comes from the scroll 1 to the channel 4, i.e. the area of the input section decreases with decrease in flow Q, constant speed, and increase in the amount of ejected gas q. The vortex scroll 1 serves as a lock for the particles larger than a set size.

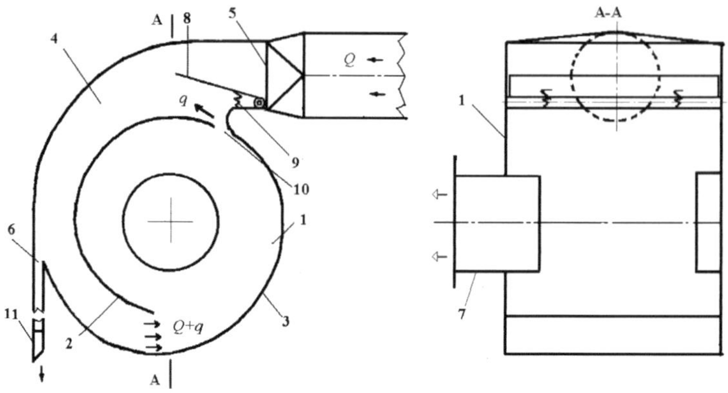

Figure 1. Separator scheme.

Thus, there are two stages of separation. The first stage takes place in the curved channel and the vortex scroll, where the separated liquid droplets pass through the curved channel and fully captured in the vortex scroll in the form of a film.

3.2 Efficiency of particle separation in curved channel

Figure 2 shows the design scheme of individual particles separation in the channel. Here, $X = R - R_в$, $H - h$ is the particle coordinate in the inlet, X_0 is the coordinate of gas discharge with separated particles flowing to the receiver *П*.

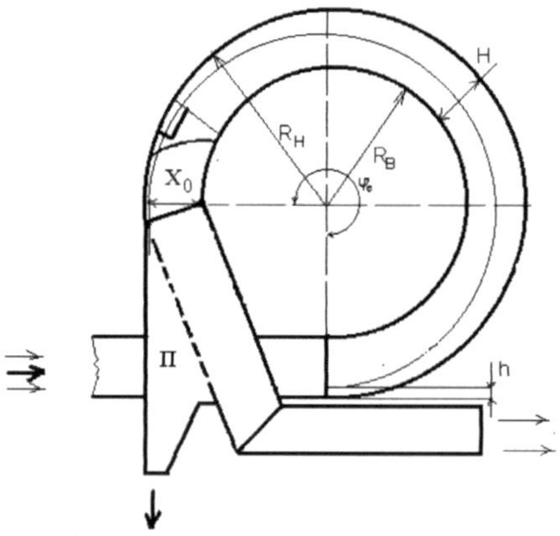

Figure 2. The scheme of particles separation in curved channel.

The speeds of the particles and gas are constant along the channel. $R_c = \frac{R_н + R_в}{2} >> R_н - R_в = H$,

$\text{Stk} = \frac{\delta^2}{18\nu} \frac{\rho_L}{\rho_G} \frac{W_0}{R_c}$, here ρ_L, ρ_G are liquid and gas density, respectively, ν is a kinematic viscosity of gas, δ is a diameter of liquid particles, W_0 is gas speed in the channel. The calculation shows that the effectiveness of particle penetration into the gap, i.e. the particle passing through the coordinate X_0, is determined by $\eta_1 = \text{Stk}\frac{R_c}{H}\varphi_c$. Here φ_c is flow deflection angle in radians. Moreover, if $\eta_1 > 1$, then η_1 will be equal to unity. There is a boundary size, and all particles which are larger than this size will penetrate into the gap of the receiver. This size can be determined by the equation $\delta_{LIM} = \sqrt{\frac{\rho_G}{\rho_L}\frac{18\nu}{W_0}\frac{H}{\varphi_c}}$, m. For example, the calculation shows that if ρ_L/ρ_G=20; ν= 1.5 $\cdot 10^{-5}$m^2/sec; φ_c =2; W_0 = 20 m/sec; H=0.1m, then δ_{LIM}=3.35$\cdot 10^{-4}$m. It means that the larger particles are separated. Consequently, in case of large liquid formations (more than 0.5 mm), complete removal of these formations occurs in the channel.

3.3. Efficiency of separation in the vortex scroll

When evaluating the effectiveness of particle separation from the flow by the method of trajectories in the vortex scroll, it is assumed that the circumferential speeds of gas and particles are remain constant and the minimum size of entrained particle is determined by the relationship

$$W_{\varphi 1}{}^2\tau/R_1 = W_{r1},\ \tau = (\rho_L/\rho_G)(\delta^2/18\nu), \qquad (1)$$

where $W_{\varphi 1}$ is a tangential velocity of gas at the radius R_1, W_{r1} is a radial velocity of gas, R_1 is a radius of the discharge connection 7(figure.1), τ is a relaxation time, ρ_L, ρ_G are liquid and gas density repsectively, δ is a diameter of liquid particles, ν is a kinematic viscosity of gas. In formula (1) it can determined the size, and the particles which are the smaller size will be submitted. Based on formula (1), it is possible to find

$$\delta_{LIM2} = \sqrt{\frac{W_{r1}R_1}{W_{\varphi 1}^2}\frac{\rho_G}{\rho_L}18\nu}\,. \qquad (2)$$

For example, when ρ_L/ρ_G=20; ν= 1,5$\cdot 10^{-5}$m^2/sec; $W_{\varphi 1}$= 30 m/ sec; W_{r1}=0,1m; R_1=0,05 m, δ_{LIM2} = 8,7$\cdot 10^{-6}$ m is obtained.

Let us calculate the efficiency of the separation process on the assumption that particle sizes are described by the integral function with the lognormal distribution of particle diameters. In this case, the amount of moisture entrainment is determined by the value of integral function in terms of particle diameters calculated according to the equation (2).

The composition of the aerosol is determined by the mass content of the particles which are smaller than a given size determined in the flow sample. The particle integral function characterized by logarithmically normal distribution has the following values (table 1):

Table 1. The integral function of size distribution X [3].

X	–2.5	–2	–1.5	–1	–0.5	0	0.5	1	1,5	2	2.5
$F(X)$%	0.62	2.28	6.68	15.87	30.85	50	69.15	84.13	93.32	97.72	99.38

Here $X = lg(\delta/\delta_m)/lg\sigma$, $\sigma=\delta_m/\delta_{16}$, δ_m is a median size. The weight of particles which is less than this size is 50% of the sample weight. Physically, the integral function is related to flow areas when sizing the sample from gas suspension particles: greater part of dispersed material of relative weight passes through a sieve with large cell size. This means that δ_m is a cell size through which 50% of the sample material passes; δ_{16} is a cell size through which 16% of the sample material passes; δ_2 is a cell size through which 2% of the sample material passes. The magnitude $d_m/\delta_2=\sigma^2$ is an indicator of particle size in small portion, i.e. the difference of cell size corresponding to the two-percent sample and the size of the cell corresponding to median value. Such an evaluation in terms of liquid particles can be performed as a part of disperse analysis using impactors [3].

Let δ_m=50 mkm, $\sigma_ч=\delta_m/\delta_{16}=2$, then the calculation of the efficiency without turbulence in cyclic trajectory is as follows $\eta_2 = 1 - F(X_y)$, where $X_y = \frac{\lg \delta/\delta_m}{\lg \sigma_ч} = \frac{\lg 8{,}7/50}{\lg 2} = -2{,}52$, $F(X_y) = F(-2.52) = 0.0062$, i.e. $\eta_2 = 1 - F(Xy) = 1 - 0.0062 = 0.9938$.

Thus, the separation process of droplets in the proposed separator is less dependent on fluctuations of gas and liquid volumes in gas-liquid mixture entering the separator.

The study of the similar method to control flow with solids is carried out in [5, 6].

The number of vortex separators and their sizes are determined by the condition that the flow rate at separator inlet should be less than 30 m/sec, and the minimum size of the entrained drops should be within the range of 2–5 mkm. However, the density of the purified gas in high-pressure devices is 2 times higher than the density at atmospheric pressure, therefore, the actual size of entrained droplets is bigger. Furthermore, taking into account the flow turbulence, the estimated droplet entrainment is 2–3 times higher than the provided calculations [5]. The vortex devices can be arranged as a battery and placed in high-capacity separators [7].

4. Conclusion

The equipment to purify gas from impurities does not operate effectively under the condition of varying impurity concentrations and gas flow rates. The vortex seperator with adjustable flow allows engineers to purify gas effectively. Usually, the qualitative dehydration of associated gas requires the large capital and operating costs, since it is required to build additional individual separation units. However, today it is possible to increase separation efficiency by upgrading equipment, precisely by using modern internal devices. One of these devices is the vortex separator. It is installed in the body of the vessel without welding, and does not change the separator technological scheme, which significantly reduces the expenses.

5. References

[1] Uzhov V N and Valdberg A U 1972 *Gas cleaning by wet filters.* (Moscow: Chemistry)
[2] Kuznetsov I E, Shmat K I and Kuznetsov S I 1989 *Equipment for sanitary cleaning of gases: Director* (K.: Technik)
[3] Birger M I, Valdberg A U, Miagkov B I and et al 1983 *Handbook of dust-and ash collection* (Moscow: Energoatomizdat)
[4] Tronov V P 2002 *Gas separation and reducing losses of oil* (Kazan: Fen)
[5] Vasilevsky M V 2008 *Dedusting gas by inertial devices* (Tomsk: Tomsk Polytechnic University)
[6] The patent of the RF 2200064 *Apparatus for separating the particulate material from the conveying gas* / Nikulchikov V K, Vasilevsky M V, Smolovik V A, Ledovskikh A K, Kosarev A E // BI 2003, №10.
[7] Lutoshkin G S 2005 *Oil, gas and water collection and treatment: Textbook for high schools.* (Moscow: Ltd. TID «Alliance»)

Integrated analysis of risks in terms of Russian Arctic zone sustainable development

Y Bolsunovskaya[1,3], L Bolsunovskaya[1,4], A Sentsov[2,5] and N Scherbinina[2,6]

[1] National Research Tomsk Polytechnic University, 30 Lenin Ave., Tomsk, 634050, Russia
[2] National Research Tomsk State University, 36 Lenin Ave., Tomsk, 634050, Russia

E-mail: [3] ju_al@inbox.ru, [4] bolsunovskl@tpu.ru, [5] sentsov@tpu.ru, [6] sapfir.19@mail.ru

Abstract. The sustainable development of the Russian Arctic zone is characterized by unique hydrocarbon and mineral resource potential and the key spheres of regional development are negatively affected by various risks. The present article addresses the issue concerning a long-term development strategy for the Arctic region growth, which is beyond the existing traditional approach. It examines a complex range of environmental, social, political, and industrial problems caused by the increasing industrial activities. The negative impact of the revealed risks has been specified. The model of the Russian Arctic zone sustainable development has been proposed.

1. Introduction

The social and economic development of a region is comprised of the agreed strategic goals and objectives which are developed and implemented by the management body with regard to total area of the region, its economic performance indexes and characteristics features, as well as its sustainable development.

The current level of Russian Arctic development shapes peculiar geographical and geopolitical factors. The geographical factors involve natural-resource, transport, economic and demographic potential. The geopolitical factors are comprised of regional geopolitical position, geopolitical interests of other countries, and geopolitical differences [1]. In general, the development of the Arctic natural resources is characterized by hard conditions, i.e. a number of unfavorable environmental, industry-related, and infrastructural factors [2, 3]. The Arctic climatic and environmental features pose the basic limitations to geophysical survey on the Arctic shelf and prompt to implement and enforce stringent requirements to machinery and equipment, infrastructure development, and environmental safety. Therefore, the expected dynamics of long-term socioeconomic development of Russian Arctic zone involves resolving a number of strategic tasks which enforce the government to develop and implement a well-planned procedures and structured set of policies to address all possible challenges including the Arctic shelf project implementation and the development of the whole region.

2. Materials and methods

Trendsetting in sustainable regional development inevitably involves assessing its weak and strong points. The present research is based on the approach that includes the following aspects: analysis of region's hydrocarbon potential, identification of key regional priorities and corresponding negative risks using a newly developed method.

The research literature includes scientific articles, normative documents and regulations, government statistics data, and the reports of various international nongovernmental organizations.

2.1. Hydrocarbon potential of Russian Arctic zone

The Russian Federation (RF) occupies especially significant position on the world market of hydrocarbons [4, 5, 6]. It takes the leading place in natural gas reserves, i.e. 23 % of world reserves, and provides 25 % of world trade; the RF ranks first in crude oil production and makes 12 % of world oil trade. Besides, the RF has tangible advantages over other Arctic countries in terms of hydrocarbon reserves, production, and consumption (tables 1, 2).

Table 1. Natural gas production and consumption in the Arctic countries [6].

Country	Country status in world gas trade	Gas production world rank	Gas consumption world rank	Gas export world rank	Gas import world rank	Gas reserves world rank
USA	Importer	1	1	10	4	4
Russian Federation	Exporter	2	3	1	28	1
Canada	Exporter	6	8	5	13	19
Norway	Exporter	8	57	3	75	18
Denmark	Exporter	51	66	38	55	65
Finland	Importer	96	69	95	38	136
Iceland	Importer	146	156	115	207	151
Sweden	Importer	195	87	184	59	196

Table 2. Oil production and consumption in the Arctic countries [6].

Country	Country status in world oil trade	Oil production world rank	Oil export world rank	Oil import world rank	Oil reserves world rank
USA	Importer	2	43	1	11
Russian Federation	Exporter	3	2	70	8
Canada	Exporter	5	7	15	3
Norway	Exporter	16	14	65	22
Denmark	Exporter	39	35	51	43
Finland	Importer	172	110	32	131
Iceland	Importer	184	130	201	147
Sweden	Importer	84	68	20	192

Russian Arctic zone holds most of hydrocarbon and mineral resources and has a huge economic potential to serve the needs of Russian economy on a long-term basis in terms of hydrocarbon and mineral resources, and to protect geopolitical interests of the country. The development of the High North and Arctic shelf hydrocarbon and mineral resources and exploitation of the Northern Sea Route are the priority projects for Russian Arctic region development. To implement these projects, it is essential to resolve a number of tasks, precisely, contribute to economic, social, scientific, and transport-related infrastructure development, as well as to guarantee environmental safety and facilitate international cooperation.

Russian Arctic zone accounts for about 12-15% of Russian gross domestic product. According to expert estimates, Arctic deposits total more than 70% (oil) and 80 % (gas) of all Russian offshore petroleum resources, with approximately 70 % being located in the Barents Sea and the Kara Sea – in the west of the Arctic [1]. The estimates of the Russian Academy of Sciences show that ultimate potential oil resources are greater in the Arctic onshore zone than in offshore area, the difference between onshore and offshore gas deposits being insignificant. However, the experts say that the Arctic territory including both onshore and offshore areas is understudied in terms of geological data. Therefore, it cannot be excluded that hydrocarbon reserves in East-Siberian region, the Laptev Sea and the Chukchee Sea, have been underestimated.

The assessment of Russian Arctic zone petroleum potential has revealed the urgency to estimate the correspondence of prospective Arctic oil and gas reserves development with the ambition for sustainable development framework.

2.2. Model of Russian Arctic zone sustainable development

The concept of sustainable development is referred to achieving overall balance between social, economic, and environmental aspects. The social aspect implies social stability in all existing cultural systems, equitable distribution of natural resources among all members of society, preservation and development of human and cultural potential. The economic aspect involves natural resource management, application of resource-saving technologies, recycling and disposal of hazardous wastes. The environmental aspect is referred to conserving biological diversity and ecosystems. The effective implementation of sustainable development plans directly depends on the possibility to adapt the above-mentioned aspects to the particular region. Compared with other Russian regions, the Arctic has a number of characteristic features that prompt to search for nonstandard solutions in implementing sustainable development plans and the Arctic shelf projects. They are as follows:

1) geographical location, climate change, icecap current conditions;
2) uncertainty in the Arctic shelf boundaries delimitation;
3) hard climatic conditions;
4) hard engineering and geological setting;
5) raw-materials export development model;
6) under population;
7) remoteness from main industrial centers, etc.

Having regarded the enumerated characteristic features in relevance to the Arctic sustainable development plans, it is also proposed to consider political and transport-related aspects (figure 1).

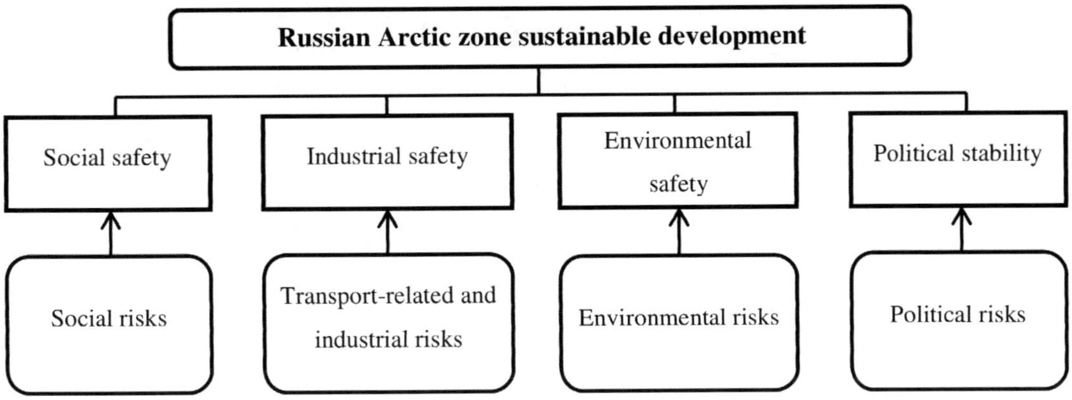

Figure 1. The Proposed model of Russian Arctic zone sustainable development.

The model of Russian Arctic zone development involves the most unstable spheres that are subjected to a great number of risks.

As a basis of further study, it is required to analyze the basic risks which adversely affect sustainable development of the Russian Arctic zone and implementation of the Arctic shelf projects.

3. Results and Discussion

Given the particularity of the Arctic and diversity of negative factors that accompany any activity within the Arctic region, it is reasonable to address the hydrocarbon production challenges on the basis of natural resource management. In addition, to maintain Arctic sustainability, elaboration of the methods to mitigate institutional risks in Arctic resource development is considered a priority task facing the state regulatory bodies.

1. Social risks.

Strategic and geopolitical importance of the Russian Arctic zone is intrinsically linked to creating favorable conditions for human potential development as a central criterion of social stability and sustainable region development [7]. Today, the RF's interests in the Arctic contradict current development trends in the Arctic region, which is manifested by negative impact of the following risks: impairment of the rights of minorities and indigenous peoples, increase in population morbidity rate [8], adaptation problems of indigenous peoples and migrated groups, lack of skilled human resources.

2. Transport-related and industrial risks.

Implementation of potentially important Arctic transport project is accompanied by a number of serious problems due to environmental and climatic characteristic features, especially ice cap conditions [9, 10]. The basic challenges in Arctic shelf projects implementation are as follows: hindrance for trans-Arctic shipping, construction and operation of offshore ice-resistant fixed platforms,insufficient provision or absence of ice-resistant sea vehicles including helicopter support facilities, supply ships, and auxiliary fleet; poor condition of coastal infrastructure, absence of service bases; stringent health and safety requirements.

3. Environmental risks.

The Arctic environment is an important indicator of changes in global climate. Assuring environmental safety as one of the key factors in sustainable development faces a number of natural and human-related challenges [3], which prevent from sustaining the balance between economic benefit of the Arctic shelf project implementation and minimized effects on the environment. As a side effect of industrial development, environmental negligence has resulted in a complex range of ecological problems [11, 12] leading to severe environmental risks.

4. Political risks.

Most of Arctic hydrocarbon and mineral deposits are within agreed boundaries of the Arctic countries. Despite this fact, the Arctic territorial disputes are still in full swing and involve not only boundary claims concerning special economic zones of the "Arctic five", but also the issues related to safety precautions, territorial and international water boundaries. The number of countries that do not directly border the Arctic Circle but are eager to participate in the Arctic problem solving is constantly increasing [13, 14]. The most serious problems are as follows: official and nonofficial claims for the North Pole, continental shelf and offshore areas; uncertainty in the maritime boundaries delimitation; conflicts in determining marine routes; disputes around the status of the Spitsbergen.

4. Conclusion

Holding most of Arctic hydrocarbon reserves and mineral deposits, Russia faces a number of problems caused by the unconformity of hydrocarbon and mineral deposit development plans and the Arctic sustainability. The characteristic features of the Arctic urge to apply new approaches toward consolidated decision making. Today, there is a huge disbalance between the above-discussed aspects of the Arctic region development, which results in high-severity risks. Taking into consideration the cost to mitigate adverse impacts, it is reasonable to solve Arctic-related problems within the framework of international cooperation. In addition, the RF should pursue initiatives to reinforce its presence in the Arctic in order to counter the increasing economic and geopolitical interests of non-Arctic countries. Besides, it is essential to actively seek to resolve inter-state conflicts, establish partnership with all participants of business activities and natural resource management, and sustain the balance between sustainable use of natural resources and economic value of Arctic hydrocarbon deposit development. The expected Arctic resource potential prompts to develop an integrated environmental safety system which would consider the negative impact of risks not only at the stage of strategy development, but also during project implementation. Overall, such a system should be integrated into all spheres of economic development of the region.

References

[1] Heininen L, Sergunin A and Yarovoy G 2014 Russian Strategies in the Arctic: avoiding a new cold war URL: http://www.uarctic.org/media/857300/arctic_eng.pdf

[2] Marchenko N 2009 Ice Conditions and Human Factors in Marine Accidents at the Arctic *International Journal on Marine Navigation and Safety of Sea Transportation* **3** 409–14

[3] Mastepanov M, Sigsgaard C, Tagesson T, Strom L, Tamstorf M P, Lund M and Christensen T R 2013 Revisiting factors controlling methane emissions from high-Arctic tundra *Biogeosciences* **10** 5139–58

[4] Piskarev A, Shkatov M 2012 Energy Potential of the Russian Arctic Seas. Choice of development strategy (Oxford: Elsevier) p 440

[5] Kontorovich A E, Burshtein L M, Kaminsky V D, Kurchikov A R, Malyshev N A, Prischepa O M, Safronov A F, Stoupakova A V, Suprunenko O I and Epov M I 2011 The potential for hydrocarbon resource development on the Russian Arctic Ocean Shelf Arctic Petroleum Geology. *Geological Society Memoir. Chapter 29* **35(1)** 443–49

[6] CIA Factbook 2013 Countries comparison: energy. URL: http://www.cia.gov

[7] Arctic Social Indicators – a follow-up to the Arctic Human Development Report 2010 Nordic Council of Ministers.
URL: http://library.arcticportal.org/712/1/Arctic_Social_Indicators_NCoM.pdf

[8] Parkinson A J 2010 Arctic Human Health Initiative.
URL: http://www.circumpolarhealthjournal.net/public/journals/32/chs/CHS_2010_6.pdf

[9] Smith L C and Stephenson S R 2013New Trans-Arctic shipping routes navigable by midcentury *Proc. Natl. Acad. Sci. U.S.A.* **110** 4871–74

[10] Arctic Shipping: Navigating the Risks and Opportunities 2014 Marsh Risks Management Research.
URL:http://www.safety4sea.com/images/media/pdf/Arctic_Shipping_Lanes_MRMR_August_2014_US.pdf

[11] Bolsunovskaya Y A and Bolsunovskaya L M 2015 Ecological risk analysis as a key factor in environmental safety system development in the Arctic region of the Russian Federation *IOP Conf. Ser.: Earth Environ. Sci.* **24** 12–16

[12] Bolsunovskaya Y A and Boyarko G Yu 2014 Osobyie ekologicheskie riski v sisteme obespecheniya ekologicheskoy bezopasnosti Arkticheskogo regiona RF *Fundamental issledovaniya* **9** 2725–28

[13] Bolsunovskaya Y A, Boyarko G Yu and Bolsunovskaya L M 2014 Political risks of hydrocarbon deposit development in the Arctic seas of the Russian Federation *IOP Conf. Ser.: Earth Environ. Sci.* **21** 250–55

[14] Sentsov A, Bolsunovskaya Y and Bolsunovskaya L 2014 Effective Planning of the Future of the Arctic *IOP Conf. Ser.: Earth Environ. Sci.* **21** 71–75

The Russian Arctic: innovative possibilities at the turn of the past and the future

A Scherbinin[1,3], E Danilova[1,4], A Sentsov[2,5], L Bolsunovskaya[2,6] and Yu Bolsunovskaya[2,7]

[1] National Research Tomsk State University, 36 Lenin Ave., Tomsk, 634050, Russia
[2] National Research Tomsk Polytechnic University, 30 Lenin Ave., Tomsk, 634050, Russia

E-mail: [3] shai@mail.ru, [4] elena.a.danilova@yandex.ru, [5] 134-43@mail.ru, [6] bolsunovskl@tpu.ru, [7] ju_al@inbox.ru

Abstract. Under the growth of the geopolitical tension a political, economic and symbolic meanings of the Russian presence in the Arctic region take on special significance. The theme of an innovation development of the Arctic and its symbolic sense for the citizens of our country are actual for the modern political discourse. The considerable scientific and technical reserve accumulated during the Soviet period and representation of the Arctic as a major symbolic resource for Russian people that helps to determine the sense of patriotism and the pride for the country can promote the growth of the authority of Russia in the region. The development of the Arctic is being conducted in two ways: expansion of the extraction of hydrocarbons and strengthening of a military and political presence. The Russian consolidation in the Arctic can potentially increase the state status in the international community. However, in order to pretend to the strong positions in the region it is absolutely necessary to answer to the arctic technological challenge by the innovative technologies and decisions.

1. Introduction

The Arctic territories that are rich in natural resources have recently become an important place of international development and geopolitical interests. In the foreseeable future "all across the Arctic, changes in climate will create new vulnerabilities for infrastructure and present new design challenges" [3]. The importance of the question encourages studies in different fields. For example, "as a part of the national Norwegian monitoring program, long-term surveys of contaminants have been carried out" [13. P. 281], another research "emphasizes a need to investigate non-stationary climate response of Alaskan coastal forests to warming in other tree species" [10. P. 211].

Also, the point is that these climate changes in the Arctic region can cause consequences around the world. That is why "given the risks posed by climate change in the Arctic, the global response to the problem is important" [4].

Currently the juridical status of the Arctic territories is settled by the system of international agreements. There are five sectors in the Arctic according to the number of state presented in the region: Russia, the United States, Canada, Denmark (Greenland), and Norway. In addition to these countries India, China, Japan and some others begin to focus theirs geopolitical interests on this region.

So, the development in the Arctic is heavily troubled by political and legal risks. In addition, "the intense cold for much of the year, long periods of near-total darkness, the potential ice-pack damage to offshore facilities, the marshy tundra dictating seasonal activity in many areas and the limited biological activity all will take a huge toll on equipment and personnel" [2. P. 5].

Natural resources of the Arctic region, especially in oil and gas, are of an exceptional quantity. Besides, recent reports show that "disagreement over the legal status of the Northwest Passage and potentially over the status of Russian Arctic waters could lead to claims that double standards are being applied" [3]. For the Russian Federation the Arctic continental shelf is a region of high geostrategic and long-term economic interests of the state. This fact makes of this territory an independent object of national policy [22]. Russian President Vladimir Putin says that "our interests are concentrated in the Arctic. And of course we should pay more attention to issues of development of the Arctic and the strengthening of our position" [12].

In this regard the problem of innovative development of the Arctic, its symbolic meaning is extremely important for the contemporary Russian political discourse.

2. The Soviet Arctic

Current political and economic possibilities of Russia in the fight for the Arctic are largely based on a strong scientific and technological potential of the Soviet period. According to Russian researchers, "a very strong industrial layer was created in the Arctic zone of Russia during the previous century, and the scale of economic activity significantly surpassed the activities of other circumpolar countries" [11. P. 6].

The leadership of the USSR in the development of the North Pole is not in doubt: legendary Russian expeditions of icebreakers in the 1930's-1940's under the direction of O. Schmidt, V. Vize and others, heroic epic rescue of Chelyuskin's associates, launch of the first polar scientific drifting station "North Pole-1" ("NP-1") under the direction of I. Papanin, heroic transpolar nonstop flights across the North Pole of V. Chkalov, landing of four Soviet aircraft on the pole, other technical records and projects.

All this created the idea of a space, where not only a feat took place, but also a qualitative synthesis of national science, engineering, and technology was presented.

We can already observe the understanding of this trinity in the memories of heroes-pilots who had taken part in the rescue of Chelyuskin's associates that were published immediately, in hot pursuit of the Arctic epic story. A complete symbolic picture gives the impression of a heavy but successful battle in which, on the one hand, "the Arctic threw all the means at its disposal: rumbling ice, polar blizzards, and terrible frost". On the other hand, "through the ice descended up to Kamchatka in winter the Soviet ships went one after another. Our glorious icebreaker were circling around the world. Soviet airships, snowmobiles, all-terrain vehicles were involved in this great struggle. Polar radio stations almost passed to a war footing…" [8. P. 32-33]. Visually it can be represented as a map of a great battle [8. P. 15].

It was very important that the Arctic had become for the Soviet society the very sacred place where appeared the first people to receive the newly established highest title of Hero of the Soviet Union. Moreover, the victory over the Arctic became a demonstration of the transition to a new era. It was believed that this was an era not only of new culture, cities, new enterprises, new technology, but also an era of millions of new people brought up by the party [8. P. 3-4].

It should be noted that the dramatic events of the rescue of Chelyuskin's associates or the drifting of Arctic station "NP-1" were actually transmitted non-stop. And the American researcher Karen Petrone compares their effect with those of a soap opera [16. P. 50]. However, such findings of Western researches could be interpreted as an instrument of information war in the geopolitical struggle for the Arctic.

The Arctic became a place where not only "the political and physical map of the country was fundamentally redrawn" [8. P. 3], but the phenomenon of its subjective measurement was fixed. So, this map was conceived by millions of citizens in terms of the Arctic, and, therefore, we could talk about the birth of the Soviet nation.

This process was not possible without the creation of the Soviet narrative. And the construction of a new socio-political reality focused not only on the heroes, but also on the new technical means of communication: telegraph, radio, film and mass distribution of print media [16. P. 49-50].

Soviet children were also involved into the sphere of this mass media. According to Karen Petrone, emulation and simple imitation mattered. Firstly, in fact, the construction of a new reality could not ignore the younger generation. So, for example, "in a nursery at the Kharkov plant women activists organized an Arctic room for a few days" [20. P. 200]. Of course, such games were regulated. So, in 1939 a scenario of mass game "In the footsteps of Papanin's associates" was published [9]. A little later the game-journey turned into one "of the most sophisticated methods of political use of geographical knowledge" [15].

Secondly, and more importantly, for many years the Arctic narrative became a theme and a place of application for children's fantasies, which raised more than one generation of Soviet citizens. Game, book, movie, magazine closely interweaved in a bunch of verbal and visual information. For example, let's consider the magazine "Murzilka" №7 of 1937, which opens with the photographs of polar explorers Schmidt, Papanin, Vodopianov, Krenkel and ends with the story of the daughter of the legendary polar explorer and radio operator Krenkel Irina about how her dad was going to the North Pole. Also, there was an announcement of the book for children "The history of our flight" written by idol of Soviet children Valery Chkalov.

Growing up, children read fiction and adventures, looked spectacular movies, thanks to which the problem of Arctic exploration became one of the central ones in the formation of a new society. It should be noted that, in school programs, and not only in geography, the Arctic theme firmly took its place as one of the supporting ones for the education of children and youth. The heroic context was constructed in compositions, songs, movies, and works of visual art. All this inseparably bounded the Arctic and the Motherland in the mind of Soviet people. Please note that even later the theme of the Arctic did not lose to the theme of space exploration.

3. Post-Soviet perception of the Arctic

The work of polar stations that provided a significant scientific and technological advance in the Arctic studies in the Soviet time, in the 1990's was ceased for a decade. An immortal expression of Bulgakov's professor Preobrazhensky that the ruin began in the heads unfortunately came true in public policy. In this case, we can not even talk about economic and technical surrender, but about a symbolic deconstruction of the Motherland. While the other Arctic powers were actively carrying out research and industrial development of the Arctic territory, Russia actually abandoned this process.

Since the early 2000's political and expert quarters began active discussions about the need to restore Russian influence in the Arctic. As Prime Minister and Acting President of the Russian Federation in March 2000 in Murmansk Vladimir Putin held All-Russia meeting on the development of the Russian Arctic sector, Eastern Siberia and the Russian Far East [14] that denoted the fundamental interest of new Russia in the Arctic issues. In 2008, Russian President Dmitry Medvedev approved the document "Fundamentals of the state policy of the Russian Federation in the Arctic up to 2020 and further perspective" [6]. Since 2014 the socio-economic development of the Arctic has been performed as part of the state program of the Russian Federation that emphasizes the priority of the public interest in regard to this region [17].

This splash of political and scientific interest is also expressed in regular holding of meetings on Arctic exploration, in speeches of country's top leadership about the need to strengthen the Russian presence in the Arctic region, in a wide coverage of this topic in mass media, in annual holding of Days of the Arctic since 2010, and other applied science conferences.

Arctic exploration has been performed by Russia in two main areas: expansion of the hydrocarbon production in the region and strengthening of its military and political presence. In the Russian Federation President's Message to the Russian Federation Federal Assembly in December 2014 the

Arctic direction is positioned as a priority for the Russian economic policy. The need for an integrated project of contemporary competitive development of the North Sea Route as an effective transit route and a stimulator of business activity on the Russian Pacific coast is emphasized [24]. The Arctic oil and gas resources present an important potential additional source of hydrocarbons for Russia. According to experts, the total reserves of fuel power resources of the Arctic part of the Russian Federation exceed 1.6 trillion tons, and the continental shelf contains about a quarter of all offshore hydrocarbon reserves in the world [23].

The resource-based structure of the Russian economy and the unstable situation on the global geo-economic market determine Russia's strategic interest in the development of Arctic fields.

4. Interpretation of the results

As a result of the political and military actions of the countries, "Arctic development is often politically contentious, with sometimes opposing interests and perspectives between local, national, and international levels. Political support for development will continue to represent an uncertainty for businesses seeking to invest in Arctic projects" [3]. This fact does not facilitate an effective modeling and a safe planning of the future of the region [21] that is absolutely indispensible.

Also, "beyond the Exclusive Economic Zone (EEZ), countries must demonstrate that the seabed is a "natural prolongation" of the continental shelf in order to claim seabed rights. These practices have led to several overlapping claims between countries. For example, Russia has made a claim that its shelf extends all the way to the North Pole along the Lomonosov Ridge, a claim that the Canadians (as well as others) vigorously reject" [2. P. 5].

In the Arctic zone the components of the Russian and American missile warning systems are located. Also there are American air-defense interceptors in Alaska, and Russian analogs located on the coast of the Arctic Ocean. There is a Russian nuclear testing area on Novaya Zemlya (an archipelago in the North of Russia in the Arctic Ocean). Some of the Arctic countries repeatedly conduct military exercises in the region. In 2013 Russia conducted military exercises using the cruiser "Peter the Great" and nuclear submarines "Orel" and "Voronezh"; launches of cruise missiles were also conducted. In turn, in 2013 the U.S. carried out the "Arctic Challenge" training exercises in the Gulf of Bothnia and in the Barents Sea with the participation of the air forces of Sweden, Finland, and the United Kingdom [22].

Hereafter we have to note that severe climate conditions require innovative approaches and technologies in the development of hydrocarbon fields. In March 2014 Deputy Prime Minister Dmitry Rogozin informed about the launch of the newest strategic project for developing of underwater (subglacial) technologies of autonomous (robotic) exploration of mineral deposits of the Arctic seas [18].

At the same time, the exacerbation of the global geopolitical situation resulting from the Ukrainian crisis could not but affect the political and economic situation in the Arctic. For example, the management of OJSC "Oil Company "Rosneft" has already announced the postponement of well-boring at least for one to two years [19]. Also, the country's political leadership declared the engaging of Indian partners from "Oil and Natural Gas Corporation "Videsh" into the development of the Arctic shelf together with Russian oil companies [25]. Such statements are also largely due to the political situation and to the attempt to change the vector of economic cooperation to the eastern one (including China).

In relation with the sharp deterioration of the geopolitical climate the country's leadership intentionally focuses the attention of the population, international community, and other geopolitical players on the subject of Russian military, political and economic presence in the Arctic. In April 2014 the Russian Federation President Vladimir Putin held a historically significant extended meeting of the Security Council of the Russian Federation where it was announced an upcoming creation of a special state body for the implementation of the state policy in the Arctic on the strategic aspects of national security [23].

We can consider as a significant event the construction and transfer for the Navy of the Russian Federation in December 2014 of nuclear strategic submarine of the fourth generation "Vladimir Monomakh" from the board of which in September 2014 a successful start of ballistic missile "Bulava" was launched as part of state test operations [5].

Russian military presence in the Arctic is linked to the political, economic, and scientific interests. According to experts, "a permanent military presence in the Arctic will permit to strengthen significantly various research and expeditions in the region developed by representatives of Russian academic and economic sectors on the high-latitude track of the Northern Sea Route" [1]. In order to achieve this goal in 2014 the Ministry of Defense of the Russian Federation made significant practical steps to improve air defense system, to return here the units of the Armed Forces of the Russian Federation, to reconstruct the transpolar military airfields, to restore the military base on the New Siberian Islands, to conducting military exercises in the region. An important step is the creation of the Joint Strategic Command "North" on the basis of the Northern Fleet [7].

5. Conclusion

At this moment the ecological, legal, political, and geopolitical risks for the Arctic region remain impressive. All these risks and political pressure, unstable global balance of power in the region slow down the development of these remarkable territories.

At the same time parallel reciprocal moves should be expected from the Arctic states of NATO that turns the Arctic into a zone of potential geopolitical conflict. In the current geopolitical realities a balanced state policy is extremely important in order to neutralize as much as possible a hotbed of tension in the north. Difficult economic situation in the country requires the mature decisions in foreign policy. Russia's presence in the Arctic is necessary and possible. The success is possible to the full extent provided that the Arctic will be recognized as an important symbolic resource for the Russians. The expansion of Russian economic, military, and political influence in the Arctic region has the potential to strengthen the position of the state in the international community. However, it is also very important what price the country should pay for this and what is the necessary responsibility on the part of key actors that make decisions.

In any case, the important condition for political and economic influence in the region is the adequate response to the Arctic technological challenge with innovative solutions and projects. Also, today it is evident that Russian political top management must create an effective system of country positioning in national and international political discourse in connection with the image of the Russian Arctic.

References

[1] An expert: a military presence of Russia in the Arctic will straighten the air-defense system (2013). RIA "Novosti". 17 October. URL: http://ria.ru/defense_safety/20131017/970666692.html (access date 13.02.2015)

[2] *Arctic oil and gas* 2013 Ernst and Young Limite

[3] Arctic Opening: Opportunity and Risk in the High North (2012). *Chatham House – Lloyd's Risk Insight Report.* 01 April 2012. URL: http://www.chathamhouse.org/publications/papers/view/182839 (access date: 15.02.2015)

[4] Ford J D 2009 Dangerous climate change and the importance of adaptation for the Arctic's Inuit population *Environ. Res. Lett.* **4** 024006

[5] From the board of atomic submarine "Vladimir Monomakh" the launch of «Bulava» was successfully executed 2014 *JSC "Sevmash"* [site]. 2014. 11 September. URL: http://www.sevmash.ru/rus/news/1769--l-r-lr-.html (access date 14.02.2015)

[6] Fundamentals of the state policy of the Russian Federation in the Arctic up to 2020 and further perspective: Direction of the Russian Federation President of 18 September 2008. (2008).

Security Council of the Russian Federation [site]. URL: http://www.scrf.gov.ru/documents/98.html (access date 03.03.2015)

[7] Gavrilov Yu 2014 "North" of the Arctic. *Rossiyskaya gazeta* 1 December. № 6546 (274), 7

[8] *How we were rescuing Chelyuskin's associates* 1934 ed O U Schmitt, I L Baevskiy, L Z Mekhlis (Moscow: Pravda)

[9] *In the footsteps of Papanin's associate*s (mass play) 1939 (Moscow: No publishing)

[10] Jarvis S K, Wiles G C, Appleton S N, D'Arrigo R D and Lawson D E 2013 A Warming-Induced Biome Shift Detected in Tree Growth of Mountain Hemlock [Tsuga mertensiana (Bong.) Carrière] along the Gulf of Alaska *Arctic, Antarctic, and Alpine Research.* **45** 211–218

[11] Melamed I I, Avdeev M I, Pavlenko V I and Kutsenko S U 2015 Russian Arctic zone in social and economic development of the country *Vlast.* **1** 5–11

[12] Nato on high alert as Putin eyes up Arctic for empire 2015 *Daily Express.* Northern and Shell Media Publications. URL: http://www.express.co.uk/news/world/508019/Nato-high-alert-Putin-eyes-Arctic-empire (access date: 05.02.15)

[13] Ørbæk J B, Tombre I and Kallenborn R 2004 Arctic-Alpine Ecosystems and People in a Changing Environment *Arctic, Antarctic, and Alpine Research* **36** 281–283

[14] Order of the Russian Federation Government of 20 January 2000 №96-r 2000 *Law portal "Poisk zakona"* [site]. URL: http://poisk-zakona.ru/169390.html (access date 03.03.2015)

[15] Orlova G 2009 «Distance journey»: managing of geographic imagination in Stalin's age. *NLO* [site]. № 100. URL: http://magazines.russ.ru/nlo%20/2009/100/or21.html (access date 10.03.2015)

[16] Petrone K 2000 *Life has become more joyous, comrades: celebrations in the time of Stalin* (Bloomington & Indianapolis: Indiana University Press)

[17] Resolution of the Russian Federation Government of 21 April 2014 №366 "About the statement of Russian Federation state program "Social and economic development of the Arctic zone of the Russian Federation up to 2020" 2014 *Government of the Russian Federation* [site]. URL: http://government.ru/media/files/AtEYgOHutVc.pdf (access date 15.02.2015)

[18] Rogozin D 2014 Let's look in abyss. *Rossiyskaya gazeta.* 14 March. **6331(59)** 17

[19] "Rosneft" will postpone well-boring at some parts of the Arctic because of the sanctions 2014 *RBK* [site]. 22 December. URL: http://www.rbc.ru/rbcfreenews/549862519a794748113309f2 (access date 23.02.2015)

[20] Scherbinin A I 2012 *Totalitarian indoctrination as a consciousness control: tutorial* (Tomsk: Tomsk State University Publishing)

[21] Sentsov A, Aleksandrov O, Bolsunovskaya Yu, Kuimova M 2015 Modeling of the Future in the Programs of Political Parties. *Proceedings of The International Conference on Research Paradigms Transformation in Social Sciences 2014* vol **166** doi:10.1016/j.sbspro.2014.12.568

[22] Sentsov A, Bolsunovskaya Yu and Bolsunovskaya L 2014 Effective Planning of the Future of the Arctic *IOP Conf. Ser.: Earth Environ. Sci.* **21** 012014

[23] Session of the Security Council about realization of the state policy in the Arctic 2014 *President of the Russian Federation* [site]. 22 April. URL: http://state.kremlin.ru/face/20845 (access date 03.03.2015)

[24] The Russian Federation President's Message to the Russian Federation Federal Assembly. 2014 *President of Russian Federation* [site]. URL: http://www.kremlin.ru/transcripts/47173 (access date 13.03.2014)

[25] Topalov A 2014 The Arctic waits for rupees. *Gazeta.ru* [site]. 12 December. URL: http://www.gazeta.ru/business/2014/12/11/6337929.shtml (access date 13.02.2015).

Assessment of Mesozoic-Kainozoic climate impact on oil-source rock potential (West Siberia)

A A Iskorkina[1,3], V I Isaev[1,4] and D A Terre[2,5]

[1] Department of Geophysics, Institute of Natural Resources, National Research Tomsk Polytechnic University, 30 Lenin Ave., Tomsk, 634050, Russia
[2] Department of Foreign Languages, Institute of Natural Resources, National Research Tomsk Polytechnic University, 30 Lenin Ave., Tomsk, 634050, Russia

E-mail: [3] iskorkina.a@mail.ru, [4] isaevvi@tpu.ru, [5] terreda@yandex.ru

Abstract. Based on paleotemperature modeling, the evaluation of the effect of Neo-Pleistocene permafrost rock thickness on geothermal regime of the Bazhenov deposits has been performed. It has been stated that permafrost about 300 m in thickness must be considered for appropriate reconstruction of geothermal history of source rocks in the south-east areas of West Siberia. This condition is relevant to a consistent consideration of oil-generation phase history and can prevent underestimation (to 25%) of hydrocarbon-in-place resources.

1. Introduction

Potential permafrost rock formation in the Late Quaternary in the area of 61°N latitude, West Siberia (Shirotnoye Priobye) has been presented in research [1]. In hydrointegrator modeling performed for a calculation period of 245 kya, A.A. Sharbatyan applied surface temperature time series as an upper-boundary condition for the stated problem (table 1). Time series for surface temperatures has been defined by time series chart of solar radiation (proposed by M. Milankovich) taking into account geographical peculiarities of the area.

Table 1. Permafrost rock thickness variation over time [1].

Time, kya (thousands of years ago)	Permafrost base depth, m	Surface temperature time series, °C
245	0	0
235	-350	-10.0
210	-450	-5.5
190	-550	-8.5
165	-450	-4.5
145	-400	-3.5
130	-350	-1.0
110	-300	-3.9
95	-300	-0.7
70	-250	-4.0
50	-250	-1.0
30	-200	-4.3
5	0	+2.5

IOP Conf. Series: Earth and Environmental Science **27** (2015) 012023 doi:10.1088/1755-1315/27/1/012023

According to recent data [2] permafrost was not only in the central and southern parts of West Siberia, but Quaternary ice cover embraced northern and northeastern parts of Kazakhstan; ice sheet can be traced in Central Kazakhstan as well.

The previous researches [3] have addressed the impact of paleoclimate (surface temperature time series) on the thermal history and oil-generation potential of Bazhenov deposits in the south-east of West Siberia (South-Siberian paleoclimate zone [4]). The studies focused on West Siberia are of particular interest provided that paleotemperature reconstruction model considers not only surface temperature time series but permafrost sequence as a peculiar lithostratigraphic unit. The purpose of the present study is to estimate the influence of Neo-Pleistocene permafrost thickness in geothermal regime of Bazhenov deposits in the south-east of West Siberia.

2. Research methods

Based on data from sedimentary cross-section of deep well №183 in Luginetskoye field (Tomsk Oblast), modeling of paleogeothermal conditions in Bazhenov deposits was performed. Hydrocarbon deposits are mainly associated with Upper-Jurassic reservoirs (J_1 horizon). The major source of hydrocarbons in J_1 traps (J_3vs) are potential oil source rocks of the Bazhenov suite (J_3bg).

Simulation of the thermal history of the Bazhenov suite deposits has been carried out on the basis of paleotectonic and paleotemperature reconstructions. In the present research paleotemperature modeling has been applied [5, 6]

The evaluation of Neo-Pleistocene permafrost rock thickness effect on the geothermal regime and degree of oil-generation potential of Bazhenov deposits is performed on the basis of result variability analysis of four optional paleotemperature reconstructions. Reconstruction 1 considers both surface temperature time series and Neo-Pleistocene permafrost sequence about 300 m in thickness. Reconstruction 2 involves surface temperature time series analysis without considering permafrost rock sequence. Reconstruction 3 gives analysis regardless of surface temperature time series and permafrost rock sequence. Reconstruction 4 refers to surface temperature time series and Neo-Pleistocene permafrost which is assumed to be up to 1000 m thick.

In Reconstruction 1 (table 2) permafrost sequence is considered to be of 300 m thick (table 1). Formalized calculation of permafrost thickness is provided beginning with 240 kya when “immediate” (by standards of geological time, over 1.5+ 3.0 ky period) replacement of “normal” sedimentary deposits by permafrost sequence with particular thermophysical parameters – thermal conductivity, temperature conductivity occurred. This sequence of permafrost rocks has overlaid sedimentary mantle for 179 ky. Hereafter, “immediately” (1.5+3.0 ky) permafrost sequence is substituted by “normal” sedimentary deposits and since that time “normal” sedimentary mantle has been retained over the recent 52 ky.

Reconstruction 4 deals with permafrost thickness of 1000 m, other procedures being the same as in Reconstruction 1. Formalized consideration of permafrost thickness has been performed in the same way as in the previous reconstruction beginning with 240 kya. Replacement of “normal” sedimentary deposits by permafrost sequence has been completed in terms of geological time over 5.0 + 3.0 ky. Later, this sequence of permafrost rocks has overlaid sedimentary mantle for 206 ky. Hereafter, permafrost is substituted by “normal” sedimentary deposits and since that time the present day section has been retained for 21 ky.

Table 2. Parametric description of sedimentation history and thermophysical properties of the sedimentary sequence tapped from well Luginetskaya №183 (NeoPleistocene permafrost thickness is 300 m).

Suite, sequence (stratigraphy)	Thickness, m	Age, Ma ago	Accumulation period, Ma	Density, g/cm^3	Thermal Conductivity, W/m K	Temperature Conductivity, m^2/s	Heat release, W/m^3
Quaternary *Q*	-	0.052–0	0.052	2.10	1.3	7e-007	1.22e-006
Quaternary *Q*	300	0.055–0.052	0.003	2.10	1.3	7e-007	1.22e-006
Quaternary *Q*	-300	0.0565–0.055	0.0015	2.10	2.09	1.05e-006	1.22e-006
Quaternary *Q*	-	0.2355–0.0565	0.179	2.10	2.09	1.05e-006	1.22e-006
Quaternary *Q*	300	0.2385–0.2355	0.003	2.10	2.09	1.05e-006	1.22e-006
Quaternary *Q*	-300	0.24–0.2385	0.0015	2.10	1.3	7e-007	1.22e-006
Quaternary *Q*	25	1.64-0.24	1.4	2.02	1.27	6.5e-007	1.1e-006
Pliocene N_2	-	1.64–4,71	3.07	-	-	-	-
Miocene N_1	-	4.71–24.0	19.29	-	-	-	-
Nekrasovskaya *nk* Pg_3	84	24.0–32.2	8.3	2.09	1.35	7e-007	1.2e-006
Cheganskaya+Lyulinvorskaya+Talitskaya *hg ll tl* $Pg_{3\text{-}1}$	173	32.2-61.7	29.4	2.09	1.35	7e-007	1.2e-006
Slavgorodskaya+ Ipatovskaya+Kuznetsovskaya *sl ip kz* K_2	364	73.2-91.6	29.9	2.15	1.4	7e-007	1.25e-006
Pokurskaya *pk* $K_{1\text{-}2}$	803	91.6–114.1	22.5	2.26	1.49	8e-007	1.25e-006
Alymskay $a_{1,2}K_{1\text{-}2}$	-	114.1-120.2	6.1	-	-	-	-
Kiyalinskay *kls* K_1	550	120.2–132.4	12.2	2.39	1.6	8e-007	1.25e-006
Tarskay *tr* K_1	74	132.4–136.1	3.7	2.44	1.62	8e-007	1.25e-006
Kulomzinskay lmK_1	237	136.1–145.8	9.7	2.44	1.64	8e-007	1.25e-006
Bazhenov bgJ_3	**16**	**145.8–151.21**	**5.4**	**2.42**	**1.62**	**8e-007**	**1.3e-006**
Georgiev *gr* J_3	-	151.2–156.6	5.4	-	-	-	-
Vasyugan $vsJ_{3\text{-}2}$	55	156.6–168.3	11.7	2.42	1.6	8e-007	1.3e-006
Tyumen *tm* J_2	115	168.3–172.0	3.7	2.46	1.64	8e-007	1.3e-006

Note. Grey shading indicates geological time intervals of "immediate" formation and degradation of NeoPleistocene permafrost sequence. Dark shading indicates time interval of existing permafrost sequence.

3. Results and discussion

A number of observations can be deduced from the analysis of computational values of mantle basement heat flow density *q* (table 3). In Reconstructions 1, 3 and 4 heat flow increases by 1.4–2.6–7.1 mW/m^2 (by 3–5–14%) relative to computational value of heat flow in Reconstruction 2 which is 52.2 mW/m^2. In Reconstructions 1and 4 the increase of computational density of heat flow *q* is due to increase of heat diffusion throughout daylight surface caused by high thermal conductivity and temperature conductivity of the permafrost sequence present in the model. In this case, more heat is dissipated through the daylight surface; therefore, higher value of computational density of mantle basement heat flow is required which, in its turn, increases calculated geotemperatures of source deposits and, consequently, volumes of generated hydrocarbon resources.

Provided that surface temperature time series (Reconstruction 3) is not taken into consideration, there is also an increase in computational heat flow – 54.8 mW/ m^2 which is due to the absence of solar source heat in the paleotemperature reconstruction model of this type.

The comparison of calculated and measured geotemperatures in the borehole is presented in table 4. Since the measured temperatures (including those defined against vitrinite reflectance) and calculated geotemperatures can have uncertainty of ±2°C, results of Reconstructions 3 and 4 cannot be regarded as admissible. In these reconstructions true error exceeds optimal rate by more than four times, while the difference between adjusted and vitrinite reflectance («maximum paleotermometer») data is 11–12°C. Thus, exclusion of paleoclimate (Reconstruction 3) does not allow producing a precise physico-mathematical model of geothermal regime of Bazhenov source rock. In the same way the hypothetical assumption about Neo-Pleistocene permafrost being 1000 m thick in the latitudes of 57-61 °is not confirmed by paleotemperature modeling.

Table 3. Calculated geotemperatures of the Bazhenov suite in Luginetskaya well №183 cross-section (South-East of West Siberia, Tomsk Oblast).

Time, million years ago	Surface temperature time series («local [3]),°C	Bazhenov suite basement depth, m	Suite geotemperatures, °C			
			Reconstruction 1	Reconstruction 2	Reconstruction 3	Reconstruction 4
0	0	2321	80	81	87	75
0.001	+1	2321	80	81	87	75
0.003	+2	2321	80	81	87	75
0.005	+3	2321	80	81	87	75
0.018	+1	2320	80	81	88	77
0.03	−2	2321	79	81	88	78
0.05	−1	2320	79	81	88	78
0.052	−1	2320	79	81	88	78
0.055	−1	2321	79	81	88	78
0.0565	−2	2320	79	81	88	78
0.07	−4	2319	79	80	88	78
0.09	−1	2319	79	80	87	78
0.11	−4	2319	79	80	87	78
0.13	−1	2319	79	80	87	78
0.15	−4	2318	80	80	87	78
0.19	−9	2318	83	81	87	79
0.21	−6	2317	84	82	87	82
0.222	−7	2317	85	82	87	82
0.225	−8	2317	86	82	87	86
0.235	−10	2317	86	82	87	91
0.2355	−9	2317	86	82	87	94
0.2385	−2	2317	86	83	87	94
0.24	0	2317	86	83	87	95
1.4	+1	2299	86	83	86	94
1.64	+1	2296	86	83	86	95
3.1	+2	2295	88	85	86	96
3.2	+2	2295	89	89	86	97
3.8	+12	2295	95	93	86	104
4.7	+3	2295	88	85	86	96
5.2	−3	2294	89	86	86	97
5.7	+7	2294	92	89	86	101
6.3	+10	2294	94	90	86	102
7	+4	2294	89	86	86	97
20	+15	2294	100	97	86	108
24	+16	2294	101	98	86	110
31.5	+17	2218	98	95	83	107
32.3	+16	2210	97	94	82	105
34	+15	2200	96	93	82	103
37.6	+14	2178	94	91	80	101
41.7	+12	2154	90	87	80	98
42	+11	2158	89	87	80	97
46	+8	2129	86	83	79	93
54.8	+19	2077	95	92	77	102
58	+24	2058	99	96	76	106
61.7	+22	2037	95	92	74	102
73	+15	1899	83	81	69	90
73.2	+16	1897	83	81	68	90
86.5	+22	1735	83	81	62	90
89.8	+22	1694	82	80	61	88
90	+23	1692	82	81	61	88
91.6	+22	1673	79	77	58	85
114.1	+21	870	50	49	29	53
118	+19	869	48	47	29	51
120.2	+19	869	47	47	29	51
132.4	+19	319	29	29	11	30
136.1	+19	245	14	27	8	18
Computational basement heat flow, mW/m^2			53.6	52.2	54.8	59.3

Note. Shaded areas indicate temperatures of major oil generation zone (OGZ) [7], dark-colour shading indicates absolute OGZ paleotemperature maximum, grey shading – relative OGZ paleotemperature maximum. Threshold OGZ geotemperature is 85 °C.

Table 4. Comparison of measured and calculated geotemperatures (Luginetskaya well №183).

Depth, m	Measured («observed») temperatures, °C	Measurement method	Reconstruction 1, °C		Reconstruction 2, °C		Reconstruction 3, °C		Reconstruction 4, °C	
			Calculated temperatures	Discrepancy	Calculated temperatures	Discrepancy	Calculated temperatures	Discrepancy	Calculated temperatures	Discrepancy
2200	77	In-place	76	–1	78	+1	83	+6	70	–7
2350	84	In-place	81	–3	82	–2	89	+5	76	–8
2345	98	VRV	101	+3	99	+1	87	–11	110	+12
Mean squared error («true error»), °C			±2		±1		±8		±9	

Note. Reconstruction 1 considers surface temperature time series, and Neo-Pleistocene permafrost sequence of 300 m thick. Reconstruction 2 deals with surface temperature time series without considering permafrost rock sequence. Reconstruction 3 gives analysis of surface temperature time series and permafrost rock sequence. Reconstruction 4 considers surface temperature time series and permafrost sequence of 1000 m thick.

Calculation of generated Bazhenov oil density R [8] (table 5) yields maximum value for appropriate Reconstruction 1 (68 cu.). This type considers presence of NeoPleistocene permafrost sequence which is 300m thick apart from surface temperature time series.

Table 5. Estimation of resource density of generated Bazhenov oils (*R*) for reconstructions considering surface temperature time series and permafrost thickness (Luginetskaya well №183).

Scenario of paleo temperature modeling	Calculated resources (*R*), cu.	Number of calculated time intervals (*n*)	Period of paleo kitchen zone activity, m years ago	Duration of kitchen zone activity, million years	Peak temperatures of paleo kitchen zone, °C
Reconstruction 1	68	24	61.7–0.222	61.5	101
Reconstruction 2	55	19	61.7–54.8 42–3.1	45.8	98
Reconstruction 3	27	29	24–0	24.0	87
Reconstruction 4	109	23	91.6–0.21	91.4	110

Note. Shaded areas indicate reconstruction types which are appropriate and consistent regarding optimal agreement of calculated geotemperatures with both measured in-place (formation) temperatures and geotemperatures determined based on vitrinite reflectance values.

4. Conclusion

By the example of Mesozoic-Kainozoic section of the south-east of West Siberia (in the latitude of Tomsk Oblast) it has been stated that neglect of surface temperature time series and NeoPleistocene permafrost thickness hinders appropriate reconstruction of thermal history of Bazhenov source rocks. To estimate hydrocarbon resources in south-east areas of West Siberia using volumetric-genetic method [9] it is advisable to apply "local" surface temperature time series [3] and deal with permafrost thickness of 300 m. The latter will allow a more consistent consideration of the history of main oil generation phase and prevent underestimation (to 25 %) of hydrocarbon-in-place resources.

References

[1] Sharbatyan A A 1974 *Ekstimalnye otsenki v geotermii i geokriologii* (Moscow: Nauka) p 123
[2] Elger Yu S 2015 K voprosu o granitsakh pokrovnogo chetvertichnogo oledeneniya v predelakh Zapadnoj Sibiri i Kazakhstana *Exploration and Subsurface Protection* **1** 30–33
[3] Isaev V I, Iskorkina A A 2014 Mezozojsko-kajnozojskij khod temperature na poverkhnosti Zemli i geotermicheskij rezhim Yurskikh neftematerinskikh otlozhenij (Uyzhnaya paleoklimaticheskay zona Zapadnoj Sibiri *Journal of Geophysics* **36**(5) 64–80
[4] Golbert A V 1987 *Osnovy regionalnoj paleoklimatologii* (Moscow: Nedra) p 222

[5] Gulenok R Y, Isaev V I, Kosygin V Y, Lobova G A and Starostenko V I 2011 Estimation of the Oil-and-Gas Potential of Sedimentary Depression in the Far East and West Siberia Based on Gravimetry and Geothermy Data. *Russian Journal of Pacific Geology* **5**(4) 273–87

[6] Isaev V I, Lobova G A and Osipova E N 2014 The oil and gas contents of the Lower Jurassic and Achimovka reservoirs of the Nyurol'ka megadepression. *Russian Geology and Geophysics* **55** 1418–28

[7] Burshtein L M, Zhidkova L V, Kontorovich A E and Melenevskiy V N 1997 Model katageneza organicheskogo veschestva (na primere bazhanovskoj svity). *Geology and Geophysics* **38**(6) 1070–78

[8] Lobova G A, Popov S A and Fomin A N 2013 Lokalizatsiya prognoznykh resursov nefti yursko-melovykh NGK Ust-Tymskoj megavpadiny *Oil Industry* **2** 36–40

[9] Kontorovich A E, Burshtein L M, Malyshev N A et al 2013 Istoriko-geologicheskoe modelirovanie protsessov nefti dogeneza v mezozojsko-kajnozojskom osadochnom bassejne Karskogo Morya (bassejnovoe modelirovanie) *Geology and geophysics* **54**(8) 1179–226

Predicted facies, sedimentary structures and potential resources of Jurassic petroleum complex in S-E Western Siberia (based on well logging data)

F Prakojo[1,3], G Lobova[1,4] and R Abramova[2]

[1] Department of Geophysics, Institute of Natural Resources, National Research Tomsk Polytechnic University, 30 Lenin Ave., Tomsk, 634050, Russia
[2] Department of Foreign Language, Institute of Natural Resources, National Research Tomsk Polytechnic University, 30 Lenin Ave., Tomsk, 634050, Russia

E-mail: [3] felix.santo@yahoo.com, [4] lobovaga@tpu.ru

Abstract. This paper is devoted to the current problem in petroleum geology and geophysics-prediction of facies sediments for further evaluation of productive layers. Applying the acoustic method and the characterizing sedimentary structure for each coastal-marine-delta type was determined. The summary of sedimentary structure characteristics and reservoir properties (porosity and permeability) of typical facies were described. Logging models SP, EL and GR (configuration, curve range) in interpreting geophysical data for each litho-facies were identified. According to geophysical characteristics these sediments can be classified as coastal-marine-delta. Prediction models for potential Jurassic oil-gas bearing complexes (horizon J_1^1) in one S-E Western Siberian deposit were conducted. Comparing forecasting to actual testing data of layer J_1^1 showed that the prediction is about 85%.

1. Introduction

Today, a highly topical problem in geology and geophysics is predicting sediment facies through formation evaluation. Well logging characteristics of sedimentary structures provide data on facies types and reservoir properties [1–3].Substrata formation conditions determining the types of sedimentary structures are generated during sedimentation [4]. Three major facies systems were described: continental including eolian, fluvial and alluvial facies; coastal-marine including delta, lagoon and shelf facies; and sea (marine) including turbidite, landslide and abyssal-marine facies [5–7].

The attributes and behavior of each facies type were determined on the basis of well logging data (SP, EL and GR) and reservoir properties (porosity and permeability). The proposed classification is based on real drilling and deep well logging data from one area in northern Tomsk Oblast (S-E Western Siberia).

2. Research methods

Sedimentary structure types, their reservoir properties, logging curve attributes in real reservoirs were analyzed via acoustic method. This method involves the systemization (based on reference data and individual observations of cross-sections and core samples) of sedimentation systems which are associated with coastal-marine sediment facies types (table1, examples).

Table 1. Examples of sedimentary structures in outcroppings (Internet sources).

Continental zone	**Coastal zone**	**Marine zone**
Eolian facies	Lacustrine facies	Turbidite facies

Gran Canaria, Spain, Duna	Colorado River, Utah, USA Mud cracks	Newfoundland, Canada Contorted (crinkled) bedding
Fluvial facies	Delta facies	Landslide facies
	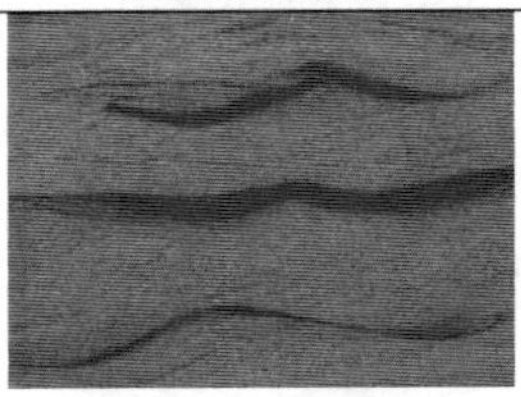	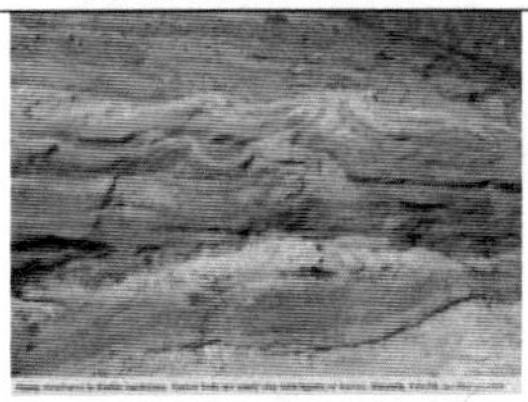
Near Kodi, Wyoming, USA Hilly oblique bedding	Pennsylvania	Bournemouth, England Landslide structure

	Kentucky, USA Flaser bedding	
Fluvial facies	Lagoon facies	Abyssal-marine facies

Baraboo, Wisconsin, USA Channel and pit	Broome Town Beach, Western Australia Linear ripples	Lester Park, Saratoga Springs, New York, USA Stromatolithic structure

3. Sedimentary structure models

Sedimentation types and sedimentary structure forms of coastal facies are complicated and diverse (table 2, examples), including continental genesis sediments, proper coastal and shelf zones and continental slope.

IOP Conf. Series: Earth and Environmental Science **27** (2015) 012024 doi:10.1088/1755-1315/27/1/012024

Table 2. Sedimentary structure models of coastal-marine sedimentation.

Continental zone	**Coastal zone**	**Marine zone**
Eolian facies	Lacustrine facies	Turbidite facies
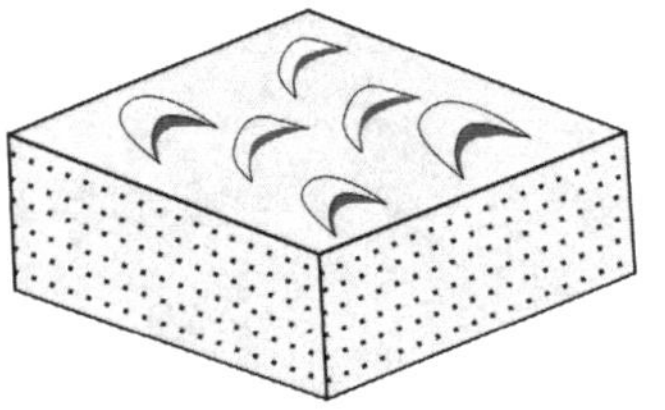	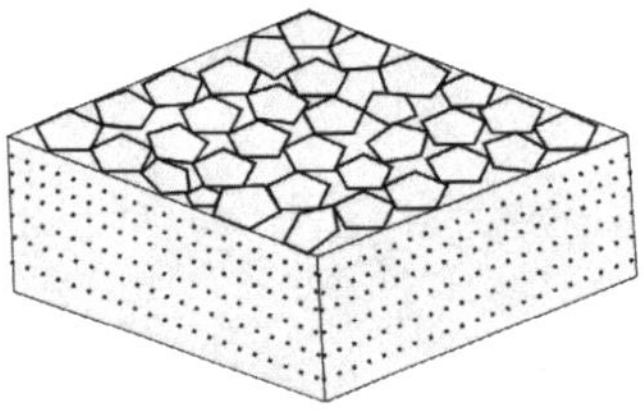	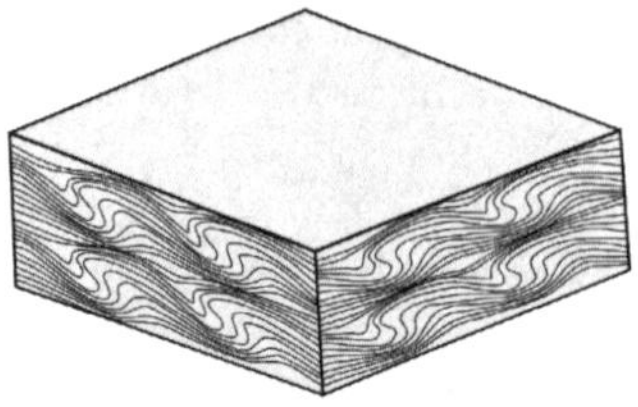
Duna	Mud cracks	Contorted (crinkled) bedding
Fluvial facies	Delta facies	Landslide facies
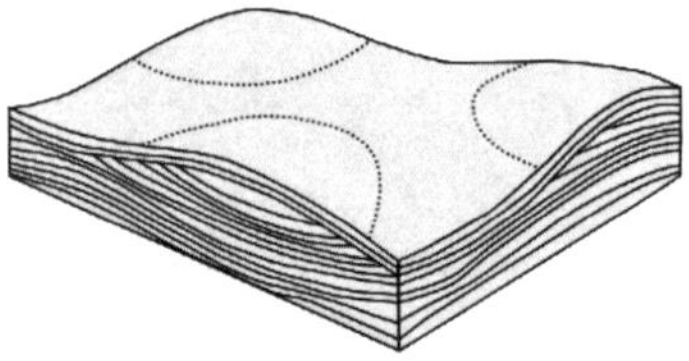	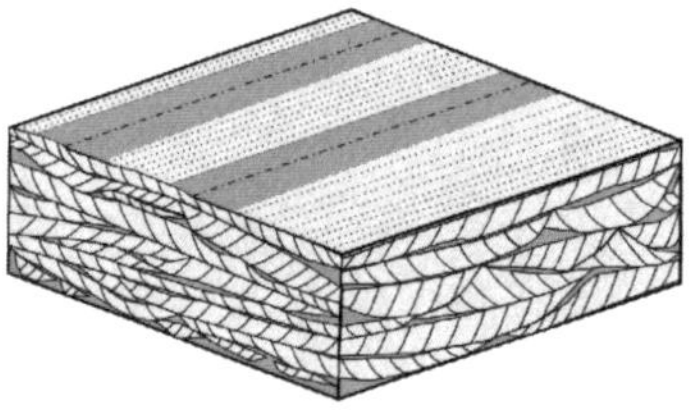	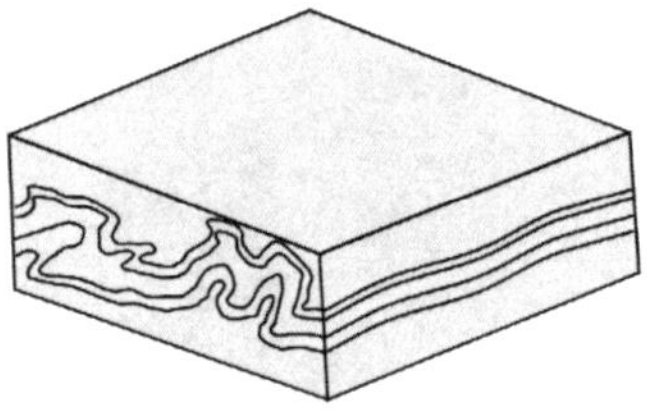
Hilly oblique bedding	Flaser bedding	Landslide structure
Fluvial facies	Lagoon facies	Abyssal-marine facies
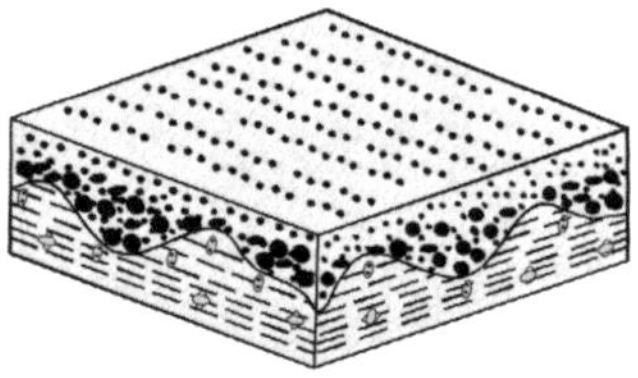	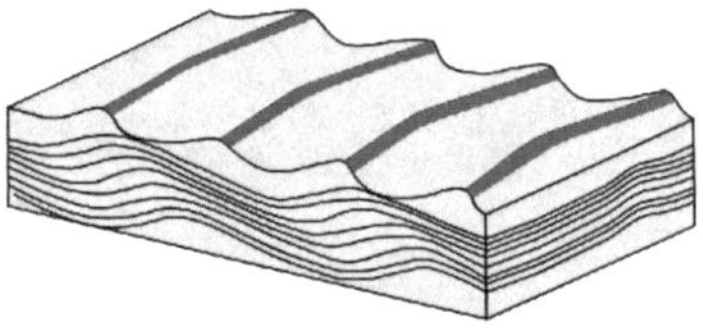	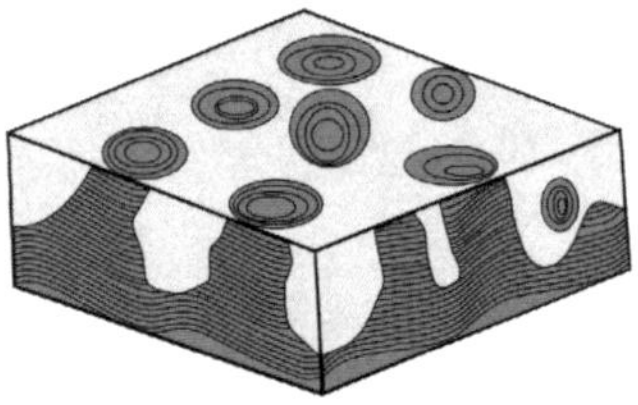
Channels and pits	Linear ripples	Stromatolithic structure

4. Summary reservoir property characteristics of sedimentation facies

Three major facies systems were embraced: continental including eolian, fluvial and alluvial facies; coastal-marine including delta, lagoon and shelf facies; and sea (marine) including turbidite, landslide and abyssal-marine facies. Sedimentary facies structure types, as a geological information feature, indicate this or that reservoir property (table 3).

Table 3. Summary of sedimentary structure characteristics and reservoir properties of typical sedimentation facies.

	Facies types	Possible sedimentary structures	Porosity (%)	Permeability (mD)
Continental	Eolian	Foreset bed, cross-bedding, bioturbation, stratification, dunas, biogenic structure	5–20	50–800
	Fluvial	Pebble bed, channel of clastics, cross bedding, hilly oblique bedding, cut-and-fill structure, occurrences, ripple marks, channels and pits	0–23	0.001–1000
	Alluvial	Mud cracks, micro-thin layers, parallel bedding, climbing ripples, flaky laminated silt and clay, columnar structure	3–15	1–50
Coastal	Delta	Lenicular bedding, swaley bedding, flaser bedding, cross bedding, herring-bone cross-bedding, linear ripples, plane stratification, foreset bed, ploughing structure traces, biogenic structure	12–34	10–1500
	Lagoon	Fine-layered structure, bioturbation abundance as a result of plant roots, lenticular, sawley, herring-bone cross-bedding	6–19	10–1500
	Shelf	Lenticular, flaser and herring-bone cross-bedding, geopetal texture	1–22	Less than 0.0001, 0.002–0.174
Marine	Turbidite	Normal sedimentary structure and reverse layers, silt-sorted sands, concretions, torch structure, contorted (crinkled) bedding	10–25	1–2400
	Landslide	Boulder sand and silt, landslide structure	10–25	1–100
	Abyssal-marine	Parallel bedding, bioturbation, micro-thin layers, carbonate silt, cupola, ball-and-pillow structure, dropstone, hilly oblique bedding, compressed-fractured structure, stromatolithic structure, biogenic structure	2–23	0.09–10

5. Logging models and testing of prediction models

Litho-facies interpretation of geophysical data was assigned to determined logging model (SP, EL and GR) for each facies. The geophysical prediction of Jurassic sediment facies in northern Tomsk Oblast was conducted. Specific characteristics of sedimentogenesis and reservoir properties of Jurassic sediments (J_1formation), the thickness of which ranged from 3 to 30m., were identified according to the integrated litho-facies analysis results and on-the-spot GIS data (SP, EL, IR and GR logging curves).

Three sediment layer types were identified in the Vasugan suite- J_1^1 layer, J_1^2 layer and J_1^3 layer based on the classification of investigated cross-sections. These layers had the following thicknesses: J_1^1 layer from 5 to 12m.; J_1^2 layer– from 3m to 13m.;and J_1^3 layer from8 m to 30m.

According to the discussed sedimentary models, logging characteristics, lithological interpretations (A Ezhova) [9], J_1^1 layer embraces predominately medium-fine grained sandstones, aleurolites, carbonaceous argillites which are in – situ oil saturated. The interpretation of logging curves showed that according to geophysical characteristics these sediments can be classified as coastal-marine-delta (table 4). It should be noted that in 7 out of 8 well models the above-mentioned facts were verified (actual productivity according to testing results).

Table 4. Example of predicted and comparable facies types in J_1^1layer.

Interval, m.	Logging	Actual productivity	Lithology (according to [9])	Facies type (author classification)	Porosity, permeability (according to author)	Productivity (according to author)
2190–2198		Oil influx rate 1.2 m^3/daily	Medium-fine grained sandstones oil-saturated	Coastal-marine-delta	12–34 %, 10–1500 mD	Productive reservoir
2190–2198		Oil influx rate 1.2 m^3/daily	Medium-fine grained sandstones oil-saturated	Coastal-marine-delta	12–34 %, 10–1500 mD	Productive reservoir
2210–2215		Dry	Medium-fine grained sandstones carbonceous argillites	Coastal-marine-delta	12–34 %, 10–1500 mD	Productive reservoir

According to the logging data of 14 wells, only in 2 wells, the facies type (coastal-marine-delta) was observed in J_1^2 layer. There is no data concerning reservoir properties and well productivity in the remaining 12 wells. In this case, the logging data of J_1^3 layer was applied in predicting in-situ facies types.

6. Conclusion

More than 100 world-wide deposits were analyzed by applying the acoustic method and the characterizing sedimentary structure for each coastal-marine-delta type was determined. The summary of sedimentary structure characteristics and reservoir properties (porosity and permeability) of typical facies were described. Logging models SP, EL and GR (configuration, curve range) in interpreting geophysical data for each litho-facies were identified.

Prediction models for potential Jurassic oil-gas bearing complexes (horizon J_1^1) in one S-E Western Siberian deposit were conducted. Layer J_1^1 embraces predominately medium-fine grained sandstones, aleurolites, carbonaceous argillites which are in – situ oil saturated. According to geophysical characteristics these sediments can be classified as coastal-marine-delta.

It should be noted that in 7 out of 8 well models the above-mentioned facts were verified. Comparing forecasting to actual testing data showed that the prediction is about 85%. Excluding logging curve analysis could result in an improper interpretation of facies based on the analysis of sedimentary structures.

References

[1] Chernova O S 2004 *Sedimentologija rezervuara* (Tomsk: TPU Publishing House) p 453
[2] Ezhova A V 2009 *Litologija* (Tomsk: TPU Publishing House) p 336

[3] Isaev V I and Nguen H B 2013 Cavitation in Oil-Gas Reservoirs of the Crystalline Basement from the Well Logging Data on the White Tiger Field in Vietnam *Russian Journal of Pacific Geology* **7**(4) 237–46
[4] Slatt R M 2006 *Stratigraphic Reservoir characterization for petroleum geologist, geophysics, and engineers* (Amsterdam: Elsevier) p 473
[5] Osipova E N, Prakoyo F S and Kudryashova L K 2014 Petroleum potential of the Neocomian deposit of Nyurolsky megadepression *IOP Conf. Ser.: Earth Environ. Sci.* **21** 012011
[6] Lobova G A, Osipova E N, Isaev V. I. and Terre D A 2015 Petroleum potential of Lower-Jurassic deposits in Nurolsk megadepression *IOP Conf. Ser.: Earth Environ. Sci.* **24** 012001
[7] Isaev V I, Lobova G A and Osipova E N 2014 The oil and gas contents of the Lower Jurassic and Achimovka reservoirs of the Nyurol'ka megadepression *Russian Geology and Geophysics* **55** 1418–28
[8] Prakojo F S 2013 Sedimentacionnye struktury peschanyh kollektorov i ih vlijanie na neftegazonosnost *Vestnik Irkutsk State University* **9** 103–10
[9] Ezhova A V 2007 Primenenie sistemnogo analiza dlja raschlenenija i korreljacii jurskih terrigennyh razrezov na mestorozhdenijah uglevodorodov Tomskoj oblasti *Vestnik Tomsk Polytechnic University* **311**(1) 59–63

IOP Conf. Series: Earth and Environmental Science **27** (2015) 012025 doi:10.1088/1755-1315/27/1/012025

Improving the accuracy of determining the character saturation reservoirs on the results of log interpretation in the oil field "K"(Tomsk region)

A A Rogov[1,2]

[1] Department of Geophysics, National Research Tomsk Polytechnic University, 30 Lenin Ave., Tomsk, 634050, Russia
[2] Department of Geology, Tomsk Research and Design Institute of Oil and Gas, National Research Tomsk Polytechnic University, 30 Lenin Ave., Tomsk, 634050, Russia

E-mail: RogovAA@nipineft.tomsk.ru

Abstract. Log interpretation has revealed a number of factors that have a significant impact on the determination of reservoir fluid content (oil saturation factor). The presented research paper is focused on the analysis of petrophysical properties of reservoir fluids considering laboratory studies of core samples. Based on core sample analysis, data of laboratory studies have been correlated to estimates of radioactive logging and high-frequency induction logging. To determine the oil saturation factor a method which considers the effect of mixed clay on reservoir fluid saturation has been proposed.

1. Introduction

Field K is located in Pudinsk oil and gas province in Western Siberia. Commercial oil content is associated with sand J_1^1 formation confined to Callovian-Oxfordian J_1 horizon of Vasyugan suite. The deposit is a complex, lithologically sealed structure located at the anticline flank. According to facies classification pay zone is related to coastal bar.

In a number of well cores from the field K the discrepancies between predicted log-derived water cut and actual water cut for tested wells have been revealed, with predicted water cut being higher than the actual one by 10% average.

The relevance of the research is that, in some cases, the current petrophysical model of field K does not determine water saturation coefficient (S_w) and reservoir fluid content (figure 1). In this respect, to refine the petrophysical model, new core sample analysis has been conducted which revealed the peculiarities of non-homogeneous field K reservoir. This data has proved the consistency of water saturation estimates and, consequently, reservoir fluid content.

The purpose of this research is to identify the factors which distort water saturation coefficient and reservoir fluid content estimates and their effects on well log interpretation.

The research target is J_1^1 reservoir rock within which reservoir properties are non-uniformly distributed from the roof to the bottom [1].

2. Selection of reference wells

Based on the comparison of predicted and actual water cut, seven wells located in the same area of the field have been selected as water cut is higher than the initial one by approximately 5-15 % (table 1).

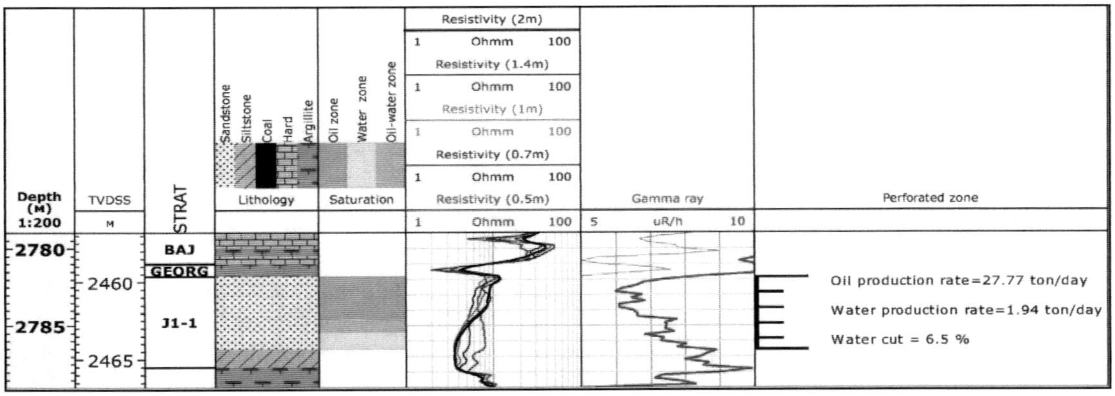

Figure 1. Ambiguity of reservoir fluid content estimates.

Table 1. Comparison of predicted and actual water cut.

Well number	Predicted water cut, %	Initial water cut, %
204	15.8	7.5
301	19.4	6.5
304	20.9	6.3
403	15.3	8.0
405	21.1	6.5
406	13.1	6.7
501	9.2	2.6

The wells are located within the zone which is not affected by production operations, as during drilling no production and injection wells were found. All the wells are located in the northeastern part of field K forming an anomalous zone on the areal map.

3. Log analysis

Well-logging suite implemented in field K involved basic methods which are used in production well testing, and included such methods as spontaneous potential log, gamma-ray logging, neutron logging, gamma-gamma density logging, lateral logging, high frequency induction logging. [5]. Such a combination makes it possible to define reservoir zones, their reservoir properties, electrical resistivity and fluid content with sufficient accuracy. It should be noted that pay zone SP curve does not indicate bottom reservoir boundary due to its regressive structure. Therefore, in the present petrophysical model, selection of effective thickness and porosity calculation was based on calculated gamma-ray relative amplitude curve; whereas porosity calculation, according to gamma-gamma density logs, has not been performed for all reference wells because of partial log recordings.

The analysis of well logs from reference wells has indicated the following effects:

1) increased shaliness of reservoir bottom according to gamma-ray logs (figure 2a);
2) high-frequency induction log inversion (figure 2b) .

Though high-frequency induction log inversion implies the presence of oil-water contact within reservoir zone, well operation tests have indicated that fluid inflow water cut does not exceed 10%, which, in its turn, provides evidence of pure oil content.

Based on qualitative analysis of well log behavior, it can be concluded that reservoir properties of J_1^1 formation bottom zone are derated. Low resistivity is probably due to overall high bulk-volume fraction of shale in the bottom zone.

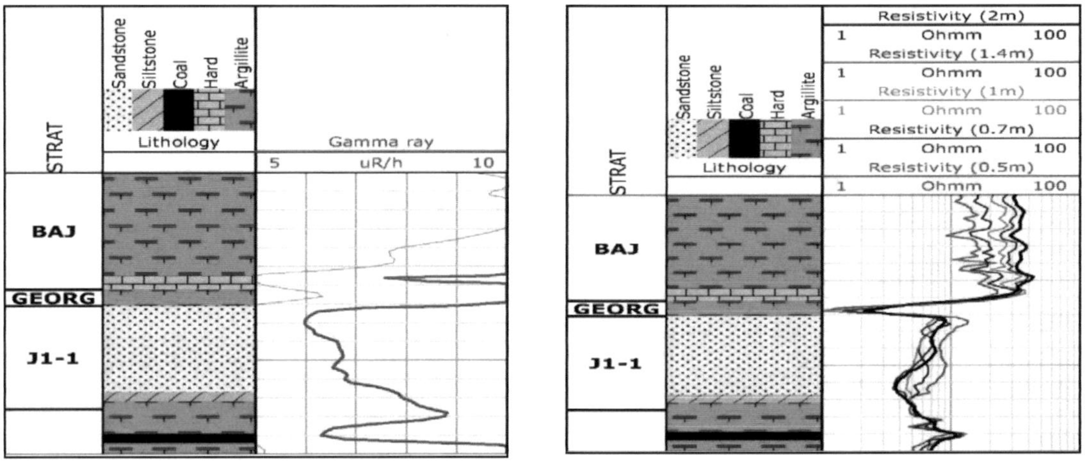

a) b)

Figure 2. Anomalous behavior of well logs (Well № 403).

4. Porosimetry, granulometry and thin section analysis

Core samples from three wells were used in analyzing reservoir properties. Porosity measurements, particle size and thin section analysis were performed to determine the lithologic composition of reservoir rocks, void pattern and mineral composition of pore-filling cement in the above-described reservoir.

According to mineralogical composition J_1^1 sandstones include greywacke-feldspar-quartz variety; rock matrix also contains mica aggregates, single grains of glauconite, as well as chlorite as pseudomorph glauconite.

Cement content in the rock is 9-18%. Cement is basically of polymineral composition, predominantly hydromica, chlorite, kaolinite and mica aggregates, and in some areas calcite, siderite and pyrite. Chlorite is a common cement constituent. Cementation is of membrane-pore type.

In terms of particle size, sandstones are fine-medium-grained; in the layer bottom where fine-grained sandstones (0.25-0.1 mm) and fine-grained siltstones (0.1-0.01 mm) are interbedded.

It can also be noted that pyrite (6-30% content) is distinguished in the mineralogical composition of the cement, which in glauconite-chlorite mineral association of cement increases the electrical conductivity, and, thereby, causes electrical resistivity decrease. Additional correction for pyritization was not applied as pyrite content was calculated on the basis of formation resistivity vs. porosity coefficient.

Porosity measurements have also been performed based on the analysis of core samples from two wells. Correlation of sampling intervals with modal pore distribution has indicated that bottom reservoir rocks are non-homogeneous and contain small-size pores with average radius of 1-2 mm [7].

5. Petrophysical properties of reservoir rocks

Considering the fact that log-derived porosity is the same as laboratory-measured formation porosity, it has been concluded that the algorithm used for calculating porosity is reliable. Consequently, the error in determination of fluid content regarding this petrophysical parameter has not been considered.

As a result of the analysis of petrophysical relationships $F_R = f(\varphi)$ and $I_R = f(Sw)$ based on core sample studies, no definite petrophysical type for clay-abundant bottom reservoir rock has been distinguished (figure 3) which, in its turn, suggests that the mineral matrix within the entire reservoir column is composed of a single rock.

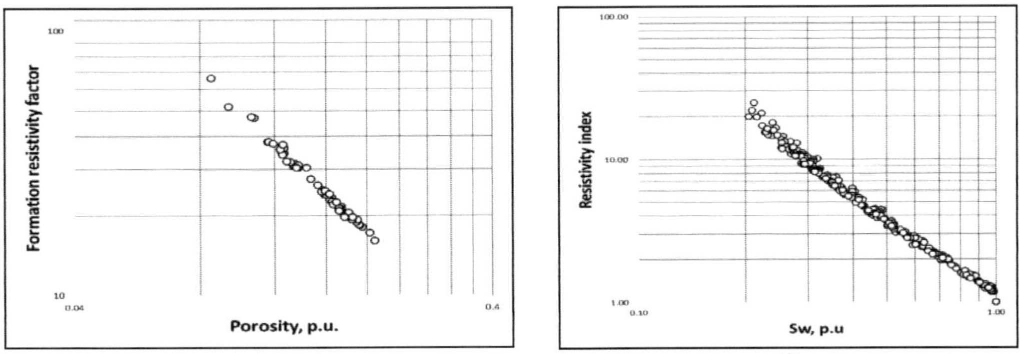

a) б)

Figure 3. Petrophysical dependency of formation resistivity factor (a) and resistivity index (b) for the studied reservoir rock.

6. Analysis of challenging factors

The study of open-hole logs, cased hole logs and core analysis has indicated that shaliness bottom J_1^1 reservoir rocks has the greatest effect on oil saturation determination. This factor causes overestimation of electrical conductivity, which, in its turn, has a significant impact on the determination of water saturation coefficient, resulting in overestimation as well.

According to facies classification, the reservoir is a regressive bar, bottom clays tend to be of a mixed type (they are mainly laminar and, to a lesser extent, structured). Therefore, to estimate its effect volumetric clay index calculated by Ellansky formula is used [2]:

$$Vcl = 1.055 - \sqrt{1.14 - 1.111 \cdot \Delta I\gamma},$$

where,

Vcl – clay volume (factor), $\Delta I\gamma$ – gamma ray delta.

To eliminate the influence of this factor the reservoir rock was divided into two types, and these types do not correspond to existing petrophysical reservoir rock types but they have been distinguished based on lithotypes that differ in void space internal arrangement and clay cement distribution. However, material which constitutes these reservoir types is the same.

Lithological types (lithotypes) are differentiated according to the following characteristic criteria:

1 type - pure medium-grained sandstone characterized by the lowest natural radioactivity, AGR> 0.90 unit fraction (figure 4)

2 type - clayey, medium-grained sandstone characterized by increased natural radioactivity, AGR≤0.90 unit fraction (figure 4), and the inversion of induction logs.

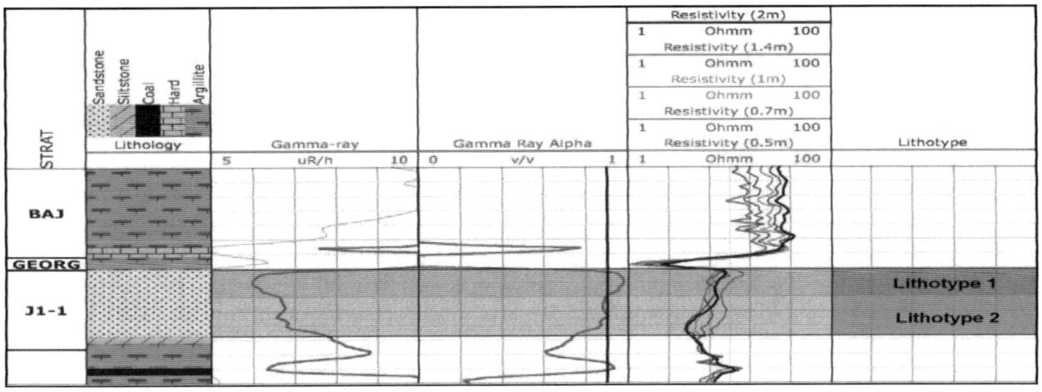

Figure 4. Well № 403 summary plot indicating lithotype classification.

7. Methods for water saturation coefficient calculation

The standard petrophysical model of water saturation coefficient calculation (Archie-Dakhnov method) is applied to homogeneous reservoirs. Provided that clay material is non-uniformly distributed within the reservoir rock, the standard method of calculation is not reliable. Calculation of water saturation coefficient using j-function is difficult, in this case, since there are no reliable data on oil-water contact in the field.

The analysis of in the laboratory core samples provides evidence of different relationships between particular mineral presence in the clay material and rock porosity. Vasyugan suite porosity factor is dependent on chlorite content. Its presence in reservoir pore-filling cement reduces effective porosity, resulting in a felted structure, and therefore, increasing the specific surface area of pore channels and amount of bound water [7].

Taking into account the regressive reservoir structure, being of mixed clay type, and the fact that cross-plots used in the interpretation were produced for cores composed of pure sandstone, a correction has been introduced into the calculation of formation resistivity factor to identify the effect of complex shaliness.

Water saturation coefficient calculation for lithotype 1 was made according to approved cross-plots for the reservoir rock composed of pure sandstone. Calculation of water saturation coefficient for lithotype 2 involved a correction for formation resistivity factor which was calculated as follows [6]:

$$F_R = \frac{a \cdot (1 - Vsh)^m}{\varphi^m},$$

where a – lithological factor,
m – cementation factor,
Vsh – shale volume factor,
φ – porosity factor.

8. Results and discussion

Based on the applied method, for wells with high predicted water cut, water saturation coefficient has been recalculated for intervals which correspond to reservoir rock lithotype 2. Recomputation of water saturation coefficient for intervals corresponding to lithotype 1 has not been performed (table. 2).

Table 2. Comparison of weighted average water saturation coefficients before and after application of the recommended method.

Well number	Sw (Archie), p.u.	Sw (Corrected), p.u.	Relative difference, %
204	0.61	0.56	-9.3
301	0.61	0.56	-7.3
304	0.60	0.54	-10.8
403	0.53	0.51	-3.7
405	0.62	0.53	-14.7
406	0.51	0.49	-3.2
501	0.64	0.57	-9.5

Recomputation of predicted water cut has also been carried out. The estimates turned out to be fairly coincident to the actual values (table 3). Prediction of water cut was carried out on the basis of capillarimetry studies of core samples. To calculate the water cut fractional flow vs. normalized water saturation coefficient cross-plot has been produced FF = f (Sw^{norm}).

Table 3. Comparison of weighted average values of predicted and initial water cut estimates before and after application of the recommended methods.

Well number	Predicted water cut, %	Inintial water cut, %	Predicted water cut including shaliness, %
204	15.8	7.5	9.4
301	19.4	6.5	13.5
304	20.9	6.3	12.5
403	15.3	8.0	13.2
405	21.1	6.5	9.9
406	13.1	6.7	8.6
501	9.2	2.6	4.8

Discrepancies of predicted and initial water cut estimates are due to log measurement error.

9. Conclusion

The analysis conducted for field K wells with high predictive water cut has identified the factor which causes errors in water saturation coefficient calculation, which, in its turn, results in misleading conclusions about fluid content in a reservoir.

To eliminate this factor effect, the method of water saturation coefficient calculation considering reservoir shale index has been applied in this field for the first time.

As a result of this technique application and based on new core data, the petrophysical model has been improved making it more reliable.

The presented method is recommended in log interpretation for step-out wells in this field, as well as in deposits with similar reservoirs. Application of this method improves the accuracy of determination of water saturation coefficient and fluid content, as well as a more accurate prediction of initial water cut.

References

[1] Dakhnov B 1985 *Geofizicheskie metody opredeleniya kollektorskikh svojstv i neftegazonasyscheniya gornykh porod* (Moscow: Nedra) (in Russian)

[2] Dobrynin B, Vendelshteyn B, Kozhevnikov D 2004 *Petrofizika* (Moscow: State University of Oil and Gas) (in Russian)

[3] Kosarev B 2009 *Kontrol za razrabotkoj neftegazovykh mestorozhdenij* (Kazan: KFU) p 145 (in Russian)

[4] Petersile B, Poroskuna B, Yatsenko G 2003 *Metodicheskie recomendatsyi po podschetu zapasov nefti i gaza objemnym metodom* (Moscow – Tver: VNIGNI, SPC "Tvergeofizika") (in Russian)

[5] Rukobodjaschij dokument №153-39.0-072-01 2001 *Tekhnicheskaya instruktsyya po provedeniju geofizicheskikh issledovanij i rabot priborami na kabele v neftyanykh i azobykh skwazhinakh* (Moscow: SSC VNIGIK VNIIgeosystem) (in Russian)

[6] Djebbar Tiab and Erle C. Donaldson 2004 *Petrophysics: theory and practice of measuring reservoir rock and fluid transport properties* (Gulf Professional Publishing is an imprint of Elsevier 200 Wheeler Road, Burlington, MA 01803, USA Linacre House, Jordan Hill, Oxford OX2 8DP, UK)

[7] Birukova L, Kolmaberg N, Loganova S 1988 *Kompleksirovanie metodov issledovaniya pri razrabotke mestorozhdenij nefti i gaza Zapadnoj Sibiri* (Tyumen: ZapSibNIGNI) (in Russian)

Structural-chemical features and morphology of glauconites in sedimentary iron ore of Bakchar prospect (Western Siberia)

M Rudmin[1,3], I Reva[1], A Gunko[1], A Mazurov[1,4] and R Abramova[2,5]

[1] Department of Geology and Mineral Exploration, Institute of Natural Resources, National Research Tomsk Polytechnic University, 30 Lenin Ave., Tomsk, 634050, Russia

[2] Department of Foreign Language, Institute of Natural Resources, National Research Tomsk Polytechnic University, 30 Lenin Ave., Tomsk, 634050, Russia

E-mail: [3] rudminma@tpu.ru, [4] akm@tpu.ru, [5] abramova_rn@mail.ru

Abstract. The research embraces the investigation results of glauconites in Bakchar iron ore occurrences to evaluate the potential diversified commercial application of this mineral. The following lab methods were used to analyze the morphology, chemical composition and structure of glauconites: granulometric analysis, optical microscopy, electron microscopy, X-ray fluorescence analysis, atomic arc-emission analysis and infrared spectroscopy. Glauconite was classified according to morphology and grain color and chemical composition and some specific characteristics were also determined (relative content of absorbed water, random distribution of smectite flakes within the grain structures). The research results showed that pistacho-green glauconite grains are less subjected to alteration than greenish-yellow grains due to the content of potassium, iron, absorbed water and organic impurities.

1. Introduction

Glauconite is iron potassium phyllosilicate nonstoichiometric mineral, $K_{<1}(Fe^{3+}, Fe^{2+}, Al, Mg)_{2\text{-}3}[Si_3(Si, Al)O_{10}][OH]_2 \cdot nH_2O$, prevailing in sedimentary rocks [1, 9, 14]. Due to its unique properties (molecular sorption, ion-exchange properties, thermal resistance, radiation stability, apparent coloring oxides, etc.), it is a diversified industrial application mineral [2-4] and a prospective geological (paleoclimatic, facies, etc.) information source [5, 7, 8, 12]. Its structure-texture properties and chemical composition determine the potential application of glauconite. Named in 1828 by Christian Keferstein, glauconite has attracted the attention of many geologists [5, 8, 12]. With the development of research and technology new and new domains of this mineral have been established: industrial application [1-4, 6] and mineral-indicator for sedimentation environments [5, 8, 11]. Recently, the processing behavior of this mineral (from the deposits in Central, Southern, North-western and Ural Federal Districts) has been investigated in details [4]. The following potential glauconite areas and provinces were determined: Central, Baltic, Volga, southern Pre-Ural, trans-Ural [3-4]. Notwithstanding the significant contribution of the following geologists E.F. Levchenko, M.L. Levchenko, N.G. Patik-Kara, L.P. Tigunov [1-4, 6, 7, 11], even, today, there is no information of the glauconite potential within the Siberian and Far Eastern territories.

This research is focused on the investigation of the glauconite structural-morphological properties and chemical composition in Upper Cretaceous sediments from Bakchar iron ore occurrences. The practical aspect of this research involves not only the possible extraction of glauconite as a by-product free-milling concentrate during the development of sedimentary iron ore in Tomsk Oblast, but also the introduction of this glauconite concentrate into different enterprise domains (agriculture, petroleum industries, etc.). Bakchar ore mineralization is located in the S-E Western Siberia iron ore basin (figure 1), 200 km north-westward from Tomsk. Ferrous iron sediments include sediments of coastal-marine facies – gritstone, oolitic ores, sandstone, aleurolite and clays. Iron ores are oolites of goethite-hydrogeothite and / or chlorite-hydrogeothite composition and confined to three horizons: Narim, Kolpashev and Bakchar [9-10]. There are three types of native iron ores within the ore mineralization itself: loose hydrogeothite, weak and medium-cemented hydrogeothite-chlorite, closely cemented siderite-hydrogeothite [9-10]. Glauconite is predominate in weak and medium-cemented hydrogeothite-chlorite ores and has the granulometric size from 0.5 to 0.1 mm. Visually, optical testing showed that the glauconite content is 15-20%. Bakchar iron ore formation embraces layers of glauconite sandstones and sands which are interbedded between iron-ore lenses and layers. Based on the research results, glauconites in iron ore deposits can be evaluated in terms of extracted sovereign concentrations under conditions of iron-ore enrichment.

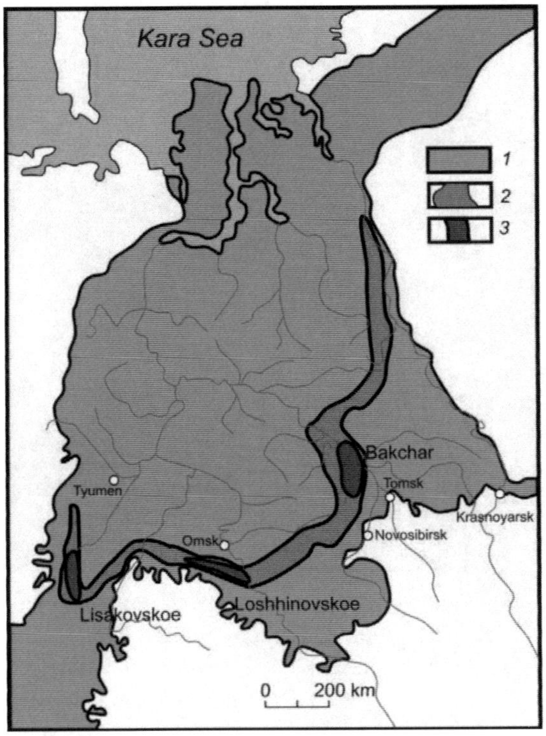

Figure 1. Map of Western-Siberia iron-ore basin [10]: 1 – Cretaceous-Paleogene extent sea area; 2 – extent coastal-marine iron-ore sediment zone; 3 – area of significant iron ore clusters (deposits).

2. Factual material and research methods

Factual material includes weakly cemented hydrogeothite – chlorite ore chip samples [10] of well cores from the depth of 180…220 meters. The following laboratory methods were used: granulometric analysis ("wet screening" method), optical microscopy, electron microscopy (TESCAN VEGA 3 SBU), X-ray fluorescence analysis (HORIBA XGT-7200), atomic arc-emission analysis (DEMO PRODIGY DC), infrared spectroscopy (Shimadzu IR Prestige-21). Glauconite monofractions of granulometric size from 0.5 to 0.1 mm were selected for binocular microscopy investigation.

3. Research results and discussion

Based on grain morphology the following types of glauconite were identified: isometrically rounded (globular, nodular) (figure 2a), elongated gobular (figure 2b), clustered (figure 2c), biomorphic (figure 2d), composite aggregates (figure 2e).The grain surfaces are rough-pit with traces of erosion (cracks, vugs) and rarely, smooth and lustrous. If increasing in volume, the internal glauconite structure changes to composite-layered randomly scattered smectite-illite plates and flakes (figure 2f) with micron eosizes (up to 2 micron). According to color, the grains were divided into two basic categories: pistachio-green and greenish yellow. However, they are also bluish-green, practically black, yellow with slightly greenish tint, as well as other color tints, which, in its turn, reflect the chemical composition variability due to the post-sedimentation changes of original glauconite assemblages.

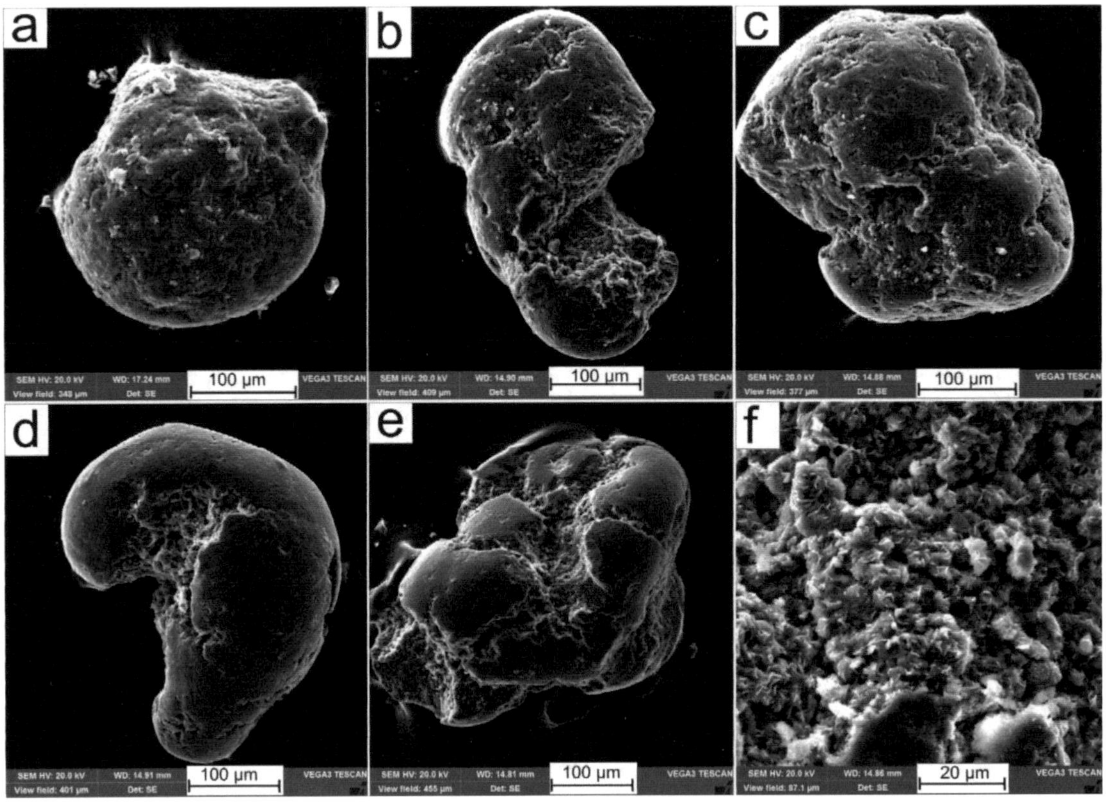

Figure 2. Morphology and internal structure of glauconite grains in Bakchar ore occurrences (electron micrograph): a –isometrically rounded; b – elongated gobular; c – clustered; d – biomorphic; e – composite aggregate; f – internal grain structure.

The microscopic investigations of glauconite surfaces and corroded vugs (cavities, cracks) revealed the following mineral impregnations (figure 3): framboid pyrite assemblages; kularite and spherulite aggregates, presumably bromin-carnallite; nickel-chrome mineral (redingtonite) adhesion. Pyrite is a framboid aggregate of up to 10 micron in size, which is cubic and /or octahedral microcrystal assemblages of up to 1 micron in size. In some cases, kularite can be found on the glauconite surfaces as microspherulite assemblages of up to 5…6 microns in size. According to the shape and chemical composition such kularite is found in hydrogeothite oolites of Bakchar ore occurrences [9]. Bromin mineral grains, comparable to bromin-carnallite ($KMg(Cl,Br)_3{\cdot}6H_2O$) in composition, are commonly found in glauconite (according to the data from www.mindat.org). These minerals (bromin-carnallite) are variform aggregates from 4…5 microns in size. Unidentified chromic mineral is found as a mineral adhesion of up to 0.5 micron thickness contouring micro- smectite flakes. Ni, Cr, Fe predominance indicates the fact that this mineral could be redingtonite, having the formula $(Fe^{2+}, Mg, Ni){\cdot}(Cr, Al)_2(SO_4)_4{\cdot}22H_2O$ (according to the data from www.mindat.org).

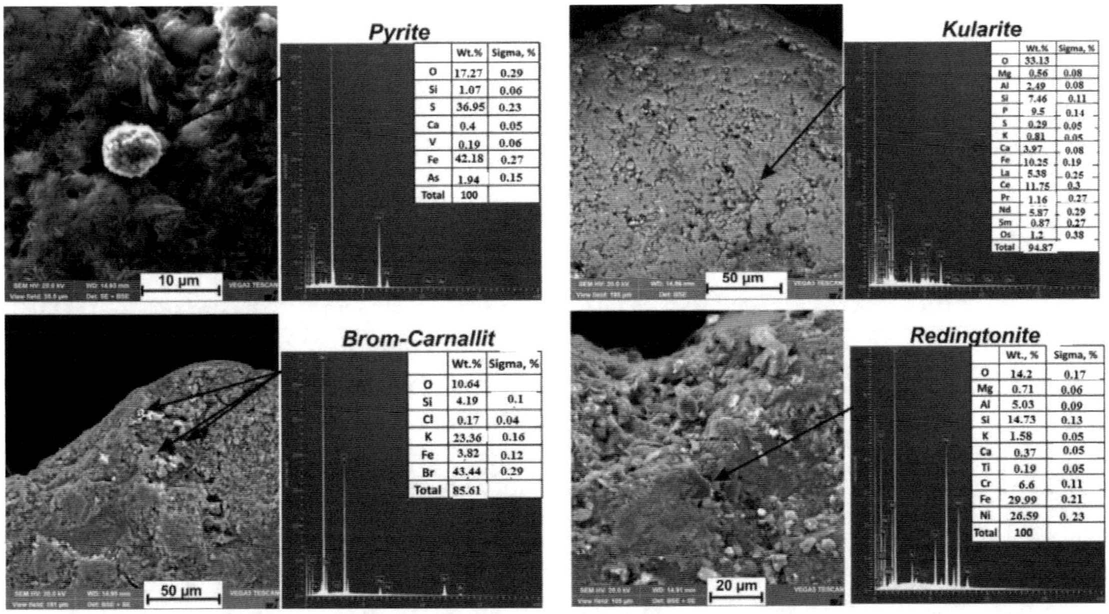

Figure 3. Electron micrographs and characteristic spectrum of mineral microphenocrysts on the glauconite grain surface.

It should be noted that the above-mentioned facts could be considered only as preliminary data, requiring further detailed investigation based on the electron microscopy analysis. Based on the X-ray fluorescence analysis the following chemical composition was defined and described in table 1. Based on these results the following conclusions can be stated: yellow and yellowish brown glauconite has a low K_2O content (up to 5.5%) and SiO_2 content (up to 39%) and a rather high $Fe_2O_{3(total)}$ content (up to 48%). The pistachio-green color of grains could be significant as a result of oxidation of the first, i.e. K_2O. The amount of impurity elements is an indicator of prevailing good glauconite sorptivity. Impurity elements precipitate in the mineral (glauconite) itself from the sedimentation environments as a consequence of its specific (random, composite-layered) inner structure. The impurity elements include Ti, V, P, Cr, Zn, Mn, As, Ni. Phosphate content varies from 0.13 to 2.04 %, while its average content is 0.42…0.83 %. Phosphate is a practical impurity for agriculture.

It is problematic to determine the absorption maximum of this or that bond in the mineral samples based on the infra-red spectrum, as the most statistically reliable absorption characteristic values, as a rule, are comparable to the equivalent bonds in reference samples. As glauconite is a mineral with alternative chemical composition, it could be also problematic to select a reliable reference infra-red spectrum. In addition, the reaction with environment bonds could possibly induce alterations of the characteristic absorption frequency itself, which, in its turn, become limiting. For example, V.I. Vigdorovich [11] described the fact that the Si-O bond in crystalline quartz is characterized by two stretching bands-1179 and 1109 cm^{-1},whereas, these bands in glauconites cannot be observed as single absorption maximums due to the fact that they could overlap PO_4 absorption maximums resulting in infrared absorption (950…1100 cm^{-1}).

Infrared spectrum (figure 4) of investigated glauconites showed stretching bands of 1570 and 1380 cm^{-1}. Stretching bands of 1625…1660 cm^{-1} reflect the inner structure of the glauconites shatter micro-smectite plates and flakes. Stretching bands of 3200…3600 cm^{-1} are related to absorbed water variations and complicated by intensive ferric stretching bands, the frequency of which is 3530cm^{-1}. Stretching band of 3600 cm^{-1} indicates a low Al^{3+} content in octahedral glauconite layers. Organic impurities show absorption maximum frequency of 2349, 667, 2850…2950 cm^{-1}.

Table 1. Average chemical composition of glauconite in iron ores of Bakchar ore occurrences.

Grain color	Size, mm	Major elements, %					
		MgO	Al_2O_3	SiO_2	K_2O	CaO	Fe_2O_3
1	0.5.....0.2	2.7	9.2	46.4	4.5	1.0	37.6
1	0.5.....0.2	2.0	9.6	52.9	5.3	1.3	29.0
2	0.2.....0.1	1.9	8.3	37.4	3.3	0.8	48.5
2	0.2.....0.1	1.6	7.7	39.5	3.9	1.0	46.0

Grain color	Size, mm	Major elements, % (+ content more than 0.01%)							
		Ti	V	Cr	P	Zn	Mn	As	Ni
1	0.5.....0.2	0.33	0.15	0.10	0.42	0.03	0.11	0.11	0.34
1	0.5.....0.2	0.69	0.11	0.27	0.83	0.03	0.15	-	-
2	0.2.....0.1	0.35	0.26	0.72	0.76	0.16	0.15	0.10	-
2	0.2.....0.1	0.28	0.24	0.13	0.67	0.08	0.13	0.10	-

*Note *Color of glauconite grains: 1 – pistachio-green, 2 – greenish-yellow*
The results in the table are based on X-ray fluorescence analysis data including 250 glauconite grains. Analyst – M.A. Rudmin

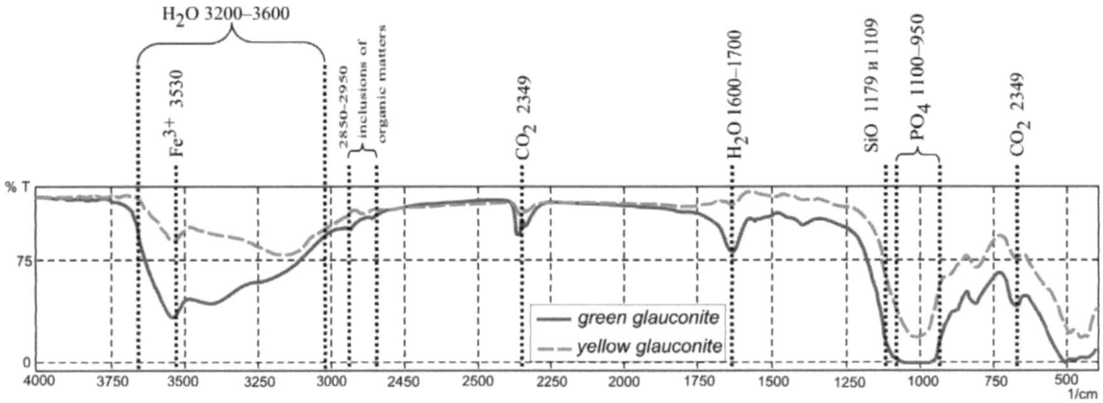

Figure 4. Infra-red spectrum of glauconite in iron ores of Bakchar ore occurrences.

Green-colored glauconite grains comparable to greenish-yellow ones predominately show intensive absorption peaks within the frequency intervals, which, in its turn, is associated with water absorption and organic impurities. This fact could be explained by the insignificant digenetic alteration of green glauconites comparable to the greenish-yellow ones, as well as the specific chemical composition of green glauconites (content of K_2O, Fe_2O_3, SiO_2).

4. Conclusion

The research results are the following: 1) according to color and morphology different glauconite grains were defined; chemical composition of glauconite was described and the content of basic impurity elements were identified; 3) structural features of glauconites were described based on the infrared spectrum; 4) micro-impregnations on the glauconite grain surface were determined and described, including bromin-carnallite and redingtonite.

Preliminary mineralogic-engineering investigation results of the glauconites in Mezozoic-Cenozoic sediments (Tomsk Oblast) were described. Thus, it could be stated that glauconite (in Tomsk Oblast) is a valuable industrial mineral with similar properties to the glauconites in Tambov and Cheljabinsk Oblasts [3-4]. It should be noted that a positive evaluation of the existing engineering properties of glauconites could further the mining efficiency of iron ores in Bakchar ore occurrences.

Acknowledgements
The investigations were supported by National Research Tomsk Polytechnic University, project No. 114-2014.

References

[1] Afanaseva N I, Zorina S O , Gubaidullina A M, Naumkina N I and Suchkova G G 2013 Kristallohimija i genezis glaukonita iz razreza «Melovatka» (Senoman, jugo-vostok Russkoj plity) *Lithosphere* **2** 065–075
[2] Grigoreeva E A 2004 Sorbcionnye svojstva glaukonita Karinskogo mestorozhdenija *Dissertation* (Chelyabinsk)
[3] Levchenko E, Patyk-Kara N and Levchenko M 2008 Glauconite deposits of Russia: perspectives of development *Abstract for the 33rd International Geological Congress (*Oslo, Norway)
[4] Levchenko M L 2008 Sostojanie syr'evoj bazy i vozmozhnosti ispol'zovanija glaukonita v Rossii *Mineral resources in Russia* **2** 27–31
[5] Meunier A and El Albani A 2007 The glauconite-Fe-illite-Fe-smectite problem: a critical review *Terra Nova* **19** 95–104
[6] Miledstkii B E 2012 Glauconite sands possible ore-metal formation *Ores and metals* **3** 69–71
[7] Nikolaeva I V 1971 Mineraly gruppy glaukonita i jevoljucija ih himicheskogo sostava *Problemy obshhej i regional'noj jekologii* 320–36
[8] Odin G S and Matter A 1981 De glauconiarium origine *Sedimentology* **28** 611–41
[9] Rudmin M, Mazurov A and Bolsunovskaya L 2014 Mineral and elemental composition features of "Loose" oolitic ores in Bakchar iron ore cluster (Tomsk Oblast) *J. IOP Conf. Ser.: Earth Environ. Sci.* **21** 012003
[10] Rudmin M, Mazurov A and Ruban A 2015 Facies and Sedimentation Model of Iron-ore Sequence in Bakchar Deposit *IOP Conf. Ser.: Earth Environ. Sci.: Scientific and Technical Challenges in the Well Drilling Progress* **24** 012028
[11] Vigdorovich V I, Filatova E Yu, Stigankova L E and Akulov A I et al. 2012 Nekotorye kristallohimicheskie parametry glaukonita Bondarskogo mestorozhdenija Tambovskoj oblasti *TSU Vestni* **17(2)** 731–34
[12] Wigley R and Compton J S 2007 Oligocene to Holocene glauconite–phosphorite grains from the Head of the Cape Canyon on the western margin of South Africa *Deep-Sea Research II* **54** 1375–95

IOP Conf. Series: Earth and Environmental Science **27** (2015) 012027 doi:10.1088/1755-1315/27/1/012027

Modeling of seismic field in porous medium: Simulation study of single pore and pore ensemble effects

A A Shatskaya[1,3], M M Nemirovich-Danchenko[1] and D A Terre[2]

[1] Department of Geophysics, Institute of Natural Resources, National Research Tomsk Polytechnic University, 30 Lenin Ave., Tomsk, 634050, Russia
[2] Department of Foreign Languages, Institute of Natural Resources, National Research Tomsk Polytechnic University, 30 Lenin Ave., Tomsk, 634050, Russia

E-mail: [3] shatsaa@mail.ru

Abstract. The article considers the effect of porous media on elastic wave field. Based on numerical modeling, diffraction pattern of the wave propagating through a single pore in carbonates has been produced. Matrix properties (calcite and dolomite) and fluid (water) are modeled based on thin core section image. The qualitative comparison with the available computational data has been performed. Provided that ensemble of pores is involved, the effect of porous medium on seismic field has been studied. For comparison with experimental data the model of porous sintered aluminum Al-6061 has been considered. The processing of numerical modeling results made it possible to estimate average velocities in the model of porous aluminum and compare them with physical modeling data. The provided estimates have indicated qualitative (single pore) and quantitative (ensemble of pores) correlation of simulation and experiment results.

1. Introduction

Porous media and their effect on the elastic wave field has been a current issue for the past few decades. This interest is due to both material engineering objectives and hydrocarbon exploration. These two different knowledge domains are interrelated in physical modeling. For example, in laboratory modeling of seismic waves in porous media several types of solids including aluminum [1, 2], soda-lime glass [3] can be used as a matrix material. In laboratory experiments both statistical and dynamic characteristics are estimated, P- and S- wave velocities being basic types among them. Considering porous media, the main disadvantage of laboratory models is that it is difficult to measure their primary properties – porosity, while elastic matrix model results in problems associated with experiment producibility.

Computer modeling does not have this disadvantage; however, it requires qualitative calibration of results and quantitative wherever it is possible. The analysis of numerical results in wave field calculation in porous media implies estimation of the wave velocity change. The comparison of these estimates with physical modeling data is referred to as quantitative calibration. For qualitative analysis one may refer to diffraction patterns, wavefront comparison, particle-motion polarization and so on.

In the present research computer modeling of total field of elastic waves (seismic field) has been performed. In solving an inverse problem an explicit second-order conditionally stable scheme is applied [4], involving a set of modifications [5] to carry out calculations for a large range of wavelengths (> 100 λ).

2. Problem setting

Modeling of total wave field is performed for two problems, two different medium models and two scenarios.

Problem one- the model of a single large-size pore filled with fluid and propagation of elastic pores is subjected to analysis. In this scenario calculations performed in this model compare the total wave field with calculation results (i.e. comparison of qualitative patterns).

Problem two- plane wave propagation in a porous layer with random pore distribution is considered. Calculations performed in this model describe how velocity of P-wave propagation changes due to porosity. Calculation involves two porosity values for which laboratory modeling data are available.

If a single pore is being considered, calculation results are compared with finite-difference modeling [6]. To calculate wave field induced in the media with ensemble of pores, the model equivalent to porous aluminium (sintered Al-6061) has been produced [2]. The choice of the material is made due to existing quantitative physical modeling data.

Linearized finite-difference method proposed by J Wilkinson is used to solve both problems [4, 5].

3. Numerical modeling

The effect of a single pore on propagating seismic wave was considered. One of the recent publications [6] is devoted to the calculation of seismic field diffraction induced by a single pore for different frequencies. It involves an image of carbonate rock section and simplified matrix of elastic characteristics for analysis.

The problem of plane wave propagation was solved through the model of medium sample introduced from the above -mentioned-research [6]. Figure 1 presents geometry of the first problem, a model containing one large-size pore, incident-wave signal. The signal (elastic wave) is represented by a damped sinusoid with wavelength which is approximately 10 times less than the characteristic "pore thickness" L (shown figure 1). In this case, calculation L is 200 computational cells; wavelength λ=20 computational cells.

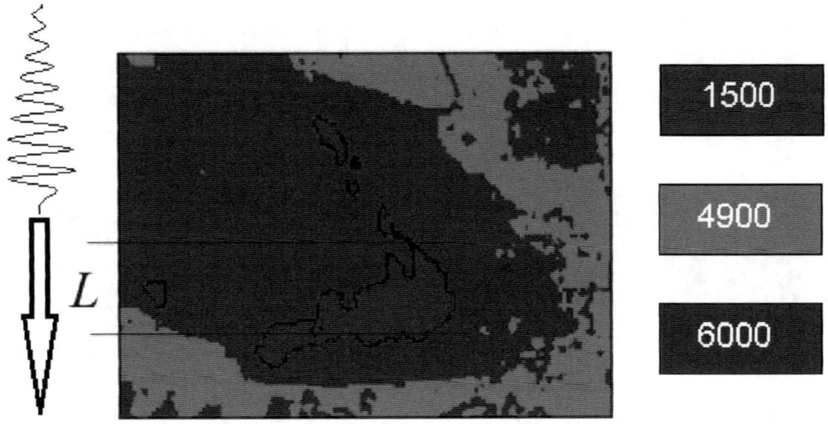

Figure 1. P-wave velocity matrix in the sample, m/s (according to [6]), an arrow indicates direction of plane-wave propagation represented by a damped sinusoid.

In this scenario the main peculiarity of diffraction is dissipation of incident wavefront around a pore (figure 2). Pore material (fluid) has substantially lower impedance. Interaction of reflections and refractions caused by a single pore results in distinctive local attenuation of wavefront is observed in the figure. Figure 2 represents a shadow-graph of wave fields. Black and white colors are extrema of alternating damped sinusoid. Grey background in figure 2 indicates areas where there are no disturbances.

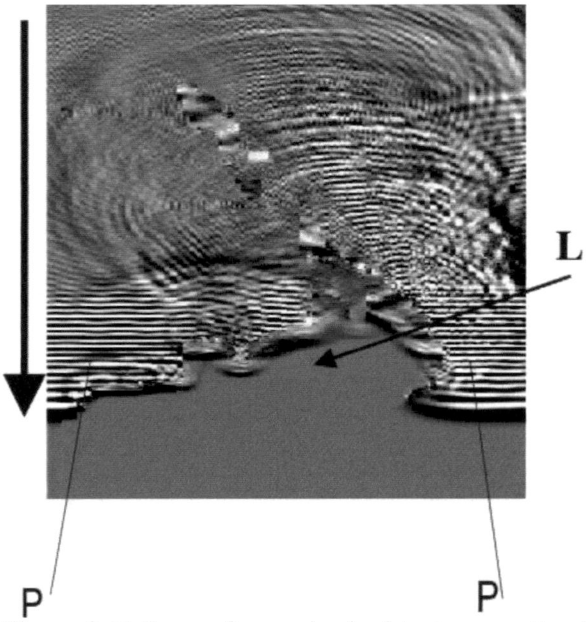

Figure 2. Full-waveform seismic data (wave pattern) resulting from impulse-pore interaction. An arrow indicates direction of P wave incidence representing a damped sinusoid. L indicates pore location.

At a later stage (figure 3 a, b) when a wave has travelled through the pore, reflected off the model boundary and travelled through the pore again, wave pattern is almost random. Small sections with P-wave trains can be distinguished (P indicates these areas in figure 3). As we compare our calculations with the results given in [6], figure 10 from the mentioned publication is presented in figure 3, b.

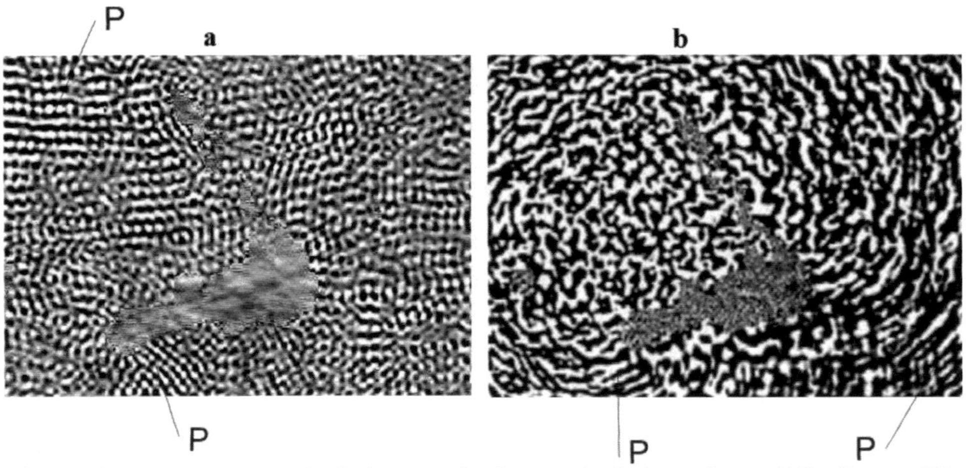

Figure 3. a – present calculation result, b – calculations from ([6], figure 10). P indicates observable P-wave trains in both calculations/

It is obvious that in both cases a pore is affected by short-period vibrations (black and white colors in the figures are wave extrema); pore margins are clearly traced in seismic profile. In figure 3, an event time corresponds to that of pseudo-random field formation which can be observed in [6]. Thus, solution to the first problem has indicated that a single pore can be displayed in a wave field due to contrasting properties, with apparent diffraction patterns being a constituent feature of the wave field. This is a kind of qualitative test for calculation program which has been further applied to the solution of porous layer problem.

Elastic wave incident in a medium which contains a layer with random arrangement of pores is being considered. Two porosity values (9% and 17 %) have been considered, since there were data of laboratory modeling for these values in publications [2]. The problem geometry is shown in figure 4.

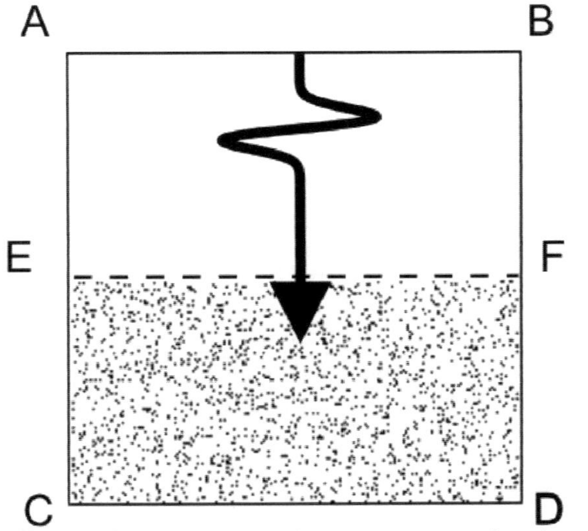

Figure 4. Geometry of wave propagation in a porous medium.

A plane compressional wave strikes AB boundary. The medium is homogenous above EF line, with aluminum parameters being Al-6061: Vp=6260 m/s, Vs =3080 m/s, ρ=2700 kg/m^3. Below EF line there is random porous medium (11% illustrated in the figure), with fluid characteristics being Vp=331 m/s, Vs =0 m/s, ρ=1.225 kg/ m^3

Contrarily to the previous problem, it is necessary to estimate wave extrema movement along the model, therefore, Riker impulse was employed as output signal.

$$\dot{U}_y|_{AB} = F(t) = -2\pi f\sqrt{e}\ (t-t_0)\cdot e^{-2\pi f(t-t_0)^2}$$

where, f – frequency, t_0 – 1/f – impulse period, $\dot{U}_y$ -Y-component of drift velocity

Pore-size distribution is presented in figure 5. Minimum size is 1 computational cell, whereas maximum size in case of 17% porosity is 10 cells.

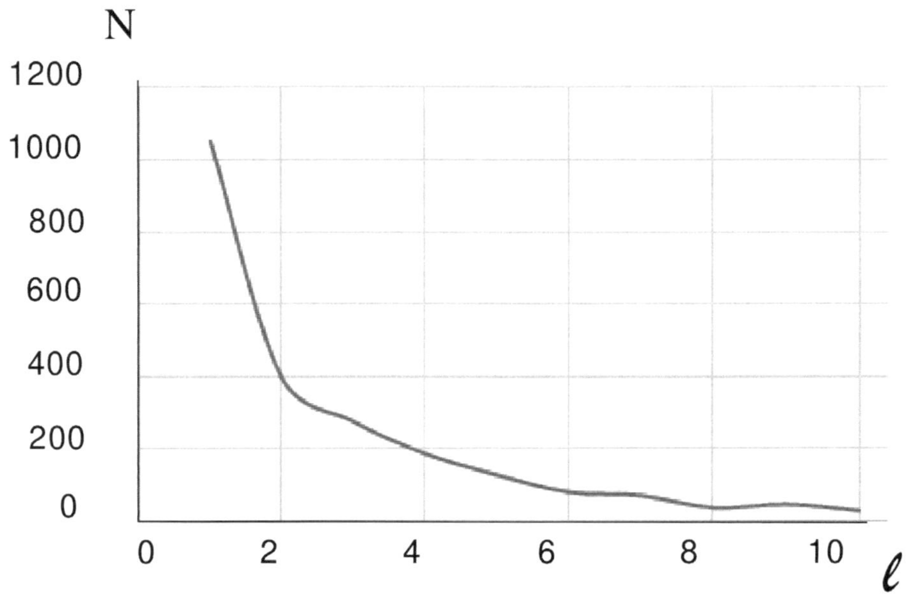

Figure 5. Pore-size distribution.

Quantitative testing algorithm for calculating elastic wave propagation in a porous medium has involved solving the problem in experimental analysis, results of which are available [2]. The mentioned experiment presents data on sintered aluminium mesostructure considering two cases: 9 % and 17 % porosity.

Pore size is about $20x10^{-6}$ m, acoustic frequency is 5MHz, wavelength ≈ 0.1 cm. Thus, pore size is merely 50 times less than the wavelength, therefore, such pores can be described by several computational cells. According to experimental results in [2] on aluminium medium with 9 % porosity, average velocity of compressional wave propagation is V_P =5125 m/s, and V_P =4120 m/s for 17% porosity sample.

Numerical modeling of plane compressional wave propagation has been carried out regarding these porosity values. During modeling average velocity of P-wave propagation was determined on the basis of the slope of time-distance graph. As matrix is dry in a medium, pores were considered to be filled with air k^{air} = 0.00014 hPa, ρ^{air} = 1.2 kg/m^3.

The digital images of seismic field throughout the model at two time points can be observed in figure 6. The record indicates that at 40 ms time point incident wave front has reached porous section of the model, while at the time point of 60 ms the wave has propagated through the calculated area. The result of pore ensemble diffraction can be clearly seen around certain pores.

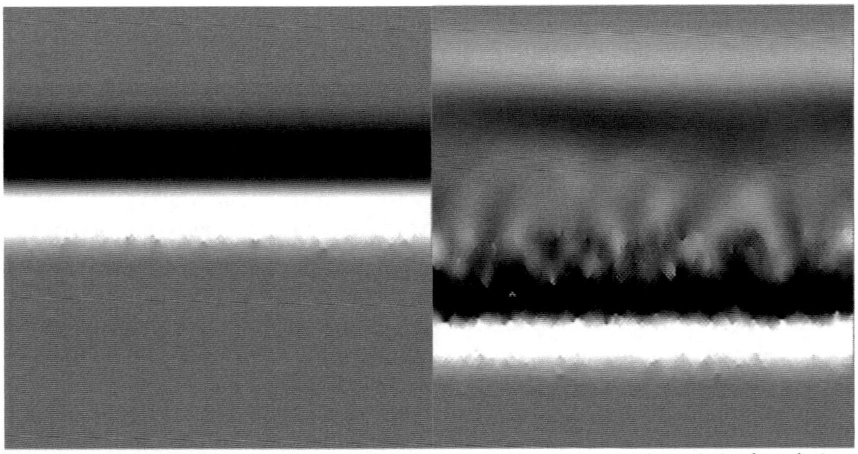

Figure 6. The left-hand shot corresponds to 40ms time point; right-hand shot – 60 ms time point.

Calculations have shown that in the case of 9% porosity compressional wave velocity is V_P=5123 m/s, while porosity at 17 %, compressional wave velocity is V_P =4074 m/s.

4. Conclusion

Thus, the numerical modeling of velocities in porous aluminium, based on the results of above-mentioned procedures differed from velocities calculated in the experiment (medium porosity 9 %) by 0.04%; whereas porosity being 17 %, discrepancy is 1.1 %.

Based on numerical modeling, the effect of porous medium on seismic field has been studied, with a single pore and pore ensemble effects being involved. In the case of a single pore simulation, calculations correspond to results of researches published by other authors. Model of medium with porous layer being produced, analysis of numerical modeling results made it possible to estimate average velocities and compare them with the data of physical modeling. It has been shown that calculations yield results are sufficiently consistent which, in its turn, is proved by qualitative (single pore simulation) and quantitative (pore ensemble simulation) testing.

References

[1] Pyrak-Nolte L J 2007 Fracture anisotropy: the role of fracture-stiffness gradients *The Leading Edge* **26**(9) 1124–27

[2] Bonnan S, Hereil P L and Collombet F 1998 Experimental characterization of quasi static and shock wave behavior of porous aluminum *Journal of applied physics* **83**(11) 5741–49

[3] Li X, Zhong L, Pyrak-Nolte L J 2001 Physics of Partially Saturated Porous Media: Residual Saturation and Seismic-Wave Propagation *Annu. Rev. Earth Planet. Sci.* **29**:419–60 419–60

[4] Nemirovich-Danchenko M M and Stefanov Y P 1995 Primenenie konechno-raznostnogo metoda v peremennykh Langranzha dlya chislennogo rascheta volnovykh polej v slozhnopostroennykh sredakh *Geology and Geophysics* **11** 96–105

[5] Melnikova N A and Nemirovich-Danchenko M M 2006 Metodika pascheta uprugykh volnovykh polej dlya mesoobema, soderzhashego neskolko raznoorientirovannykh crystallov s razlichnymi svojstvami *Physical Mesomechanics* **9**(1) 103–10

[6] Endar B, Nurhandoko B, Wardaya P D, Adler J, and Siahaan K R 2012 Seismic wave propagation modeling in porous media for various frequencies: A case study in carbonate rock Citation: *AIP Conference Proceedings* 1454 109; doi: 10.1063/1.4730699

Petrophysical zoning elements of Chertovo Koryto gold-ore deposit (Patom Upland, Eastern Siberia)

S V Sokolov[1,3], Y V Kolmakov[1] and D A Terre[2]

[1] Department of Geophysics, Institute of Natural Resources, National Research Tomsk Polytechnic University, 30 Lenin Ave., Tomsk, 634050, Russia
[2] Department of Foreign Languages, Institute of Natural Resources, National Research Tomsk Polytechnic University, 30 Lenin Ave., Tomsk, 634050, Russia

E-mail: [3] sokolovsv@sibmail.com

Abstract. The paper considers magnetic susceptibility (χ) and electrode potentials (EP) of rocks in the Chertovo Koryto deposit. Carbon-bearing substance is found in all the studied samples, but in some cases, this substance supplies EP (-150 ÷ -400 mV). In these samples χ rarely exceeds $40 \cdot 10^{-5}$ SI units, while, in other samples χ is 8-10 (up to 30) times higher. Less intensive EP (-20 ÷ -240 mV) is furnished due to the sulfides in this deposit. Rocks with polarized carbon-bearing substance do not contain magnetic pyrrhotine and are negative linear EP anomalies. Rocks in which carbon-bearing substance is associated with pyrrhotine are revealed as magnetic anomalies. The adjacent rocks determine petrophysical zoning of the Chertovo Koryto deposit. The combination of negative linear EP anomalies and magnetic anomalies is a potential indicator and can define the multi-stage formation of the deposit itself.

1. Introduction

Central Patom Upland hosts Bodaibo ore-placer area where gold-mineralization has been explored completely. Today this area embraces precious metal reserves of thousands of tons including the unique Sukhoi Log deposit. There is the underexplored Tonodsk ore-placer area in the northern Patom Upland. Being currently placed under production, Chertovo Koryto deposit, containing more than 100-ton reserves, confirms the gold mineralization potential of this area (figure 1). The deposit is located in carboniferous shales of Kevaktinsk ore zone where numerous gold placers, the ore bodies of which, are still to be detected.

Efficiency of the magnetic survey in Chertovo Koryto deposit [3] makes it possible to apply geophysical methods in prospecting similar deposits.

The article presents data on the ore occurrences in the Chertovo Koryto deposit via natural electric and magnetic fields. Investigating the matter within this deposit makes it possible to identify the petrophysical zoning of the deposit and explain the structure and disposition of its geophysical anomalies.

2. Geological structure

The extended ores in the Chertovo Koryto deposit are located in Low-Proterozoic terrigeneous-sediment formations of Mikhailovsk suite, which, in its turn, includes carbon-bearing substance content of up to 2%. Metasandstones and metasiltstones account for up to 90 % of the host rocks and

IOP Conf. Series: Earth and Environmental Science **27** (2015) 012028 doi:10.1088/1755-1315/27/1/012028

according to transformation grade are referred to as initial alterations of albite-epidote facies regional metamorphism [2].

The gently sloping structure, hosting the ore deposit, is confined to the innermost part of the sublateral Mikhailovsk syncline; while eastward to high-angle fault interarea which could have been an ore conduit. These displacements trending west -south-west plunge at the angles of 15-20° and 60°, respectively.

Basic ore deposit minerals are pyrite, pyrrhotine, arsenic pyrite, while galena, sphalerite, chalcopyrite are of smaller proportions. The ground mass of sulfides formed as inclusions and veinlets are concentrated in aposhale metasomatites, the rear zones of which are associated with the beresite formation. Gold is associated with sulfides and occurs often in quartz.

3. Research methods

The natural electric field (EF) of Kevaktinsk ore zone, including Chertovo Koryto deposit, was surveyed on the scale of 1:10000 during 1978-1980 under the guidance of M.M.Bazhenov. In 2006 magnetic survey with proton magnetometers was carried out, applying the above-mentioned survey grid (figure 1).

To understand and explain the relationship between magnetic field (ΔT) and EF anomalies, 139 core samples from exploratory wells № 301-308 and 190-199 were selected study the physical properties (figure 1). The electrode potential (EP) and magnetic susceptibility (χ) was also studied. EP measurements were conducted in sulfides and carbon-bearing substance to a copper unpolarized electrode. All the samples contained carbon-bearing substance. Visible sulfide grains suitable for EF observations were found in 23 samples.

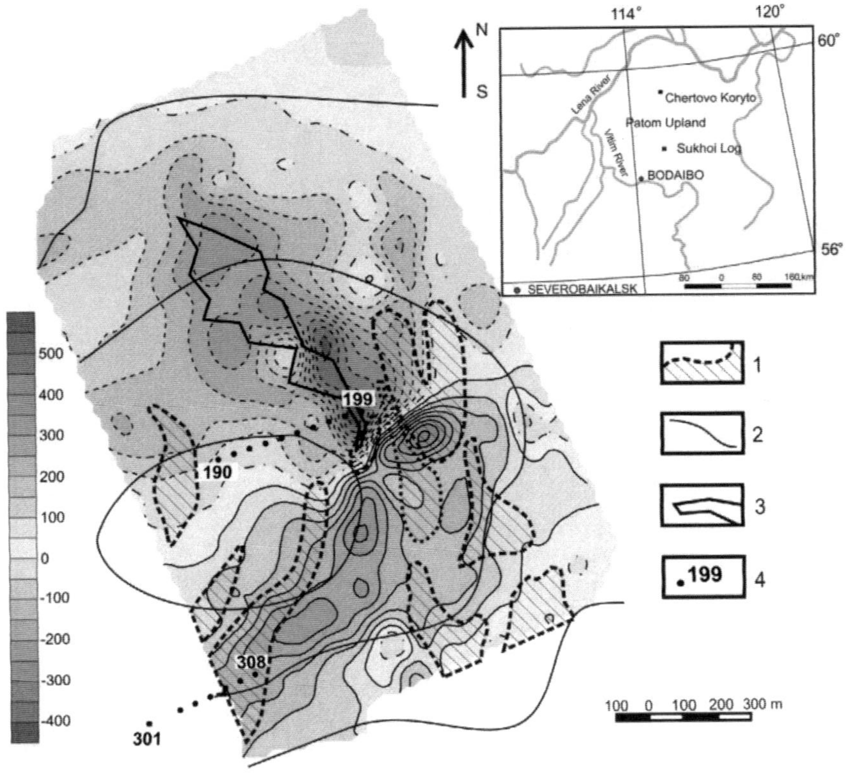

Figure 1. Scheme of magnetic anomaly field (ΔT, nT) in Chertovo Koryto deposit: 1 – anomalies of natural electric field (-300 ÷ -850 mV); 2 – contours; 3 – ore body outcropping (C_{Au}>1g/ton); 4 – wells.

4. Research results

With reference to carbon-bearing substance EP was detected only in 42 samples from 301-307 wells. Their values vary within -150 ÷ -400 mV. χ values of these samples seldom exceed $40 \cdot 10^{-5}$ SI units (figure 2 A, B). In the remaining samples, containing unpolarized carbon-bearing substance, χ is 8-10 (up to 30) times higher.

The most intensive EP was detected in pyrite comparable to other sulfides. Their values range from -100 to -245 mV (figure 2, D). Pyrrhotine and arsenic pyrite occurrences furnish EP values within -20 ÷ -100 mV range.

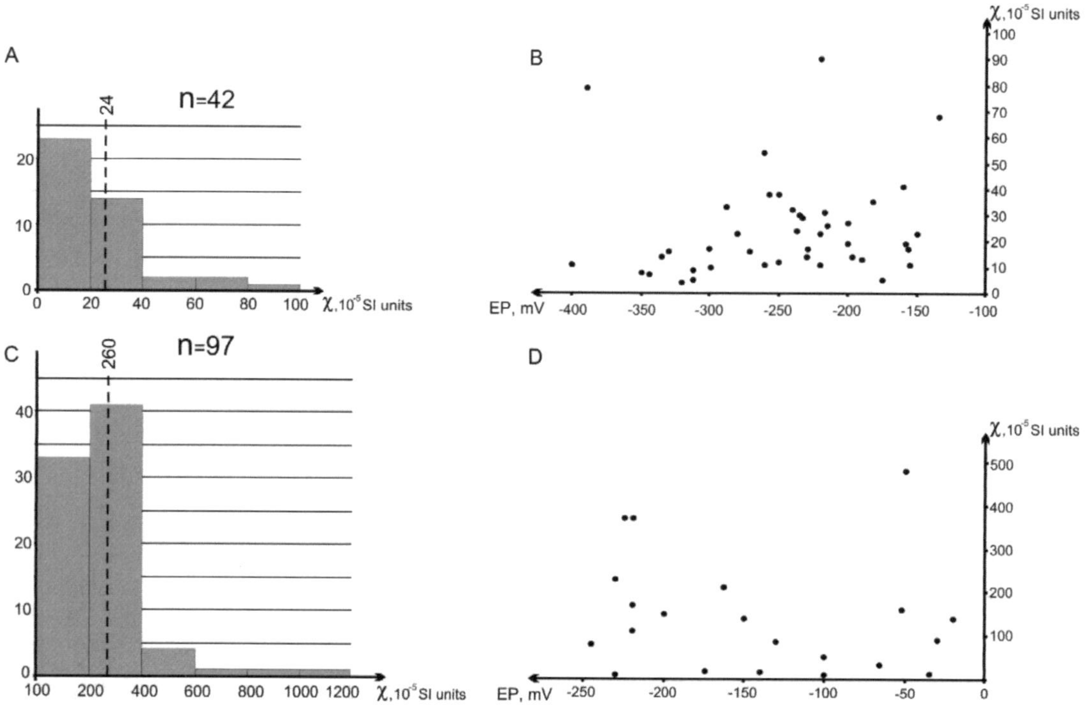

Figure 2. χ and EP measurement results in samples with polarized (A, B), unpolarized (C) carbon-bearing substance and visible sulfides (D). Sample number and average χ values in this selection (dotted line) are indicated at the top of the bar charts.

5. Discussion

The χ value of Chertovo Koryto deposit rocks is directly associated with the concentration of pyrrhotine - the only ferromagnetic ore mineral. It is obvious that it is observed in samples with polarized carbon-bearing substance and, vice versa- in samples where carbon-bearing substance does not produce EP.

Intensive magnetic anomaly is evident on the surface of pyrrhotine concentrations. Pyrite, arsenic pyrite and other sulfides are distributed comparatively the same as pyrrhotine within this deposit. Sulfide concentration is rather high in commercial gold deposit zones, however, polarized carbon-bearing substance is absent (samples from 190-199 wells) and, consequently, there are no negative EP anomalies.

Petrophysical measurements of 1-2meter-interval cores (301-307 wells) indicate the fact that there are cases of alternating non-magnetic samples with lenses of polarized carbon-bearing substance and magnetic samples with unpolarized carbon-bearing substance. This could explain the fact why linear negative EP anomalies in the southern and eastern parts of the deposit are adjacent to magnetic anomalies.

Thus, a distinctive petrophysical zoning can be observed in this deposit, which involves areal isolation, and contiguous two carbon-containing rock types: 1) rocks with intensive EP to carbon-bearing substance and low χ value; and 2) rocks with zero EP to carbon-bearing substance and high χ value. The same tendency can be traced in EF and ΔT anomalies. Although they are contiguous, their epicenters are not coincident (figure 1).

The main source of above-described regularities could be the multi-stage formation of the Chertovo Koryto deposit itself. Apparently, polarized carbon-bearing substance was formed during the pre-ore stage. Later, during the ore stage beresitization occurred due to the intensive hydrothermal solubility in ore-hosting thicknesses. Carboniferous minerals with dielectric properties are typical of this process, overlapping the polarized carbon-bearing substance and disintegrating the aggregates at the micro-level. As a result, beresitization responses to a conductivity loss and further becomes depolarized. In this case, commercial gold mineralization zone are more intensively subjected to these processes. The southern and eastern field areas were less affected and carbon-bearing substance was preserved retaining its electrochemical activity. This can also be observed in Blagodatnoe gold-ore deposit located within Kordinsk suite carboniferous shales (Yenisei Ridge) [1].

6. Conclusion

Petrophysical zoning of Chertovo Koryto deposit is apparently due to its multi-stage formation. The combination of linear negative EP anomalies and magnetic anomalies is a significant indication of ore grade mineralization similar to the ore types of Chertovo Koryto deposit and this fact can be applied in mineral exploration.

References

[1] Kolmakov Yu V 2014 Blagodatnoe gold-sulfide deposit (Yenisei Ridge, Russia): the nature of geophysical anomalies and the succession and causes of formation of petrophysical zoning. *Russian Geology and Geophysics* **55** (11) 1321–32

[2] Kucherenko I V, Gavrilov R Yu, Martynenko V G and Verkhozin AV 2009 Strukturno-dinamicheskaya model zolotorudnykh mestorozhdeniy obrazovannykh v neslantsevom i chernoslantsevom substrate -Mestorozhdenie Chertovo Koryto. *Bulletin of the Tomsk Polytechnic University* **314** (1) 23–38

[3] Sokolov S V, Kolmakov Yu V and Gusev E V. 2011 Tektonicheskaya obuslovlennost raspredeleniya sulfidnoj mineralizatsii i magnitnoj anisotropii porod na zolotorudnom mestorozhdenii Chertovo Koryto *Subsurface prospecting and protection* **12** 19–26

Impact of oil on groundwater chemical composition

N N Brakorenko[1]

[1] National Research Tomsk Polytechnic University, 30 Lenin Ave., Tomsk, 634050, Russia

E-mail: brakorenko@sibmail.com

Abstract. The objective of the paper is to characterize the chemical composition of groundwater samples from the monitoring wells drilled in the petrol station areas within the vicinity of Tomsk. The level of contamination has increased since many macro – and microcomponent concentrations (such as petroleum products, chlorine, sulphates, carbon dioxide and lead, etc.) in groundwater samples of the present study is higher than that in previous period samples.

Introduction

Currently, in the territory of Tomsk there are 119 petrol stations. The intense construction of filling stations, oil refineries, and oil depots within urban areas caused one of the major environmental problem, which is pollution of the geological environment with petroleum products (PP). A great number of works addressed the issue of assessing the degree of groundwater contamination with oil products. However, these studies do not have information on the hydrogeochemical analysis of the contaminated groundwater. In this regard, the assessment of the impact of petroleum products on the chemical composition of the groundwater is topical.

Materials and methods

To clarify the characteristics and the hydrogeochemical state of the groundwater contaminated with oil products, the materials obtained in the course of the present study were analyzed. The author studied chemical composition of the groundwater, sampled from the monitoring wells drilled in the areas of the filling stations in the city of Tomsk.

The background data used for comparison with the present study results are the average values of components contained in the groundwater of the river Tom floodplain, terraces, and watershed, which are represented in the materials by Tomskgeomonitoring [2, 3] and the paper by E.M. Dutova et al. [4].

Results and discussion

The values of major elements concentrations in contaminated groundwater are shown in tables 1 and 2. In the course of the research it was found out that the values of Quaternary horizon waters exceeded the limit values of oil products in all studied samples.

The maximum level of pollution (>7.0 mg/dm^3) is characteristic of the oldest petrol station, namely, petrol station 3 and petrol station complex 28, with service life more than 30 years. Moreover, the PP concentration in the groundwater is determined by the lithological composition of the rocks composing the section of the unsaturated zone, as well as by the depth of the groundwater level, which ensures a certain of protection. According to the method developed by V. M. Goldberg, protection of the groundwater in the areas of filling stations corresponds to categories I–III (from the lowest to medium).

In some samples, the values of lead and sodium exceeded the limit concentrations (table 2).

In case the petrol station and petrol complex were put in line after 2000 (for example, petrol complex 50 works since 2002), the content of lead in the groundwater is less than the maximum contaminant levels, which can be explained by the fact that since 1 July 2003, according to the national law, the production and distribution of leaded petrol are banned in the territory of the Russian Federation.

The results obtained are significantly different from the values characteristic for the groundwater of the river Tom floodplain, terraces, and watershed (beyond the areas of filling stations) (table 1). In the waters contaminated with oil products, pH values decreased in comparison with those in previous research samples, which signifies that the acidity increased. Chloride, sulphate, and free carbon dioxide are the components whose concentrations in the present samples are significantly higher than in the previous research ones.

Table 1. Groundwater chemical composition.

Components		Content (minimum – maximum)			
		Floodplain and river terraces		Watershed	
		Background data (number of samples)	Contamination with PP	Background data (number of samples)	Contamination with PP
pH	mg/dm^3	7.2 (1262)	6.68-6.76	7.64 (488)	5.91-6.15
Na^+	mg/dm^3	7.6 (1249)	4.15-46	8.5 (483)	5.27-40
K^+	mg/dm^3	1.11 (137)	2.23-5.43	0.94 (87)	2.22-5.1
Ca^{2+}	mg/dm^3	28.6 (1264)	60-148	44.9 (487)	18-24.2
Mg^{2+}	mg/dm^3	7.3 (1240)	9.76-80.52	13.3 (483)	1.22-3.54
NH^{4+}	mg/dm^3	0.65 (1151)	1.58-5.69	0.7 (473)	0.53-2.44
Cl^-	mg/dm^3	5.3 (1263)	19.5-101.2	3.5 (482)	19.52-62.5
SO_4^{2-}	mg/dm^3	4.51 (513)	<2-27.6	2.69 (92)	<2-2.04
NO_2^-	mg/dm^3	0.3(170)	0.01-0.025	0.2 (91)	0.025-0.112
NO_3^-	mg/dm^3	2.0 (269)	0.51-2.46	1.0 (57)	0.32-0.37
HCO_3^-	mg/dm^3	158.6 (1265)	180-810	274.6 (487)	42.7-98
$CO_{2\ СВ}$	mg/dm^3	17.1 (40)	132-176	19.6 *	35.2-61.6
Total hardness	$mg\text{-}EQ/dm^3$	3.0(1256)	3.8-14	4.4 (486)	1.0-1.5
Mineralization	mg/dm^3	97.5(1256)	303-1191	224 (461)	91-233

Conclusion

The results of the present study show that the groundwater in the areas of filling stations is contaminated with oil products due to poor environmental protection.

The concentrations of many components (such as petroleum products, chlorine, sulfates, free carbon dioxide and lead, etc.) in the present samples are significantly higher than those in the previous research samples.

Today, the monitoring wells are in the areas of 21 petrol stations and complexes, out of total 119 located in Tomsk. It is recommended that monitoring wells should be drilled in the areas of all Tomsk filling stations for groundwater sampling no less than twice per year.

Table 2. Chemical composition of groundwater contaminated with petroleum products (present study data).

Components	Petrol station №28	Petrol station №37	Petrol station complex №4, well H-1	Petrol station complex №4, well №9	Petrol station complex №4, well №6	Petrol station complex №4, well №3	Petrol station №60	Petrol station complex №3	Petrol station complex №50	Petrol station №4	Maximum contaminant level (in compliance with Sanitary rules and regulations 2.1.4.1074-01)
The index of the aquifer	$^{1}Q_{III}$	$Q_{I-II}fd$	$^{3}Q_{III}$				Q_{IV}	$Q_{E}kc$	$Q_{E}kc$		
pH	6.68	6.27	6.48	7.57	7.59	7.81	6.76	5.91	6.15	8.67	6-9
Electrical conductivity, mS/cm	0.42	0.128	0.338	0.244			1.292	0.301	0.1		
Bicarbonate – ion, mg/dm^3	180	61	165	166	315	335	810	98	42.7	480	
Carbonate ion, mg/dm^3	<3	<3	<3	<3	<3	<3	<3	<3	<3	36	
Carbon dioxide, mg/dm^3	176	35.2	48.4	3.52			132	61.6	37.84	<3	
Ammonium ion, mg/dm^3	5.69	0.1	0.33	0.066	0.57	0.04	1.58	2.44	0.53	0.41	
Nitrite ion, mg/dm^3	<0.01	0.037	<0.01	0.25	0.19	2.43	0.025	0.025	0.112	0.09	3.0
Nitrate ion, mg/dm^3	2.46	0.25	0.14	0.59	9.2	4.04	0.51	0.37	0.32	1.41	
Sulfate ion, mg/dm^3	27.6	2.23	<2	<2	90	27.6	<2	<2	2.04	<2	
Chloride ion, mg/dm^3	19.5	19.5	64	23.1	510	49	101.2	62.5	19.52	254	
Phosphate ion, mg/dm^3	0.03	<0.01	<0.01	0.01	0.003	0.005	0.014	<0.01	<0.01	0.01	
Total hardness, mg-EQ/ dm^3	3.8	1.1	3.7	1.4	6.6	2.44	14	1.5	1.0	8.3	7.0
Calcium, mg/dm^3	60	20	44	24	95	39	148	24.2	18	80	
Magnesium, mg/dm^3	9.76	1.22	18.3	2.44	22.73	7.83	80.52	3.54	1.22	52.46	
Sodium, mg/dm^3	4.15	10.3	17.1	40	330	111	46	40	5.27	197	200
Potassium, mg/dm^3	2.23	1.33	2.5	9	19	7.34	5.43	5.1	2.22	4.1	
Iron, total, mg/dm^3	681	124.5	69.1	31.65	18.87	257.7	137.7	55.86	27.2	46.55	0.3
Mineralization on the amount of salts, mg/dm^3	303	116	311	264	1382	574	1191	233	91	1104	1000
Oil, emulsion, mg/dm^3	2250	1.75	0.54	0.298	0.601	0.915	0.833	7.41	4.8	27.5	0.1
Oil, vol %	7.98							8.42			
Lithium, mg/dm^3	0.003	0.002	0.007	0.001	0.0012	0.0009	0.005				0.03
Zinc, mg/dm^3	0.22	0.055	0.5	0.22	0.11	0.58	0.36	0.089	0.068	0.26	5.0
Cadmium, mg/dm^3	<0.0002	<0.0002	0.00021	<0.0002	<0.0002	0.00022	0.00056	<0.0002	<0.0002	0.0009	0.001
Lead, mg/dm^3	0.21	0.018	0.05	0.038	0.0045	0.029	0.076	0.0045	0.0059	0.0092	0.03
Copper, mg/ dm3	0.52	0.029	0.11	0.06	0.022	0.19	0.056	0.2	0.024	0.57	1.0

References

[1] Chetvergov D N 2009 *Monitoring the condition of natural resources in the territory of Tomsk Oblast'. Newsletter 2008* (Tomsk: Tomskgeomonitoring) **14** 22

[2] Chetvergov D N 2010 *Monitoring the condition of natural resources in the territory of Tomsk Oblast'. Newsletter 2009* (Tomsk: Tomskgeomonitoring) **15** 9

[3] Dutova E M, Kuzevanov K I, Pokrovsky D S, Nalivayko N G, Bezrukova O A and Fedorova V P 2000 Chemical and microbiological composition of groundwater and surface water sources in the territory of Tomsk *300 years of mining and geological service of Russia: Regional conference of the geologists from Siberia, Far East and North-East of Russia: Materials (Tomsk, 18–23 September 2000)* **1** 373–75

The Relationship of Physical Property Indicators and Clay Soil Structural Strength of Tomsk Oblast Territory

V V Kramarenko[1], A N Nikitenkov[2], V Y Molokov[3], I A Matveenko[4] and A V Shramok[5]

[1] Associate professor, National Research Tomsk Polytechnic University, 30 Lenin Ave., Tomsk, 634050, Russia
[2] Senior Teacher, National Research Tomsk Polytechnic University, 30 Lenin Ave., Tomsk, 634050, Russia
[3] Post-graduate, National Research Tomsk Polytechnic University, 30 Lenin Ave., Tomsk, 634050, Russia
[4] Associate Professor, National Research Tomsk Polytechnic University, 30 Lenin Ave., Tomsk, 634050, Russia
[5] Student, National Research Tomsk Polytechnic University, 30 Lenin Ave., Tomsk, 634050, Russia

E-mail: [1] kramarenko-v-v@mail.ru; [2] nik@tpu.ru; [3] vik3011347@yandex.ru; [4] mia2046@yandex.ru

Abstract. The article deals with the characteristic of initial condition in fine-grained soils – its structural strength – p_{str}. Estimation and measurement of this factor at soil testing are of primary importance for defining its physical and mechanical properties as well as for subsequent calculation of foundation settlements that is insufficiently covered in Code of practice, national standard and inefficiently applicable in practice of engineering geological investigations. The article reveals the relationship between soil physical property, its occurrence depth, which will make possible to forecast p_{str} over the given territory.

1. Introduction

By structural strength p_{str} we mean the strength conditioned by the presence of structural bonds and characterized by pressure up to what soil sample is not or lightly deformed at its vertical loading [10]. It is one of the characteristics defining the soil initial conditions in natural environment that is of great significance for determination of its physical-mechanical properties, state of consolidation of sediments, calculation of foundation settlements and estimation quality of selection sample with undisturbed structure. It is obvious that clay soil with mixed microstructure has a pronounced discontinuity point related to their structural strength. Consequently, at low loads, coefficient of volume compressibility equals the hundredth parts of MPa $^{-1}$, whereas at higher loads its values increase to the tenth parts of MPa $^{-1}$. Structural strength influences the soil shear parameters as well. At loads increasing its angle of internal friction and cohesion sufficiently decrease in case of undrained shear strength, particularly. The value of residual strength amounts not less than 30-40 %. There often occur week or overconsolidation soils among them. The former are developed due to early cementation, but the latter are formed at lithified rock tipping with partial content of cement phase

[23]. According to [3] "… *all misunderstandings of soil mechanical properties and calculations based on them are connected with complete neglecting the initial condition of soil and reluctance of recognition that soil bearing capacity is defined by its initial state*". «The *initial conditions* can be established in terms of two sets of parameters: (a) indices; and (b) state parameters. *The index parameters provide a measure on the types, shapes, ranges, and descriptive values on the particles that make up the composition. Indices can be performed on remoulded or disaggregated samples from the field. The state parameters represent quantification on how the particles are arranged or packed with respect to each other, thus leanding to concepts such as "loose" vs. "dense" and "soft" vs. "hard", as well as special facets concerning bonding, cementation, fracturing, sensitivity and structural fabric. The initial state reflects the long-term geostatic conditions, often over many thousands to millions of years, and the value of these state parameters is given a subscript "nought" to indicate its beginning, such as initial void ratio (e_0), initial effective overburden stress (σ_{v0}'), initial hydrostatic porewater pressure (u_0), and initial lateral stress state (K_0)*» [16].

The value of soil structural strength depends on age, lithification, genesis, structural bonds and disturbances, such as loading or sampling. It is believed that the least structural strength is typical for relatively young clay soils with weak coagulative contacts, whereas the highest one is inherent to relatively old lithified ones with strong mixed and phase cement contacts that to some extent reflect the basic characteristics of soil physical properties such as density, dry soil density, and moisture. As a result, relationship of p_{str} with soil composition and dependence of physical properties from soil stress history and formation conditions is of great interest for researchers. Therefore, the purpose of the given article is to estimate state of consolidation and reveal the relations of structural strength and characteristics of physical property of cohesive soils over the territory of Tomsk Oblast to forecast p_{str} preliminarily.

The research tasks include sampling, laboratory investigation of main soil physical and mechanical properties, and statistic processing of the results obtained to verify the relationships between soil composition and its properties.

2. Methods

Soil testing was made in accordance with the methods given in the Russian national standard [8, 10, 11]. Sampling was performed in 60 points; they were mostly located in the central, most developed part of Tomsk Oblast as well as its under-developed Northern and Northern-West regions. Sampling was carried out up to the depth of 230 m from sediments of Quaternary system, Paleogene and Cretaceous periods. In total about 220 samples of clay soils were treated. The following parameters were defined: grain size composition, density *(ρ)*, solids density (ρ_s), dry soil density (ρ_d), water content (w), plastic and liquid limits (w_L and w_p), deformation and strength characteristics; the following characteristics were calculated: porosity, void ratio, degree of saturation, plasticity and liquidity indexes, over consolidation ratio *(OCR)* and others.

Structural strength obtained in compression tests is defined in the same way as pressure of preconsolidation stress σ_p' by the methods of Casagrande, Burmister, Schemertmann, Akai, Janbu, Christensen, Sellfors, Pacheco Silva, Butterfield, Tavenas, Oikawa, Jose, Sridharan, Burland, Jacobsen, Van Zelst, Becker, Lebert and Horn, Senol, Saglamer, Nagaraj, Thøgersen, Einav, Carter , Gregory, Grozic, Imhoff, Dias Junior and others [1, 2, 4, 5-7, 9, 10, 12-14, 17-33, 24]. The most common ones are graphical techniques of Casagrande, Butterfield, Becker, Strain Energy-Log Stress, Sellfors and Pacheco Silva; in Russia, the Casagrande method is widely applied. There are a lot of interesting papers dealing with existing methods of estimation of the preconsolidation stress, history of mechanical stress, factors and characteristics defining state of consolidation of sediments [23-25].

Factor p_{str} determined in this work by the standard technique described in GOST 12248 [10]. In compression test for determination of p_{str} the first and subsequent stages of loading were taken as equal to 0,0025 MPa up to the compression point of soil sample, for which relative vertical soil deformation is taken ε >0,005. *Structural strength* of soil was defined in terms of the initial portion of compression curve $e_i = f(lg\ \sigma')$, where e_i – the void ratio at load σ_i. The break point after initial linear portion of curve corresponds to soil structural compression strength. Graphical processing of the results was performed by the traditional methods of Casagrande and Becker, as well.

3. Results and discussion

The research was made for clay soil of mostly solid and semisolid consistency. The results showed that in general dry density changes from 1,16 to 1,77 g/cm^3 for lean soil; from 1,18 to 1,65 g/cm^3 for fat and silty clay increasing with depth. On average, solid particle density for all types amounted 2,68-2,69 g/cm^3, at minimum 2,50-2,58 g/cm^3 and maximum 2,83-3,00 g/cm^3 and more – for rock and clay soil cemented with iron oxides and hydroxides. The statistic results showed that the values of structural strength change depending on the magnitudes close to 0 kPa to 300 kPa and higher – for rock and solid clay. Distribution of p_{str} depends on grain size composition (hydrometer analysis was performed for 60 samples) and the differences of clay soil are presented in figure 1. It is seen that the maximum values are obtained for silty clay: 130 kPa on average (at maximum - 177 kPa); on the whole the values 93-95 kPa prevail for fat and lean clays at maximum 170-188 kPa.

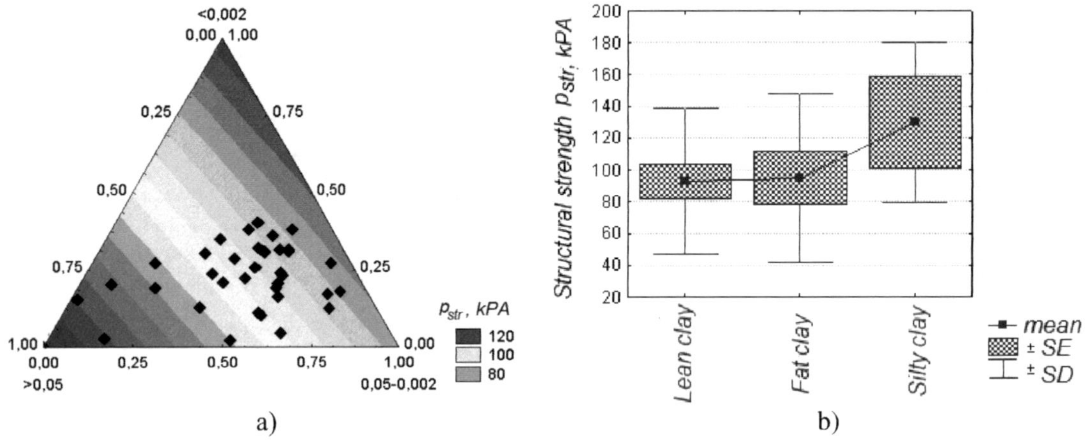

Figure 1. Distribution values of structural strength depending on a) grain size composition; b) different clay soils/

OCR change analysis with depth revealed that soils lower 20 m are normally consolidated; it should be noted that in the upper part of cross-section the *OCR* values range is wider – there occur overconsolidation soils. Such a picture is typical, since aerated zone soils occur in solid and semisolid state, whereas with increase water content, maximum water capacity and degree of saturation, the structural strength decreases, which was supported by correlation analysis.

The closest relationship was noted between structural strength and density parameters (figure 2): p_{str} and dry soil density ρ_d (the ratio coefficient r = 0,53); p_{str} and solids density ρ_s (r = 0,50); p_{str} and density ρ (r = 0,41); as well as between the following characteristics: p_{str} and water content w (r=-0,48); p_{str} and maximum water capacity w_n (r = -0,43); p_{str} and liquidity index I_L (r=-0,42). The insignificant relationships were obtained for plasticity index I_p, liquid limit w_L and plastic limit w_p. In correlation analysis the rock data were not used. But, as the graph shows, some samples had high density and structural strength values a well. Based on the stated relationships, the following equations of regression were obtained allowing prediction of structural strength (in kPa) depending on soil density and moisture:

p_{str} =211,6·ρ- 301,4;
p_{str} = 210,8·ρ_s - 465,5;
p_{str} = 195,5·ρ_d-187,7;

$p_{str} = 222{,}7 - 4{,}1 \cdot w;$
$p_{str} = 221{,}6 - 387{,}8 \cdot w_n.$

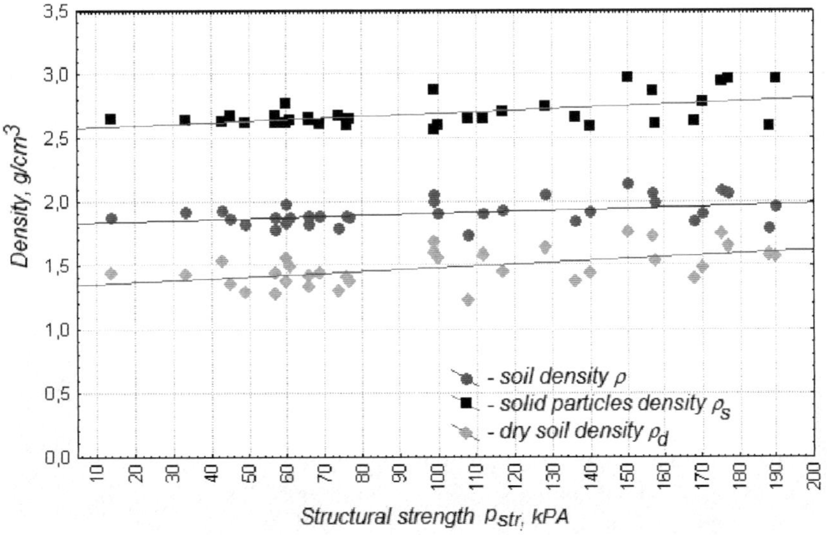

Figure 2. Dependence of soil density from its structural strength.

Besides, the relationship of indicator change with depth h ($r = 0{,}79$) was revealed.

$p_{str} = 19{,}6 + 0{,}62 \cdot h.$

According to [15], it is enough for one value of σ_p' to determine *OCR,* but to obtain characteristics of deformation necessary for settlement calculation at spotting the curve straight line section before and after p_{str}. It is desirable to obtain the two key points describing minimal $p_{str/min}$ and maximal $p_{str/max}$ structural strength. These can be points of tangent separation to the initial and final curve sections, or additional points obtained at processing of compression curves by Casagrande, Sellfors and Pacheco Silva methods [7, 19, 20]. As a guideline for research of deformation parameters we can recommend to determine soil physical properties as: void ratio, density, and water content, which correspond to minimum and maximum structural strength.

3. Summary

The results of laboratory research in composition, structural strength, and physical soil properties of Tomsk Oblast have permitted us to characterize their changes numerically and evaluate the state of soil consolidation up to 230 m depth. The *OCR* change analysis showed that soils lower 20 m are normally consolidated, i.e. structural strength does not exceed effective pressure, in the upper part of the cross-section there occur both week and overconsolidation soils. The stated closest relationships between p_{str} and soil density and moisture parameters allowed the regression equations to forecast the structural strength values. Such a forecast is useful for its express evaluation before soil shear and consolidation test for the given area, as well as estimation quality of undisturbed samples for laboratory testing.

References

[1] Akai K 1960 *Die structurellen Eigenshaften von Schluff* Mitteilungen, Heft 22 (Die Technishe Hochchule, Aachen) p 116

[2] Becker D B, Crooks J H A, Been K and Jefferies M G 1987 Work as a criterion for determining in situ and yield stresses in clays *Canadian Geotechnical Journal* **24** 549–64
[3] Bogdanov E N 2012 About mechanical properties of clay soils *Soil* **1** 64–72
[4] Burland J B 1990 Thirtieth Rankine Lecture: On the compressibility and shear strength of natural clays *Géotechnique* **40** 327–78
[5] Burmister D M 1951 The application of controlled test methods in consolidation testing. Symfosium on Consolidation testing of soils *ASTM*. STP **126** 83–98
[6] Butterfield R 1979A natural compression law for soils (an advance on e–log p') *Geotechnique* **24** 469–79
[7] Casagrande A 1936 The determination of the preconsolidation load and its practical significance *Proc. First Int. Conf. Soil Mechanics and Foundation Engineering* (Harvard Printing Office, Cambridge, Mass, UK) vol 3 pp 60–64
[8] Code of practice 47.13330.2012 Engineering survey for construction. Basic principles.
[9] Dias Junior M S and Pierce F J 1995 A simple procedure for estimating preconsolidation pressure from soil compression curves *Soil Technology* **8** 139–51
[10] GOST 12248-2010. Soils. Laboratory methods for determining the strength and strain characteristics.
[11] GOST 30416 -2012. Laboratory testing. General requirements.
[12] Gregory A S et al. 2006 Calculation of the compression index and precompression stress from soil compression test data *Soil and Tillage Research* **89** 45–57
[13] Imhoff S, Silva A P, Dias Junior M S and Tormena C A 2001 Quantificação de pressões críticas para o crescimento das plantas *Revista Brasileira de Ciencia do Solo* **25** 11–18
[14] Janbu N 1969 The resistance concept applied to deformation of soils *Proc.7th Int. Conf. Soil Mechanics and Foundation Engineering (Mexico City, 25–29 August 1969)* ed A A Balkema (Rotterdam, the Netherlands) vol 1 pp 191–96
[15] Kramarenko V V, Nikitenkov A N and Molokov V Y 2014 About the preconsolidation pressure of clay soils in Tomsk region *Modern problems of science and education* **5** URL: http://www.science-education.ru/119-14703
[16] Mayne P W, Coop M R, Springman S, Huang A-B and Zornberg J 2009 GeoMaterial Behavior and Testing *Proc. 17th Int. Conf. Soil Mechanics and Geotechnical Engineering (ICSMGE, Alexandria, Egypt, 5-9 October 2009)* vol 4 pp 2777–872
[17] Oikawa H 1987 Compression curve of soft soils *Journal of the Japanese Geotechnical Society, Soils and Foundations* **27** 99–104
[18] Onitsuka K, Hong Z, Hara Y and Shigeki Y 1995 Interpretation of oedometer test data for natural clays *Journal of the Japanese Geotechnical Society, Soils and Foundations* 35(3) 61–70
[19] Pacheco Silva F 1970 A new graphical construction for determination of the preconsolidation stress of a soil sample // *Proc/4th Brazilian Conf. Soil Mechanics and Foundation Engineering (Rio de Janeiro, August 1970 1995)* vol. 2 pp 225–32
[20] Sallfors G 1975 *Preconsolidation pressure of soft, highplastic clays* (Goteborg: Geotechnical Department of Chalmers University of Technology) pp 231–237
[21] Schmertmann J H 1953 Undisturbed Consolidation Behavior of Clay *Transaction, ASCE* **120** 1201
[22] Senol A and Saglamer A 2000 Determination of Preconsolidation Pressure with a New Strain Energy-Log Stress Method *Electronic Journal of Geotechnical Engineering* **5** URL: http://www.ejge.com/index_eige.htm
[23] *Soil* ed. V T Trofimov (Moscow: Moscow State University Press) 2005 p 1024
[24] Tavenas F, Des Rosier J P, Leroueil S et al. 1979 The use of strain energy as a yield and creep criterion for lightly overconsolidated clays *Géotechnique* **29** 285–303
[25] Tomasz S 2007 OCR versus YSR– a discussion of classic and today's views on preconsolidation of clayey soils *Prz. Geol.* **55** 405–10

Hydrogeological Conditions Changes of Tomsk, Russia

V D Pokrovsky[1,3], E M Dutova[1,4], K I Kuzevanov[1,5], D S Pokrovsky[2,6] and N G Nalivaiko[1,7]

[1] National Research Tomsk Polytechnic University, 30 Lenin Ave., Tomsk, 634050, Russia
[2] Tomsk State University of Architecture and Building, Solyanaya sq., 2, Tomsk, 634003, Russia

E-mail: [3] vdp@sibmail.com, [4] dutova@sibmail.com, [5] kki@tpu.ru, [6] dsp@sibmail.com, [7] corestone@mail.ru

Abstract. The hydro-geological conditions of Tomsk are determined by both natural factors and the impact of the urban infrastructure. Important impact on subsurface water flows involves the complex hydraulic relationship of several geological layers and the ancient and modern relief. Increasing groundwater abstraction has generally led to lowered piezometric heads in the deeper aquifer horizons, while in the uppermost horizons, rises in the water table and formation of new perched water tables are experienced due to leaking pipes and impedance of groundwater flow by deep foundations. In this paper special attention is paid to the Quaternary aquifer complex. Barrage effects of pile foundations and the intensive development of perched water distributed on flat surfaces of the watersheds and high terraces, complicated conditions for the construction and operation of facilities, leading in some cases to emergency situations.

1. Introduction

One of the actual problems of the urbanisation is a technogene underflooding of the newly developing territories. It is manifested in the raise of the subsoil waters level, the formation of new aquifers and upper layers of ground water. These processes lead to getting wet and stability loss of the soils of the foundations, flooding of underground components of buildings and constructions, changing exploitation conditions, breakdowns and out -of – action engineering communications. They are the reason of the ecological and social situation degradation.

From the beginning of the 70-s (20th century) great attention has been paid to underflooding in Russia. Since 1972 regular conferences on this problem have been conducted, in which the representatives of scientific design, prospecting and building-exploitation organizations took participation. During the last decades a large amount of factual material has been stored, theoretical conceptions and investigation methods of hydro-geological conditions within urban territories have been worked out, as well as prevention and straggle with negative underflooding consequences [1–3]. At the same time, this problem remains open because of the variety of natural and technogene circumstances, complication of different and often diverse direct factors. At present, in Russia hundreds of large cities are under underflooding influence by underground waters, as well as many rural settlements and large industrial enterprises within the European territory, in the Urals, in Siberia, in the Far East. Tomsk is not an exclusion.

Since 1932 the hydro-geological conditions in Tomsk have been studied by M.I. Kuchin, later on by his followers V.A. Nudner, B.V. Plotnikov et al. The authors turned to this problem in 1979, generalizing the materials of the engineering-geological prospecting for building purposes. The scientific report of 1981 was the result of this work, which became the basis for future hydro-geological prospecting on the territory of the city by different organizations. Later on, the works on this theme were continued by such authors as [4–8]. The formed ideas about the directionality of the hydro-geological processes are given below.

2. Situation and Natural Conditions

Tomsk, an ancient student, cultural and industrial city, is situated in a redundantly humidified taiga zone of West Siberia, on the river Tom, the right tributary of the Ob River (figure 1). Its area embraces 295 km2, including a weakly developed left-bank part and a green zone. The W-N-W relief is complicated by terracing complexes of the Tom River and Ushaika, Kirgizka and Basandaika valley tributaries. The relief involves watersheds in the southern part of up to 190-210 m, lowering to 120-150 m in the northern part. The gently sloping terrace ledges smoothly accompanied by the watershed and often confined by slopes of different steepness inclining from 5-10 up to 50-60 degrees, and exceeding up to 50-60 m. The minimum relief is marked by the water cut in the Tom River of up to 68-70 m (absolute height). The historical central part of the city is mainly of ancient two- and three-storey wooden houses, neighbouring with five- and fifteen- storey buildings in residential micro-districts, being densely populated.

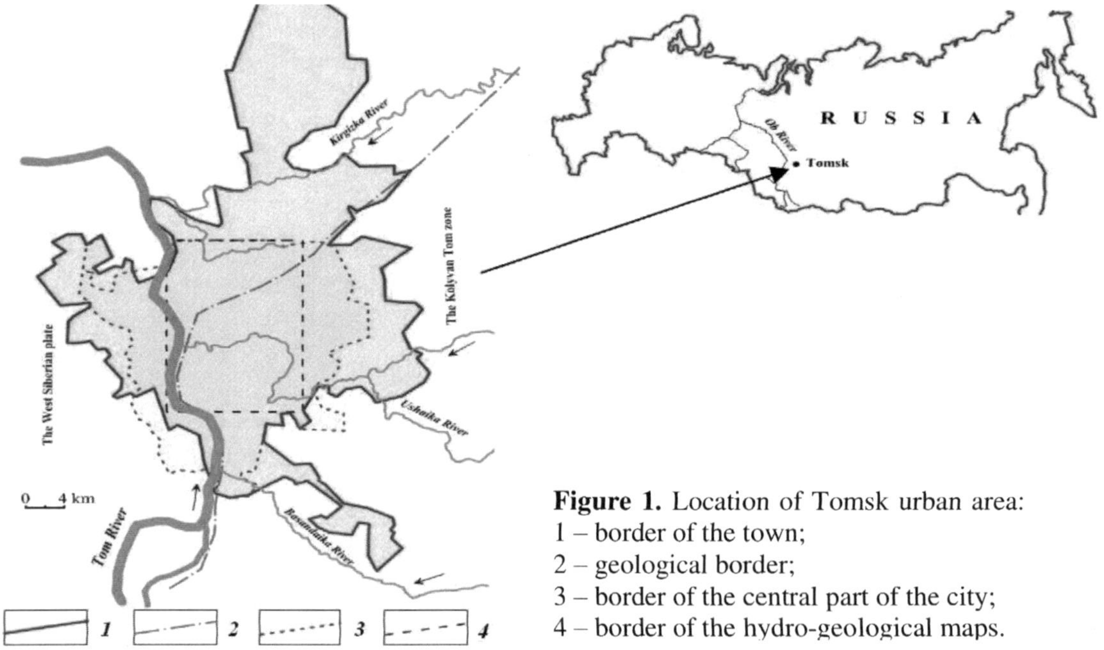

Figure 1. Location of Tomsk urban area:
1 – border of the town;
2 – geological border;
3 – border of the central part of the city;
4 – border of the hydro-geological maps.

3. Hydro-geological Conditions

Hydro-geological features of the territory are determined by its location in the joint of the Kolyvan-Tomsk folded zone and West Siberian platform. The peculiarities of this geological area are defined as a two-level hydro-geological territorial structure.

Water-bearing rocks of the Palaeozoic folded substructure, forming the low hydro-geological level mainly include agrillaceous slates, to a less extent, sandstones. The rocks of the substructure are exposed in the course of the River Tom, in the southern part of the city and in the central area within Ushaika River valley. Water-bearing zones of regional crushing with a thickness of 20-80 m, developed in the upper parts of this section. Within the top of Palaeozoic deposits, the clay weathering crust is developed, which serves as a reliable aquifuge and separates the lower hydro-geological level from the upper one. Waters are of a confined ground nature. Water tables are fixed at the depths of 30-35 m. Water abundance of the deposits is irregular and, on the whole, it is not significant. Specific discharges of the boreholes vary from thousands` up to 0.2 l/sec. Areas with the highest water abundance are within the zones of crushing, also in zones of structural eluvium developed by the widths of a sandy content. In the northern and north-western directions the rocks of the substructure are immersed. They occur at the depths of 80-100 m and more within the bounds of the city.

Loosen formations of the cover, forming the upper geological level, involve aquiferous complexes of Quaternary, Palaeogene and Cretaceous deposits. The Quaternary complex deposits include aquifers (flood-plain and the first terrace), high terraces (the second, the third and the fourth on flood-plains) of the river Tom and its tributaries, watershed and its slopes. The levels of the flood lands are widely developed

4. Changing hydro-geological conditions

The territory of Tomsk is exposed to dangerous engineering-geological processes, where flooding is the most significant one. The development characteristics of this process are determined by the complex influence of both artificial and natural factors. The most important of which are intensity of additional groundwater supply above the natural conditions, structure of the geological section in the zone of influence and the drainage of the area. One of the leading places sources is the additional groundwater recharge to the water supply systems. The urban area receives volumes of water from the external surface and underground sources that are comparable to the intensity of rainfall. In addition, the city has about 200 wells, which capture a considerable amount of water from the regular aquifer and moving to the upper level, promoting the development of flooding. The intensity of technogenic impact on the geological environment is largely determined by density and condition of engineering networks. An extremely widespread engineering infrastructure via water communications and their long-term exploitation contribute to the formation of large quantities of leaks. Taken into account the nature of the construction and condition of the water communications, the calculations show that the intensity of the additional infiltration, in ordinary cases, is 150-200, and in some places up to 1000 – 1500 mm per year, i.e. a higher order of magnitude than in the case of natural ones.

Analysis of the geological structure and lithological variability rocks allows to select one-, two-, and three-layer filtration slit models defining a predisposition of the territory to the flooding process development. This selection is based on the ratio of good and low permeable layers.

One layer section: F-type I-1 classified the soils with high filtration properties (gravel-pebble deposits, sand, sandy loam, bulk soils, as well as loess-like loam top loessial soil). These soils exist in various relations and have a total capacity of more than 15 m.

F-type I-2 classified sections consisting of soils with low filtration properties (loamy soils in all categories except for the area of the upper horizon, clay, uliginous difference between rocks and argillaceous products of the weathering crust) to a depth of 10-15 m.

Two-layer sections: F-type II-1 are soils with high filtration properties, deposits with a width from 2 to 10-12 m, low permeable undersoils and waterproof horizons.

F-II- 2 is characterized by the development of low permeable soils at the top of the section. At depths of 5, sometimes more meters, soils with high filtration properties are deposited.

Three-layer sections: F- type III-1 includes soils with high filtration properties, which have a discrete layer of aquitard with the width 2-6 m. F-type III-2 involves soils with low filtration properties, including good permeable layer with the width 5-6 m.

According to generally accepted criteria, taking into account the ratio of hypsometric location of the underground water supply and discharge areas and the extent of the relief dissection, this territory is well – drained. In general, underground water of the lower hydro-geological and Quaternary alluvial deposit horizons have good drainage. Top soils of sandy and clayey horizons, flooded once in a while, are characterized by very low drainability. This determines the possible active flooding development. Increased levels caused by the development of new areas and reconstruction of old ones are identified by the geomorphologic elements. They are visible in the watershed areas composed of different types of cross-sections, inclined to the formation of flooding processes. First of all, the formation of perched water should be noted in a number of areas which it has not been previously described.

The technogenic impact results on the water imbalance (drainage deterioration, filtration losses from water supply and sewage networks, regimentation of surface runoff, etc.), formation of perched water, and developed flooding. The worsening drainage conditions caused by the barrage effect during the construction of buildings on pile foundations cause the flooding of watershed and transit sites of ground flow. Pile foundations of planned and existing buildings can play an important role in the formation of the damaged structure of seepage flows, especially in the case of mass building. Barrage effect of pile foundations, without taking account underground water movement direction, was repeatedly described in literature. In 1981 this effect was described concerning the possible underground water level rise in some areas of Tomsk. This was also confirmed recently. The scheme of building foundations interaction on one of Tomsk construction sites is described. This site is located on the surface of the second terrace over flood-plain of Tom River, directly near its border. Two major aquifers are allocated in the cross-section. The non-pressure aquifer is located from the surface to depths of 2.3-3.7 m. Pressurized aquifer is opened at depth 16.4 m. The depth of perched water in the backfill soil of near-surface cross-section ranges from 1.2 to 2.8 m.

Based on field studies results aquifer of Quaternary deposits located within reach of the pile foundation is sited in the zone of engineering structure interaction. In the area of the construction site, there is a regional groundwater flow, N-W oriented to the Ushayka River valley which is a natural drain. Analysis of hydro-geological conditions showed that the building foundations existing in the area of the construction site are under the influence of the groundwater flow. This creates favourable conditions for the barrage effect. Modeling of the hydro-geological conditions of this site was performed (figure 2a).

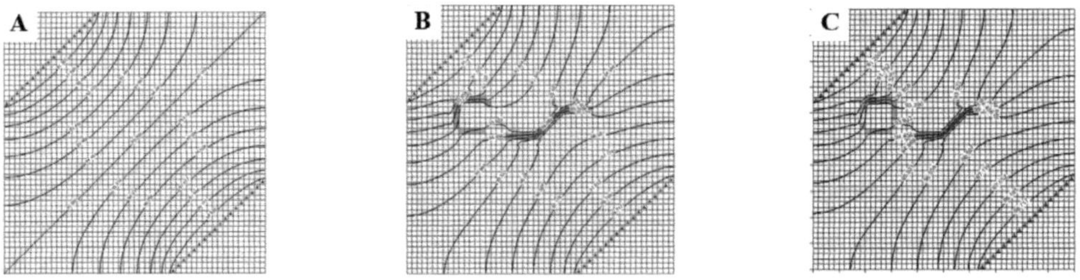

Figure 2. Modeling results of groundwater flow:

A – under natural conditions of the aquifer with homogeneous filtration properties;
B – under disturbed conditions, reducing the filtering coefficient of piles field in 100 times
C – under disturbed conditions, reducing the filtering coefficient of piles field in 100 times. The planning design of building site was changed.

An illustrative example is the design of piles field configuration disregarding the hydrodynamic characteristics of the geological cross-section (figure 2-B). The courtyard represents a "trap" for the regional filtration flow. Within this area the drainage will be completely impossible, but the flow of underground water from the South-East will continue. It can be stated that construction site on a pile foundation in combination with the existing buildings could drastically worsen the conditions of natural drainability, leading to a redistribution of filtration flows and the rise of groundwater levels.

After analysing the results of numerical simulation (figure 2-C), the proposed design solution dramatically reduces the risk of a sharp rise of groundwater levels. Alternative construction project of the residential house was adopted to implement the proposed recommendations. This example shows the importance of considering the characteristics of hydro-geological conditions in the development planning of any urban territory. The study of other technogenic influenced areas in the analysis of the water flows structure can also be very useful.

This can be illustrated by examining the pollution processes. Zoning of the city territory under terms of run-off was proposed. This scheme is as following: the eastern part of the subdistrict A-I-1 (figure 3) embraces modern multi-storey buildings, which affect the groundwater discharge along Tom River high terraces. The western part of the subdistrict embraces old two-three storey buildings in so-called private sector and characterized by the underdevelopment of the sewerage and water supply. The northern part of the subdistrict A-I-2 experiences a complex impact of mixed ancient urban building, including the low-rise part of the private sector along the western border. Technogenic load is reduced to a minimum southward. Area B-II is subjected to the complex nature of pollution. The effect of high-rise modern building is prominent along the southern boundary. In the northern part of the territory the influence of the industrial zone is increasing.

The effect of various kinds of technological factors is pronounced within valleys of permanent watercourses, which are tributaries of the first order Kirgizka River and are treated as separate sections (not shown on the map).

Subdistrict C-I-1 on the right bank of the Ušajka River valley is mainly influenced by high-rise multi-storey building construction development. Subdistrict C-I-2 combines three tributaries of the Ušajka River, forming separate areas within the valleys, vary by relative shares of watershed development. The types of economic development are approximately similar for all three valleys. Subdistrict C-I-3 is characterized by Northwest direction of runoff and has a direct impact on modern buildings. Subdistrict C-I-4 is defined by South-East direction of surface run-off and reflects a mixed impact of modern building with elements of individual construction.

Subdistrict C-II-1 undergoes a mixed influence of urban building. The underground water discharge from the bed of Ušajka River is more prone to the old two-three storey construction. Subdistrict C-II-2 is influenced by two-three storey construction development. Subdistrict C-II-3 is characterized by mainly North direction of surface run-off and by limited influence of technogenic load because of the weak development of the territory in the areas of groundwater supply.

Area D corresponds to the drainage basin of Basandjak River and occupies an insignificant area within the administrative boundaries of the central part of the city.

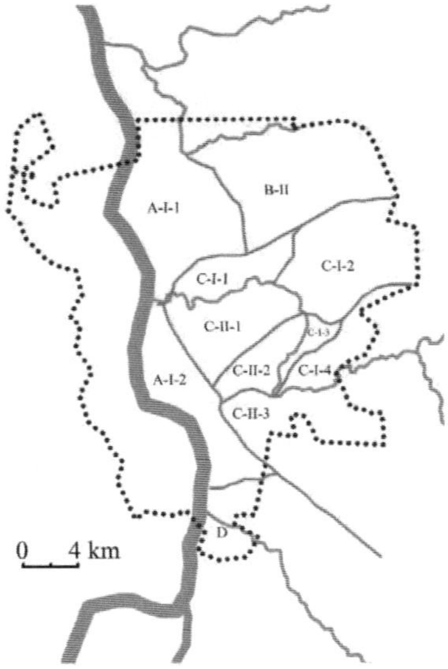

Figure 3. Zoning of the central part of Tomsk territory according to the surface discharge conditions.

Performed zoning allows obtaining analysis basis showing the development of pollution processes. The possible differentiated approach in analysing the potential sources of pollution in urban area became evident. It is possible to determine the major type of pollution, emerging as a result of new multi-storeyhigh-rise or old two-three storey buildings. This is crucial for Tomsk, which was historically developed without considering zoning in the construction residential and industrial areas.

5. Conclusion

The territory of the city as a whole is not prone to the intensive development of the groundwater flooding processes. High hypsometric marks of the area, rugged river network and logs, the presence of powerful horizons of sandy soil create favourable conditions for drainage. At the same time, the intensive development of perched water distributed on flat surfaces of the watersheds and high terraces, complicate conditions for the construction and operation of facilities and, in some cases, to emergency situations. Barrage effects of pile foundations in areas with near deposition of ground waters are pronounced.

1. structure of the regional natural underground water flow does not change in time. The supply and the drainage areas keep their position in space.
2. urbanized territories local seepage flows structure is formed not only under control of natural factors but largely has been restructuring under the influence of technogenic impact, among which, the most important are the additional supply of underground water and direct impact, that are expressed in terms of barrage effect of underground engineering constructions.
3. impact of cities on the deep aquifers can be seen in the lowering levels of underground water; their hydrodynamic regime has no effect on the conditions of construction.
4. areas of the city within the development of low terraces are the natural flooded condition.
5. main reasons of flooding are the rising value of additional underground water supply through leaks from water communications and the deterioration of drainage in the terms of the accepted system of the construction.

Acknowledgements

Research is executed with financial support of the RFBR "Spatial and time research geophysical and hydrogeological fields changes of Altai Mountains seismically active territories"; contract N13-05-98072

References

[1] Dzektser E.S. 1994. The methodological aspects of the problem of geological danger and risk *Geoecology* **3** 3–10 (in Russian)

[2] Yang Y, Lerner D N, Copperthwaite N H and Barrett M H 1999 Assessing the impact of Nottingham on groundwater using GIS *Groundwater in the Urban Environment* 281–87

[3] Pokrovsky D S, Dutova E M, Kuzevanov K I 2008 Application of geoinformation technologies for estimation of hydrogeoecology conditions of developed territories *News of higher educational institutions. Construction* **3** 107–12 (in Russian)

[4] Pokrovsky D S, Kuzevanov K I and Sereda N P 1983 An assessment of the potential underflooding of the territory of the city (the example of Tomsk) *Geology and mineral resources of the West-Siberian plate and its folded margin* (Tyumen: ZapsibNIGNI) 226–28

[5] Pokrovsky D S and Kuzevanov K I 1987 Hydrogeological conditions and underflooding processes in the territory of Tomsk *Underground waters of the south-eastern Siberia* (Novosibirsk: SB Nauka) 146–53

[6] Dutova E M, Nalivaiko N G, Kuzevanov K I and Kopylova J G 1998 The chemical and microbiological composition of urban groundwater, Tomsk, Russia *Groundwater in the Urban Environment: Proceedings of the XXVIIth Congress IAH (Nottingham, UK)* **2** 125–30 (in Russian)

[7] Pokrovsky D S, Rogov G M and Kuzevanov K I 1998 The impact of urbanisation on the hydrogeological conditions of Tomsk, Russia *Groundwater in the Urban Environment: Proceedings of XXVIIth Congress IAH (Nottingham, UK)* **2** 217–23

[8] Pokrovskiy V D 2008 Antropogenicunderflooding as a rational restoration planning factor of Tom architecture monuments *International scientific symposium in honor of academician M.A. Usov Problems of geology and subsurface development (Tomsk: TPU Publ.)* 919–20

Karst hazard assessment in the design of the main gas pipeline (South Yakutia)

L A Strokova[1,2], E M Dutova[1], A V Ermolaeva[1], I N Alimova[1] and A B Strelnikova[1]

[1] National Research Tomsk Polytechnic University, 30 Lenin Ave., Tomsk, 634050, Russia

E-mail: [2] sla@tpu.ru, alyona7@inbox.ru

Abstract. The paper represents the description of the zonal and regional geological factors of geoengineering conditions which characterize the territory in South Yakutia crossed by the designed main gas pipeline. Cryogenic processes and karst are considered to be the most dangerous hazards for gas pipeline maintenance. Karst hazard assessment of the gas pipeline section made in the course of the research has involved a complex of geological methods: geoengineering, geophysical, hydrogeological, and mapping. Sections prone to karst development have been identified. The authors have suggested the measures to protect potentially hazardous sections and to ensure timely informing on sinkhole collapses.

1. Introduction

1.1. Significance of the problem
Constructing gas pipelines in the areas characterized by unfavourable environmental conditions, such as boggy areas or zones of ever-frozen subsoil, is a current challenge in the field of hydrocarbons transportation. The issue is particularly important when the pipelines are constructed in the territories with frozen karst, for instance, the Lena tableland in South Yakutia. To prevent gas pipeline failures, it is necessary to develop the measures which ensure timely informing on sinkhole collapses.

The karst phenomenon has been researched in the works of many Russian and foreign scientists. This is due to the fact that karst zones are varied and widely spread. In the works by N.A. Gvozdetsky [3, 4], G.A. Maksimovich [8] the following questions were discussed: solubility of karst rocks under different physiographic and geological conditions, the mechanism of karst collapse, stability assessment in karst areas, the methods efficient to detect karst cavities in the course of land development, etc.

The objective of this paper is to assess karst hazards for the first section of the main gas pipeline «The Power of Siberia» with the complex implementation of different geological methods: geoengineering, geophysical, hydrogeological, and mapping. On the basis of the data analysis sections prone to karst development have been identified and the appropriate measures have been suggested to protect potentially dangerous sections and to ensure timely informing on sinkhole collapses.

1.2. Object of study
The study area is located in Lensk Ulus, the Republic of Sakha (Yakutia). The description of geoengineering conditions is provided on the basis of the archive materials, which contain the data

obtained by VostSibAGE, Promneftegazproyekt, Ingeokom, Fundamentproyekt, SPF DIEM, Schmidt Institute of Physics of the Earth of the Russian Academy of Sciences (IPE RAS) in the course exploration work for pipeline route in 2010-2012. The length of the track was 160 km. The region is poorly studied and the lands are poorly reclaimed.

The climate of the region is sharp continental, with lukewarm summer and severe dry winter. The average annual air temperature is –5.6-6.2 °C. The annual amount of precipitation is 346-409 mm, 75% of which fall during the warm period.

In terms of geomorphology, the gas pipeline route stretches through the Prilensk erosive tectonics plateau bounding with Patom highland in the southeast and with the Leno-Vilyuysk basin in the north and northeast. The surface of the plateau is 400-600 m above sea level and is tilted gently to the northwest. The highest point above the sea level is Byugyuekh watershed divide (554 m) and the lowest point is the valley of the river Nuia (243 m). The erosion and denudation landform is predominant. It is represented by Lower-Middle Cambrian sediments of steeply-sloping plateau and Upper Cambrian and Ordovician sediments of the ripple-semi-sloping plateau.

In terms of tectonics, the region is located in the southern part of the Siberian platform, beginning from the Nepsky arch of the Nepsko-Botuobinsk anticlise and moving to Predpatomsk depression. The area is composed of Cambrian and Ordovician sediments as comb-shaped folds extending to the north-east along the border of Baykal-Patom mantle folded formation. The folds are fractured predominantly with south-east thrusting. There are also transverse steeply dipping submeridional strike fractures. Two sections of the track cross the zones of Jurassic sediments overlying Lower Paleozoic sediments with angular unconformity. The summary information on the region geology is represented in table 1.

Table 1. The geology of the region.

№	Sediment age *period* – thickness, m	Lithologic composition of sediments
1	Cenozoic *Quartenary*	Low ice-content clay loam, sand clay, pebble, gruss, peat
2	Mezozoic *Jurassic* – 90	Diversely-grained sand with sandstone and conglomerate lamina
3	Paleozoic *Ordovician* – 300 *Cambrian* – 1300	 Aleurolite, argillite, Limestone, dolomite

On the seismic zoning map the track crosses the zone with seismic intensity of 6 points of MSK-64 scale.

Within the zone, country rocks and quaternary sediments are identified. Country rocks are of two formations: Lower Cambrian carbonate and Middle Cambrian and Ordovician terrigenous-carbonate. The following genetic types are found among Quaternary sediments: alluvial, alluvial-diluvial, diluvial-proluvial, diluvial-colluvial, colluvial, eluvial, eluvial-deluvial, and biogenic.

The region is within the area of insular permafrost. The lowest annual permafrost temperatures (up to –1.7 °C) are characteristic for valley bottoms and the lower parts of slopes, while the highest temperatures (0 – minus 0.6 °C) are characteristic for watershed and tops of smooth slopes. Thermal radiation subgelisols are typical for high watershed and fractured carbonate rocks slopes. Channel subgelisols are typical for stream valleys, especially within the zones of tectonic fractures and karst zones within the carbonate rocks of the channels.

Preferentially, island nature of distribution of permafrost within the territory and limited distribution of icy soil predetermined nature of development of cryogenic processes – dominance of the one-year, seasonally failing cryogenic educations – river sleets, one-year hillocks of a frost heaving, arising not annually and changing places of the education depending on freezing conditions. Cryogenic processes have local distribution and are coordinated generally for the bottoms of valleys.

The insular permafrost and scarce icy ground of the region result determined the character of cryogenic processes: one-year cryogenic formations, which break on-season, are predominant – flood icing and one-year frost mounds, which emerge from time to time and change the place of emergence.

Karsts are widely spread in carbonate formations. Intensive karst development takes place due to fissured lands and fractured rocks. Karst development is caused by dissolution of carbonate rocks by surface and ground waters. The process is intense in fractured zones of smooth carbonate kettle back, which is the anticlinal core. Cambrian carbonates include sulphate of lime, cube spar, limestone, and as a result, both the thickness and the extension of karst zones vary significantly.

In the course of exploration different forms of karst were identified – karst windows and sinking streams. The diameter of the karst window is from 10 to 50 m and the depth is from 0.5 to 3.3 m. The process of karst development within some windows is quite intense which is indicated by recently fallen trees and sinks 5 m in width and 3 m in depth. The 10 m well drilled within the karst window revealed uncovered loam soils and sandy medium gravel clay, while no hard limestone, supposed to occur at the window periphery at the depth of 1.5– 3.0 m, were identified. Ground waters were neither found.

Sinking streams can be observed in slightly sloping depression which stretches along the kettle back slope for 1 km. The well drilled at the edge of the depression uncovered homogeneous formation of medium gravel loam soils while country rocks were not identified. Neither were ground waters found.

The channel karst was identified on the left bank of the river Tokunda. At the interval between 10 and 13 m below 3 m layer of medium hard limestone, a decompressed zone, was found. It was filled with sandy clay with noncoherent limestone clasts. Ground waters were not found. The zones of potential karst hazard are smooth carbonate formations (watershed divides, gentle slopes, and valley bottoms). Slope and erosion exogenous process, as well as bogging and suffusion, are less intense within the study area.

2. Methodology

2.1. Choice of methods and work sequence

One of the key issues in karst research is karst related zoning. The average annual intensity of karst sinkhole development (number of sinkholes / km^2 per year) was suggested by Z.A. Makeev [7] in 1948 and has been the conventional measure to assess karst hazard for a long time. We support the idea of V.S. Lukin that the appropriate methods to assess karst hazard are different for different zones; for instance, to assess karst hazard in Kungur territory, the basic criterion is the diameter of the sinkhole [10].

Since the data on intensity of karst sinkhole development, sinkhole diameter, and sinkhole age are not enough to assess karst hazard in the study area, we suggested assessment on the basis of the integral criterion of potential karst development.

To provide rationale for the karst hazard criteria, we analysed the data collected from various karst areas. The data analysis allowed recognizing the evidences of karst hazard – lithological composition of rocks, massif's hydrogeological characteristics and fracturing (table 2).

Table 2. Evidences of potential karst sinkhole development.

№	Evidences
1	Surface karst landforms
2	Dissolving capacity of natural waters
3	Zones of loosened ground, identified through geo-electrical and velocity anomalies
4	Intensive vertical filtering of underground waters
5	Existence/absence of the aquaclude superposing soluble rocks

To assess the karst hazard of the area, firstly, the well columns indicating data on karsting rocks and aquacludes were identified. The zones of surface karst and water sampling points were reflected in the base map.

Secondly, equilibrium and non-equilibrium states of the system "natural waters – soluble rocks" were studied to assess the dissolving power of natural waters.

At the next stage, geophysical survey data were analyzed and the zones of loosened ground were identified through geo-electrical and velocity anomalies. Finally, the karst related zoning map of the study area was designed.

2.2. Hydrogeological research

The assessment of karst hazard and the intensity of karst development was done on the basis of laboratory study of waters sampled in 2010-2012 with the application of hydrogeochemical methods.

The underground waters are characterized by sporadic distribution. The waters within thin and marginal diluvial formations at slopes are found at the depth of 3-5 m; in terms of content, they are hydrocarbonate, magnesium-calcium, and sodium-magnesium-calcium. Aquaclude is hard and medium hard soil, sometimes multi-year.

The waters within diluvial-proluvial formations at slopes are associated with over- and inter-permafrost subgelisols. Water-bearing soil are medium gravel clay loams and brash.

The calculation of water equilibrium was done with the method by Garrels and Christ [2] for the temperature of 25°C and the pressure of 0.1 MPa. Thermodynamic modeling is done on the basis of HydroGeo software [1]. In the process of modeling only the minerals widely spread within the study area were considered. The issue of mineral matter transformation under the given hydrogeochemical conditions was considered on the basis of non-equilibrium index (SI). As the saturation with any chemical bonding increases, the non-equilibrium index decreases tending to zero; in case of over-saturation the index value is positive; zero value signifies equilibrium.

Table 3. Average value macro component composition (мg/l), pH and total salt content (g/l).

Component	Types of water	
	surface	ground
pH	7.2	7.1
$(HCO_3)^-$	182	335
$(SO_4)^{2-}$	69	84
Cl^-	7.5	11
Ca^{2+}	53	78
Mg^{2+}	18	29
Na^++K^+	10	16
M	0.2	0.54
$CO_{2\ aggresive}$	15.4	14.7
Number of test	7	23

The data of chemical analysis made on the basis of HydroGeo software allowed identifying equilibrium and non-equilibrium states of the system calci-spar – underground water. On the basis of data obtained, non-equilibrium index dependency graph of $CaCO_3$ mineralization was designed. The dependence was described through logarithmic equation with the coefficient of correlation R^2 = 0.41. The waters with mineralization higher than 0.6 mg/l are saturated with calci-spar, and within these zones, there is no hazard of karst development or the process in not intense.

2.3. Geological survey

On the basis of vertical electrical sounding survey geoelectric models of the upper part of the section up to the depth of 15-20 m over the winter period were studied. In this depth interval geoelectric sections are approximated with three-four-layer models. The first geoelectric layer within the upper interval of the section is of high-resistivity, with the thick seasonally-frozen interlayer. The specific electric resistance of the soil (ρ) is from 400 to 800 Ohm·m. The geoelectric layer thickness varies from 0.5 to 4 m, with the increase in the depressed areas. According to the data of the seismic survey (SRM), the velocity of P-waves within the interval preceding the first rigid border (V_1) is from 500 to 1800 m/s. The zones, where according to seismic survey data the refractor velocity is abnormally low ($V_2 \leq$ 1800 m/s,) are thick layers of friable eluvial ground (loosened ground). There are 48 zones of this type on the track. Geoelectric layer characteristics are represented in table 4.

Table 4. Geoelectric layer characteristics.

Layer	Characteristics
1	Seasonal-frozen layer – peat, loams, sandy loams of different consistence and cryostructure Thickness 0.5-4 m ρ_1= 400-800 Ohm·m V_1=500-1800 m/s
2	Rubbly, crushed, clumpy eluvial soil, loams of different consistence and cryogenic characteristics Thickness 1-7 m, ρ_2= 80 -200 Ohm·m V_2=1100 - 3900 m/s.
3	Hard carbonate rocks (limestone, dolomite, marl) Thickness 5-10 m, ρ_3= 600 Ohm·m V_3=2500 – 3900 m/s.
4	Zones of interbedding of the dense lithological varieties of dolomite, limestone and marls, clay and soapstone ρ_4= 80-220 Ohm·m V_4=2000 – 4600 m/s.

2.4. Karst hazard mapping

To design the karst related zoning map, it was necessary to identify the values of the model parameters on the basis of characteristics represented in table 2. The data points were grouped through calculation of the integral criterion for a characteristic presence or absence. If the characteristic was absent, the zone was considered potentially dangerous.

3. Results

3.1. Data analysis

The pipeline route is split into sections which are considered hazardous, potentially hazardous, and unhazardous in relation to karst development. Hazardous zones are those with karst landforms, with no aquaclude superposing soluble rocks or with the thickness less than 5 m. Potentially hazardous zones are those composed of loosened rocks, with the aquaclude superposing soluble rocks or with the thickness from 5 to 10 m. Unhazardous zones are those with the aquaclude superposing soluble rocks.

3.2. Recommendations on gas pipeline protection

To protect potentially hazardous sections and to ensure timely informing on sinkhole collapses, the following measures recommended:
- shutting off karst cavities and loosened ground zones with high-strength and low-deformation geosynthetic materials;

- soil stabilization by biocement injection fixing; the advantage of this method in comparison with other methods of chemical stabilization is low viscosity of the slurry and its deep penetration into porous ground;
- constructing bored piles with lodgement and steel cable rope;
- implementing a reliable alarm system to indicate the changes in the stressed-deformed state of the pipeline.

4. Conclusion

The lack of study area data makes karst related zoning conditional and requires further exploration. To update the karst related zoning map, continuous monitoring of karst development is relevant. To describe geological objects, the advantages of GIS technologies should be used, namely, spline functions, which are flexible, relatively simple and easy to use.

References

[1] Bukaty M B 1997 Development of the software in the field of oil and gas hydrogeology *J. Investigation and protection of a subsoil* **2** 37
[2] Garrels R M and Christ C L 1968 *Solutions, minerals, equilibria* (Moscow: Mir) p 368
[3] Gvozdetsky N A 1981 *Karst* (Moscow: Myisl) p 214
[4] Gvozdetsky N A 1988 *Karst landscapes* (Moscow: Moscow State University Press) p 112
[5] Zverev V P 2007 *Underground waters of crust and geological processes*. (Moscow: Scientific world) p 256
[6] Katayev of V N 2001 *Methods and practice of comparative and evaluation karst division into districts* (Perm: Perm State University Press) p 85
[7] Makeev Z A 1948 Principles of engineering-geological division into districts *J. Karstovedeniye*. **4** 43
[8] Maksimovich G A 1963 *Karst bases.* (Perm: Perm State University Press) vol 1 p 443
[9] Falcons D S 1962 *Main conditions of development of a karst* (Moscow: Gosgeoltekhizdat) p 322
[10] Tolmachev V V and Royter F 1990 *Engineering karst* (Moscow: Publishing house Subsoil) p 151
[11] Homenko V P 2012 Normative assessment of karst danger: a crisis situation *Geotechnical problem of design of buildings and constructions on the karst danger territories: Proc. Int Conf. (Ufa).* (Ufa: Bashniistroy) pp 240–45
[12] Scherbakov S V and Katayev V N 2011 An integral assessment of karsto-danger of the urbanized territories (on the example of Kungur) *Scientific notes of the Kazan university. Series Natural sciences* **153** (1) 203
[13] Fleury S 2009 *Land use police and practice on karst terrains: living on Limestone* (Dordrecht: Springer Science) p 153
[14] Smith P 2007 *The fundamentals of piping design: drafting and design methods for process applications* (Houston: Gulf Publishing Company) p 214

IOP Conf. Series: Earth and Environmental Science 27 (2015) 012033 doi:10.1088/1755-1315/27/1/012033

Methods for karst hazard forecast and pipeline protection in South Yakutia

L A Strokova[1,2] and E A Teterin[1,3]

[1] National Research Tomsk Polytechnic University, 30 Lenin Ave., Tomsk, 634050, Russia

E-mail: [2] sla@tpu.ru, [3] badscrool@gmail.ru

Abstract. The objective of this paper is to consider the ways of monitoring and protecting the pipeline «The Power of Siberia» from karst hazards. The method of protection with bored piles has been recommended and the forecasting methods have been overviewed. Thus, ground penetrating radar allows discovering a soil cavity when the overburden layer is not very deep and predicting sinkhole collapse when used in combination with a balance arch model. Monitoring of triggering factors is widely used to forecast karst collapse when the opening is caused by pumping as the data on dynamic groundwater conditions can be obtained in real-time. However, the suggested protection method is expensive and some forecasts might be difficult to make. The authors have suggested the ways to improve the current situation and to reduce the damage caused by karst collapse.

1. Introduction

1.1. Problem statement

Currently, in the field of hydrocarbon transportation, there is a challenge of constructing the pipeline through the areas of severe climatic and environmental conditions. These areas are the regions with a great number of bogs and ever-frozen ground. However, the most complicated challenge is to construct the pipeline through karst areas. As for the geologically unfavorable regions described above, we have found the ways to solve problems possible in oil and gas pipeline operation, but the karst areas are an exception since there is a lack of applicable and reliable to prevent and eliminate probable accidents. Often transporters do not have an opportunity to build a pipeline through a more preferable route, because it is economically inadvisable or due to environmental requirements, etc. Therefore, firstly, we have to study karst areas and then identify the factors, which can reduce karst risks in the potentially dangerous regions the pipeline runs through.

1.2. Rationale of the research

Karst is dangerous especially because of possible underground cavities, which might represent a challenge in terms of engineering geological assessment. The process of voids development starts under the three co-occurring conditions:

1) soluble fraction;
2) capability of water to dissolve;
3) diverting of ground water.

IOP Conf. Series: Earth and Environmental Science 27 (2015) 012033 doi:10.1088/1755-1315/27/1/012033

Subsequently, due to the growth of cavity space at the threshold tension, the surface may collapse and form a sinkhole. This destroys the base of the pipe reducing the pipe resistance and its load-bearing capacity [1]. As the diameter of the sinkhole can be from a few meters to more than 150 meters and the depth is greater than 20 meters [2], the most unfavorable effect in the loss of pipeline integrity. Figure 1 represents the pipeline in the area of karst cavities which correspond to thermokarst (upper picture), in the sinkhole (lower picture), [3].

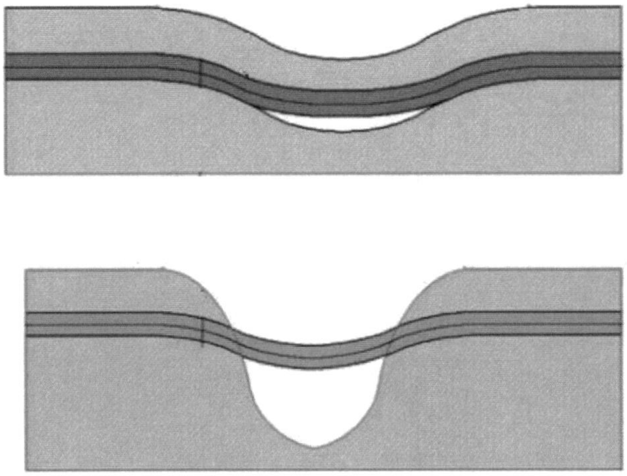

Figure 1. Condition of the pipeline.

1.3. Object of study

The Russian gas pipeline system is one of the largest in the world. In 2012, the length of the main pipeline was more than 175 thousand kilometers. According to the official statistic data represented by «Gazprom», 42% of all gas pipeline accidents are directly or indirectly caused by natural factors.

One of the major problems is «The Power of Siberia» pipeline. The length of the potentially dangerous part is 160 kilometers. This section starts from Chayandinskoye deposit. The pipeline stretches through the territory of Lensky Ulus of the Sakha Republic (Yakutia). The climate conditions are severe, for example, the temperature in the winter can changes from -25 °C to -62 °C, the spring comes quickly, with significant differences between the night and day temperatures. Moreover, many karst regions are located within this area. Every year new sinkholes are developing. Development of karst is connected with rock fracturing. Ground water circulates through the rock dissolving and removing soluble minerals. Therefore, the area is characterized by severe climatic and environmental conditions and tends to be even more dangerous because of the karsts [4].

To start solving this problem, firstly, it is necessary to predict the deformations of the pipelines on the basis of the karst hazard forecast. However, often it is impossible to predict when exactly a sinkhole will open, so secondly, it is important to monitor bending pipeline sections to prevent the pipeline failures and accidents. The efficient solution is to apply the advanced methods of pipeline monitoring and protection.

2. Methods of detection and protection

2.1. Pipeline protection with bored piles

The invention is relevant for pipeline construction and operation and can be used to prevent accidents in pipelines caused by karst failures (figure 2).

IOP Conf. Series: Earth and Environmental Science 27 (2015) 012033 doi:10.1088/1755-1315/27/1/012033

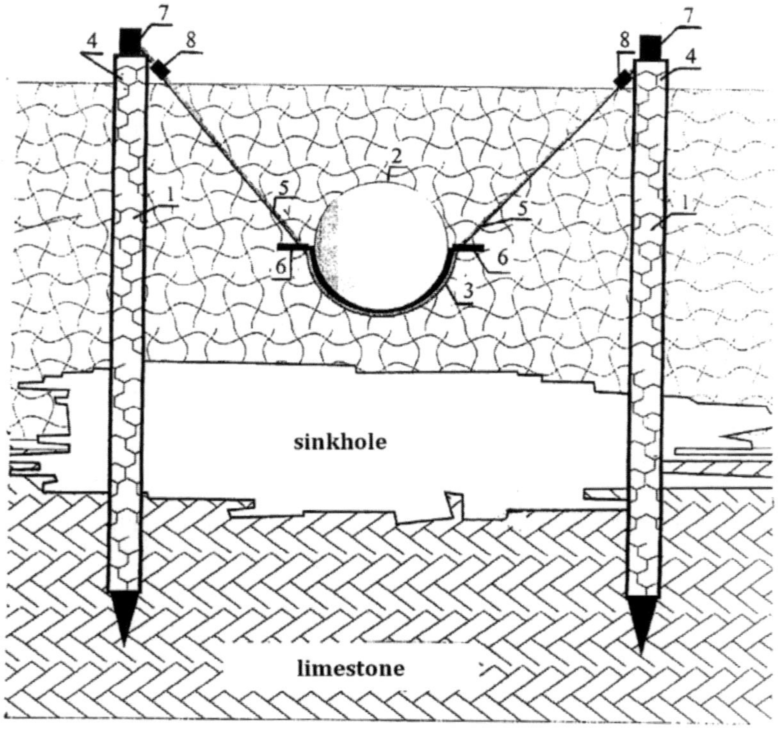

Figure 2. Protection with bored piles.

The bored piles 1 are placed on the both sides of the pipeline 2 and connected with the metal lodgement 3. The construction is fixed in such a way that the pile heads are above the earth's surface, the metal lodgement and the bored piles are connected with the steel cable rope 5, which bypasses the pipe, and the ends of the rope are fixed on the tip piles 4 with the help of the tension clutches regulating the tension of the cable rope. At the ends of the rope before the tension clutches there are the compression couplings sensors 8 (figure 2 [5]).

The disadvantage of this method is that the prototype is difficult to construct since the metal beam is connected with the bored piles through welding. Another disadvantage is the large quantity of metal per structure and the absence of visual control over the construction during the operation.

The advantage of the invention is that the pipeline is independent from any movements, which might occur in the ground.

2.2. Ground Penetrating Radar (GPR) Surveys

This simplest and most popular method for forecasting karst collapse is very popular in China. GPR can directly identify soil cavities. The Ground penetrating radar modeling for predicting the sinkhole is based on an idealized balance arch as shown in Figure 3, where σ_v is the natural vertical stress; h is the height of the arch; b is the half span; and f is the Protodyakonov coefficient.

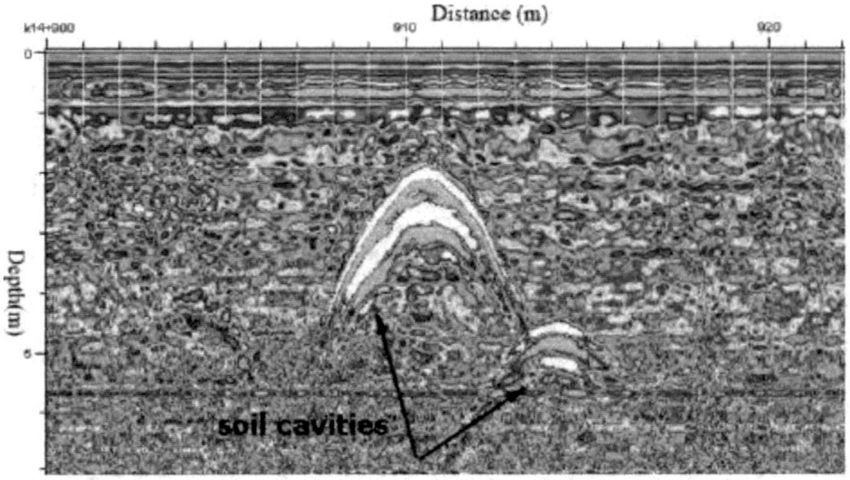

Figure 3. GPR image of a highway and the idealized model of its balance arch [6].

Basically, the equilibrium equation of the arch is

$$— \quad (1)$$

Thus, if

$$h \quad — \quad (2)$$

then the arch will be stable, but the hole will develop further when

$$h \quad — \quad (3)$$

until a sinkhole opens.

The disadvantage of this method is as follows: frequent scanning is necessary to monitor changes in the arch, which might result in high costs when the survey is large. Moreover, GPR fails to be accurate in case of deep cavities because of technological limitations.

The advantage of this method is the time since it is the fastest way to identify the potentially dangerous zones [6].

2.3. Monitoring triggering factors

The aim of this method is to monitor changes in the groundwater through the karst conduit (Figure 4).

Determining the forecasting threshold is the most important stage of this method. First, the soil samples are obtained to be tested in the lab. The main purpose of test is determination of the hydraulic gradient.

If the hydraulic gradient in the laboratory tests is greater than in the field, deformation will generate in the soil, and a karst collapse will open at the surface.

The disadvantage of this method is that the accuracy of forecasting is low due to discrepancies in the structures and physical-mechanical properties of soil even within the same layer, and the test results may be different from the real data.

However, to improve the accuracy, a new method was developed. This method is based on the residual analysis of groundwater pressure. It is important to control relationships between the time that anomalous monitoring data were recorded and the time of karst collapse. The times of maximum, minimum and most anomalous values and the time of the sinkholes opening have a linear correlation. Between the time of the most anomalous value and the time of the sinkhole opening was

. Basically, when an anomalous value fits the equation following monitoring then the time of the sinkhole opening can be predicted.

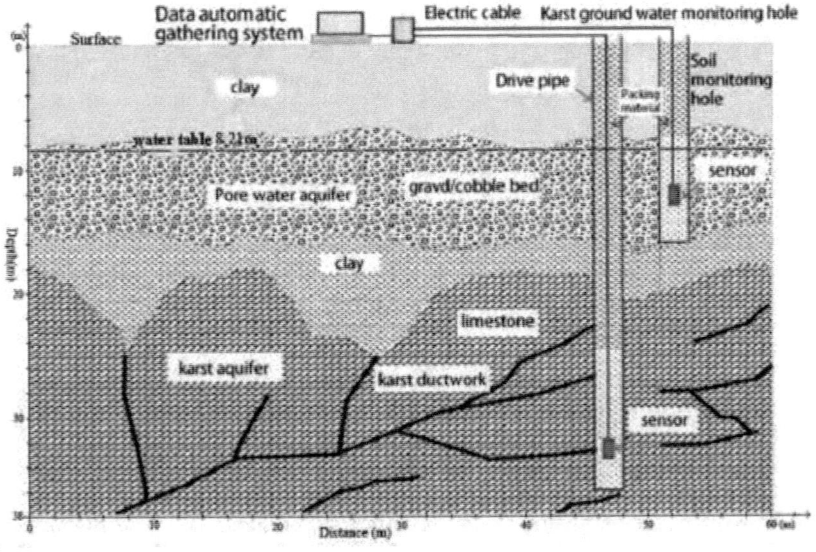

Figure 4. Monitoring of groundwater pressure [6].

The advantage of this method that it is more accurate and provides additional data to analyze the situation: for example, this sensor can obtain the data on the chemical composition of the ground water, etc. [6]

2.4. Barkhausen noise sensor

This method is one of the most reliable and efficient way to identify deformations. The Barkhausen effect is an indirect evidence of the magnetic domains existing within the ferromagnetic materials. When domains grow, under the applied magnetic field, the movement of the domain walls occurs by discontinuous and abrupt Barkhausen jumps. The jumps in magnetization of a ferromagnetic material can induce a voltage in a winding coil of wire that in turn can produce Barkhausen noise [7].

The overhead sensors are used for excitation and detection of magnetic Barkhausen noise (figure 5).

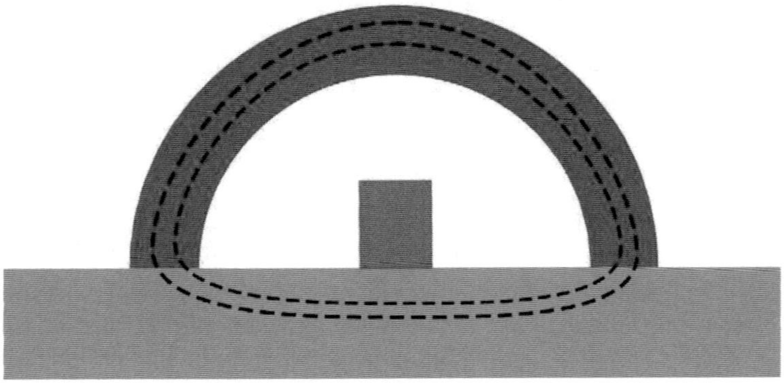

Figure 5. A set-up for non-destructive testing of ferromagnetic materials: arc – magnetising yoke, square – inductive sensor, rectangle – sample of pipeline under test.

IOP Conf. Series: Earth and Environmental Science **27** (2015) 012033 doi:10.1088/1755-1315/27/1/012033

Transverse field dipole magnet of the sensor generates varying magnetic field in the study area. This magnetic field creates jumps of magnetization resulting in noise signals in the receiving coil of the sensor which are registered by the device. The level of the magnetic noise depends on the properties and state of the crystal lattice and mechanical tension.

In most types of steel under the tension the intensity of the Barkhausen noise increases while under compression it decreases. The control of the pipe's deformation is provided on the basis of this property.

Moreover, this kind of sensors may be used in hard conditions. In particular, the working temperature is from -70 °C to +70 °C. It is very important for Chayandinskoye region.

The disadvantage is that the method cannot be applied for the pipelines made of non-metal materials.

3. Conclusion

There are different methods to detect sinkholes, as well as the ways of predicting and monitoring the growth of cavities and protecting pipelines against their negative impact. However, karst hazard is still a great challenge for particular regions. Unfortunately, there is no optimal solution that satisfies all needs of the humanity yet. Karst problem is a serious challenge for the safe oil and gas transportation and as a result, it is still the issue researched all over the world. The impact of natural factors on the stability of pipelines in karst areas influences the development of petroleum industry, as well as contributes to investment in the knowledge-intensive field. The analysis of all methods overviewed in the paper demonstrates that the sensors based on Barkhausen effect is the best way to monitor pipelines deformations since the method is applicable at different temperatures.

References

[1] Shammazov A M 1999 *Calculation of gas pipelines in a karst area* (Ufa: Gilem)

[2] Taviani M, Angeletti L, Campiani E et al 2012 Drowned karst landscape offshore the Apulian Margin (Southern Adriatic sea, Italy). *J. Cave and Karst Studies* **74**(2) 197–212

[3] Teterin E A and Strokova L A 2015 Karst development in the cambrian limestones, Jakutia, Russia. *Proc. XIX intern. symp. in Honour of Academician M. A. Usov: Problems of Geology and Subsurface Development (Tomsk, 8–10 April 2015)* (Tomsk: TPU) (in prep)

[4] Strokova L A and Ermolaeva A V 2015. Natural features of construction of the main gas pipeline «The Power of Siberia» on a site Chayandinskoye oil and gas field - Lensk *Bulletin of the Tomsk Polytechnic University* **326** (4) 41–55

[5] Al-Sayyal E B and Shiryaeva T S 2008 *Method for pipeline protection against emergency state caused by karst holes.* Russian Patent 2316630

[6] Meng Y, Dai J, Jia D and Lei M 2013 Typical methods for forecasting karst collapse in China. In Land L, Doctor DH, Stephenson JB, (ed.) *Sinkholes and the Engineering and Environmental Impacts of Karst: Proc. of the Thirteenth Multidisciplinary Conference (Carlsbad, New Mexico, 6–10 May 2013) NCKRI Symposium 2. Carlsbad (NM)*: National Cave and Karst Research Institute pp 239–245

[7] Cullity B D 1972 *Introduction to magnetic materials* (USA: Addison-Wesley Pub. Co)

Uranium and thorium behavior in groundwater of the natural spa area "Choygan mineral water" (East Tuva)

Y Kopylova[1], N Guseva[1,3], A Shestakova[1,4], A Khvaschevskaya[1] and K Arakchaa[2,5]

[1] National Research Tomsk Polytechnic University, 30 Lenin Ave., Tomsk, 634050, Russia
[2] Research Institute of Medical and Social Problems and Management of Tuva, 48 Lenin Street, Kyzyl, 667000, Russia

E-mail: [3] guseva24@yandex.ru, [4] stysha@sibmail.com, [5] chodura@yandex.ru

Abstract. The natural spa area "Choygan mineral waters", a unique deposit of natural carbon dioxide mineral waters in Siberia, is located in the Eastern Sayan Mountains. There are 33 spring discharges in this area. Spring waters are mainly of HCO_3-Na-Ca type. TDS varies from 300 mg/L to 2600 mg/L, the temperature ranges from 7 °C (in spring 33) to 39 °C (in spring 12), pH varies from 5.9 to 8.3, and the value of the oxidation-reduction potential is from -170 mV to 236 mV. All studied waters were divided into two groups according to their temperature and geochemical conditions: cold fresh water in oxidizing environment and warm slightly brackish water in reducing environment. The uranium concentration varies from 0.7 to 14 μg/l and the thorium concentration varies from 0.001 to 0.33 μg/l in the studied waters. The predominant uranium complexes are $(UO_2(CO_3)_3)^{4-}$, $(UO_2(CO_3)_2)^{2-}$, UO_2CO_3, $(UO_2(PO_4)_2)^{4-}$ in the waters in oxidizing and reducing environments. It was found that acid-alkaline and oxidizing-reducing conditions were the determining factors for uranium behavior and speciation in the studied waters. The pH conditions are determining factors for thorium behavior and speciation in the studied waters. In slightly acidic water the predominant thorium species is negatively charge complex $(ThCO_3(OH)_3)^-$ (more than 95 %).

1. Introduction

The natural spa area "Choygan mineral waters", a unique deposit of natural carbon dioxide mineral waters in Siberia, is located in the Eastern Sayan Mountains. Its formation is related to the latitudinal fault in the area with current and recently extinct volcanic activity in the East Sayan Mountains. The Choygan warm and cold springs are a potential deposit of natural carbon dioxide medicinal waters. The discharging cold and thermal natural carbon dioxide groundwaters have a unique set of bioactive components: Rn, SiO_2, CO_2, Fe, etc. This unique site is of abiding interest for researchers. The first studies of these groundwaters were performed by I.S. Kryzhin (1858), S.V. Obruchev (1945), V.G. Tkachuk (1955), E.V. Pinneker (1966–1967).

The attention focused on the study of these waters has contributed to the development and improvement of the research and getting new knowledge about chemical composition, processes and mechanisms of formation, balneology, etc [1, 2].

However, the abundance of most trace elements, including uranium and thorium, in groundwaters of the natural spa area Choygan has not been studied in detail. Therefore, the aim of this research is to study the abundance and species of uranium and thorium in groundwaters of the natural spa area "Choygan mineral waters".

Uranium can be found in nature both in geological and aqueous environment. The concentration of U in groundwater is less than 15 μg/L. The U concentration in groundwater and surface water can reach 1 mg/L [3]. Uranium exists as various chemical species in natural waters, including the free metal ions (UO_2^{2+}), and complexes with inorganic and organic ligands (both anion and cation) [4].

A lot of works are devoted to the study of the uranium behavior in the environment. A speciation of uranium depends on several factors, including pH, Eh, concentrations of organic matter, calcium carbonate or other ligands. Phosphate, carbonate, and humic substances are the most common ligands in natural waters. Hexavalent (U (VI)) and tetravalent (U (IV)) are the main oxidation states of U in nature waters.

In an open system, under oxidizing conditions, pH<5, lack of natural organic matter and potential ligands, uranyl ion (UO_2^{2+}) dominates in waters with pH below 5 or hydroxy complexes in waters with pH below 6,5. Increase of pH (> 7) and the CO_3^{2-} and HCO_3^- ions contribute to the formation of the di- and tri-carbonate complexes [3]. Moreover, as referred to [5, 6] phosphorus with a relatively high concentration (6×10^{-4} mol/L) and low pH contribute to the sorption of U (VI).

The subsurface uranium environment could be strongly adsorbed on iron hydroxides, clay minerals and dissolved organic matter, although desorption may occur due to competition between mineral surface and ligands for binding of uranyl [3]. The last data show that complexes with dissolved organic matter (DOM) and uranyl complexes of calcium (Ca-uranyl carbonate complexes) play an important role in water chemistry of uranium. These complexes are still omitted from many speciation calculations and have minor importance in the alkaline pH range. However, Gustafsson et al. (2009) included dissolved organic C (DOC) in their simulation of U species and showed that the dominance of DOM-UO_2 complexes at pH < 7.2 under atmospheric CO_2 pressure. When partial CO_2 pressure (P_{CO2}) in that study was increased 10-fold, the pH threshold, at which DOM-UO_2 complexes dominated decreased to 6.6, illustrating that carbonate species are predominantly complex-forming ligands [7].

2. Materials and methods

The natural spa area "Choygan mineral waters" is located between 52°34'45.98"N and 52°34'59.95"N, and between 098°45'17.12"E and 098°45'25.27"E in the absolute height of 1550 m in the north-eastern part of the Tuva region on the western slope of the Eastern Sayan Mountains. The groundwaters discharge in 33 springs within a relatively small area of the Arzhaan-Khem river valley.

These mineral springs are confined to a large E–W striking fault in Precambrian rocks (gneisse, marble, schist) intruded by Paleozoic granites and diorites. The Upper Proterozoic Sailygskaya water-bearing sequence embraces marble, schist and marbled limestone interbedded with shale, quartzite, conglomerate and intruded by Devonian granite intrusions of Brensky sequence. Discharge of hot and cold springs are accompanied by the deposition of recent travertine. The travertine dome has a brown color due to Fe oxides and hydroxides.

The study of Choygan groundwater was performed in the summer of 2013. 33 water samples were collected. Electrical conductivity, temperature and pH were measured *in situ* using a portable device Water Test. Water chemical composition analysis was performed in Fundamental Research Laboratory of Hydrogeochemistry, Education and Research Centre "Water" (Tomsk Polytechnic University, Tomsk, Russia). Water samples were investigated by the following methods: titrimetric (HCO_3^-, CO_2, CO_3^{2-}), ion chromatography (SO_4^{2-}, Cl^-, Ca^{2+}, Mg^{2+}, SO_4^{2-}, Na^+, K^+). The concentrations of trace elements, including uranium and thorium, were measured by ICP-MS (NexION 300D, Perkin Elmer, USA).

The geochemical modeling program HydroGeo (author M. Bukaty) was used to predict uranium and thorium speciation in the studied waters. For the calculations, the following variables were used: pH, Eh, temperature, 17 based ions: H^+, OH^-, H_2O, e^-, Na^+, Mg^{2+}, K^+, Ca^{2+}, Cl^-, $(SO_4)^{2-}$, $(HCO_3)^-$, Fe^{3+},

Al^{3+}, $(UO_2)^{2+}$,$(PO_4)^{3-}$, SiO_2 and 72 species, including 49 uranium species and 12 thorium species: UO^+, U^{2+}, $(UO_3)^-$, $(UOH)^{2+}$, UO_3, $(UO_4)^{2-}$, $(HUO_4)^-$, $(HUO_3)^-$, UO_2OH, $((UO_2)_4(OH)_7)^+$, UO_2OH^+, $UO_2(OH)_2$, $((UO_2)_2(OH)_2)^{2+}$, $((UO_2)_3(OH)_4)^{2+}$, $((UO_2)_3(OH)_5)^+$, $((UO_2)_3(OH)_7)^-$, $(UO_2(OH)_3)^-$, $((UO_2)_2(OH))^{3+}$, $(UO_2(OH)_4)^{2-}$, $(U(OH)_3)^+$, $U(OH)_4$, $(U(OH)_2)^{2+}$, HUO_2, $(HUO_2)^+$, U^{4+}, U^{3+}, $(UOH)^{3+}$, UO_2, $(UO_2)^+$, $(USO_4)^{2+}$, $U(SO_4)_2$, $UO_2S_2O_3$, $(U(CO_3)_5)^{6-}$, $(U(CO_3)_4)^{4-}$, $((UO_2)_2CO_3(OH)_3)^-$, $(UO_2(CO_3)_3)^{5-}$, $(UO_2(CO_3)_3)^{4-}$, $(UO_2(CO_3)_2)^{2-}$, UO_2CO_3, $UO_2(HCO_3)_2$, $(UO_2HCO_3)^+$,$((UO_2)_3(CO_3)_6)^{6-}$, $(UO_2(PO_4)_2)^{4-}$, $(UO_2)_3(PO_4)_2$, $(UO_2H_2PO_4H_3PO_4)^+$, $(UO_2H_3PO_4)^{2+}$, $(UO_2H_2PO_4)^+$, $(UHPO_4)^{2+}$, $U(HPO_4)_2$. $(Th(OH)_5)^-$, $(Th(OH)_3)^+$, $(Th(OH)_2)^{2+}$, $(ThOH)^{3+}$, $(Th(OH)_6)^{2-}$, $Th(OH)_4$, $(ThCl)^{3+}$, $(ThSO_4)^{2+}$, $Th(SO_4)_2$, $(Th(SO_4)_3)^{2-}$, $(ThCO_3(OH)_3)^-$, $(Th(CO_3)_5)^{6-}$.

3. Results and Discussion

According to the hydrogeological and hydrogeochemical features, the studied groundwaters are divided into several types: natural carbon dioxide fissure-vein waters in reducing and oxidizing environments and groundwater of the regional-jointing zone (background waters) (table 1). The natural carbon dioxide fissure-vein waters in reducing environment are the water of springs № 1, 6–13, 15–17, 19–20, 22, 31–32. The natural carbon dioxide fissure-vein waters in oxidizing environment discharge in springs № 2–5, 21 and 23–28, 30. The groundwaters of the regional-jointing zone discharge in springs 33, 29, 29a [1, 2].

Within the studied area, the warm natural carbon dioxide fissure-vein waters in reducing environment are located in the central part of the Argens-Khem river valley, including the left bank. The temperature varies from 24 °C to 40 °C. This type of waters is characterized by slightly acidic conditions (mean value of pH 6.4). The mean value of Eh is 21 mV. TDS values are up to 1969 mg/L. Warm waters belong to the HCO_3-Ca-Na type. The TDS values and the concentrations of such major ions as HCO_3, Cl, Ca, Na, Mg, K and Si increase to the water temperature.

Table 1. Chemical composition of Choygan mineral waters, mg/L.

Parameter	Mean content		
	Natural carbon dioxide fissure-vein waters		Groundwaters of the regional-jointing zone
	Reducing environment	Oxidizing environment	
T, °C	30	18	9
Eh, mV	21	196	203
pH	6.4	6.3	7.6
CO_2	759.5	880.3	22.8
HCO_3^-	1619.4	959.6	229.5
SO_4^{2-}	14.2	28.4	10.1
Cl^-	23.8	11.0	3.3
Ca^{2+}	221.6	185.4	66
Na^+	290.4	109.2	2.3
Mg^{2+}	32.5	20.3	6.1
K^+	44.6	16.8	2.7
PO_4^{3-}	0.11	0.03	0.04
Th	0.00003	0.000004	0.000003
U	0.003	0.005	0.001
Th/U	0.008	0.002	0.014
TDS	1969	1212	320
Chemical type	HCO_3-Ca-Na	HCO_3-Na-Ca	HCO_3-Ca
Number of points	18	12	3

The natural carbon dioxide fissure-vein waters with oxidizing conditions are cold where the temperature ranges from 11 to 20 °C. The waters are fresh and slightly acidic with the mean value of TDS of 1212 mg/L and the pH value of 5.9–6.7. The free carbonic acid concentration reaches 1488 mg/L (spring 4). The values of Eh vary from 170 to 236 mV. The mean value is 196 mV. Besides, the concentrations of sulfate ions in the waters from 5 to 59 mg/L underline the oxidizing conditions in the studied waters (table 1). According to the ratio of major ions the studied waters refer to the HCO_3-Na-Ca type.

The groundwaters of the regional-jointing zone are characterized by oxidizing conditions. The mean value of the redox potential is 203 mV. These waters are cold with the mean temperature of 11 °C. Acid-alkaline properties of the water are slightly alkaline. The studied waters belong to the HCO_3-Ca chemical type. The TDS values vary from 290 to 350 mg/L.

The studied water types differ from each other by the concentration of uranium and thorium. The uranium concentrations vary from 0.7 to 12 μg/L in the natural carbon dioxide fissure-vein waters in reducing environment. The highest concentrations of uranium are observed in springs 16, 17, 19 and 31. From reducing to oxidizing conditions, the uranium concentration changes and increases with an increase of the TDS values. High uranium content varying from 0.16 to 14 μg/L is observed in the natural carbon dioxide fissure-vein waters in oxidizing environment. The high concentration of uranium is typical for the waters with positive values of Eh. This is caused by the high solubility of U^{6+} in comparison to U^{4+} [8, 9].

Thorium accumulates mainly in the natural carbon dioxide fissure-vein waters in reducing environment. The mean concentration of thorium is 0.00003 mg/L in this type of waters. Its concentration is 8 times higher than those in the waters in oxidizing environment, where the mean concentration of thorium is 0.000004 mg/L.

The world average concentration of uranium in the rocks is 1–4 mg/kg while the content of thorium is three times higher and reaches 13 mg/kg. The concentrations of U in groundwater are usually low, typically in the range of 0.1–1 μg/L, although they can reach up to several tens or hundreds of μg/L as a result of reactions with minerals in aquifers. The concentration of thorium is quite low and reaches up to 0.24 mg/L, which can be explained by the weak migration capacity of thorium in groundwater [3]. Uranium and thorium tend to occur together in the geological unit formation, due to the similarity in their ionic radius. The U^{4+} is easily oxidized and migrates during crustal evolution, whereas Th is stable in the oxidation zone. Therefore, Th is considered as a reference concentration and can be used to study the state of the original uranium concentration [10].

According to the abundance of uranium and thorium in granitic rocks, where the investigated waters occur, the thorium-uranium ratio is not less than 4. In the studied waters, the thorium-uranium ratio decreases from 0.0027 (an average) in the natural carbon dioxide fissure-vein waters in reducing environment to 0.0003 in natural carbon dioxide fissure-vein waters in oxidizing environment. This tendency is caused by uranium and thorium behavior in natural waters. The concentration of thorium in the natural carbon dioxide fissure-vein waters in reducing environment exceeds those in the natural carbon dioxide fissure-vein waters in oxidizing environment. At the same time, reverse behavior is characteristic for uranium: the concentration of uranium in the waters in the oxidizing environment is two times higher than those in the waters in the reducing environment. This is the main difference between the abundance of these elements in waters and rocks, which is determined by the peculiarities of the uranium and thorium speciation in natural waters [8, 9].

According to the results of physicochemical modeling of uranium speciation, bicarbonate complexes $(UO_2(CO_3)_2)^{2-}$ (up to 67 %), $(UO_2(CO_3)_3)^{4-}$ (up to 65 %), and UO_2CO_3 (up to 2 %) are the most abundant uranium species in the natural carbon dioxide fissure-vein waters with the Eh values varying from 0 to 142 mV and pH 6,4–6,9 (figure 1A). In the waters with Eh <0 and pH 6,1–6,4, the proportion of UO_2 and $U(OH)_4$ complexes increases up to 48% and $(U(OH)_3)^+$ (up to 5 %) and $((UO_2)_3(CO_3)_6)^{6-}$ (up to 5%). With the decrease of Eh value up to -90 – -170 mV and the pH value up to 6, the proportion of UO_2^0 complex increases up to 48 % and becomes predominant, whereas the proportion of $(U(OH)_3)^+$ complex reaches 5 %.

In the spring waters with PO_4^{3-} content is more than 0.05 mg/L, uranyl phosphate $(UO_2(PO_4)_2)^{4-}$ emerges. The content of this complex reaches 10 %. In the water with the concentration of PO_4^{3-} is 0.46 mg/L this complex prevails, its content reaches up to 76%.

The predominant uranium complexes are $(UO_2(CO_3)_2)^{2-}$ (up to 80 %), $(UO_2(CO_3)_3)^{4-}$ (up to 54 %), and UO_2CO_3 (up to 3 %) in the slightly acidic natural carbon dioxide fissure-vein waters in oxidizing environment. With the values of pH up to 5.9–6.2, the proportion of $(UO_2(CO_3)_2)^{2-}$ complex increases, whereas the percent of $(UO_2(CO_3)_3)^4$ complex decreases, and the UO_2CO_3 and $((UO_2)_3(CO_3)_6)^{6-}$ complexes emerge. The proportions of $(UO_2(CO_3)_2)^{2-}$ and $(UO_2(CO_3)_3)^{4-}$ complexes are equal under neutral conditions. Such behavior of uranium complexes is typical for the investigated type of waters [7, 11–13].

The formation of uranyl-carbonate complexes is caused by the occurrence of HCO_3^- and CO_3^{2-} in the groundwaters and pH 4.5–6.5. In this case a sufficient concentration of the ligand (CO_3^{2-}), and, consequently, the displacement of equilibrium toward the formation of carbonate complexes are provided [7, 9, 11–13].

The migration of uranium with carbonate compounds reduces, whereas the migration of uranyl ion with phosphate complexes increases in the waters with low pH. The percent of migration $(UO_2(PO_4)_2)^{4-}$ is 30 %, $(UO_2H_2PO_4)^+$ – 10 %, $(UO_2H_2PO_4H_3PO_4)^+$ – 35 % at pH 5.9.

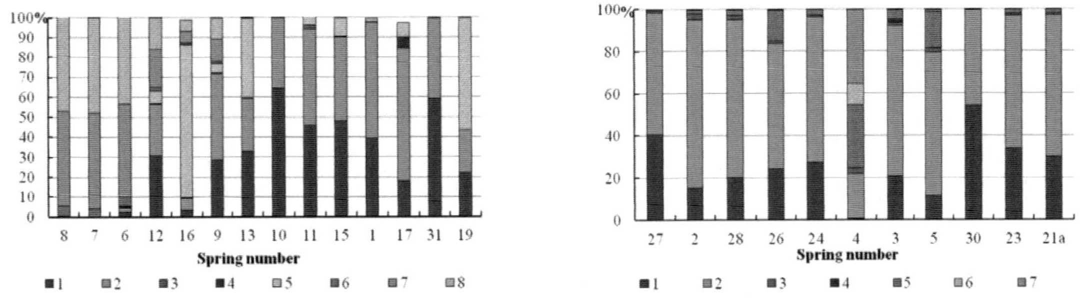

A B

Figure 1. Proportions of uranium complexes as a function of Eh: (A) in the natural carbon dioxide fissure-vein waters with reducing conditions (1 – $(UO_2(CO_3)_3)^{4-}$, 2 – $(UO_2(CO_3)_2)^{2-}$, 3 – UO_2CO_3, 4 – $(UO_2)_3(CO_3)_6)^{6-}$, 5 – $(UO_2(PO_4)_2)^{4-}$, 6 – $(U(OH)_3)^+$, 7 – UO_2, 8 – $U(OH)_4$); (B) in the natural carbon dioxide fissure-vein waters with oxidizing conditions (1 – $(UO_2(CO_3)_3)^{4-}$, 2 – $(UO_2(CO_3)_2)^{2-}$, 3 – UO_2CO_3, 4 – $(UO_2)_3(CO_3)_6)^{6-}$, 5 – $(UO_2(PO_4)_2)^{4-}$, 6 – $(UO_2H_2PO_4)^+$, 7 – $(UO_2H_2PO_4H_3PO_4)^+$).

The acid-alkaline conditions in the waters have a great influence on thorium migration as element-hydrolyzate. The negatively charge $(ThCO_3(OH)_3)^-$ is predominant thorium complex (more than 95 %) in slightly acidic waters of all studied types. The percent of this complex increases up to 100 % under neutral conditions. The proportion of the $(ThCO_3(OH)_3)^-$ complex decreases due to the formation of the positively charge thorium hydroxocomplexes ($(Th(OH)_3)^+$ and $(Th(OH)_2)^{2+}$) if pH-value decreases to 5.9–6.2.

4. Summary and Conclusions

The features of the water chemical composition in the natural spa area "Choygan mineral waters" were studied. The following water types were identified: warm natural carbon dioxide waters in reducing environment, cold natural carbon dioxide waters in oxidizing environment and groundwaters of the regional-jointing zone (background water).

The natural carbon dioxide waters in reducing environment belong to the HCO_3-Ca-Na chemical type. The mean value of TDS is 1916 mg/L. The uranium concentrations vary from 0.7 to 12 μg/L. The thorium concentrations vary from 0.001 to 0.33 μg/L.

The second group of the waters is cold natural carbon dioxide waters in oxidizing environment. They are fresh with the average value of TDS is 1212 mg/L. Their chemical type is HCO_3-Na-Ca. The free carbonic acid concentrations vary from 200 to 1488 mg/L. The geochemical conditions of this water type are favorable for the uranium accumulation, the concentration of which is from 0.93 to 14 μg/L. The uranium concentration is higher in this type of water than that in the natural carbon dioxide waters in reducing environment. This is caused by the high concentration of free carbonic acid in the waters and the high positive values of Eh. The thorium concentrations in these waters range from 0.001 to 0.03 μg/L.

The predominant uranium complexes are $(UO_2(CO_3)_3)^{4-}$, $(UO_2(CO_3)_2)^{2-}$, UO_2CO_3, $(UO_2(PO_4)_2)^{4-}$ in the waters in oxidizing and reducing environments. However, in the waters in oxidizing environment there are also such uranium complexes as $(UO_2H_2PO_4)^+$ and $(UO_2H_2PO_4H_3PO_4)^+$, the total percent of which is less than 2 %. In the waters in reducing environment the total proportion of these complexes increases up to 10 % and 35 %, respectively. In addition to the carbonate and phosphate complexes the UO_2, $U(OH)_4$ and $(U(OH)_3)^+$ compounds occur in the water in reduction environment. It was proved that acid-alkaline and oxidizing-reducing conditions are the determining factors for uranium behavior and speciation in the studied waters.

The acid-alkaline conditions are determining factors for thorium behavior and speciation in the studied waters. In the slightly acidic water the predominant thorium species is a negatively charge complex $(ThCO_3(OH)_3)^-$ (more than 95 %). The proportion of the $(ThCO_3(OH)_3)^-$ complex decreases due to the formation of positively charge thorium hydroxocomplexes $((Th(OH)_3)^+$ and $(Th(OH)_2)^{2+})$ when pH-value decreases up to 5.9–6.2.

Acknowledgements
This work was supported by the State Assignment "Nauka" № 5.1931.2014/K.

References

[1] Kopylova Y G, Guseva N V, Shestakova A V, Khvashchevskaya A A and Arakchaa K D 2014 Geochemistry of Carbon Dioxide Mineral Waters of the Choigan Natural Complex (Northeastern Tuva) *Russian Geol. and Geoph.* **55** (11) 1295–1305

[2] Kopylova Y G, Guseva N V, Shestakova A V, Khvashchevskaya AA and Arakchaa K D 2014 Dissolved Gas Composition of Groundwater in the Natural Spa Complex "Choygan Mineral Waters" (East Tuva) *IOP Conf. Ser.: Earth Environ. Sci.* **21** 012022

[3] Wu Y, Wang Y and Xie X 2014 Occurrence, behavior and distribution of high levels of uranium in shallow groundwater at Datong basin, northern China *The Science of the total environment* **472** 809–17

[4] Zhao J, Fasfous I I and Murimboh J D 2009 Kinetic study of uranium speciation in model solutions and in natural waters using Competitive Ligand Exchange Method *Talanta* **77** 10150–102

[5] Wang X, Ni S and Shi Z 2014 Uranium distribution in the sediment of the Mianyuan River near a phosphate mining region in China and the related uranium speciation in water *Chemie der Erde – Geochemistry* **74** (4) 661–69

[6] Guo Z, Li Y and Wu W 2009 Sorption of U(VI) on goethite: Effects of pH, ionic strength, phosphate, carbonate and fulvic acid Original Research Article *Applied Radiation and Isotopes* **67**(6) 996–1000

[7] Norrström A C and Löv Å 2014 Uranium Theoretical Speciation for Drinking Water from Private Drilled Wells in Sweden – Implications for Choice of Removal Method *Appl. Geoch.* **51** 148–52

[8] Lisitsin A K 1975 *Hydrogeochemistry of Ore Formation (as Example Exogenic Epigenetic Uranium Ore)* (Moscow: Publishing House «Nedra») (in Russian)

[9] Kraynov S R, Ryzhenko B N and Shvets V M *2012 Geochemistry of Groundwater. Theoretical, Applied and Environmental Aspects* (Moscow: CentrLitNefteGas, 2012) (in Russian)

[10] Asfahani J, Aissa M and Al-Hent R 2007 Uranium migration in a sedimentological phosphatic environment in Northern Palmyrides, Al-Awabed area, Syria. *Applied radiation and isotopes: including data, instrumentation and methods for use in agriculture, industry and medicine* **65** (9) 1078–86

[11] Hoyer M, Zabelt D, Steudtner R, Brendler V, Haseneder R and Repke J-U 2014 Influence of Speciation During Membrane Treatment of Uranium Contaminated Water *Separation and Purification Technology* **132** 413–21

[12] Semião A J C, Rossiter H M and Schäfer A I 2010 Impact of Organic Matter and Speciation on the Behaviour of Uranium in Submerged Ultrafiltration *Journal of Membrane Science* **348** (1–2) 174–80

[13] Rossiter H M, Graham M C and Schäfer A I 2010 Impact of Speciation on Behaviour of Uranium in a Solar Poweredmembrane System for Treatment of Brackish Groundwater *Separation and Purification Technology* **71** (1) 89–96

IOP Conf. Series: Earth and Environmental Science **27** (2015) 012035 doi:10.1088/1755-1315/27/1/012035

Microbiological composition of river waters in the Ob' basin (West Siberia) and its associations with hydrochemical indices

O G Savichev[1,2], O G Tokarenko[1,3], E Yu Pasechnik[1,4], N G Nalivaiko[1], E A Ivanova[1] and L V Nadeina[1]

[1] National Research Tomsk Polytechnic University, 30 Lenin Ave., Tomsk, 634050, Russia

E-mail: [2] osavichev@mail.ru, [3] tog@tpu.ru, [4] paseyu@yandex.ru

Abstract. Chemical and microbiological composition of the Mid-Ob' and its feeders' waters has been studied. Swampiness of the area is the cause of significant organic and biogenic substance content in streams, and it is also responsible for a large variety of organotrophic microflora. Microbiological composition of studied streams characterizes them as contaminated. Settlements are the main sources of the investigated area water pollution.

1. Introduction

Microorganisms are of vital importance in ecologo-geochemical formation of biosphere components, including aquatic ecosystems. Therefore, quantitative and group compositions of microorganisms are one of the most important criteria of surface water body condition. It is impossible to objectively estimate recent and future condition and plan effective aquatic protection activities without direct knowledge of river water microbiological composition, and this fact determines the relevance of these investigations. The research of microbiocoenosis in Siberian aquatic ecosystems is carried out by scientists (with the participation of authors) of National Research Tomsk Polytechnic University during 1994–2014. The results of one of its stages, which dealt with the Ob' and its inflows water microbiological composition (West Siberia, the Russian Federation), are given below. The objective of the study is to prove the fact that microorganisms are both an indicator of natural water condition and one of the factors, affecting on formation of their chemical composition.

2. Materials and methods

2.1. *Subject of research*

Microbiocoenosis of river waters in the Ob' basin in the section of its middle reach (figure 1) is the subject of research. The investigated area is located in forest part of physiographic West Siberia and in continental West-Siberian climatic region of temperate climatic zone. This area is mainly characterized by distribution of plain sub-taiga, mid and south taiga landscapes. The intense quantity of bogs which determines natural degree of flow control and high concentration of organic and biogenic substances in waters is the main feature of this area. Quite harsh weather conditions determine long winter season. During this period there is stable ice cover on rivers and the rivers are mainly fed by ground water. In spring snow melting results in heavy flood and, in turn, under plain relief it results in ordinary flooding and underflooding of a lot of areas. When the flood is going down the erosion process intensity increases, and these processes have a very negative effect on economic entities and apartment houses. Therefore, Tomsk region's natural conditions, on the one hand, stimulate water resource formation, and, on the other hand, they determine unfavorable conditions of water use caused by low quality waters of boggy areas and necessity of considerable financial and material inputs to make arrangements for organizing flood-protection works and erosion control measures [1, 2].

Figure 1. Map of the studied area.

2.2. Sampling and research methods

The considerable part of material used in this paper, especially microelement, organic substance, microorganism content data, was obtained by authors in 1994 – 2014 yrs. This material was received due to field researches of Tomsk polytechnic university (TPU), Tomsk branch of the institute of petroleum geology and geophysics the Siberian Branch of Russian Academy of Sciences (SB RAS), Tomsk state university (TSU), Institute of water and ecology problem SB RAS, OJSC "Tomskgeomonitoring", LLC "INGEOTEH". And it was also obtained due to a number of researches, and surface-water body, water facilities monitoring in the territory of Siberian Federal District. According to working paper (WP) 52.24.353-94 "Recommendations. Sampling of land surface waters and treated sewerage" and GOST P 51592-2000 "Water. General requirements for sampling", field works were carried out by conjugate geochemical testing of natural environment components according to WP 52.24.353-94 "Recommendations. Sampling of land surface waters and treated sewerage" and GOST P 51592-2000 "Water. General requirements for sampling".

The field works included the reconnaissance study of the water body sites and surroundings which should be investigated, surface and underground water sampling, atmosphere precipitation, peat and bottom sediment sampling. If it is necessary, other environment components will be investigated and associated works will be carried out. Samplers or special containers were used for surface water sampling. The sampler was submerged at a depth of 0.3–0.5 m below surface. After reaching the required depth, the sampler was lifted to the surface and sample was poured in especially prepared containers made of chemically inert materials. To determine microelements, petroleum products, phenols, anionic surfactants, nitrogen compounds and phosphorus, the samples were preserved according to method of analysis. To carry out microbiological analysis, the water samples were taken without preservation in especially prepared glass containers. In transit and in the lab the samples were kept under conditions which excluded the possibility to contaminate them.

Methods of separation and interpretation of results are given in authors' and other researchers' publications [3–5].

3. Hydrochemical characteristic of river water

3.1. Microcomponents and pH

Large river waters of investigated area are at the average defined as low and mean mineralization freshwaters; they are also defined as hydro-carbonate calcium (according to classification by O.A. Alekhin), neutral and weakly alkaline ones. Minor and small river waters are characterized as from low to high mineralization freshwaters; in terms of pH, they range from weakly acidic to weakly alkaline ones. Main ions contents vary in a rather wide range, and they reach a maximum in south-western part of Tomsk region, in left-bank part of the Mid-Ob' basin, at the interface between a forest steppe zone and a south-taiga zone, which is in line with the Chaya and the Shegarka basins. There are the lowest and medium macro-component concentrations in river waters of right-bank taiga part of the Mid-Ob' basin, in the Tym and the Ket' drainage basins, and also in the Tom' basin (its upstream and midstream parts).

3.2. Dissolved organic substance

Both indirect indicators (chemical oxygen demand COD – indicator of resistant to oxidation substance content, 5-day biochemical oxygen demand BOD_5 – indicator of easily oxidizable substance content) and specific substance concentrations data were used to determine the level of organic substance contents. Total organic carbon in the region river waters is strongly varied a wide range from 3–4 mg/dm^3 and less in the Tom' basin to 18–20 mg/dm^3 and more in water of rivers of the Ob'- Irtysh interfluve with waterlogged drainage basins.

The swamp area surface runoff which gives at an average over 17 000 tons of carbon of organic compounds per year is the most important source of resistant to oxidation organic substances according to COD. Sufficiently large concentrations of substances identified as hydrocarbons of the oil sequence are in waters of all investigated rivers. It can be associated with not only anthropogenic effect but specific natural conditions as well, for example, carbonaceous rock distribution in the Tom' basin [7, 8]. Swamps can be a source of hydrocarbon intake in river waters. It is confirmed by data of organic substance composition of peat and organic compound content findings in the immediate swamp waters [9]. Analysis of organic compounds brings us to the conclusion that there is predominance of odd carbon atom n-alkanes, and this clearly demonstrates their natural origin. In addition to hydrocarbons, a number of different organic compounds, including carboxylic acids, phthalates, chlorine-bearing substances, is found in river waters in Tomsk region.

3.3. Iron and biogenic substances (compounds of nitrogen, phosphorus and silicon)

As noted above, surface and subsurface runoff from swamps, containing a lot of organic substances, come into drainage network [10]. Organic substances in the swamp waters derive directly from the formation and evolution of swamps. Excessive moistening and low drainage conditions of watersheds, in turn, contribute to the swamps. Some of organic substances mineralize to form ammonia, and in aqueous medium they mineralize to form ammonium ions. If there is oxygen deficiency, nitrification process is less intensive, and it results in accumulation of ions NH_4^+ and NO_2^- in swamp waters. Moreover, some organic acids (for example, fulvic acids) can form very stable iron, manganese and different metal compounds, accumulating in aqueous medium (mainly in the form of suspensions and colloids). As a result, the elevated concentrations of Fe, NH_4^+ and rare NO_2^- are identified not only in swamp waters but in river waters as well. Moreover, amounts of them exceed specified water quality standards very often (with probability more than 20 %). Swamp waters are diluted and decomposed in more oxygen-rich medium (it concerns organic and biogenic substance contents) in the process of swamp water inflow in river net. Taking this into account, it seems clear that, as a rule, the content of iron, ammonium nitrogen, nitrite nitrogen and organic substances decreases in river waters.

3.4. Microelements
River waters of investigated area are characterized by extremely wide range of microelements variation. As a rule, well-defined regularities of spatial microelements distribution in river waters are not identified. Microelements content in waters were described in detail in papers [12, 13].

As compared with maximum permissible concentration for water in a river (MPCr) the contents of Cu, Mn, Zn and Al, and it is rare Hg, Mo, V, Sr, Ni are higher in Tomsk region's river waters, though Mn standard violations in the Ob' and its plain feeders with heavy swampy watersheds are defined more often than in the Tom'. The highest concentrations of Cd, Hg and Li are confined to the river Tom' and its feeders flowing in large industrial centers area [13; 1].

4. Results and discussion

4.1. Microbiological composition of river waters
Due to researches carried out in TPU and Tomsk branch of the institute of petroleum geology and geophysics SB RAS, the various groups of heterotroph and lithotroph microorganisms were identified in region river waters. They are 1) saprophytic bacteria, mineralizing dead organic material; 2) oligotrophic bacteria which are able to develop at low concentrations of organic substances; 3) chemo-organotrophic bacteria; 4) denitrifying bacteria which are able to reduce nitrates to gas; 5) nitrifying bacteria – autotrophic microorganisms receiving energy due to oxidation of deoxidized nitrogen compounds; 6) ammonifying bacteria, which can decompose complex nitrogen containing organic compounds (proteins, amino acids and a number of various substances), resulting in ammonia, hydrogen sulphide and carbon dioxide isolation; 7) thionic bacteria receiving energy due to oxidation of sulphur and its reduced compounds; 8) sulphate-reducing bacteria – anaerobic microorganisms which are able to reduce sulphates to sulphides when in use metabolism [3, 1]. Total characteristic of identified ecologo-trophic groups of microorganisms, contained in waters of the Ob' and its feeders, is placed in table 1.

Table 1. Average contents of certain groups of bacteria in waters of the Ob', the Tom' and their feeders.

Bacteria	Unit of measurement	The Ob'	The Tom'	Feeders of the Tom'	Plain feeders of the Ob'
Saprophitic	cell/ml	12131	12532	7576	21227
Oligotrophic	cell/ml	41891	37653	65096	33890
Thionic	number	4	6	2	0
Nitrifying	number	7	5	3	–
Denitrifying	number	9	10	10	12
Ammonifying	cell/ml	8646	4451	–	–
Sulfate-reducing	point	<1	1	0	–
Oxidizing:					
Heptane	c.u.	326	235	169	–
Octane	c.u.	332	279	300	–
Decane	c.u.	386	338	344	–
Phenol	c.u.	229	245	309	–
Number of samples		20	9–32	3–11	6

The obtained data analysis has shown that in the Ob' there is oligotrophic bacteria predominance in group composition of microorganisms adapted to scattered organic material uptake. At the average saprophitic bacteria concentrations are more than twofold lower than oligotrophic ones, and saprophitic and oligotrophic bacteria concentration ratio is known to decrease according to movement of water masses from South to North [3]. Thionic bacteria were noted to develop in the Ob' high and low of the Tom' mouth, and near the villages (Krivosheino, Molchanovo, Mogochino) in September, 1991[11]. Their presence and high intensity of evolution allow us to make an assumption that there were reduced sulphur compounds in river waters in these sites. At the same period of time (September, 1991) thionic bacteria were not identified further northward, starting with the point located upwards of Kolpashevo (town). Their presence in the Ob waters, at least in 1991, is very likely to deal with inflow of the Tom' contaminated waters [3].

All in all, a decrease in the number of bacteria, which oxidize nitrogen containing organic substances, is observed northward. The similar tendency was not identified in the content change of other heterotrophic bacteria groups. As for hydrocarbon-oxidizing bacteria their total intensity of evolution in the Ob waters is mainly stable. The exception is provided by the special case of some points where the sharp increase of oil oxidizing bacteria (from 0…5000 cell/ml in the most of sections to 9000…28000 cell/ml downstream from the village Kargasok) was observed in summer 1999. Taking into account the fact that whole high concentration samples were selected at the river transport parking places or downstream from them, it may be concluded that there is pollution of the river waters with oil products, and it is anthropogenic pollution [3, 1].

Saprophytic microorganism contents mineralizing dead organic material are an average higher in waters of the Tom' than in the Ob', though their highest concentrations are identified not only in the sections of cities, which are potential microbiological pollution sources of water bodies (62000 cell/ml near Tomsk), but in Kemerovo region the river upwards from Mezhdurechensk (town) (69000 cell/ml 5 km upwards). In whole, bacterial contamination of the river is occasionally observed along the river.

Contrary to saprophytes, the level of oligotrophic bacteria content which are able to develop at low organic substance concentrations is higher in the Ob' but not in the Tom'.

Taking into account the fact that it is difficult for oligotrophic bacteria to develop in medium with increased concentration of nitrogen-containing organic compounds, it can be said that there are significant differences in qualitative composition of organic substances dissolved in waters of two rivers. Sulfate-reducing oxygen deficit bacteria should be noted among groups of microorganisms identified in waters of the river Tom'. Such kind of microorganisms in quantity 1 cell/ml were identified 0.3 km upwards Tomsk in March, 1998. In addition, it was noted that there were thionic bacteria in Kemerovo and Tomsk sections. This allowed us to make an assumption that there were reduced sulphur compounds in investigated river waters. Oil oxidizing bacteria contents in the Tom' waters as in the Ob' change slightly, and they account about 2000 cell/ml; this is in agreement with regular hydrocarbon distribution in river waters. Chemo-organotrophic bacteria distribution is characterized by a higher level of irregularity [3].

5. Conclusions

In general, according to microbiological indices waters of the Middle Ob' are characterized as contaminated. In accordance with requirements of GOST 17.1.2.04–77 "Indices of condition and assessment regulations of fishery water bodies" the category of saprobity vary greatly – from xenosaprobic ("clear") to polysaprobic ("dirty"). The worst condition is characteristic of minor rivers located in settlements or in the near vicinity of them. As for microbiological indices, the certain river water quality impairment is identified in sections of the river Tom', which are located within settlements or downstream from them, and sewage disposal of housing and utility companies.

Probably, considerable part of microorganisms, biogenic and organic substances, some microelements enter the region's rivers from natural and anthropogenic sources distributed in watersheds area. Swamps and wet parts of floodplains and valleys play an important role among natural sources; and as for anthropogenic sources, the main role belongs to the runoff from settlement and industrial areas. The purpose of microorganisms are also supposed to be the most important in the process of formation of nitrogen compound content and runoff; and to be of little importance when it deals with macroelements.

Acknowledgements

The investigations were supported by Russian Foundation for Basic Research (Project No 14-05-31518).

References

[1] Savichev O G 2010 *Vodnye resursy Tomskoy oblasti* (Tomsk: Izdatel'stvo Tomskogo politekhnicheskogo universiteta) p 248
[2] Akhmed-Ogly K V, Savichev O G, Tokarenko O G, Pasechnik E Yu, Reshetko M V, Nalivajko N G and Vlasova M V 2014 *IOP Conf. Ser.: Earth Environ. Sci.* **21** 12021
[3] Savichev O G, Nalivajko N G and Trifonova N A 2002 Mikrobiologicheskiy sostav rechnykh vod basseyna verkhney i sredney Obi *Sibirskiy ekologicheskiy zhurnal* **2** 173
[4] Rodina A G 1965 Metody vodnoy mikrobiologii (Moscow: Nauka)
[5] Kuznetsov SI and Romanenko V I 1963 Mikrobiologicheskoe izuchenie vnutrennikh vodoemov (Moscow: AN SSSR)
[6] Savichev O G 2000 Prostranstvennye i vremennye izmeneniya geokhimicheskogo sostoyaniya rechnykh vod basseyna Sredney Obi *Geografiya i prirodnye resursy* **2** 60
[7] Rasskazov N M, Shvartsev S L, Trifonova N A and Nalivajko N G 1995 Neletuchie organicheskie veshchestva i mikroorganizmy v podzemnykh vodakh rayona Krapivinskogo vodokhranilishcha na reke Tom' (Kuzbass) *Geologiya i geofizika* **4** 30
[8] Rogov G M and Popov V K 1985 Gidrogeologiya i katagenez porod Kuzbassa (Tomsk: Izd-vo Tomsk. gos. un-ta)
[9] Shvartsev S L, Serebrennikova O V, Zdvizhkov M A, Savichev O G and Naimushina O S 2012 Geochemistry of wetland waters from the lower Tom basin, Southern Tomsk oblast *Geochem. International* **50** 367
[10] Savichev O G 2009 Khimicheskiy sostav bolotnykh vod na territorii Tomskoy oblasti (Zapadnaya Sibir') i ikh vzaimodeystvie s mineral'nymi i organomineral'nymi soedineniyami *Zhurn. Izv. Tomskogo polit. univ.* **314** 72
[11] Shvartsev S L, Savichev O G, Vertman G G, Zarubina R F, Nalivaiko N G, Trifonova N G, Turov Yu P, Frizen L F and Yankovskii V V 1996 The ecological and geochemical state of the water in the middle course of the Ob River. *Water Resources* **23** 673
[12] Shvartsev S L and Savichev O G 1997 Ekologo-geokhimicheskoe sostoyanie krupnykh pritokov Sredney Obi *Zhurn. Vodnye resursy* **6** 762
[13] Shvartsev S L, Rasskazov N M and Savichev O G 1997 Contents and migration forms of elements in natural waters of the mid-Tom' basin *Geologiya i Geofizika* **12** 1955

Influence of peat formation conditions on the transformation of peat deposit organic matter

O V Serebrennikova[1,2], E B Strelnikova[2], Yu I Preis[3] and M A Duchko[1,2,4]

[1] National Research Tomsk Polytechnic University, 30 Lenin Ave., Tomsk, 634050, Russia
[2] Institute of Petroleum Chemistry SB RAS, 4 Akademichesky st., Tomsk, 634021, Russia
[3] Institute of Monitoring of Climatic and Ecological System SB RAS, 10/3 Akademichesky st., Tomsk, 634055, Russia

E-mail: [4] maria.duchko@gmail.com

Abstract. The paper studies the individual composition of n-alkanes, polycycloaromatic hydrocarbons, steroids, bi-, tri-, and pentacyclic terpenoids of two peat deposits of rich fen Kirek located in Western Siberia. Considering the individual n-alkanes concentrations, some indexes were calculated to estimate the humidity during peat formation. It was shown that the pH of peat medium primarily affects steroids, tri- and pentacyclic terpenoids transformations.

1. Introduction

The world's largest wetland systems are situated on the territory of Western Siberia. However, despite many years of wetlands research carried out in the region, this issue is still understudied. n-Alkanes, steroids, bi-, tri- and pentacyclic terpenoids presented in peat are biomarkers. Their composition shows the contribution of different peat-forming plants, peat formation conditions and the processes occurring in peat deposit to the peat composition. The study of individual composition of peat organic compounds allows drawing conclusions about the humidity and the pH impact on the direction of peat organic matter transformation [1].

2. Samples and analysis

Fen Kirek is located in Tomsk region, on the south-eastern shore of lake Kirek. It is a lowland mire, which is supplied by hydrocarbonate groundwater and lake water. Two fen Kirek peat deposits were chosen as the study objects. The first one (sedge peat), bing of lake genesis and 6.0 m in depth, was formed by carbonate sapropel deposit, followed by forested grass-sedge floating and fen sedge, grass and sedge-moss peat deposition. Lake waters are alkaline, so peat accumulated in an alkaline environment in constant high humidity conditions. In the lower part of the sedge peat deposit pH is 7.5, decreasing up to 6.9 with depth decrease.

The second deposit (wood-grass peat) is 4.2 m in depth. It was formed due to lake shore paludification, followed by carbonate sapropel deposition, forested sedge floating formation, wood and wood-grass peat deposition accompanied by lake drawdown, with conifers being predominant [2]. Peat was formed in neutral environment with sharp change in water regime. pH values lower from 7.0 to 6.3 with depth decrease in the wood-grass peat deposit.

Peat macrofossil composition was determined by microscopic method. All peat samples were dried and powdered before the analysis. Bitumens were concentrated by extracting with 7% methanol in chloroform at 60^0C. Group and individual composition of n-alkanes, polycycloaromatic hydrocarbons, steroids and terpenoids in peat samples were determined by gas chromatography-mass spectrometry (GC-MS). Thermo Scientific gas chromatograph with mass detector fused silica capillary column (30 m in length, 0.25 mm bore, 0.25 μm in film thickness) was used for biomarker molecule analysis. The operating conditions were as follows: temperature was held at 80°C for 2 min ramped to 300°C at 4°C/min, with He acting as carrier gas. The ionization energy of the mass spectrometer was set to 70 eV, with a scanning range being from 50 to 550 amu. The compounds were identified by comparison of their mass spectra with that of reference compounds.

3. Results and discussion

The following groups of organic substances were detected in peat bitumen: n-alkanes, polycycloaromatic hydrocarbons (PAHs), steroids, bi-, tri- and pentacyclic terpenoids. n-Alkanes with the odd homologues C_{21}-C_{31} base are the main representatives of organic compounds in the studied peat samples. Sedge peat is specified by high relative content of n-C_{23} alkane, the sample S-1 contains almost equal concentrations of n-C_{27}. The C_{27} homologue dominates in all samples of wood-grass peat, as well as in wood material from fen Kirek. The molecular mass distributions of n-alkanes in all samples of sedge peat and the sample W-3 of wood-grass peat are shown in figure 1.

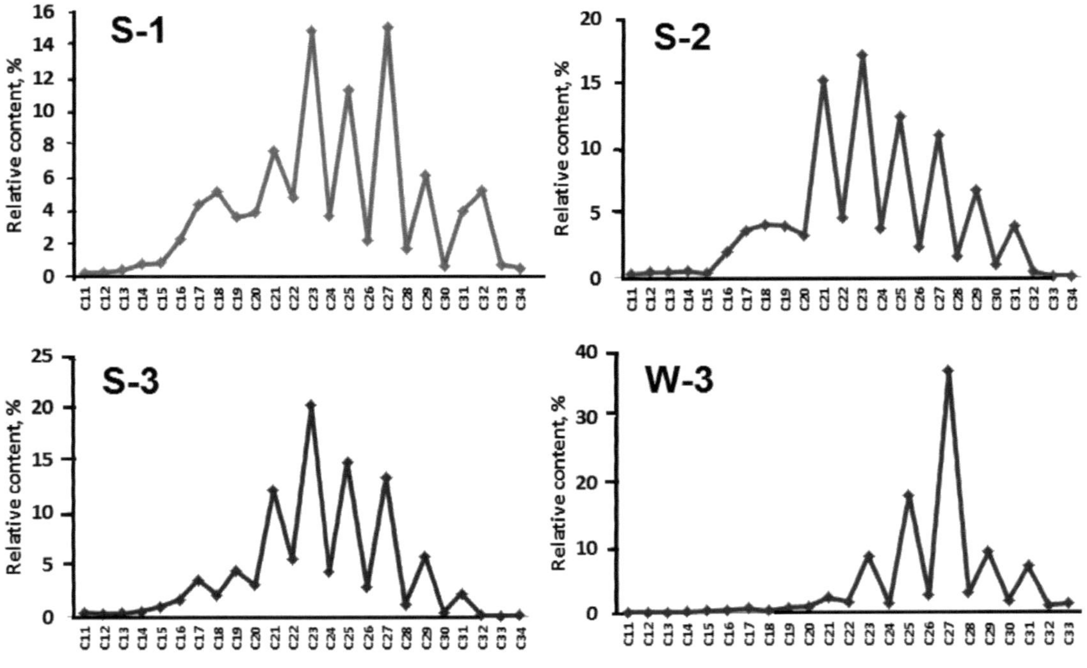

Figure 1. The molecular mass distributions of n-alkanes in samples S-1, S-2, S-3 and W-3.

Considering individual n-alkanes content, one can calculate several indices: Carbon Preference Index (CPI), humidity index (P_{aq}) and the concentrations ratio of n-alkanes, specific for different types of plants.

The predominance degree of molecules with an odd number of carbon atoms can be estimated using index CPI, which is a weight ratio of odd to even homologues [3, 4]. Peat formation conditions can be estimated using the humidity index (P_{aq}) [5, 6].Considering n-alkanes concentrations ratio, other indices can be calculated to gain more information about mire water balance. Strong relationship between distribution of plants and humidity conditions allows using the relative content of peat-forming plants biomarkers for mire paleohydrologic history reconstruction [7]. The n-alkanes n-C_{23}/n-C_{25} concentrations ratio shows the predominance of marsh plants that prefer wet conditions over the plants that grow in dry conditions [8].The dominant n-alkane, CPI, humidity index (P_{aq}), and the values of the ratio n-C_{23}/n-C_{25} are presented in table 1.

Table 1. The dominant n-alkane, CPI, P_{aq} and n-C_{23}/n-C_{25} in fen Kirek peat.

Sampling depth, m	Sample code	Dominant n-alkane	CPI	P_{aq}	n-C_{23}/n-C_{25}
		Sedge peat			
0.3	S-1	C_{23}, C_{27}	3.38	0.72	1.32
1.35	S-2	C_{23}	4.32	0.73	1.38
2.0	S-3	C_{23}	4.35	0.82	1.37
		Wood-grass peat			
0.25	W-1	C_{27}	7.36	0.77	0.71
1.0	W-2	C_{27}	6.39	0.83	1.10
1.3	W-3	C_{27}	6.96	0.62	0.48
1.5	W-4	C_{27}	21.40	0.88	0.60

CPI values increase with depth and, respectively, peat maturity, which is consistent with the published data [4]. The values of P_{aq} and ratios of n-alkanes n-C_{23}/n-C_{25} concentrations do not change in sedge peat. The values of these parameters for wood-grass peat change abruptly and correlate with each other well – an increase in n-C_{23}/n-C_{25} and n-P_{aq} values in the samples W-2 and W-4 might indicate a change of mire water regime due to lake Kirek periodic floods.

Bicyclic (naphthalene and its methyl-, dimethyl-, trimethyl-, tetramethylderivatives and cadalene), tricyclic (phenanthrene, its methyl, dimethyl, trimethyl and tetramethyl derivatives) and tetracyclic (fluoranthene and pyrene) structures were identified among PAHs in studied samples. Bicyclic PAHs predominate in all samples of sedge peat and in the samples W-2 and W-4 of wood-grass peat, tricyclic – in the samples W-1 and W-3. Tetracyclic PAHs are presented in low concentrations in all peat samples. Trimethylnaphtalenes dominate in the sample S-1 of sedge peat, phenanthrene – in the sample S-2. Cadalene content substantially exceeds the concentration of other PAHs in the sample S-3 as in sedge from fen Kirek [1]. Trimethylphenantrenes prevail in the samples W-1 and W-3 of wood-grass peat, cadalene – in the samples W-2 and W-4. Cadalene has a very stable relic structure and is genetically connected with bicyclic terpenes.

Bicyclic partially unsaturated structures, naphthenoaromatic compounds presented by calamenene, α- and β-calacorenes were identified among bicyclic terpenes in fen Kirek peat. All these structures are presented in a wide range of plants (in most conifers) [9]. γ-Cadinene prevails in the sample S-1 of sedge peat, δ-cadinene – in the samples S-2 and S-3, in all samples of wood-grass peat, as well as in sedge and wood material from fen Kirek. The concentrations of cadinene isomers changes slightly with depth decrease. It proves the resistance of these compounds and explains their high content in peat.

The prevalent square of fen Kirek is forested by coniferous trees. That is why tricyclic terpenoids are mainly presented in peat by abietic acid transformation products which are the part of conifers resin, abietic acid monoaromatic analogue, which can be found in pine bark, and 8,13R-epoxylabd-14-ene, distinctive for grass plants and shrubs. Abietic acid derivatives include naphthenic, naphthenoaromatic, and aromatic hydrocarbons formed by decarboxylation, hydrogenation, demethylation or dehydrogenation of the parent acid. Table 2 lists relative concentrations of tricyclic terpenoids and pentacyclic terpenoids groups in the investigated peat.

Table 2. The content of tricyclic terpenoids and pentacyclic terpenoids groups in the peat of fen Kirek, % rel.

	Sedge peat			Wood-grass peat			
Sampling depth, m	0.3	1.35	2	0.25	1	1.3	1.5
Sample code	S-1	S-2	S-3	W-1	W-2	W-3	W-4
Tricyclic terpenoids							
18-Norabietane	52.17	89.66	76.71	81.99	92.29	88.33	39.32
18-Norabieta-8,11,13-triene	3.19	2.36	6.51	5.08	4.55	3.56	5.7
Abieta-8,11,13-triene	0.43	0.28	0.39	0.28	0.19	0.15	0.32
10,18-Bisnorabieta-5,7,9(10),11,13-pentaene	2.54	1.75	8.25	5.16	1.27	3.16	28.6
Methyl-6,8,11,13-abietatetraene-18-oat	0.12	0.18	0.1	0.05	0.04	0.1	0.04
Methyl-8,11,13-abietatriene-18-oat	0.37	0.27	0.53	0.28	0	0.12	0.37
Abieta-8,11,13-triene-18-ol	0	0	0	0.07	0	0.02	0.08
Dehydroabietic acid	0	0	0	0.11	0	0.98	3.92
8,13 R -Epoxylabd-14-ene	38.2	3.06	4.32	1.22	0.78	0.63	0.86
Retene	3	2.44	3.18	5.75	0.88	2.94	20.8
Pentacyclic terpenoids							
Unsaturated structures	92.23	89.65	59.7	86.99	75.47	83.9	98.64
Saturated structures	7.77	10.36	40.3	13	24.54	16.1	1.37
Hydrocarbons	39.41	22.75	82.09	27.06	62.84	54.84	20.41
Ketones	60.59	77.26	17.91	61.31	37.17	24.34	28.96
Alcohols	0	0	0	11.62	0	20.82	50.64

The samples S-2 and S-3 are specified by a significant predominance of saturated hydrocarbon – 18-norabietane, which dominates also in the sample S-1. The content of 8,13R-epoxylabd-14-ene is also rather high in this sample. 18-Norabietane prevails in all samples of wood-grass peat while high concentrations of 10,18-bisnorabieta-5,7,9(10),11,13-pentaene and retene were found in the sample W-4.

The content of 18-norabietane, reduction product of molecular biological source, increases whereas the concentration of bi- and triaromatic diterpene hydrocarbons, dehydrogenation product of initial biomolecules declines with depth decrease in sedge and wood-grass peat profiles. pH of the peat medium changes in the same way [10].

Monoaromatic dehydroabietic acid is absent in sedge peat while its methyl ester (methyl-8,11,13-abietatrien-18-oat) was detected in all samples. The acid content is low in wood-grass peat, it declines with depth decrease. Low concentration of dehydroabietic acid in the upper part of the peat profile can be explained by its rapid esterification. The small amount of free protons in fen Kirek bottom waters made it possible to preserve diterpenic acid and prevented its esterification. The content of methyl-8,11,13-abietatrien-18-oat in wood-grass peat varies similarly.

Triterpenoids are presented by hopane, oleane, ursane and lupane derivatives with different lateral substituents and unsaturated bond positions in molecules. Structural analogues of oleane and lupane include C_{30} ketones and alcohols with one or two unsaturated bonds in the molecule. Only urs-12-ene-3-one was identified in the sample W-4 of wood-grass peat among ursane derivatives. The distribution of oleane, lupane and hopane derivatives in fen Kirek peat is presented in Figure 2A.

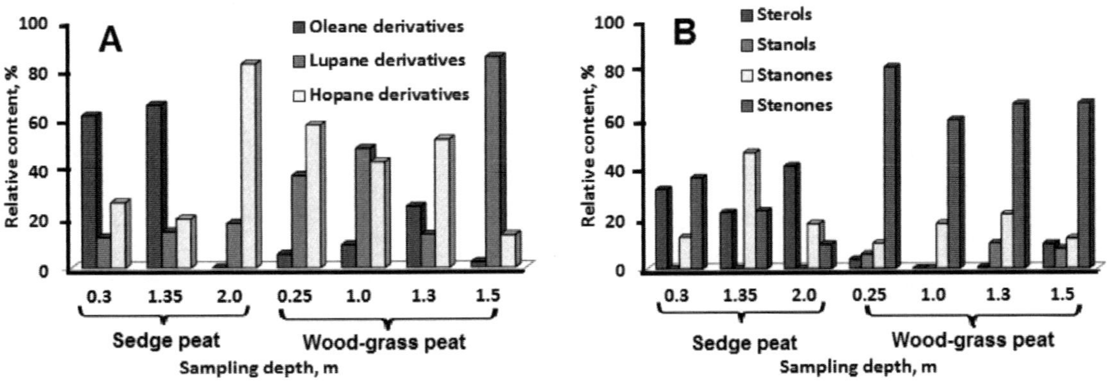

Figure 2. The content of pentacyclic terpenoids structural (A) and steroid (B) groups in fen Kirek peat, % rel.

Oleane derivatives, reflecting angiosperms contribution in peat formation, dominate in the samples S-1 and S-2 of sedge peat. It was detected that D-friedoolean-14-en-3-one has the highest concentration. Oleane derivatives completely disappear in the sample S-3 while hopenes begin to dominate. It may indicate high degree of peat bacterial processing.

The content of pentacyclic structures with double bonds is significantly higher than the one in saturated compounds. The contribution of triterpenic saturated structures grows with depth decrease in the wood-grass peat deposit alongside reducing pH values. The content of saturated structures declines with depth decrease in the sedge peat deposit.

The concentration of ketonic derivatives of pentacyclic terpenoids increases with depth decrease in both peat profiles alongside the transition to a more acidic environment during peat formation. The alcohol content decreases while the concentration of hydrocarbons increases with depth decrease in wood-grass peat. Alcohol derivatives are completely absent in sedge peat while hydrocarbons content reduces with depth decrease.

Steroids in fen Kirek peat are presented by the set of cholestane, ergostane, stigmastane and cycloartane derivatives. Total concentrations of these compounds are presented in table 3. Stigmastane derivatives with C_{29} composition dominate among the steroids in all studied peat samples. Stigmast-4-ene-3-one prevails in the sample S-1 of sedge peat, α-stigmastan-3-one – in the sample S-2. Stigmast-4,6-dien-3-ol was detected in high concentrations at these depths. It dominates in the sample S-3. Stigmast-4-ene-3-one prevails in the samples W-1 and W-2 of wood-grass peat, stigmast-5-en-3-ol (sitosterol) – in the samples of W-3 and W-4.

Table 3. The content of steroid groups in fen Kirek peat, % rel.

Sampling depth, m	Sample code	Cholestane derivatives	Ergostane derivatives	Stigmastane derivatives	Cycloartane derivatives
			Sedge peat		
0.3	S-1	0.94	4.47	93.47	1.12
1.35	S-2	0	12.1	87.89	0
2.0	S-3	2.24	5.23	88.8	3.73
			Wood-grass peat		
0.25	W-1	2.68	5.65	90.63	1.05
1.0	W-2	0	6.2	93.96	0
1.3	W-3	0	3.04	96.97	0
1.5	W-4	3.91	2.79	87.86	5.42

Cholesterol and its derivatives (C_{27}), reflecting microalgae contribution in peat formation, were found in the upper and lower part of both peat profiles in low concentrations. The distribution of sterols, stanols, stanones and stenones in fen Kirek peat is presented in Figure 2B. Stenones prevail in the sample S-1 of sedge peat, stanones – in the sample S-2, sterols – in the sample S-3. High relative content of sterols in the sedge peat deposit might be explained by their good preservation in alkaline conditions. Stenones prevail in all samples of wood-grass peat. Low concentrations of sterols in this peat deposit may be related to the specific composition of the original plants which often contain saturated sterol derivatives.

4. Conclusion

n-Alkanes with dominance of the odd homologues C_{21}-C_{31}, bi-, tri- and tetracyclic aromatic hydrocarbons, partially unsaturated bicyclic and naphthenoaromatic compounds, tricyclic transformation products of abietic acid and also 8,13R-epoxylabd-14-ene were identified in the studied samples of fen Kirek peat. Pentacyclic terpenoids are presented by hopane, oleane, ursane and lupane structural analogues, steroids – by a set of cholestane, ergostane, stigmastane and cycloartane derivatives with various lateral substituents and unsaturated bond positions in molecules. The studied peat largely inherits the composition of initial peat-forming plants.

It was found that the pH of the medium affect to the greatest extent the transformation of steroids, tri- and pentacyclic terpenoids during peat formation. The content of ketonic derivatives of pentacyclic terpenoids and 18-norabietane increases with depth decrease in sedge and wood-grass peat profiles whereas the content of bi- and triaromatic diterpene hydrocarbons declines alongside pH of the peat medium. The small amount of free protons in the wood-grass deposit bottom waters prevents diterpenic acid esterification.

Thus, it was shown that peat formation conditions have a significant impact on the conversion of peat deposits organic matter. Therefore, considering detailed data about the individual composition of organic compounds, the rough reconstruction of peat formation conditions can be carried out.

Acknowledgements

This work was supported by RFBR under grant 15-05-03910 and by RSF grant 15-17-10003. Results were supported by the funds of Strategic Programme on National Research Tomsk Polytechnic University Competitiveness Enhancement in the Group of Top Level World Research and Academic Institutions.

References

[1] Serebrennikova O V, Strelnikova E B, Preis Y I, Gulaya E V and Duchko M A 2013 *Bul. of Tom. St. Pol. Univ.* **323** 40
[2] Serebrennikova O V, Strelnikova E B, Preis Y I and Duchko M A 2014 *Bul. of Tom. St. Pol. Univ.* **325** 80
[3] Bush R T and McInerney F A 2013 *Geoch. et Cosmoch. Acta.* **117** 161
[4] Zhanga Z, Zhaob M, Eglinton G, Lud H and Huange C 2006 *Quat. Sc. Rev.* 575
[5] Ficken K J, Li B, Swain D L and Eglinton G 2000 *Org. Geochem.* **31** 745
[6] Lopez-Días V, Urbanczyk J, Blanco C G and Borrego A G 2013 *Int. J. of Coal Geol.* **116-117** 270
[7] Nichols J E, Booth R K, Jackson S T, Pendall E G and Huang Y 2006 Org.Geochem. **37** 1505
[8] Ronkainen T, McClymont E L, Väliranta M and Tuittila E S 2013 Org. Geochem. **59** 1
[9] Lu Y, Hautevelle Y and Michels R 2013 *Org. Geochem.* **2** 175
[10] Serebrennikova O V, Strelnikova E B, Preis Y I and Duchko M A 2014 *Bul. of Tom. St. Pol. Univ.* **325** 80
[11] Serebrennikova O V, Strelnikova E B, Averina N G, Kozel N V and Duchko M A 2014 *Fund. Res.* **12-1** 112

Size fractionation of trace elements in the surface water and groundwater of the Ganjiang River and Xiushui River basins, China

E A Soldatova[1,3], N V Guseva[2], Z Sun[2] and I S Mazurova[1]

[1] National Research Tomsk Polytechnic University, 30 Lenin Ave., Tomsk, 634050, Russia
[2] East-China Institute of Technology, 418 Guanglan Avenue, Nanchang, 330013, China

E-mail: [3] 2a61@mail.ru

Abstract. The research of trace element behaviour and size fractionation is significant to understand the processes in the water–rock system. Moreover, the aqueous speciation of trace elements is strongly related to their distribution and toxicity. In this research, trace elements behaviour and their size fractionation are investigated in the natural water (the surface water and shallow groundwater) of the Ganjiang River and Xiushui River basins. Trace element concentrations were measured by ICP-MS. The authors estimated the size fractionation of the trace elements between suspended forms (>0.45μm) and dissolved fractions, which included truly dissolved (1 kDa) and colloidal particles (0.45 μm–1 kDa)), after filtration through a 0.45 μm acetate cellulose membrane filter. The distinct features of the trace element size fractionation were identified. The size fractionation of the trace elements under different conditions (in the surface water and shallow groundwater) was studied.

1. Introduction

The Ganjiang River and the Xiushui River are the main rivers of the Poyang Lake basin which disembogued into Poyang Lake from the west. This territory is the part of a unique wetland system, which is the habitat of rare animal species. Moreover, it is an important part of economy of Jiangxi Province (China). Wide alluvial plains make this area one of the main agricultural regions in China; these territories are often irrigated for rice and other crops cultivation. In such rural regions as the Ganjiang River and Xiushui River basins, where population density is quite high, intensive agricultural activity is developed, and often, there are no sewer systems, the vexed problem is shallow groundwater and surface water pollution by fertilizers, domestic sewage, manure and other livestock waste. Therefore, the investigation of the shallow groundwater and surface water chemical composition of agricultural regions is important to support the ecosystem sustainability.

Trace elements are the essential part of natural water chemical composition. They are usually contained in natural water at low concentrations but have a great significance to human health [1]. Some of them have a healthy influence. The others, on the contrary, are not required for human body function, and, as numerous works have showed, they may provoke various pathological processes [2, 3]. However, there is a lack of information about the concentrations and behaviour of the trace elements in the natural water of the Ganjiang River and Xiushui River basins.

The distribution and toxicity of trace elements are strongly related to their aqueous speciation. Thus, both the total concentrations of trace elements and their chemical forms determine their bioavailability and potential environmental pollution [4]. Trace elements in natural water may migrate in different forms: particles (>0.45μm), colloids (0.45 μm – 1 kDa), and dissolved forms (<1 kDa) [5]. Saito et al. [6] and McNaught and Wilkinson [7] reported that colloids in natural water have at least one of their dimensions from 1 nm to1 μm. Nystrand and Őstерholm [4] have demonstrated that the dissolved forms of trace elements are the most mobile and available to biota; hence, the dissolved fraction of trace elements may be identified as the most toxic fraction. However, there is the increasing number of evidences that natural colloids are carriers for toxic substances [6, 8]. It is also important that the concentrations of trace elements should be considered in the investigation of a redistribution process in the water-rock system. In most cases, the precise evaluation of processes in the water–rock system is impossible without data on trace element content in natural water. The size fractionation of trace element species influences the element capability to transport and react with other water compounds.

The present work is focused on the investigation of trace element behaviour and size fractionation in the natural water (the surface water and shallow groundwater) of the Ganjiang River and Xiushui River basins, along with the study of the main ion composition and geochemical conditions of this area. The authors of the article have compared the concentrations of the trace elements after filtration through a 0.45 μm acetate cellulose membrane filter. This procedure was used to separate a suspended fraction (>0.45μm) from a dissolved fraction (<0.45μm). The dissolved fraction includes colloidal (0.45 μm – 1 kDa) and truly dissolved particles (1 kDa) and is called "semi-dissolved" according to Nystrand and Ősterholm [4]. It allows identifying the distinct features of behavior and the size fractionation of the trace elements under different conditions in the natural water of the Ganjiang River and Xiushui River basin.

2. Materials and Methods

2.1. Study area

The study area is situated in Jiangxi Province (southeast China). It is the part of the Poyang Lake catchment, in particular, the Ganjiang River basin and the Xiushui River basin, which are two of the five major rivers feeding Poyang Lake.

The territory belongs to the subtropical humid climate zone with the average annual temperature 17.5 °C [9]. The annual precipitation is 1400–2400 mm [10]. Irregular rainfall distribution during the year is controlled by the East Asian monsoon. The wet season lasts from March – April to June, then rainfall decreases from July to September, with evapotranspiration reaching its maximum point [11]. After September the dry season begins and lasts till December–February. The study area has a relatively flat relief, with alluvial plains being predominant [9].

Shallow aquifers within the study area are porous water-bearing rocks composed of Quaternary fluviolacustrine and deltaic sand-clay sediments, including red earth.

2.2. Sampling and analytical procedures

Field work in the study site was conducted in October, 2013, during the dry season. Sampling points were located to the west of Poyang Lake in the lower course of the Ganjiang River basin and the Xiushui River basin (figure 1). In the course of research, the following waters were studied: the shallow groundwater, which was sampled from public and private household wells (18 samples), and the surface water from the main rivers, the Ganjiang and the Xiushui, and from Poyang Lake (5 samples). The surface water samples were collected near the riverbank (the Xiushui River) or using a boat in the middle of the stream (the Xiushui River, the Ganjiang River, Poyang Lake).

Two sterile 50 mL polyethylene vials were used to collect water at each sampling point. The water sample for the first vial was collected without any preparation; another water sample was immediately filtered through a 0.45 μm acetate cellulose membrane filter of 30 mm in diameter using a pre-cleaned syringe. About 5 mL of the filtered water was eliminated at the beginning of filtration.

The unfiltered and filtered water samples were analysed for 42 elements using ICP-MS (PerkinElmer NexION 300D) in the Fundamental Research Laboratory of Hydrogeochemistry of Education and Research Centre "Water" (Tomsk Polytechnic University, Russia).

At each sampling point 600 mL of water was collected using polyethylene bottles and analysed for main ions (HCO_3^- by titration with 0.1M HCl; SO_4^{2-}, Cl^-, Ca^{2+}, Mg^{2+}, Na^+, K^+ by ion chromatography (ICS 1000, ICS 2000)), nitrogen compounds (NO_3^-, NO_2^-, NH_4^+) by photocolorimetry (KFK-2). Electrical conductivity, temperature and pH were measured *in situ*.

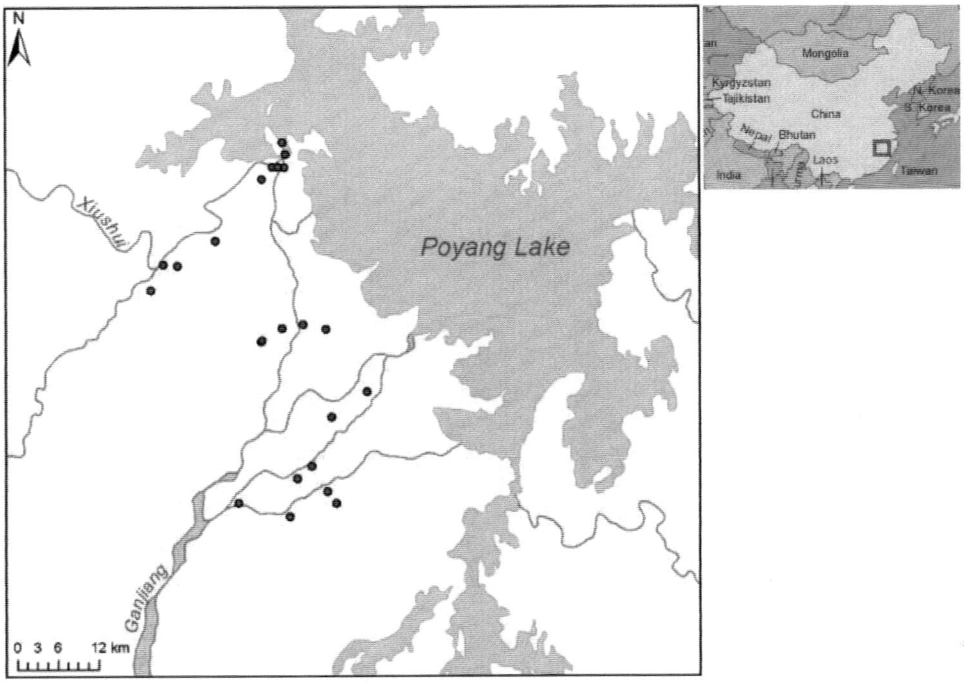

Figure 1. The scheme of the sampling points location.

3. Results and Discussions

The shallow groundwater of the study area is basically fresh with TDS up to 500 mg/L. The average value of TDS is about 196.9 mg/L (table 1). The acid-alkaline properties of the shallow groundwater vary significantly from 5.90 to 7.05. Geochemical conditions change from acid to neutral. An average pH value is 6.36.

As mentioned above, the study area is characterized by subtropical humid climate, where water exchange is active and there are no significant salt deposits in the upper part of the earth crust. According to the theory of the water–rock system evolution [12, 13] the shallow groundwater of such area should be HCO_3–Ca and HCO_3–Ca–Mg. In fact, the main ions in the shallow groundwater are HCO_3^- and Ca^{2+}. However, we observed the increased role of chloride and nitrate and, less often, of sodium, potassium and sulphate in the shallow groundwater of the Poyang Lake basin, which changes the chemical type of the groundwater [14, 15].

The surface water of the study area is ultrafresh (TDS<200 mg/L) with the average TDS value 92.3 mg/L (table 1). Geochemical conditions are usually neutral with the average pH value 7.05. The main ions of the chemical composition are HCO_3^- and Ca^{2+}. However, the average concentrations of the main ions are lower than that in the shallow groundwater. In particular, it concerns NO_3^-, which is no longer significant in the chemical composition of the surface water, whereas Cl^-, SO_4^{2-} and Na^+ remain the important components in the chemical composition and make a significant contribution to the TDS value. Especially, it concerns Cl^- and Na^+, which reach 30% of the anion and cation concentrations, respectively.

Table 1. The chemical composition of the natural water of the Ganjiang River and Xiushui River basins, mg/L, Eh – mV.

Values	pH	Eh	HCO_3^-	SO_4^{2-}	Cl^-	Ca^{2+}	Mg^{2+}	Na^+	K^+	NH_4^+	NO_2^-	NO_3^-	TDS
	Shallow groundwater												
Min.	5.90	-91	18.3	0.67	1.38	2.18	1.18	8.30	0.74	<0.05	<0.02	0.15	49.5
Max.	7.05	176	170.8	111.4	58.9	47.5	9.70	43.4	76.0	6.40	0.3	93.9	523.7
Average	6.36	26.2	86.2	20.9	20.4	20.6	5.72	16.5	8.14	1.32	*NA*	17.2	196.9
	Surface water												
Min.	6.70	51	32.3	8.50	6.10	10.0	1.59	4.80	1.32	0.15	<0.02	0.93	67.3
Max.	7.30	140	43.9	16.5	17.2	16.7	2.70	10.9	2.61	0.58	3.15	5.4	115.9
Average	7.05	85	38.9	12.9	12.0	13.6	2.28	8.27	2.00	0.30	*NA*	2.36	92.3

Note: *NA* – not available.

The investigated waters are characterized by the low Eh-value, which varies from -91 to176 mV. The average Eh-value is no higher than 85 mV even for the surface water. Usually the Eh values are much higher in shallow groundwater and surface water.

We identified two groups of the trace elements characterized by the peculiar features of size fractionation between a suspended phase and dissolved phase (table 2). On the one hand, Li, Be, B, Si, Sc, V, Mn, Fe, Co, Ni, Zn, As, Se, Rb, Sr, Mo, Cd, Cs, Ba, W, Re, Tl, U were found basically in the dissolved form in accordance with the fact that their average concentrations in the filtered and unfiltered samples are distinguished slightly. On the other hand, the concentrations of some elements (Al, Ti, Cu, Sn, Ga, Y, Sn, Pb, rare earth elements) significantly decreased after filtration through a 0.45 μm membrane filter. This demonstrates that this group of the trace elements tends to migrate in the suspended phase.

Table 2. The average concentrations of the trace elements in the unfiltered and filtered water samples, mg/L.

Element	Water sample		Element	Water sample		Element	Water sample	
	Unfiltered	Filtered		Unfiltered	Filtered		Unfiltered	Filtered
Li	0.00268	0.00265	**Ga**	0.00009	0.00004	**Nd**	0.000329	0.000089
Be	0.00007	0.00006	**As**	0.01430	0.01308	**Sm**	0.000071	0.000018
B	0.0075	0.0077	**Se**	0.0008	0.0007	**Eu**	0.000022	0.000013
Al	0.100	0.008	**Rb**	0.00728	0.00751	**Gd**	0.000078	0.000021
Si	10.06	9.64	**Sr**	0.0973	0.0978	**Tb**	0.000012	0.000004
Sc	0.0003	0.0003	**Y**	0.000354	0.000148	**Dy**	0.000056	0.000017
Ti	0.009	0.002	**Mo**	0.0013	0.0014	**Ho**	0.000012	0.000005
V	0.0021	0.0019	**Cd**	0.000085	0.000091	**Er**	0.000032	0.000012
Mn	0.915	0.925	**Sn**	0.00006	0.00002	**Yb**	0.000031	0.000011
Fe	9.32	8.88	**Cs**	0.000122	0.000083	**W**	0.000181	0.000162
Co	0.00113	0.00097	**Ba**	0.0837	0.0839	**Re**	0.000006	0.000006
Ni	0.0015	0.0013	**La**	0.000371	0.000111	**Tl**	0.000027	0.000025
Cu	0.00075	0.00035	**Ce**	0.000690	0.000125	**Pb**	0.00187	0.00029
Zn	0.012	0.006	**Pr**	0.000084	0.000023	**U**	0.000092	0.000057

However, average concentrations do not always reflect all peculiarities of trace element behaviour. The case in point is Fe behavior. In several sampling points where the Eh-value is below 0 mV, the Fe concentrations are high in both filtered and unfiltered samples. The oxidation degree Fe^{2+} results in the absence of significant differences between the concentrations of Fe in the filtered and unfiltered samples. The Fe ion with this oxidation degree is formed under specific reduction conditions. It leads to the accumulation of Fe in the dissolved phase. At the sampling points with the positive Eh-values the concentrations of Fe are much lower and tend to decrease significantly after filtration procedure. The authors suppose that this decrease occurs due to Fe hydroxide precipitation under oxidizing conditions.

It is typical that in the surface water, unlike the shallow groundwater, the transport of most trace elements occurs in the suspended form (figures 2a, b). Obviously, the high content of suspended matter in the surface water contributes to the trace element transport with large sized particles. In the research described, Be, V, Co, Cd, Cs and U, which usually migrate in the dissolved form, tended to be transported in the suspended phase (figure 2b).

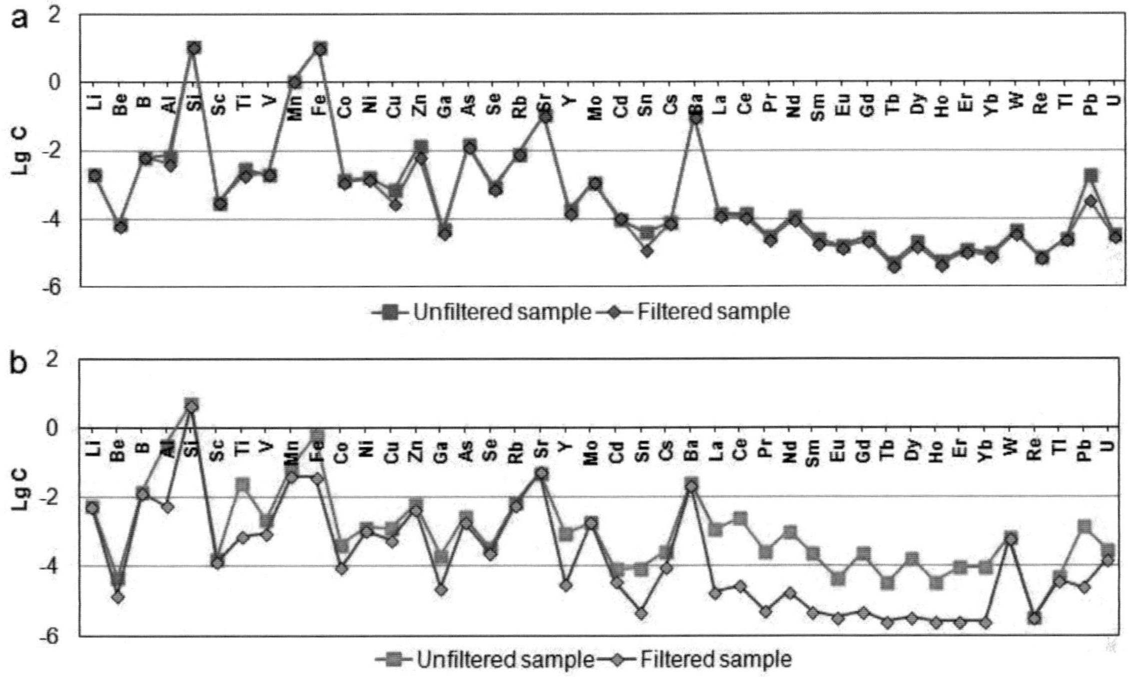

Figure 2. The concentrations of the trace elements in the most typical unfiltered and filtered samples: (a) the shallow groundwater; (b) the surface water.

4. Summary and Conclusions

In the investigated waters the authors have identified two groups of the trace elements characterized by peculiar features of the size fractionation between the suspended phase and the dissolved fraction. The first group includes Li, Be, B, Si, Sc, V, Mn, Fe, Co, Ni, Zn, As, Se, Rb, Sr, Mo, Cd, Cs, Ba, W, Re, Tl, U, which were found in the dissolved form. In the second group, Al, Ti, Cu, Sn, Ga, Y, Sn, Pb and rare earth elements migrate in the suspended fraction.

It has been found that the trace elements in the surface water tend to migrate in the suspended phase, whereas the trace elements in the shallow groundwater usually migrate in the dissolved form.

The size fractionation of Fe basically depends on geochemical conditions rather than the presence of the suspended matter in the natural water. Reduction conditions lead to accumulation of Fe in the dissolved phase. Under oxidizing conditions, Fe tends to migrate in the suspended phase, which might be caused by the hydroxide precipitation process.

Acknowledgements

The investigation was supported by The Russian Foundation for Basic Research (Project No 14-05-31267) and State Assignment "Nauka" № 5.1931.2014/K.

References

[1] Fraga C G 2005 Relevance, essentiality and toxicity of trace elements in human health *Molecular Aspects of Medicine* **26** 235–44
[2] WHO 2011 *Guideline for Drinking Water Quality* fourth ed. (Geneva: World Health Organization (WHO))
[3] ATSDR 2008 *Draft Toxicological Profiles* (Atlanta: US Department of Health and Human Services, Public Health Service, Agency for Toxic Substances and Disease Registry (ATSDR)).
[4] Nystrand M and Ősterholm P 2013 Metal species in a Boreal river system affected by acid sulfate soils *Appl. Geochem.* **31** 133–41
[5] Ren H, Liu H, Qu J, Berg M, Qi W and Xu W 2010 The influence of colloids on the geochemical behavior of metals in polluted water using as an example Yongdingxin River, Tianjin, China *Chemosphere* **78** 360–67
[6] Saito T, Suzuki Y and Mizuno T 2013 Size and elemental analyses of nano colloids in deep granitic groundwater: Implications for transport of trace elements *Colloids Surf., A* **435** 48–55
[7] McNaught A D and Wilkinson A 1997 *Compendium of Chemical Terminilogy*, second ed. (Oxford: Blackwell Scientific)
[8] McCarthy J F and Zachara J M 1989 Subsurface transport of contaminants – mobile colloids in the subsurface environment may alter the transport of contaminants *Environ. Sci. Technol.* **23** 496–502
[9] Ye X, Zhang Q, Liu J, Li X and Xu C Y 2013 Distinguishing the relative impacts of climate change and human activities on variation of streamflow in the Poyang Lake catchment, China *J. of Hydrology* **494** 83–95
[10] Wang Q, Riemann D, Vogt S and Glaser R 2014 Impacts of land cover changes on climate trends in Jiangxi province China *Int. J. Biometeorol.* **58** 645–60
[11] Li X and Zhang Q 2011 Estimating the potential evapotranspiration of Poyang Lake basin using remote sense data and Shuttleworth-Wallace model *Proc. Environ.Sci.* **10** 1575–82
[12] Shvartsev S L 2008 Geochemistry of fresh groundwater in the main landscape zones of the Earth *Geochemistry Int.* **46** 1285–1398
[13] Shvartsev S L 2012 The internal evolution of the water-rock geological system *Herald of the Russian Academy of Sciences* **82** 2 134–42
[14] Soldatova E, Guseva N and Wang G 2014 Characteristic features of groundwater pollution in the Poyang Lake catchment *IOP Conf. Ser.: Earth Environ. Sci.* **21** 012023 doi:10.1088/1755-1315/21/1/012023
[15] Sun Z, Soldatova E A, Guseva N V and Shvartsev S L 2014 Impact of human activity on the groundwater chemical composition of the south part of the Poyang Lake Basin *IERI Procedia* **8** 113–18

Modeling results of calcium-containing minerals precipitation in the alkaline hydrotherms of Baikal Rift Zone: calcite and dolomite

O G Tokarenko[1,2] and E V Zippa[1,3]

[1] National Research Tomsk Polytechnic University, 30 Lenin Ave., Tomsk, 634050, Russia

E-mail: [2] tog@tpu.ru, [3] zev-92@mail.ru

Abstract. The calculation modeling results of the nitric hydrotherms saturation in Baikal Rift Zone with calcite and dolomite are presented. The calcite and dolomite make the carbonate barrier to thermal waters equilibrium with primary minerals of igneous rocks. In the research territory, there are three main types of geochemical thermal waters which are characterized by the saturation degree with the calcite and dolomite and the proportion of precipitating minerals phases. It has been established that nitric thermal waters-rock system has equilibrium with these minerals, which leads to bonding migrated from the rocks calcium and magnesium by the secondary formed minerals – calcite and dolomite.

1. Introduction

Previously, the assessment of preliminary results of thermal waters equilibrium with a variety of carbonate minerals was considered [1]. The assessment of water equilibrium degree with the water-bearing rocks is currently widely used by the hydrogeochemists [2]. The equilibrium degree assessment of water with water-bearing rock helps to determine the conditions of chemical composition formation [3, 4]. It is shown in the research [5] that the thermal water of Baikal Rift Zone (BRZ) is the product of long-term interaction of ancient and modern infiltration waters with the host rocks followed by large-scale redistribution of chemical elements between the water and secondary formed minerals. In the solution, the equilibrium with a variety of minerals is achieved (carbonate and aluminosilicate) which results in formation of various secondary formed minerals.

The work purpose is to determine quantitative proportion of new secondary formed mineral in order to prove the hypothesis characterized peculiar of BRZ hydrotherms chemical composition. More preciously, the study considers formation of calcite and dolomite, which formation of them is the main part in the evolution of water-rock system.

2. The geologic and thermal condition of hydrothermal systems formation

The temperature of aqueous solutions also exerts a significant effect on the chemical composition of thermal waters: when heated, the water-rocks interaction accelerates and the products of aluminosilicate hydrolysis intensely pass into the solution. At the same time, carbonate precipitation accelerates. It is supposed that the character of water composition transformation is much influenced by the temperature gradient. When temperature drastically changes, the chemical transformation is mainly determined by the reactions in the solution. At a slow temperature change, the water–rock interaction is a determining factor.

The geochemical composition of rocks in the area where hydrotherms originate definitely influences the chemical composition of the waters. The BRZ is mainly composed of similar intrusive and metamorphic types. However, there are also geochemical traces of the existed Barguzin–Vitim ocean basin and conjugate Kelyana and Meteshikha island-arc systems. The deep fault rocks that occured during the formation of the Sayan–Baikal Fold Belt [6] show a great geochemical diversity. These ancient geologic objects and associated geochemical redistribution of substance influence the chemical composition of modern surface and underground waters.

2. Materials and Methods

2.1. The nitric thermal waters concise description

The detailed analysis of the hydrotherms chemical composition peculiarities is presented [7]. It should be taken into account that low mineralization is of the main peculiarities of thermal water. The TDS exceeds 1 g/L, but it is usually bellow 0.5 g/L. Besides the TDS, the thermal waters have the same differential peculiarity as the high alkalinity (pH from 8.2 to 10), with high concentration of silicium (sometimes 120 mg/L) and fluorum (to 47 mg/L) being identified. As it has been defined, the content of fluorum perfectly correlates with carbonate ions but poorly – with sulfates. In general, five chemical types of thermal waters have been formed within BRZ – HCO_3–Na, HCO_3–F–Na, HCO_3–SO_4–Na, SO_4–HCO_3–Na и SO_4–Na, which differ in chemical composition transportation, correlation with total mineralization, temperature, pH, contents of fluoride and sulfate ions. Therefore, the alkaline thermal waters of sodic type are fresh (M<0.5 g/L). This fact has not been scientifically explained yet.

2.2. Investigation methods

The chemical composition data of 47 springs were used in the modeling process. The saturation index was used to define thermal waters saturation degree to individual minerals. The index of saturation is

$$L = \lg Q/K$$

where Q – ion activity product; K – solubility product. As solution is saturating in relation to any mineral, the saturation index increases to zero (equilibrium state). In case of oversaturation, its values become positive. The parameter calculation and modeling of mineral phases solution/precipitation processes were made according to methods which were developed on the basis of thermodynamics hydrogeochemical processes [8] using computer program [9]. The modeling process of minerals formation is based on the free energy minimization within the program. The method principle is the calculation of studied geochemical system equilibrium composition based on the data of its initial elemental composition according to the fundamentally principles of the equilibration chemical thermodynamics. Thus, the most probable of development system is its transition to state that corresponds to the minimum of total Gibbs thermodynamic system potential (Gibbs free energy) under given thermodynamic conditions. In this particular case, the real temperature of water and ranging from 40 to 80°C was taken into account in the calculations. When setting a model, 14 minerals which contain calcium were included in the system: $CaCO_3$ (к) – calcite, $CaMg(CO_3)_2$ – dolomite, $CaCO_3$ (a) – aragonite, CaF_2 – fluorite, $Ca_{0,15}Al_{1,9}Si_4O_{10}(OH)_2$ – montmorillonite, $Ca_{0.187}Na_{0.0205}K_{0.0205}Fe^{3+}{}_{0.141}Mg_{0.336}Al_{1.59}Si_{3.93}O_{10}(OH)_2$ – montmorillonite, $CaAl_2Si_2O_6(OH)_4$ – lawsonite, $CaAl_2Si_4O_8(OH)_8$ – laumontite (zeolite), $CaAl_4Si_2O_{10}(OH)_2$ – margarite, $CaMg_3(CO_3)_4$ – guntit, $CaSO_4$ – anhydride, $CaSO_4(H_2O)_{0.5}$ – bassanite, $CaSO_4(H_2O)_2$ – gypsum, $CaAl_2Si_4O_{12}(H_2O)_2$ – wairakite (zeolite).

3. Results and Discussion

3.1 The index of saturation for calcite and dolomite

In spite of nitric thermal waters low salinity (1–7 mg/L) and low calcium concentrations in the thermal waters, the equilibrium with calcite has been established. It was found that most springs are saturated to calcite and dolomite (figure 1, figure 2).

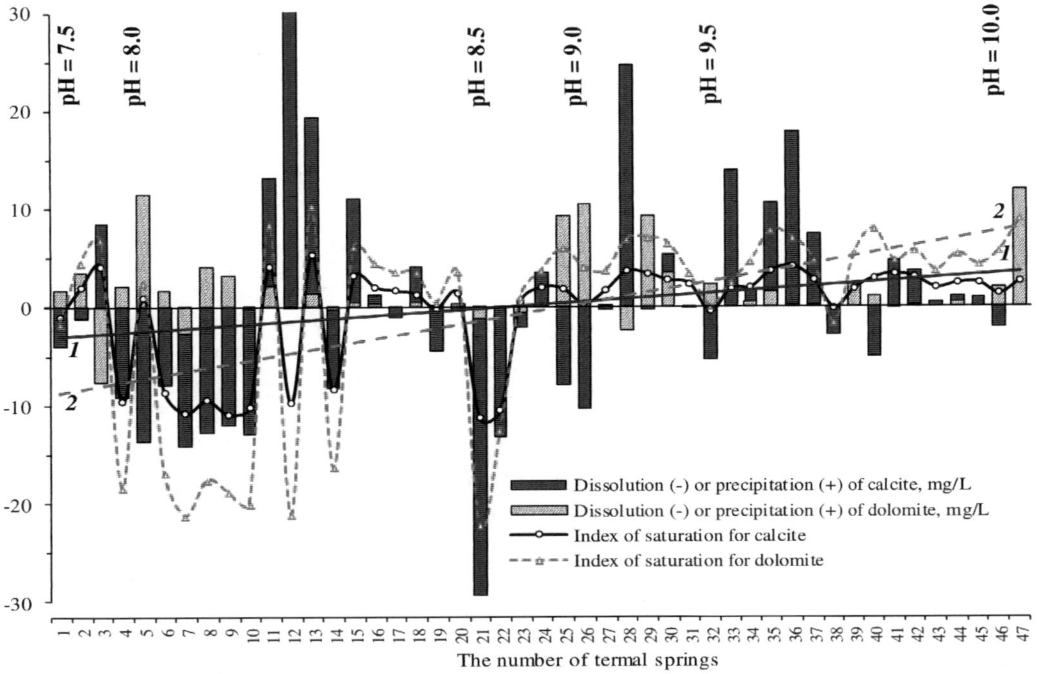

Figure 1. Dependence of saturation index values and quantity of mineral dissolution/precipitation on pH of springs.

Trend lines: 1 – for precipitate calcite quantity; 2 – for saturation index to dolomite. The number of termal springs: 1 – Engorboyskii; 2 – Garginskii; 3 – Zmeinyi; 4 – Teplyi Klyuch; 5 – Nilova Pustyn'; 6 – Irkaninskii-2; 7 – Alginskii; 8 – Korikeiskii; 9 – Sartinskii; 10 – Busanskii-2; 11 – Irkaninskii-1; 12 – Gusikhinskii; 13 – Tochinskii-2; 14 – Busanskii-1; 15 – Tochinskii-1; 16 – Bauntovskii-1; 17 – Bauntovskii-2; 18 – Mogoiskii-1; 19 – Mogoiskii-2; 20 – Mogoiskii-3; 21 – Mogoiskii-4; 22 – Mogoiskii-5; 23 – Shurindinskii-1; 24 – Shurindinskii-2; 25 – Verkhneangarskii; 26 – Kotel'nikovskii; 27 – Bylyrinskii; 28 – Pitatelevskii; 29 – Pitatelevskii (well); 30 – Dzelindinskii; 31 – Kyrinskii-1; 32 – Dzelinda well; 33 – Kyrinskii-2; 34 – Kulinye Bolota; 35 – Tolstikhinskii; 36 – Allinskii-2; 37 – Allinskii 1; 38 – Goryachinskii; 39 – Angarkan-Sartinskii; 40 – Uluriiskii-1; 41 – Uluriiskii-2; 42 – Kuchigerskii-1; 43 – Umkheiskii-1; 44 – Umkheiskii-2; 45 – Kuchigerskii-2; 46 – Solnechnyi; 47 – Bol'sherechenskii.

The similar results were previously obtained by using calculation methods based on the stability fields of carbonate minerals. However, the data of quantitative minerals formation were obtained for the first time.

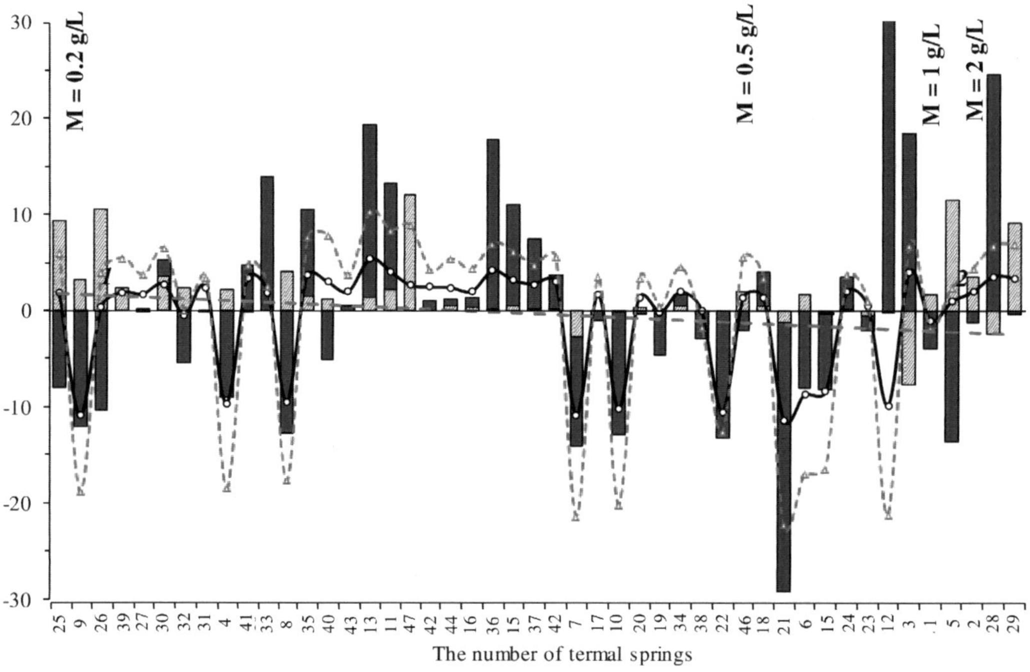

Figure 2. Dependence of saturation index values and quantity of mineral dissolution/precipitation on TDS of springs (the legend is in the figure 1).

Three types of hydrochemical springs are distinguished: 1 type – the springs with high degree of undersaturation to calcite and dolomite which are consequently dissolved due to chemical reaction; 2 type – springs that have reached the equilibrium state or oversaturation to calcite and dolomite, where both or one of the minerals are dissolved due to chemical reaction; 3 type – springs that are oversaturated to calcite and dolomite which precipitate as solid secondary formed minerals. Alongside, there are springs where one more geochemical combination can be observed: the springs are nonsaturated to calcite or dolomite but dolomite precipitates as a solid. This fact has no clear explanation and has to be more explored.

With springs pH growth calcite precipitation is likely to occur. With pH<8.5, calcite dissolves while precipitating, with the amount ranging from 2 to 18 mg per liter of water solution with pH>8.5. The maximum quantity of calcite is formed in the Pitatelevskii spring (25 mg/L), with the TDS values being more than 2 g/L.

It should be noted that thermal water-mineral equilibrium can be achieved only under particular geochemical conditions, for example, right temperature which is necessary for calcite concentration, definite pH, etc. (figure 3). The saturation degree to calcite increases with the growth of mineralization (TDS) (figure 3*a*). It is related to accumulation of calcium (figure 3*c*) and carbonates in the solution. However, the saturation growth occurs till the particular values. Thus, the thermal waters with TDS 0.2 g/L reach equilibrium and oversaturate by calcite in the high temperature conditions. Then it is formed as a solid phase. Under standard conditions, the equilibrium with calcite is reached under slightly higher thermal water mineralization – from 0.32 g/L (figure 3*d*).

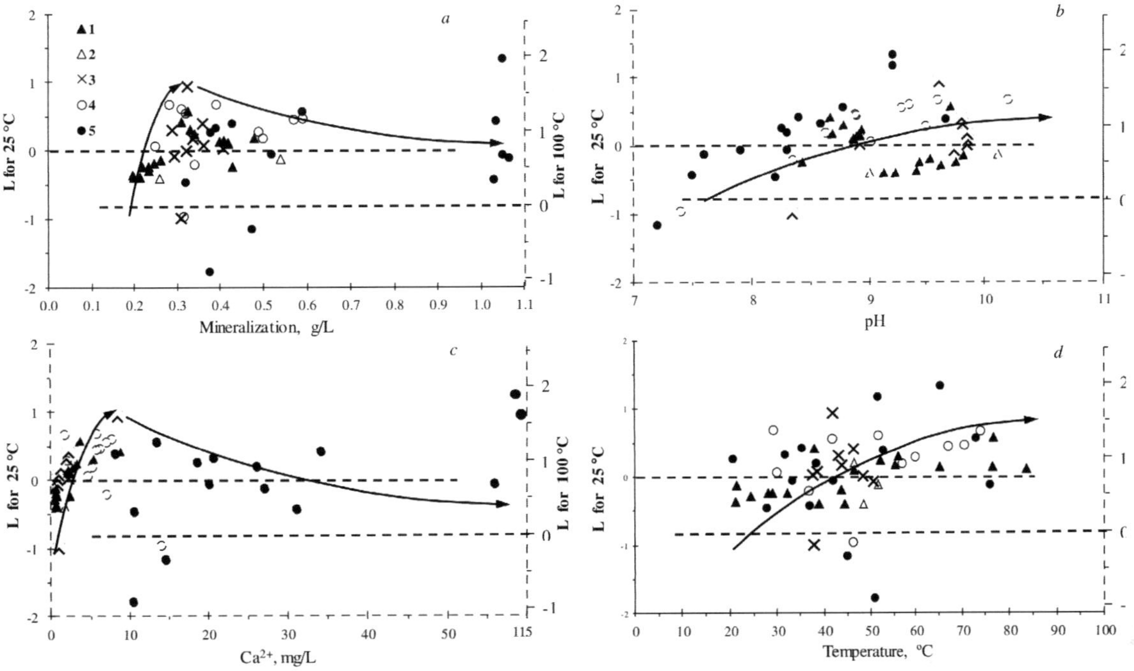

Figure 3. Dependence of thermal water saturation index (L) to calcite on general check parameters of thermal waters composition.

The chemical type of thermal waters: 1 – HCO_3–Na; 2 – HCO_3–F–Na; 3 – HCO_3–SO_4–Na; 4 – SO_4–HCO_3–Na; 5 – SO_4–Na.

This group includes the thermal waters with high concentration of sulfate-ion, i.e. SO_4–HCO_3–Na and SO_4–Na types of thermal waters. The index saturation value is lower in these thermal waters types due to less importance of carbonate component.

As carbonate dissolubility decreases due to the temperature increase, the saturation degree of nitric thermal waters to calcite and dolomite grows with depth increase. According to this, we believe that all nitric thermal waters have equilibrium with carbonate mineral at the shallow depth. Disturbance of equilibrium occurs only when they rise to the surface, cool and are diluted by fresh groundwater.

The sulfate type of thermal waters can reach the equilibrium with calcite when pH 7.5, under deep conditions when pH 8.4 – for standard conditions (figure 3*b*). For HCO_3–Na и HCO_3–SO_4–Na types of thermal waters the pH values are significantly higher (9.9) to reach equilibrium to calcite.

The equilibrium with calcite is established even at the low spring temperature. Most of HCO_3–SO_4–Na thermal waters types in the field of the mineral saturation. As for SO_4–HCO_3–Na, it is represented by all types. The gradual decrease in saturation to calcite occurs with higher concentrations of Ca^{2+} (from 10 to 115 mg/L) and with lower values of pH. This can be observed in the SO_4–Na thermal waters. It can be explained by balancing calcium with sulfate-ion in the thermal waters and insufficient amount of carbonate ion under these pH conditions.

Thus, the equilibrium and oversaturation of nitric thermal waters by calcite and dolomite lead to its formation as secondary forms minerals (authigenous). In this case, definite set of secondary formed minerals corresponds to each stage of water-rock interaction. The composition of authigenous minerals can change over time. Where being formed, the secondary formed minerals, in its formation process, bind the particular chemical elements of the solution, therefore, the content of these chemical elements is not grow. As a result, a variety of geochemical barriers appear. The geochemical barriers prevent the equilibration with primary (endogenous) minerals. First of all, it concerns Ca-, Mg-aluminosilicates of anorthite, forsterite, fayalite types, and etc., as moving elements of these minerals form the carbonates (calcite, dolomite and others) which completely prevent the equilibrium of solutions with igneous minerals.

4. Conclusion

To sum up, the water equilibrium with the majority of host rock minerals occurs at a certain stage of water-rock system evolution. Instead of being dissolute, minerals of host rocks begin to form. Thus, not only chemical equilibrium but also dynamic balance is reached in the system. The latter ensure the equality of entering elements to solution and precipitating elements. The consequence of this equilibrium is low mineralization in the thermal water which can slightly increase over time or does not increase.

One of the main peculiarities of water-rock system is due to the fact that in the course of evolution the ratio between chemical elements is constantly changing, which, in its turn, leads to constant change of groundwater chemical composition and increases the difference between groundwater and host rock composition[10]. This is explained by the fact that one part of elements is bond by precipitated secondary minerals and the second one which remains in solution, can concentrate redundant. The water is the medium where the chemical bond between ionic pairs breaks down, redistribution of chemical bonds between the ions occurs and new ion pairs form. According to hydrodynamics laws, the new ionic pairs form other solids that are different from primary one.

Due to above, there is another peculiarity of secondary minerals formation. A mineral forms under the conditions of equilibrium with solution. Therefore, the sudden change (increase or decrease) in concentration of elements that form the mineral, is not possible in the solution. Only elements which enter the solution from nonequilibrium with water minerals are precipitate. But the solution does not change, of course, if the system keeps equilibrium-nonequilibrium state.

Acknowledgements
The investigations were supported by The Russian Foundation for Basic Research (Project No 14-05-31518).

References

[1] Tokarenko O G 2013 Thermal nitric water equilibrium from Baikal rift zone with carbonaceous minerals of country rock. *Mineral resources development in Siberia: from Obruchev V.A., Usov M.A., Urvantsev N.N. to present* (Tomsk: TPU) p 569 (in Russian)
[2] Shvartsev S L 2008 *Geoch. International* **46**(13) 1285
[3] Kopylova Y G, Lepokurova O E, Shvartsev S L and Tokarenko O G 2011 *Doklady Earth Sciences* **436** 284–289
[4] Timoshenkova A N, Pasechnik E Yu and Tokarenko O G 2014 *IOP Conf. Ser.: Earth Environ. Sci.* **21** 12024
[5] Shvartsev S L 1998 *Hydrogeochemistry of hypergenesis zone* (Moscow, Nedra) p 366 (in Russian)
[6] Gordienko I V 2008 *The Earth's Evolution* (Novosibirsk: Akad. Izd. "Geo") (in Russian)
[7] Plyusnin A M, Zamana L V, Shvartsev S L, Tokarenko O G and Chernyavskii M K 2013 *Russian Geol. and Geoph.* **54**(5) 495
[8] Garrels R M and Christ Ch L 1965 *Solutions, Minerals and Equilibria* (New York: Harper & Row). P 450
[9] Bukaty M B 2002 *Bulletin of Tomsk Polytechnic University* **305**(6) 348 (in Russian)
[10] Shvartsev S L 1994 *Geologiya Rudnykh Mestorozhdenii* **36**(3) 261

Comparative analysis of forest lands cadastral appraisal estimated with regards to wood and food resources

V Kovyazin[1], A Romanchikov[2] and O Pasko[3]

[1] Doctor of Biology, professor, National Mineral Resources University (Mining University), 21 Line V.I., 2, St. Petersburg, 199106, Russia
[2] Post-graduate student, National Mineral Resources University (Mining University), 21 Line V.I., 2, St. Petersburg, 199106, Russia
[3] Doctor of Agricultural Science, professor, National Research Tomsk Polytechnic University, 30 Lenin Ave., Tomsk, 634050, Russia

E-mail: [1] vfkedr@mail.ru, [2] romanchicov@inbox.ru, [3] helgapas@mail.ru

Abstract. Cadastral appraisal of forest fund is one of the topical challenges of modern natural resource management. The paper delivers comparison of different approaches to cadastral appraisal of forest lands. The authors suggest a uniformed model to compare objectively and choose the most effective use of parcels.

1. Introduction

Forest lands prevail in the structure of Russian land funds [2, 3]. However, their use is inefficient, which leads to both profit loss and forest damage caused by lack of funding. Cadastral appraisal of forest lands plays an important role in efficient forest use and forest taxation. In Soviet time cadastral appraisal was conducted according to all-union standards, which was convenient if a country's economy is centralized. In market conditions, the Russian forest funds appeared to be in a tricky position – on the one hand, private ownership of forest land has not been permitted [5], on the other, logging industry has started developing intensively on the state lands. Thus, a forest user has no right to choose the optimal way to use forest land, and the state cannot collect full statistical data on economic efficiency of logging and its basic parameters, which makes general condition of the industry vague and unclear. All mentioned above proves that cadastral appraisal of forest lands in Russia is a challenging task that needs special approach. Market value comparison, which is used for other land categories, cannot be applied since there is no forest land market due to the absence of private forest land ownership. Cost approach does not take into account the quality of obtained forest resources. There is the last approach left – income approach that has been developed by Russian legislators for 15 years.

The first attempts to apply the income approach were made in 2000. The RF government developed a Guideline for economic assessment of forests [9] that determined a basic assessment concept: resource cost is income minus expenditure. It was for an appraiser to determine income and expenditure, which made the procedure unclear.

In 2002 a new Guideline for cadastral appraisal of forest lands was approved [6]. According to it, cadastral values were to be defined by discounting flows of money in three levels: federal constituent entity, regional forestry and a forest ranger station. The classical Faustmann model was a source of basic concepts to appraise forest lands [1]. However, the fact that the model had been developed for forest owners with all their rights and responsibilities was not taken into account.

In 2010 the Guideline was abolished to provide more effective legal framework, but nothing was introduced instead. Nowadays there is a specific indicator of cadastral value (SICV) [8] which is unified for a whole constituent entity of the Federation (Federal subject). This indicator does not differentiate land areas according to their quality, which makes it useless. Thus, it is not studied in the research.

For some decades forest has been regarded not only as a wood source but also as a source of non wood forest products (NWFP) or as an area used for recreational purposes. Therefore, it is NWFP that is studied in the article. The guideline [6] suggests appraising NWFP resources by rent capitalization. Such approach is quite vague: it may be difficult to determine a capitalization rate, since it changes from lumber company to lumber company and also depends on forest growth conditions. Another factor that makes such approach doubtful is the fact that a significant part of NWFPs is obtained from forest for non-commercial, personal use. Thus, such forest use cannot be regarded as business, and the capitalization rate can hardly be applied in this case.

Moreover, the Guideline [6] implies using this way of appraisal only for forest lands producing NWFP. NWFPs are not implied being gathered from areas meant for clear cutting, thus integrated cadastral appraisal of such areas is not provided in the guideline. It may be explained by the fact that discounting method and capitalization method of forest resources appraisal give incomparable results. For example, the cadastral value of blueberry-birch forest in Leningradskaya oblast based on mushroom resources can be three times higher than the cadastral value of the same area based on wood resources appraisal. Therefore, these appraisal results can hardly be compared and united for integral assessment. However, development of multifunctional cadastral appraisal is still one of the modern challenges of land management. It is necessary to take into account all possible ways of forest use, and find balance between them while making cadastral appraisal of an area. The results of cadastral appraisal should be the base for choosing the most efficient forest use of an area.

The aim of the research is to align forest appraisals made according to different methods, which would make it possible to compare them and to determine the most efficient and sustainable way of using particular forest area.

2. Objects and research methods

The data to study is mensuriation description data of 259 compartments of 11 units of "Pesochinskoye" district forest ranger station of "Curort lesopark" forest-park, St. Petersburg [7]. A forest park refers to urban protective forests. However, the species content and forest growth conditions refer to taiga biogeocenosis.

To compare incomes from clear cutting and MWFP (we chose mushrooms for calculations: cepe, milk mushroom, orange agaric, aspen mushroom, annulated boletus, rough boletus, sharp agaric, russula) we should make evaluation by unified methods. We suggest changing capitalization approach to average annual specific NPV. Thus, it is necessary to estimate income obtained from different forest services and goods for a particular period then discount it and divide it by the period.

Firstly, let us define a period. Income from clear cutting is received at any one time at the end of cutting cycle. However, over mature stands should be cut in the moment of evaluation. It is reasonable to consider a period as period before the end of cutting cycle plus a period of one cutting cycle. Let us take a period of 60 years for deciduous species and 80 years for coniferous ones. It should be noted that while changing continuous cutting to a certain limited period, the value is lost to a certain extend. However, preliminary estimation showed that the loss will not exceed 10%. Moreover, the current prices can hardly be taken into account to calculate costs in more than 60 years.

Let us use mensuriation description to estimate the income from forest harvested mushrooms. According to the Guideline [4] on NWFP appraisal, there is a relation between a forest type and an average annual mushroom yield in forests of Russia's North-West. The data show only the average yield from areas with the same forest growth conditions, that is why the yield value is quite low. We should underline that appraisal of every unit can hardly be implemented in this case, since the average values can not reflect a yield capacity of this or that compartment. Thus, an average unit yield will be taken as the most reliable for further cadastral evaluation.

To estimate total income from mushroom yield, it is necessary to multiply mushroom yield capacity by income from one kilo of particular mushroom species and by mushroom season period. The mushroom season period, as it is described above, equals the number of years before cutting plus one cutting cycle for a dominating wood species. The fact that mushroom yield is low during the first 10 years after clear cutting should be taken into account. Industry profitability is taken as 34% [10]. To bring cost values to the estimated period, let us discount the obtained value of total income. According to Russian agricultural ministry's regulation, the discounting rate is 0.02. Let us divide the obtained value by mushroom season period and we have a specific average annual cadastral value of a unit with mushroom yield. When the unit estimations completed, it is possible to evaluate the average cadastral value for a whole research object.

The general formula of calculation is as follows:

$$P^{mush}{}_{cad} = \frac{\dfrac{\sum_{1}^{i} P_i^{mush} W^{mush}{}_i \cdot t}{(1+d)^t} + \dfrac{\sum_{1}^{i} P_i^{mush} W^{mush}{}_i \cdot (T-10)}{(1+d)^{(t+10)(T-10)}}}{t+T},$$

where

$P^{mush}{}_{cad}$ – specific average annual cadastral value, ruble/ha;

$i = 1\ldots12$ – species of gathered mushrooms;

P_i^{mush} – income from one kilo of mushrooms, ruble/kg;

$W^{mush}{}_i$ –average annual mushroom yield capacity, kg/ha;

t – period before the end of cutting cycle for main wood species, years;

T – cutting cycle, years;

d = 0.02 – discounting rate [6].

To make further comparison and choice of alternative forest use, the average annual cadastral value was estimated for the same period without clear cutting:

$$P^{mush}{}_{cad} = \frac{\sum_{1}^{i} P_i^{mush} W^{mush}{}_i}{(1+d)^{(t+T)}}$$

To estimate the cadastral value with regards to wood cutting, we use the same discounting method:

$$P^{wood}_{cad} = \frac{\dfrac{P_{hardw} W_{hardw}}{(1+d)^t} + \dfrac{P_{hardw} W_{hardw}}{(1+d)^{t \cdot T}} + \dfrac{P_{softw} W_{softw}}{(1+d)^t} + \dfrac{P_{softw} W_{softw}}{(1+d)^{t \cdot T}}}{T+t},$$

where

$P^{wood}{}_{cad}$ – specific average annual cadastral value, ruble;

P_{hardw} – income from 1 m^3 of hardwood timber production is taken as 1272 ruble/m^3 according to the average 7-year cost [11] and the profitability of 6% [12];

W_{hardw} – stand of hardwood timber per hectare, m^3/ha;

P_{softw} – income from 1 m^3 of softwood timber is taken as 1475 ruble/m^3 according to the average 7-year cost [9] and the profitability of 6% [12];

W_{softw} – stand of softwood timber per hectare, m^3/ha.

3. Results and discussion

The research resulted in average annual specific cadastral values for each unit with regards to different ways of forest use: only timber production, mushroom production with clear cutting and mushroom production without clear cutting. As a result, the average annual specific cadastral value with regards to mushroom production with clear cutting on all compartments is 29 ruble/ha, without clear cutting – 75 ruble/ha.

The average annual specific cadastral value with regards to timber production is 208 ruble/ha. Taking into account complementary forest yields (timber and mushroom yields), the average annual specific cadastral value increases up to 231 ruble/ha. With the average productive period of the units being 89 years, the cadastral value with the complimentary forest use for the whole period is 20,559 ruble/ha vs 18,512 ruble/ha of the cadastral value without mushroom yield. Thus, the average cadastral value should increase by 2,047 ruble/ha or 23 ruble/ha annually.

While making forest land cadastral appraisal, it is advisable to use average annual parameters rather than general cadastral values, since the average values are easy to compared. To calculate the charge for forest use, the cadastral value is of little importance since the charge value can be controlled by an adjusting factor.

Further research is to be focused on other NWFPs, such as tree biomass components (fir-needles, foliage, tanner's bark, and birch bark), birch juice, oleoresin, berries and herbal medical products and starting materials. Integration of the cadastral values with regards to different ways of forest use will allow us to assess true cadastral value of forest as a natural resource.

References

[1] Amacher G, Ollikainen M and E Koskela 2009 *Economics of Forest Resources*. (Massachusetts: The MIT Press) Date Views 1.05.2015 URL: http://iopscience.iop.org/1755-1315/21/1/012026

[2] Kovyazin V, Belyaev V, Pasko O and Romanchikov A 2014 Taxation indices of forest stand as the basis for cadastral valuation of forestlands *IOP Conf. Ser.: Earth Environ. Sci.* **21** 012026

[3] Pasko O A The forest resource of Tomsk Region and its use 2013 *Applied and fundamental Studies: Proc. 3rd Int. Academic Conf. (Publishing House and Science and Innovation Center: St. Louis, Missouri, USA, 30-31 August 2013)* p 102–9

[4] Gryaz'kin A, Kondratenko N and Pona D 2006 *Nedrevesnaya produktsiya lesa* (*Non-wood forest production*) (St. Petersburg: Polytechnic University publishing house) p 338

[5] Kovyazin V and Skachkova M 2013 *Osnovy lesnogo, sadovo-parkovogo i priusadebnogo khozyaystva (Fundamentals of forest, park and garden industry)* (St. Petersburg: Polytechnic University publishing house) p 135

[6] *Forest Code of the Russian Federation: Federal Law dated 04.12.2006. № 200-ФЗ* (edit. 12.03.2014) Date Views 1.05.2015 URL: http://www.consultant.ru/popular/newwood (in Russian)

[7] *Ob utverzhdenii metodiki gosudarstvennoy kadastrovoy otsenki zemel' lesnogo fonda Rossiyskoy Federatsii: prikaz Roszemkadastra 17.10.2002 № P/336* Date Views 1.05.2015 г. URL: http://www.bestpravo.ru/federalnoje/zk-zakony/q3k.htm (in Russian)

[8] *Ob utverzhdenii metodiki ekonomicheskoy otsenki lesov: prikaz Federal'noy sluzhby lesnogo khozyaystva Rossii 10.03.2000 № 43* Date Views 1.05.2015 URL: http://docs.cntd.ru/document/901828190 (in Russian)

[9] *Ob utverzhdenii rezul'tatov gosudarstvennoy kadastrovoy otsenki zemel' lesnogo fonda na territorii Leningradskoy oblasti: postanovlenie pravitel'stva Leningradskoy oblasti №76 30.04.2010* Date Views 1.05.2015 URL: http://www.bestpravo.ru/leningradskaya/xg-gosudarstvo/h4p.htm (in Russian)

[10] *Taksatsionnye opisaniya nasazhdeniy Pesochinskogo lesnichestva Kurortnogo lesoparka* 2013(*Mensuriation description of district forest ranger station of the forest-park "Curort lesopark"*) (St. Petersburg: Lesproyekt) p 586 (in Russian)

[11] *Uroven' rentabel'nosti (ubytochnosti) rabot, uslug v oblasti lesnogo khozyaystva i lesozagotovok, vsego po Rossiyskoy Federatsii v protsentakh na 1 yanvarya issleduemogo goda (Profitability/loss ratio in forest and timber industry of the Russian Federation for the first of January)* Date Views 1.05.2015
URL: http://www.umocpartner.ru/assets/files/Analitika/Uroven% 20rentabelnosti% 20lesnogo%20hozyaistva%20i%20lesozagotovok%20za%202005-2010%20godi.pdf (in Russian)

[12] *Federal State Statistics Service* Date Views 1.05.2015 *http://cbsd.gks.ru/* (in Russian)

Comparative analysis of land management in the world

A Pomelov[1], O Pasko[2] and A Baranova[3]

[1] Director of Republican Unitary Enterprise "BelNITszem", candidate of economic sciences, Belarus research land management center, Minsk, the republic of Belarus
[2] Professor of National Research Tomsk Polytechnic University, Doctor of Agricultural Science, 30 Lenin Ave., Tomsk, 634050, Russia
[3] Senior lecturer of Foreign Language Department, Tomsk Polytechnic University, 30 Lenin Ave., Tomsk, 634050, Russia

E-mail: [1] pomelov@mail.bn.by, [2] helgapas@mail.ru, [3] baranskikh@yandex.ru

Abstract. The article studies and compares agricultural land management performance criteria applied in different countries worldwide. It is stated that there are different definitions and classifications of agricultural lands in different countries, which implies different indicators of land use efficiency. Sometimes these indicators cannot be correlated or compared with each other. It is suggested using such criteria as arable land area and fodder yield per 1 ha as the most unified and appropriate ones.

1. Introduction

Agricultural land covers about 38 % of the world's land area, arable land makes 11 %. This means that the ploughness level of agricultural land reaches 30 %. There is less that 0.7 ha of agricultural land and 0.2 ha of arable land per each of the world's inhabitant [7, 12, 16]. Though, since 1993, there have been taken a number of attempts to standardize national systems and to develop a unified international land accounting system, different approaches to the issue still exist [1, 3, 6].

The aim of the research is to make a comparative analysis of basic performance indicators of agricultural land use management in the world.

To achieve the aim, the following objectives are to be put:

- to collect necessary data;
- to compare, correlate the basic performance indicators and choose the most optimal ones;
- to use them for the analysis of agricultural land condition and agricultural land use performance.

2. Objects and methods of research

The research objects are agricultural lands in Europe (including Russia and Belarus), the USA and Canada. Analytical and comparative methods are used to achieve the aim of the research. According to Food and Agricultural Organization of the United Nations (FAO) [17], agricultural lands are divided into crop areas, permanent meadows and pasture areas. Crop areas are divided, in its turn, into permanent crops and arable areas; arable areas are subdivided into temporary crops, temporary meadows and fallow areas.

3. Results and discussions

According to Belarus State Land Cadastre data, dated the 1st of January, 2014[5], agricultural land area is 8,726 M ha (42%) including 5,560 Mha (26.8 %) of arable land area, with the total area of the country 20,760 M ha. The level of ploughness of agricultural lands reaches 63.7 %. There is 0.9 ha of agricultural land area, with 0.6 ha of arable land area per each resident of Belarus.

According to the cadastral data, 1 January 2015, the total area of Russia (with Crimea) is 17151442 sq. km, the area of productive lands is about 9 million sq km; i.e. 11% of the world's productive land area. The number of productive soil is 6ha per capita, with arable land area 130.3 million ha and the level of ploughness is 7.7% [7, 14].

In the frame of land accounting system "EC CORINE Land Cover" agricultural lands are classified into arable land, permanent crops, pasture area and other agricultural areas.

Analyzing national and international agricultural land classification and accounting systems, we pay attention to the following: in other countries but Belarus and Russia, arable lands are considered to be those that in Belarus are classified as arable and permanent meadow cultivated. It covers the area of 7,618 M ha (36.7 % of the total area and 87.3 % of the agricultural area). Such "arable" area per one resident of Belarus is 0.8 ha.

Arable lands in Russia are defined as regularly cultivated agricultural lands, plowed and sown, including planting perennial grasses, emergency fields, areas under glass and fallow areas [9].

In Belarus and Russia, permanent meadows and pastures, which cover vast areas in some countries, are considered to be agricultural lands used to cultivate perennial grasses and lands with natural grass stand [13]. The area of such lands is only 974.4 Mha, or 11.1 % of total agricultural land area, with this number constantly decreasing. Meadow lands are not regarded as agricultural lands in cadastral systems of some countries. This approach was often suggested being implemented in Belarus national agricultural politics.

Thus, taking into account all mentioned above, to compare land resource potential of different countries it is reasonable to consider arable land area as a more informative indicator. According to FAO (2009) [2, 16], the countries with the most amount of arable lands are China (220 million hectares), the USA (163), India (158), Russia (122), Brazil (61), Australia (47) and Canada (45). Area of arable land per one inhabitant in a country should also be taken into account. Thus, in Australia, the value is 2.2 ha /person, Canada – 1.5, Kazakhstan – 1.5, the Republic of Niger – 1, Russia – 0.9, Argentine – 0.8, Ukraine – 0.7. The value in Japan is only 0.03 to compare.

It is desirable to compare these values with those of the countries having the same climatic, natural, social and economic conditions. For example, the European Union countries neighboring Belarus have the following values: Poland -0.3 ha/person, Lithuania -0.6, Latvia -0.5. In Northern Europe the values are the following: Estonia - 0.5 ha/person, Denmark – 0.4, Finland – 0.4, Sweden – 0.3 and Norway – 0.2. The most developed European countries have less arable area per inhabitant: France – 0.3 ha/person, Germany, Great Britain, the Netherlands and Switzerland – 0.1. Thus, in terms of area of arable land per one inhabitant, Russia and Belarus are among the world leaders.

However, the question is if it is possible to compare land quality and land use efficiency in different countries.

Unfortunately, there are no unified approaches to compare quality of lands in different countries, including agricultural and arable lands. In most countries, there are national methods to assess and classify land quality according to internal purposes.

Typically, to make a cross-country comparison, specialists apply indirect correlation of agricultural land use efficiency rates in value terms and/or physical terms (figure 1).

Potentially, the best relative indicator of agricultural land use efficiency might be land productivity, for example, fodder yield per hectare. However, such approach is not widely spread world wide. International organizations more often use methods of arable land assessment based on average grain and leguminous crop capacity, especially wheat. It is quite reasonable, since grain and leguminous crops are of strategic value all over the world and cover more than a half of the whole cultivated area of the world.

Wheat plays the leading role in gross grain yield structure of the world and it determines food safety policy and agriculture economy both at the global and local levels.

The leading European countries (Germany, Great Britain, France, Denmark and others) have reached wheat crop capacity of 7000-8000 kg/ha [2, 3, 6], which is considered to be close to physiological limit, according to some specialists [6]. Belarus, as well as Lithuania, Latvia and Poland, have yielded more than 3300 kg/ha (average value of 2005-2012 years [15]). Russia have gained 2390 kg/ha. It should be noted that it is not quite correct to compare wheat crop capacity in different countries if we take into account the difference in climatic and environmental conditions that influence the agriculture structure of a country [10].

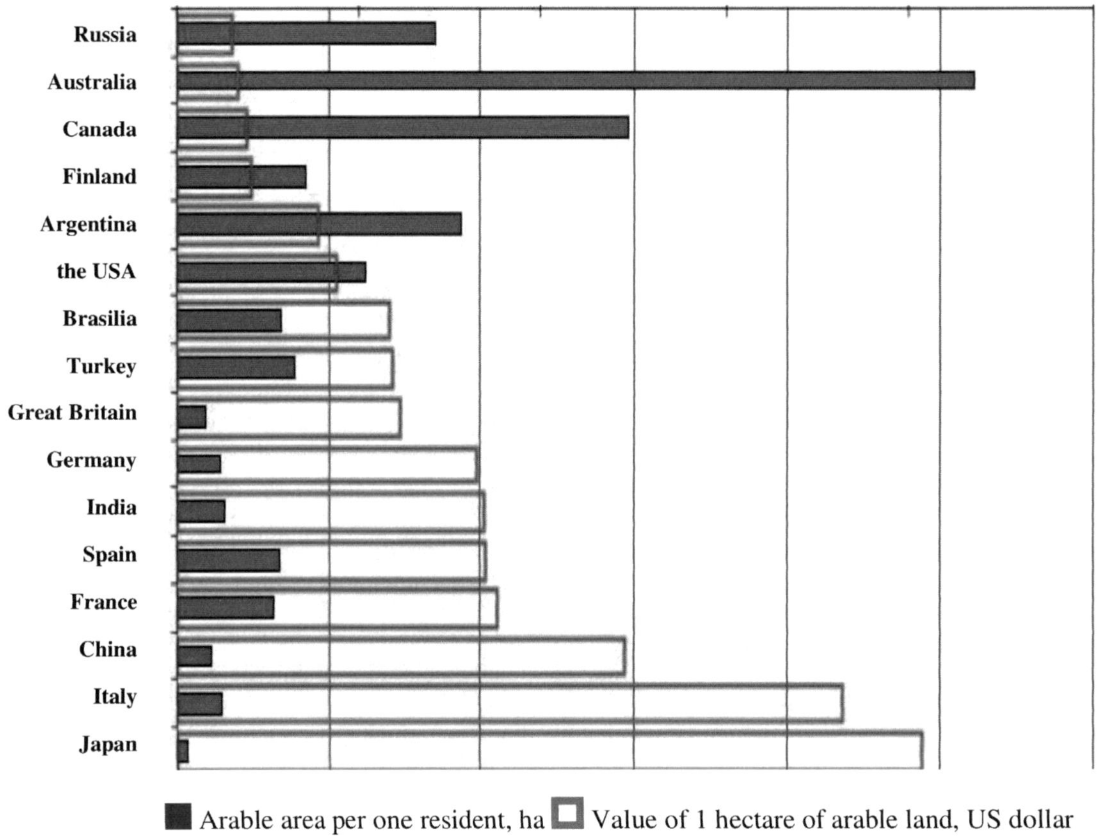

Figure 1. Value of 1 ha of arable land and the amount of arable land per one resident in some countries.

However, the climatic and environmental conditions of the southern part of Sweden, which is most agriculturally developed, and social and economic priorities are rather comparable with those in Belarus. Nevertheless, the average wheat crop yield in Sweden is 6000 kg/ha.

Thus, taking into account the interconnection of crop capacity of different agricultural plants and the fact that crop capacity conditions animal agriculture productivity, it is possible to say that Belarus and Russia have significant, almost double potential of national agriculture. It is necessary to fulfill the potential in the nearest future by using modern techniques and technologies of land management, agro chemistry, genetics, mechanical operations etc.

References

[1] Baumane Vand Pasko O 2014 Comparison of land reform of Latvia and Russia in conditions of transition period. *Baltic Surveying* **1** 40-44 Date Views 1.05.2015*: http://www.lif.llu.lv/print_pdf.php?article=269*
[2] *Land availability*. Date Views 1.05.2015 *www.fao.org*
[3] Pasko O 2013 Vergleichende Analyse des effizienten Flaechenverbrauchs in der Landwirtschaft in Deutschland und Russland *Economic development and perspectives of cooperation between the USA, Europe, Russia and CIS states* **1** (New York : CIBUNET Publishing) 45–60
[4] Berezovskaya O 2011 *Kadastrovaya Otsenka Zemel' Naselennykh Punktov Natsional'nyy Doklad o Sostoyanii, Ispol'zovanii i Okhrane Zemel'nykh Resursov Respubliki Belarus' (Cadastral Appraisal of Urban Areas. National Report on Condition, Use and Protection of Land Resources in Belarus)* (Minsk: RUP "Bel-NITszem") 57–62
[5] *Gosudarstvennyy Zemel'nyy Kadastr Respubliki Belarus' 2014 (State Land Cadastre of Belarus)* Date Views 12.05.2015 *www.gki.gov.by.*
[6] *Evropa Istoshchila Svoy Potentsial Uvelicheniya Urozhaynosti Zernovykh (European Capacity to Grow Food Is Plateauing)* Date Views 1.05.2015 *www.inosmi.ru.57*
[7] *Land resources of the world (*only in Russian*)* Date Views 1.05.2015 *www.land-in.ru*
[8] *Land Resources of Russia* (only in Russian) Date Views 1.05.2015 *http://old.rgo.ru/2010/07/zemelnye-resursy-rossii-—-struktura-i-mirovoe-znachenie/*
[9] *Land Code of the Russian Federation* Date Views 1.05.2015 *http://kodeksy-by.com/dictionary/p/pahotnye_zemli.htm*
[10] *Land of Russia and of the World* Date Views 1.05.2015 *http://maxpark.com/community/3135/content/810065*
[11] *Instruktsiya o Poryadke Vedeniya Gosudarstvennogo Ucheta Zemel' i Sostavleniya Otcheta o Nalichii, Kachestvennom Sostoyanii i Ispol'zovanii Zemel' 1993 (Guideline on State Land Accounting and Making Reports on its Quality and Use)* (Minsk: Belgiprozem) 46
[12] *Land Use in the World*. Date Views 1.05.2015 *www.ru/worldstat.info/World/Land.*
[13] *Land Classification*. Date Views 1.05.2015 *http://elc.baa.by/kadastr.doc*
[14] *Total Area of Russia* Date Views 1.05.2015 *http://fb.ru/article/163763/obschaya-ploschad-rossii-obschaya-ploschad-rossii-s-kryimom*
[15] Kuznetsov G, Moroza G and Tsytron G 2010 *Pokazateli Kadastrovoy Otsenki Zemel' Sel'skokhozyaystvennykh Organizatsiy i Krest'yanskikh (Fermerskikh) Khozyaystv (Indicators of Cadastral Assessment of Farmlands)* (Minsk) 127
[16] *Agriculture and Food Production* Date Views 1.05.2015 *www.unep.org.*
[17] *System of Agricultural Census and Surveys. World Program for the Census of Agriculture 2010. Theme 01:Land // Statistic database of FAO*. 2005. Part 11. (Rome: FAO) 95–103

IOP Conf. Series: Earth and Environmental Science **27** (2015) 012041 doi:10.1088/1755-1315/27/1/012041

Evaluation of metal content in perch of the Ob River basin

N A Osipova[1], K D Stepanova[1] and I A Matveenko[1,2]

[1] National Research Tomsk Polytechnic University, 30 Lenin Ave., Tomsk, 634050, Russia

E-mail: [2] mia2046@yandex.ru

Abstract. The geochemical features of river perch in the River Ob basin have been studied (the upper and middle reaches of the Ob River and the lower reach of the Tom River). The contents of Ag, Bi, Cd, Co, Cr, Cu, Fe, Mn, Mo, Ni, Pb, Sn, W, Zn, Hg in perch's soft tissue are defined by the methods of ICP AES and stripping voltammetry, that of mercury in bones – by the atomic absorption method using mercury analyzer PA-915+. The distribution series of metal absolute concentrations in perch's soft tissue from the Ob River basin are plotted: Fe > Zn > Cu > Mn, typical for uncontaminated or slightly metal contaminated water bodies. In soft tissue of the studied samples the metal content does not exceed the permissible values. The mercury content in bones of studied samples is in the range 0,036-0,556 mg/kg. The mercury concentration is higher in bones in comparison with soft tissue in all samples.

1. Introduction

Interest in the content of the chemical elements in fish of Ob basin is related to the anthropogenic impact on natural water bodies that disturbs the natural cycle of chemical substances in the biosphere, and to the direct impact of those chemicals on human health.

The information on composition and amount of the metals can be found in the research of the fish (e.g. perch) muscle and bones chemical composition. Fish take the upper trophic level in aquatic ecosystems biocenosis and have a pronounced ability to accumulate elements [1]. Elevated levels of metals in fish indicate their significant concentration in the aquatic medium.

The average metal content in perch soft tissue of the Upper Ob River according to [2] was respectively: Cu – 0.94; Mn – 2.24; Fe – 61.87; Zn – 9.29; Pb – 0.28; Cd – 0.12; Hg – 0.015 mg/kg. According to the data of 1990, 1991, and 2000, the changes in the element content in the River Tom ranged in the following way: Cu – 0.73- 0.2; Zn – 8.5-14; Pb – 0.08-3.92; Cd – 0.01-0.23; Hg – 0.11-0.76 mg/kg [3]. In 2000, in contrast to 1990-1991, in the samples of fish soft tissue from the Tom River the contents of metals (Co, Ni, Cu, Cd, Hg and Pb) were considerably less, which is likely to be a consequence of decreasing the concentrations of these metals in abiotic river ecosystem components within this period.

Thus, there is an objective need for systematic data analysis on the same metals in the same time period in the lower reaches of the Tom River and the upper and middle reaches of the Ob River, where there are densely populated settlements of the Tomsk region.

The purpose of the work is as follows:

- To identify the geochemical characteristics of perch of the Ob River basin by examining its muscle and bone

- To estimate the quality of the river perch, as one of the staple food of the Ob River basin residents.

The stages of the work are: sampling the fish; sample preparation for atomic emission spectrometry ICP, voltammetry and atomic absorption analysis; determination of the elements contained in the bone and muscle tissue of the fish; data analysis.

For a number of reasons the river perch was chosen as an object of the analysis [2]: it accumulates the toxicants in much higher concentrations compared to those in the environment; its migration is of low intensity; it has large population; it has a relatively long life cycle; it can be of a relatively large size that makes possible to take enough quantity of samples; it is suitable enough for catching and sampling the organs and tissues; it is representative in terms of determining the degree of contamination according to the certain morphophysiological and environmental parameters.

2. Materials and methods

2.1. Study area, Sampling and preparation

The samples of perch were taken in June – July, 2014 in the upper and middle reaches of the Ob River and in the lower reach of the Tom River (figure 1). The point observation network was used during the sampling (15 points in total). The total mass of one sample was 500 to 700 grams, it included at least 7 to 10 fish. The muscle tissue was separated and crushed with a plastic tool. The bone tissue was taken out of the fish and subjected to drying and grinding.

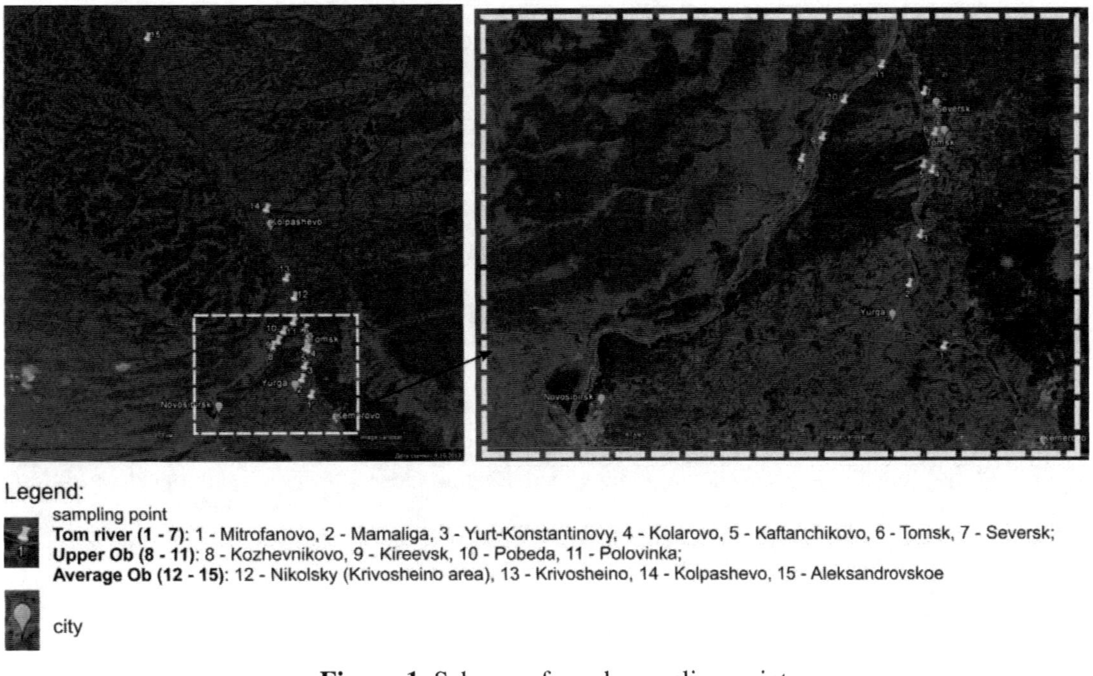

Figure 1. Scheme of perch sampling point.

2.2. Analysis

The quantitative analysis of muscle tissue samples for the content of Ag, Bi, Cd, Co, Cr, Cu, Fe, Mn, Mo, Ni, Pb, Sn, W, Zn was performed by atomic emission spectrometry with inductively coupled plasma «iCAP 6300 Duo» (ThermoScientific Company), with sensitivity being 10^{-8}-10^{-9} per cent and with accuracy results up to 20 to 50 per cent depending on the element detected. The content of Hg, As, I, Se, Zn, Cd, Pb, and Cu in the perch muscles was determined by stripping voltammetry method. There is satisfactory convergence of these methods.

The content of Hg in the perch bone tissue was examined by the absorption method using the mercury analyzer RA-915+ equipped with the auxiliary device PYRO-915.

3. Results and discussion

The elements content analysis of the fish species muscle tissue revealed the following: most of the metals concentrations (Ag, Bi, Cd, Co, Cr, Mo, Ni, Sn, W) is lower than the sensitivity of the procedures being used, the average content of Cu, Fe, Mn, Zn, Se, I, As, Pb, and Hg in perch muscle does not exceed the permissible values for food products [4].

The content of iron (Fe) in the perch muscle tissue ranges from 6.44 to 24.2 mg/kg, the maximum concentration of the element was detected in the samples 1 and 14 (figure 2). It is impossible to draw a definitive conclusion about the causes and sources of elevated concentrations of Fe in the above mentioned sites of the Ob River basin without carrying out some additional research.

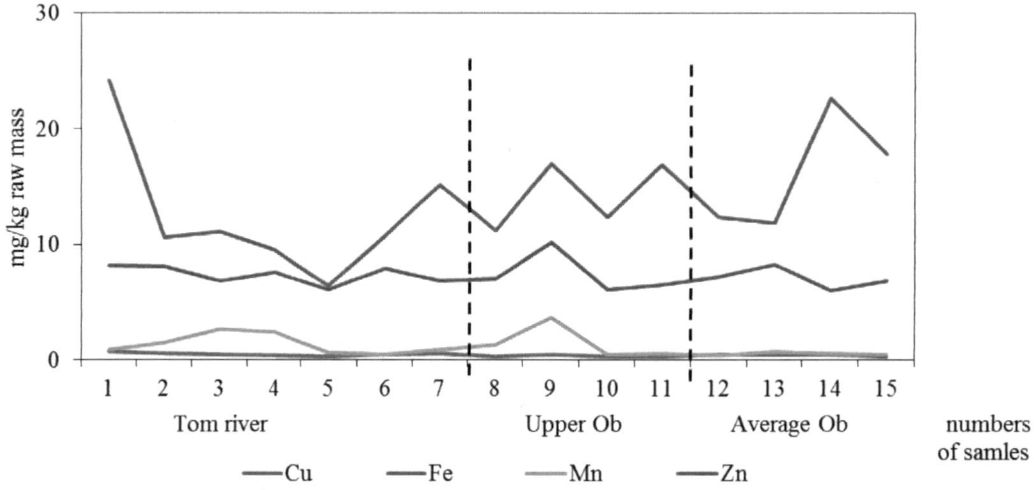

Figure 2. The content of metals in muscle tissue of perch caught in the Ob River basin, according to the results obtained by ICP AES.

In terms of decrease in metal absolute concentration in the perch muscle tissue the distribution series are built (figure 3). The first series corresponds to the fish tissue samples No. 1, 2, 3, 4, 5, 7, 8, 9, 10, 11, 13, 14, and 15. For samples number 6 and 12, the second series of element distribution is typical.

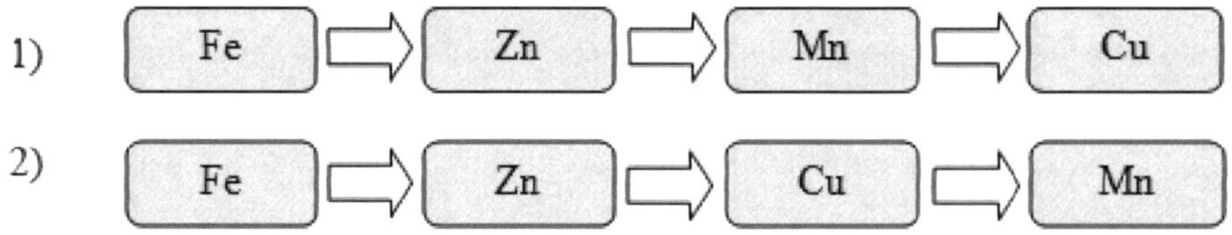

Figure 3. Series of metal absolute concentration in the perch muscle tissue, the Ob River basin.

In general, in all the investigated samples of muscle tissue, iron (Fe) and zinc (Zn) were predominant. This confirms the available literature data on high concentrations of these elements in fish, and significant accumulation coefficients compared to those of other metals. This fact is explained by the intensive accumulation of microelements (Fe, Zn) in the organism, actively involved in a number of physiological processes (breath, blood formation, deposition, excretion, etc.).

The important fact is that the resultant distribution series correspond to the fish caught in uncontaminated or slightly metal contaminated waters [3]. In such species of ichtyofauna the content of Fe, Zn, Co, and Mn are predominant, zinc in almost all samples being an absolute leader after iron. Cu and Mn can interchange with each other. Co, Ni, Cd, Hg, are, as a rule, in the right side of the series, Hg and Cd being the rightmost.

In the samples of perch muscle tissue taken in the Tom River in, Pb follows Zn, thus reflecting its high accumulation in fish (table 1) [3]. However, the research of 1991 showed that copper follows zinc, not lead.

Table 1. Metal Content Ranking in Muscle of Perch in the Tom River.

Year	Ranked sequences
1990	Zn>Pb>Mn>Hg>Cu>Cd>Ni
1991	Zn>Cu>Mn>Pb>Co>Ni>Cd>Hg

One of the most harmful elements polluting the biosphere, possessing the highest toxicity index rate among heavy metals is mercury (Hg), due to its ability to block the protein molecules, break down their biosynthesis, cause mutagenic changes in DNA [5]. No positive effect of mercury on living organisms is established.

The analysis revealed that the content of mercury in the samples studied is in the range 0.036 to 0.556 mg/kg, the content of the element in all the test samples being higher in bone tissue in comparison with that in muscle tissue, which can be explained by the property of mercury to accumulate toxic elements (figure 4). Many researchers revealed similar trend of toxic elements accumulation in bones of living organisms [6-7].

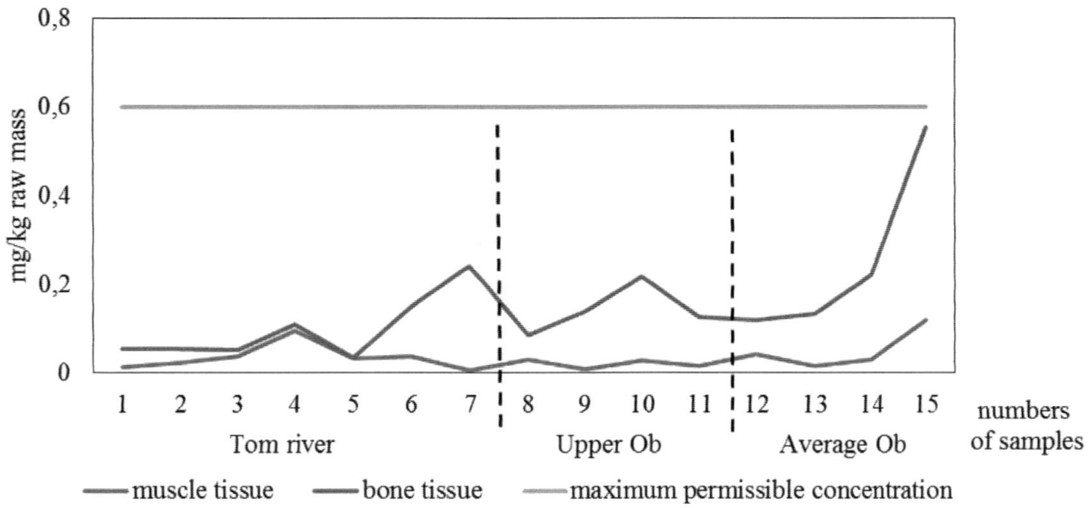

Figure 4. Comparison of mercury content in muscle and bone of perch of the Ob basin.
Note: Muscle tissue was investigated by stripping voltammetry, bone tissue was investigated by atomic absorption method.

In addition, the results of the study carried out has shown that the mercury levels in samples of the river fish of the Ob River basin do not exceed the maximum permissible concentration, except for the area of intensive oil production - 0.556 mg/kg, wet weight. This fact is not an unambiguous evidence of anthropogenic factor influence, but we cannot exclude that the extraction and the use of oil and gas make a substantial contribution to mercury emission into the environment [7].

Mercury can pollute the atmosphere when gas is burnt in flares, and then, can settle on the water surface with precipitation. Fish, involved in the upper trophic level of the hydroecosystem, actively accumulate various elements, including mercury from the water surface.

4. Conclusion

As a result of the study, the following conclusions are made: in muscle tissue of perch of the Ob River basin the content of the following elements can be established: Cu, Fe, Mn, Zn, the maximum permissible concentration not being exceeded. The obtained distribution series of metals in perch muscle tissue correspond to the fish from uncontaminated or slightly metal contaminated waters. As compared to the mercury content in the soft tissue, it is higher in the bone tissue of the river perch. The maximum permissible concentration is not exceeded. All the studied samples of perch of the Ob River basin do not pose any hazard to the human health as a consumer of fish.

References

[1] Moiseenko T I and Kudryavtseva L P 2005 Evaluation of the geochemical background and anthropogenic load on the bioaccumulation of trace elements in fish *Water resources* **6** 700–11

[2] Glazunova A 2007 The heavy metal content and character of distribution in the some fish of the High Ob River *The news of Altay State University*3 20–22

[3] Popov P A 2002 *Assessment of the ecological status of water bodies means bioindication* (Novosibirsk: Novosibirsk State University press) p 270

[4] *Sanitary rules and norms* 2.3.2.560-960 1997 Hygiene requirements for the safety and nutritional values of food 269

[5] Gordeeva O N, Belogolova G A and Grebenshikova V I 2010 The distribution and migration of heavy metals and arsenic *Problems of regional ecology* **3** 108–113

[6] Jezierska B and Witeska M 2006 The metal uptake and accumulation in fish living in polluted waters *Soil and water pollution monitoring, protection and remediation* **12** 107–114

[7] Mercury emissions in the oil and gas industry: The Intergovernmental Negotiating Committee for the preparation of a legally binding global instrument on mercury 2011 *The United Nations Environment Programme* 1–7

Mineralogical and geochemical characteristics of drinking water salt deposits

B R Soktoev[1,3], L P Rikhvanov[1,4] and I A Matveenko[2,5]

[1] Department of Geoecology and Geochemistry, Institute of Natural Resources, National Research Tomsk Polytechnic University, 30 Lenin Ave., Tomsk, 634050, Russia
[2] Department of Foreign Languages, Institute of Natural Resources, National Research Tomsk Polytechnic University, 30 Lenin Ave., Tomsk, 634050, Russia

E-mail: [3] bulatsoktoev@gmail.com, [4] rikhvanov@tpu.ru, [5] mia2046@yandex.ru

Abstract. The article presents the research results on the features of element and mineral composition of salt deposits (limescale) formed in household conditions in heat exchanging equipment. The major part of limescale is represented by two species of calcium carbonate - calcite and aragonite. We have shown that high concentrations of chemical elements in the limescale promote the formation of their own mineral forms (sulphates, silicates, native forms) in salt deposits. Detecting such mineral formations suggests the salt deposits of drinking water to be a long-term storage media which can be used in the course of eco-geochemical and metallogenic studies.

1. Introduction

According to several encyclopedias and reference books, salt deposits, or limescale, are "... solid deposits that form on the inner walls of steam boilers, water economizers, super-heaters, evaporators and other heat exchangers in which the evaporation or heat of water containing certain salts takes place" [1]. The study of the salt deposits formed in drinking water systems at various stages of water treatment and transportation is the relevant issue all over the world. Usually most of the studies deal with methods of treatment and removal of salt deposits. The issues discussed there are: 1) development and influence of various factors on these processes [2-4]; 2) their mineral composition [5-9]; 3) disposal processes and methods. However, all these studies characterize the processes taking place in industrial production systems: water treatment, water heating in boilers of thermal power plants, etc.

Our research is based on the data of the studies of drinking water salt deposits which are formed directly in household conditions (enamelled and electric kettles, pots, pans) in the process of multiple boiling and cooling of water. In our opinion, such data could be a source of information on chronic, "long-term" input of chemical elements in a human body. Investigations of the Department of Geoecology and Geochemistry (GEGC) (Tomsk Polytechnic University (TPU)) carried out for over 20 years have shown that the study of limescale element composition is an informative way to assess the quality of drinking water as well as environmental and geochemical conditions of the sites. Furthermore, element composition of these formations can be used in metallogenic forecasting [10-16].

According to our data, the information on the mineral composition of limescale formed directly in the household conditions appears to be absent. Therefore, the aim of this work is to study the characteristics of the mineral composition of drinking water salts.

2. Materials and methods

So far, the research covers the territory of southern part of Urals and Siberia: Irkutsk, Tomsk, Chelyabinsk Oblasts, the Republic of Altai and Buryatia (Russia), as well as Pavlodar Oblast (Kazakhstan).

We collected samples of limescale from different utensils, which were repeatedly used to boil drinking water (enamelled and electric kettles, pans, pots, samovars). In most cases, the selected materials presented powdered mass or separate plate aggregates on the walls of utensils. If limescale was firmly fixed on the walls, the samples were taken with a knife made of stainless steel: salt deposits were carefully removed from the walls of the heat exchanging household utensils. Selected samples were packed in plastic bags. Each sample was labeled with the following information: kind of vessel where water was boiled and the depth of the aquifer. If we had the opportunity to take samples from different type of dishes (plastic pots, metal teapots and samovars) in the same yard, we necessarily did it. In the laboratory, samples were dried at room temperature and prepared for analyzes according to the relevant standards.

At the first stage, all samples were sent to an instrumental neutron activation analysis (INAA) to determine the concentrations of 28 chemical elements (Na, Ca, Sc, Cr, Fe, Co, Zn, As, Br, Rb, Sr, Ag, Sb, Cs, Ba, La, Ce, Sm, Eu, Tb, Yb, Lu, Hf, Ta, Au, Th, U). The analysis was performed in the nuclear geochemical laboratory of the International Innovation Research and Educational Center (IIREC) "Uranium Geology" (TPU) (accreditation number ROSS RU.0001.518623, October 10, 2011, analysts are senior researcher A.F. Sudyko, and L.V. Bogutskaya).

The samples with elevated concentrations of chemical elements were studied in detail by other methods for the mineral composition analysis: 1) X-ray diffraction (DRON-3M, GEGC, TPU, head of the laboratory is G.A. Babchenko; D2 Phaser, Bruker AXS GmbH, Moscow, analyst is Dr. S.N. Putilin); 2) scanning electron microscopy (Hitachi S-3400N, IIREC "Uranium Geology", analyst is S.S. Il'enok). U occurrence modes were determined using the method of fission radiography (f-radiography).

3. Results and discussion

Analysis of more than 600 samples showed that the element composition of limescale is changed within the wide ranges. The leading group includes such elements as Ca, Na, Fe, contents of which in all samples exceed n•100 ppm. In our opinion, these elements are structure-forming for the minerals of drinking water salts. This is also reflected in the composition of the limescale matrix. Our studies show that most of them have a composition identical to travertine – natural carbonate rocks formed by the deposition of Ca, Mg, Fe, Na carbonates out of carbonated springs water [17]. Both travertine and limescale consist of two modifications of calcium carbonate (90-95 %) – calcite and aragonite – with impurities of magnesium and iron species (figure 1). It should be noted that whereas calcium carbonate in travertines is almost always crystallized in the form of calcite, this trend is not observed in limescale. For instance, in some cases it crystallizes in the form of pure calcite, sometimes aragonite and calcite are formed together in various proportions.

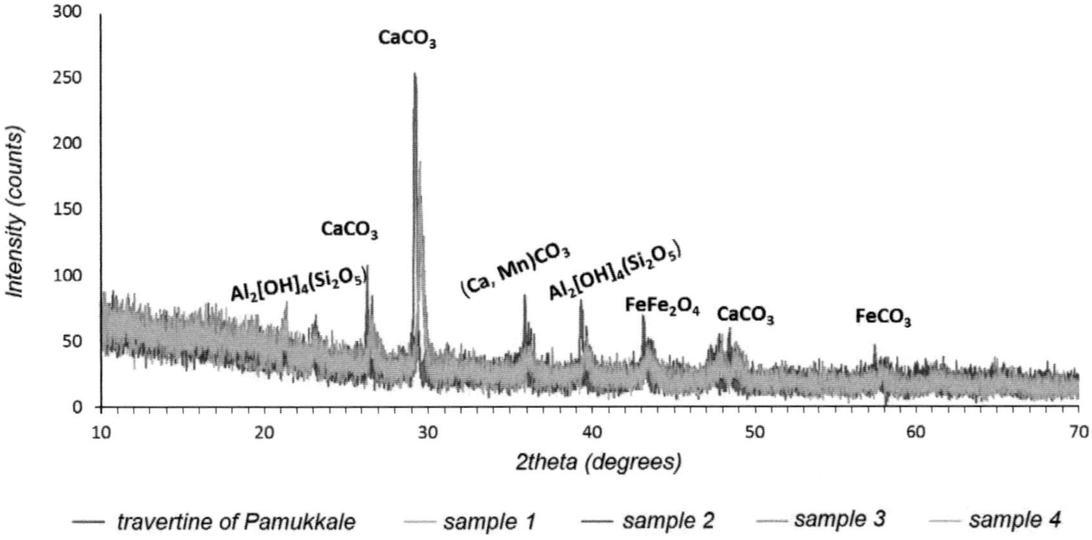

Figure 1. Comparison of mineral composition of Pamukkale travertine and drinking water salt deposits.

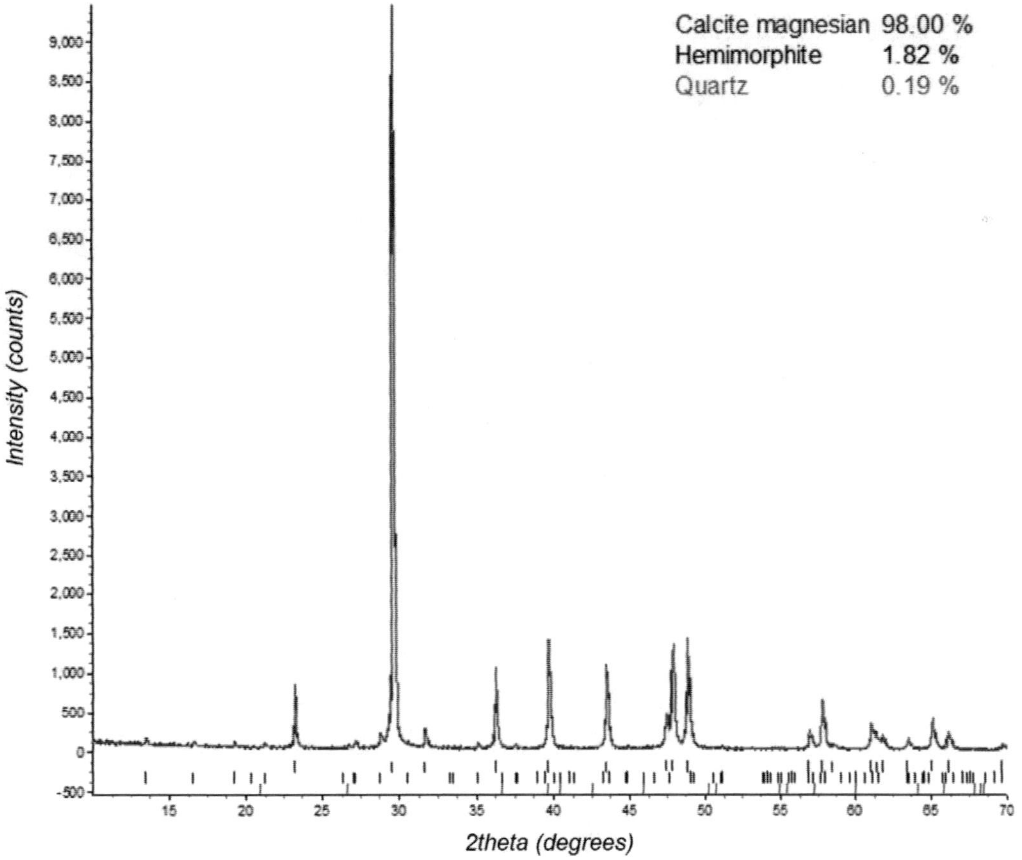

Figure 2. Diffractogram of drinking water salt deposits sample (content of Zn – 3%, Ba – 0.8%).

Issues of formatting aragonite and calcite in travertines are widely discussed in literature. According to various researchers, among the possible factors influencing the formation of aragonite in travertine, there are water temperature [18, 19], the ratio Mg/Ca [18, 20], saturation coefficient [21], Sr^{2+} and SO_4^{2-} [22], pCO_2 and CO_2 rate [23], Fe^{2+}, Ba [21]. On drinking water limescale we found only one research [6], where the high content of Fe ions in water is stated as the factor preventing formation of aragonite.

The elements with high concentrations (> n•10 ppm) in salt deposits are also Zn, Sr, Ba. They seem to isomorphically substitute Na, Ca, Fe ions in the structure of aragonite-calcite, and form their native mineral phases due to their similar geochemical properties. We observed mineral native forms of Zn (hemimorphite) (figure 2), Ba (barite), etc.

Elevated concentrations of some trace elements are also reflected in the mineral composition of limescale. For instance, native minerals of Ag, Au, Ba, Co, Cu, Ni, Ta, Th, Ti, REE (particle sizes range from n•10 to n•100 μm) (figure 3) were revealed by scanning electron microscopy (SEM).

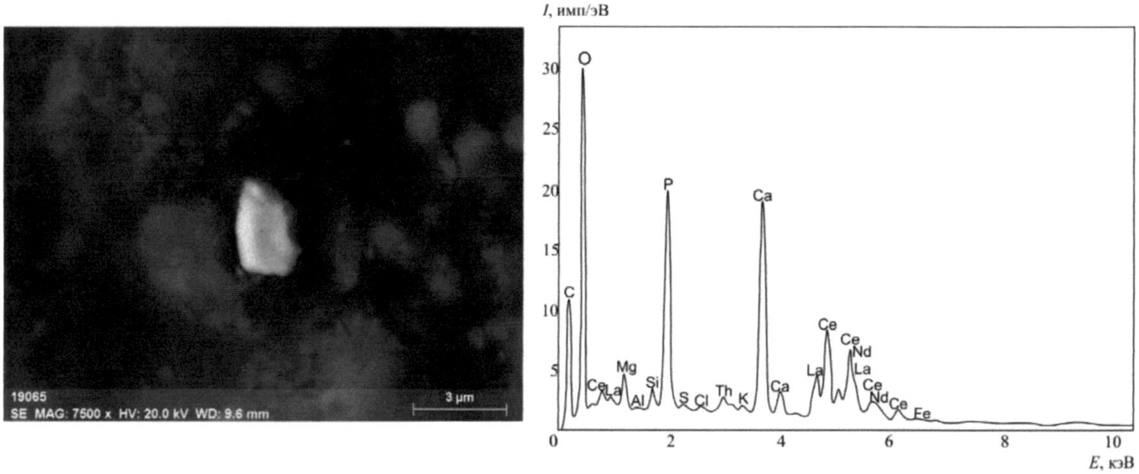

Figure 3. Rare-earth elements phosphate in drinking water salt deposits according to SEM.

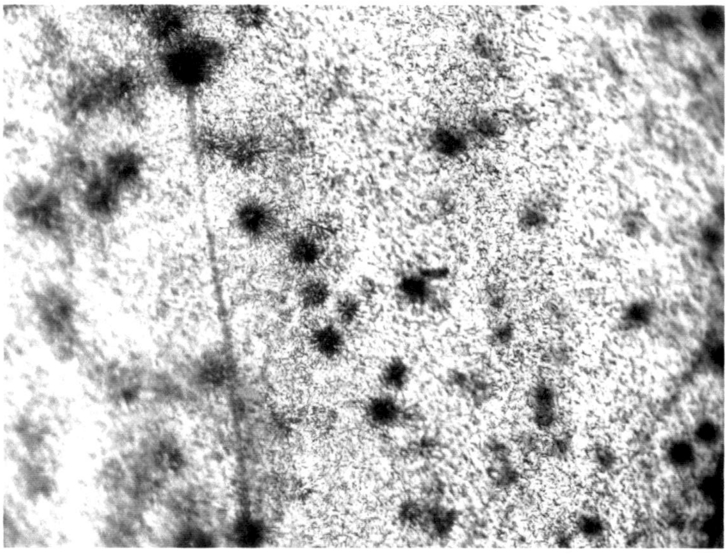

Figure 4. Uranium modes of occurrence in drinking water salt sediments according to fission radiography.

The study in U modes of occurence in limescale performed by the f-radiography indicates that even at the U concentrations of 1.8 ppm it can be found both in the form of native uranium microminerals (observed as "stars") and molecular form of nonstructural scattering, uniformly distributed in the matrix of salt deposits (Fig. 4).

4. Conclusion

The study in mineral composition of the drinking water salts shows that main matrix minerals are calcite and aragonite in various ratios. Elevated concentrations of several chemical elements allow revealing their own mineral forms using methods of X-ray diffraction and scanning electron microscopy. It is also of importance that the elevated contents of elements and minerals in limescale are not always confirmed by the analysis of water chemical composition. This suggests that the drinking water salt formations are depositing media that can provide information on the long-term accumulation of chemical elements.

References

[1] Amjad Z (ed.) 2014 *Mineral scales in biological and industrial systems* (Boca Raton, FL: Taylor & Francis Group)

[2] *Encyclopedia of Desalination and Water Resources*, 2012. URL: http://www.desware.net (date of access 12.12.2013).

[3] Gal J-Y, Fovet J and Gache N 2002 Mechanisms of scale formation and carbon dioxide partial pressure influence. Part I. Elaboration of an experimental method and a scaling model *Water Research* **36** 755–63.

[4] Gal J-Y, Fovet J and Gache N 2002 Mechanisms of scale formation and carbon dioxide partial pressure influence. Part II. Application in the study of mineral waters of reference *Water Research* **36** 764–73

[5] Pokrovskiy D S (ed.) 2002 *Mineral growths on the water intakes in Tomsk region* (Tomsk: Publishing House of Scientific and Technical Literature) (in Russian)

[6] Potapov S S, Votyakov S L and Borisov D R 1998 Mineralogy and spectroscopy of technogenic and anthropogenic (domestic) limescale *Ural'skiy mineralogicheskiy sbornik* **8** 151–70 (in Russian)

[7] Cowan J C and Weintritt D J 1976 *Water-formed scale deposits* (Houston: Gulf Publishing Co)

[8] Clark L M 1948 The identification of minerals in boiler deposits. Examples of hydrothermal synthesis in boilers *Mineralogical Magazine* **28** 359–66

[9] Candeais J P, de Oliveira D F, ods Anjos M J and Lopes R T 2014 Scale analysis using X-ray microfluorescence and computed radiography *Radiation Physics and Chemistry* **95** 408–11

[10] Mongolina T A, Baranovskaya N V and Soktoev B R 2011 Element composition of drinking water salt sediments in Tomsk Oblast *Bulletin of Tomsk Polytechnic University* **319(1)** 201–11 (in Russian)

[11] Soktoev B R, Rikhvanov L P, Taisaev T T and Baranovskaya N V 2014 Geochemical characteristics of drinking water salt deposits of Baikal region *Bulletin of Tomsk Polytechnic University* **324(1)** 209–23 (in Russian)

[12] Soktoev B R, Rikhvanov L P, Usenova S Z, Mongolina T A and Baranovskaya N V 2014 Drinking water salt sediments as an indicative media in ecogeochemical and metallogenic investigations *Herald of Irkutsk State Technical University* **1(84)** 40–45 (in Russian)

[13] Yazikov E G, Rikhvanov L P and Baranovskaya N V 2004 Indicative role of salt sediments from water in geochemical monitoring *Izvestiya vuzov. Geologiya i razvedka* **1** 67–69 (in Russian)

[14] Yazikov E G, Rikhvanov L P, Baranovskaya N V and Yankovich E P 2009 Featires of element composition of drinking water salt sediments in the south of Tomsk Oblast *Environmental Geoscience* **4** 375–81 (in Russian)

[15] Rikhvanov L, Baranovskaya N, Soktoev B and Mongolina T 2011 Evaluation of drinking water

according to geochemical composition of its salt deposition *Proc. of 8th Int. Conference Environmental Engineering (19-20 May 2011) (Vilnius: Vilnius Gediminas Technical University Press)* pp 337–42
[16] Tapkhaeva A E, Taisaev T T, Rikhvanov L P, Yazikov E G and Baranovskaya N V 2010 Geochemical specialization of limescale of water sources illustrated by two regions in Siberia *Contemporary Problems of Ecology* **3** 498–507
[17] Pentecost A 2005 *Travertine* (Berlin Heidelberg: Springer-Verlag)
[18] Folk R L 1994 Interaction between bacteria, nannobacteria, and mineral precipitation in hot springs of Central Italy *Geographie physique et Quaternarie* **48** 233–46
[19] Fouke B W, Farmer J D, Des Marais D J, Pratt L, Sturchio N C, Burns P C and Discipulo M K 2000 Depositional facies and aqueous-solid geochemistry of travertine depositing hot springs (Angel Terrace, Mammoth Hot Springs, Yellowstone National Park, U.S.A) *Journal of Sedimentary Research* **70** 565–85
[20] Leitmeier H 1915 Zur kenntnis der carbonate. II. Teil *Neues Jahrbuch fur Mineralogie Beilageband* **40** 655–700
[21] Buczynski C and Chafetz H S 1991 Habit of bacterially induced precipitates of calcium carbonate and the influence of medium viscosity on mineralogy *Journal of Sedimentary Petrology* **61** 226–31
[22] Malesani P and Vanucci S 1975 Precipitazione di calcite o di aragonite dalle acque termominerali in relazione alla genesi e all'evoluzione dei travertini *Accademia Lincell, Rendiconti Scienze fisica, matematica e naturale* **58** 761–76
[23] Kele S, Demeny A, Siklosy Z, Nemeth T, Maria T and Kovacs M B 2008 Chemical and stable isotope compositions of recent hot-water travertines and associated thermal waters, from Egerszalok, Hungary: depositional facies and non-equilibrium fractionations *Sedimentary Geology* **211** 53–72
[24] Meyer H J 1984 Influence of impurity on the growth rate of calcite *J. Cryst. Growth* **66** 639–46

IOP Conf. Series: Earth and Environmental Science **27** (2015) 012043 doi:10.1088/1755-1315/27/1/012043

Fluorine concentration in snow cover within the impact area of aluminium production plant (Krasnoyarsk city) and coal and gas-fired power plant (Tomsk city)

A V Talovskaya[1], N A Osipova[1], E A Filimonenko[1,2], S A Polikanova[1], N P Samokhina[1], E G Yazikov[1] and I A Matveenko[1]

[1] National Research Tomsk Polytechnic University, 30 Lenin Ave., Tomsk, 634050, Russia

E-mail: [2] filimonenkoea@mail.ru

Abstract. The fluorine contents in snow melt water find in the impact areas of aluminum production plant and coal and gas-fired power plant are compared. In melt water, soluble fluoride is found in the form of fluoride ion, the content of which was determined by the potentiometric method using ion-selective electrode. According to the measurements of 2013-2014, fluoride content in melt water ranges 10.6-15.4 mg/dm^3 at the distance 1-3 km from the borders of Krasnoyarsk aluminum plant with the mean value 13.1 mg/dm^3. Four-year monitoring from 2012 to 2015 in the impact area of Tomsk coal and gas-fired power plant showed that fluoride content in melt water in vicinity of the thermal power plant is significantly lower than in the samples from the impact area of the aluminum plant. But higher content of fluoride ion (0.2 – 0.3 mg/dm^3) in snow samples in vicinity of coal and gas-fired power plant was revealed in winter of 2015. Intake of soluble fluoride is mostly explained by dust-aerosol emissions of study plants and deposition of fluorine compounds from air.

1. Introduction

Fluoride with its high chemical reactivity occupies a special place among other elements, since its physiologically necessary amount is close to the damaging effect doze. A distinguishing feature of the element is a narrow range of its optimal concentrations in drinking water varying 0.7 – 1.2 mg/L [1-4]. On the one hand, fluoride is included in the group of microelements participating in the processes of growth, development, reproduction, bone tissue metabolism; therefore, it is required for an organism life. On the other hand, even in small concentration it has toxic properties [1-5].

Fluoride clark amounts to 350 g/t for biosphere, 1.3 g/t for hydrosphere [1, 2]. The data on fluoride content in water differs greatly, however, the following common empirical pattern is observed: a positive correlation of fluoride concentrations and Na/Ca ratio [1, 2, 5]. The industrial sources of fluoride and its compounds in the environment are chemical, metallurgical, glass, ceramic, cement plants, coal-fired power plants, and nuclear fuel cycle enterprises.

Fluoride content in the environment is studied in a significant number of works. Conventional natural forms of fluoride in soil are unavailable for plants in contrast to man-made F, water-soluble compounds of which can be absorbed by plants, participate in intensive bioaccumulation, and pass along food chains [1, 2].

Yanchenko (2013) [6] showed that fluoride content in snow melt water can serve as an site indicator of environmental conditions. Snow is used to study the pollution level of urbanized areas in cold climate conditions, where snow accumulates pollutants for several months [7]. The urgent research in snow is also conditioned by the fact that precipitations are a part of the surface water balance influencing the conditions of soil, plants, and ground waters, as well as ecosystem, in general [7, 8].

Aluminum smelters sufficiently contribute to environmental fluoride pollution. In manufacturing practice, to produce aluminum the initial raw material – cryolit (sodium hexafluoroaluminate) - is added, as a result of which fluoride gaseous compounds and inorganic fluorides are released in the process of electrolysis. In such conditions significant losses of feedstock is observed. In the impact area of Bratsk aluminum plant fluoride content in melt water filtrate amounted to 18 mg/L at a distance of 3 km [6]. In the impact area of aluminum smelters in Canada fluoride content in melt water was determined as up to 13 mg/L [9].

The contribution of thermal power industry in fluoride emissions was studied by the example of some regions [10, 11], however, this involves insufficient information. In the process of coal combustion fluoride-containing compounds can enter the environment in the form of hydrogen fluoride. A certain part of fluorine-containing emissions can be transformed into salts. Depending on wind direction and distance from the thermal power plant (1.7-4.5 km), fluoride concentration ranges from 0.56 to 1.94 mg/L at the mean value of 1.21 mg/L according to [12].

Based on the literature [6, 11, 13-17] fluoride content in snow melt water from the impact areas of aluminum production in Western Siberia was compared (figure 1). These data show that fluoride content in melt water sufficiently exceeds the background values – from 10 to 80 times. But the level of fluoride in filtrated melt water within 1-3 km from site of aluminum plants is 3-7 times higher as compared to fluoride content at the distance of 5-10 km. Fluoride content in melt water at the distance of 29-30 km amounts to 0.5 – 1 mg/L, which is lower than the fluoride content in the samples taken in the vicinity of the plant – 20 – 50 times (figure).

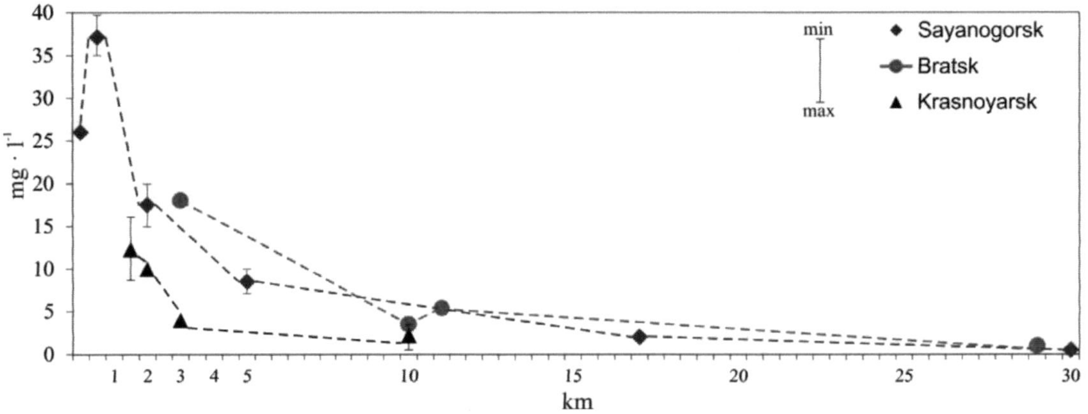

Figure. Spatial dynamics of fluoride content in snow melt water at the distances from aluminum plant sites in Western Siberia (according to the data [6, 11, 13-17]).

Spatial dynamics of fluoride content in melt water depends not only on the distance from the source of emission and wind pattern, but also the areal landscape. Moreover, in Krasnoyarsk, according to the data [13], over the last 15 years the wind pattern has changed. It is the result of the major emissions of aluminum production plant fall in Krasnoyarsk [13]. Therefore, the study issue in fluoride content of melt water in the sites of industrial enterprises as an environmental condition indicator of the area, identifying spatial and temporal variability of its content is still topical. This paper is aimed at studying the content of water-soluble fluoride in filtrated snow melt water in the impact areas of aluminum production plant (Krasnoyarsk) and coal and gas-fired power plant (Tomsk).

2. Materials and methods

To determine the fluoride content in snow melt water in the vicinity of Krasnoyarsk aluminum production plant northern-eastern site of the impact area was chosen. In our case, the choice of sample points with regard to landscape, meteorological conditions, and urban development [18] is a special task. Snow sampling was performed in northern-eastern site of aluminum plant as pollutants are transported here in the prevailing wind direction. Besides, there are not any other pollution sources around the plant. In 2013 in the vicinity of Krasnoyarsk aluminum production plant samples were taken at the distance of 1, 2, 3 km to the north-east from the borders of the plant, since the previous research had shown the most intensive accumulation of fluoride in melt water just within the radius 1-3 km from production sites of this aluminum production plant [6, 11, 13-17]. In 2014 snow samples were additionally taken at the distance of 8 and 13 km from the plant. In total, 8 snow samples were taken over the monitoring period.

The previous research in snow pollution of Tomsk territory showed that the dust distribution zone with chemicals from stacks of coal and gas-fired power plant is located in the north-east at the distance of 2 km [19]. Therefore, in the territory of Tomsk, snow sampling (the number of samples is 5) was performed in the northern-eastern impact area of Tomsk coal and gas-fired power plant at the distances of 0.7; 1.0; 1.3; 1.6 and 2.0 km from coal and gas-fired power plant stacks. The thermal power plant operates on Kuznetsk basin coal and natural gas. The main share of coal (up to 80 – 90 %) of the total annual amount is burnt during the winter period – from November till March. The height of stacks is 100 m that contributes to emission distribution over the town territory.

Each sample was taken according to [18] snow in-depth with the exception of 5 cm layer over the soil to avoid sample contamination with soil particles. To take samples plastic bags and non-metal bucket were used. In the course of sampling the width and length of pits were measured, the sampling dates were registered. Then samples were taken to the laboratory, where snow was melted in plastic bowls at room temperature. After melting samples were filtered, separating solids, and content of fluoride ion content in filtered melt water was determined.

Fluoride ion content in the samples of snow melt water was determined by potentiometric technique using the device Anion 4100 with fluoride-selective electrode according to [20]. Satisfactory convergence of fluorescent analysis results is shown. The low detection limit amounts to 0.05 mg/dm^3.

3. Results and discussion

The data analysis has shown that in 2012 in the impact area of Tomsk coal and gas-fired power plant the maximal value of fluoride ion content was revealed in samples taken 2 km from the stacks of the plant (table 1). In 2013 the values of fluoride ion content in snow melt water did not exceed the detection limit in all sampling points. In the samples taken in 2014 and 2015 fluoride ion content did not sufficiently change with the distance from the plant. Fluoride ion content in the samples taken in 2015 is 2-5 times higher than that of the samples taken in the period from 2012 to 2014. During the whole monitoring period the fluoride ion loading rate in the study area changed from 0.46 to 8.27 mg/m$^{2}\cdot$ per month, the maximum values being recorded in the samples of 2015. Remarkable changes in fluoride ion content in melt water and fluoride ion loading rate at the distance from the stacks of coal and gas-fired power plant were not observed over the whole monitoring period.

In 2015 the elevated concentrations of fluoride ion content in the samples and fluoride ion loading rate in the study area could be explained by the climatic conditions of that year. The winter of 2015 was characterized by heavy snow falls that could result in more precipitation of fluoride from the air. All values obtained within the four-year period did not exceed maximum permissible concentration in drinking water (1.5 mg/dm^3 [21]).

Coal combustion is also one of the principle sources of fluoride in the air [4]. As mentioned above [10], the main part of coal at Tomsk coal and gas-fired power plant is used mostly in winter; hence, water-soluble fluoride compounds in the melt water samples taken in vicinity of the plant are likely to be associated with coal combustion. When coal is burnt, coal can release fluoride compounds which can enter the environment in the form of hydrogen fluoride. The other gaseous compounds, which can be present in the gaseous flow (SiF_4, H_2SiF_6), are transformed into salts or hydrogen fluoride [4].

Table 1. Dynamics of fluoride ion content (mg/dm^3) in snow melt water and its loading rate (mg/m^2·month) in the impact area of Tomsk coal and gas-fired power plant between 2012 and 2015.

Distance, km*	2012		2013		2014		2015	
	C_{F^-}	H	C_{F^-}	H	C_{F^-}	H	C_{F^-}	H
0.7	0.03	n.d.	<0.05	0.67	0.09	1.88	0.24	8.27
1.0	0.14	n.d.	<0.05	0.87	0.07	1.22	0.20	7.25
1.3	0.07	n.d.	<0.05	0.59	0.07	1.18	0.26	6.68
1.7	0.08	n.d.	<0.05	0.66	<0.05	0.46	0.24	5.01
2.0	0.20	n.d.	<0.05	0.85	<0.05	0.48	n.d.	n.d.

Note: * - distance from the stacks of coal and gas-fired power plant to the sampling point; C_{F^-} – content of fluoride ion in melt water, mg/dm^3; *H* – loading rate of fluoride ion in the territory, mg/m^2·per month; *n.d.* – no data

Table 2 shows fluoride ion content in snow melt water samples from the impact area of Krasnoyarsk aluminum production plant. It was determined that in the samples of 2013 fluoride ion content exceeded maximum permissible concentration in drinking water in 9 times on average. In the samples fluoride ion content exceeded maximum permissible concentration in 5.5 times in 2014. The data on fluoride ion content in melt water are comparable with the previously published data (figure). The results of two-year monitoring show that fluoride ion content in the samples did not sufficiently change at the distance from 1 to 3 km from the borders of aluminum production plant. In 2014 additional study at the distances of 8 and 13 km from the plant borders showed the increase fluoride ion content in 4 times at this distance.

Table 2. Dynamics of fluoride ion content in melt water in the impact area of Krasnoyarsk aluminum production plant between 2013 and 2014, mg/dm^3.

Year of sampling	Distance from the plant borders to the sampling point				
	1 km	2 km	3 km	8 km	13 km
2013	13.13	13.76	15.38	n.d.	n.d.
2014	13.83	10.64	11.62	3.88	1.64

Note: *n.d.* – no data, sampling was not performed

In 2014 the loading rate of fluoride ion in the impact area of aluminum production plant ranges from 13 to 172 mg/m^2·per month, the maximal loading rate is observed in the sample taken at the distance of 3 km from the plant borders (table 3).

In addition, according to the data obtained it can be stated that water-soluble compounds of fluoride are recorded at significant distances from the emission sources. According to [6, 7, 11, 13-17], the impact radius of large aluminum production plants emissions can reach 30 km.

In snow fluoride content can be conditioned by the peculiarities of water-soluble fluoride compounds transportation in the environment. According to [22] in air gaseous hydrogen fluoride is absorbed by air moisture forming aerosol or smog of hydrofluoric acid solution. Fluoride is removed from the air in the process of wet or dry deposition.

Table 3. Loading rate of fluoride ion in the impact area of Krasnoyarsk aluminum production plant, mg/m^2·per month.

Year of sampling	Distance from the plant borders to the sampling point				
	1 km	2 km	3 km	8 km	13 km
2014	140	58	172	56	13

4. Conclusion

Thus, fluoride contents in melt water and its loading rates in the impact area of Tomsk coal and gas-fired power plant and Krasnoyarsk aluminum production plant are different, which is explained by the plants' specificity. It was shown that the aluminum production plant is a powerful pollution source of snow with fluoride. We have determined that fluoride ion content in snow melt water at the distance of 0.7 to 1 km from the stacks of coal and gas-fired power plant did not change sufficiently over the monitoring period, which can be conditioned by specific air micro-circulation in the city territory because of urban development. In the impact area of Krasnoyarsk aluminum production plant elevated fluoride ion content were identified at the distance of up to 3 km from the plant borders, which could indicate the transportation of fluoride compounds over long distances. The study area is a plain without urban construction that contributes to emission distribution. The data obtained show that fluoride enters the town territory with the study plants emissions with high probability degree. Taking into account high solubility of some fluorides, mainly sodium fluoride, those compounds can pass into soil from snow during snow melt. Therefore, further study should be aimed at the research in content of water-soluble fluoride compounds in soil in the impact area of study plants.

Acknowledgment

This work was partly financially supported by the Presidential Grant for young Russian scientists (MK 951.2013.5).

References

[1] Ivanov V V 1994 Ecological geochemistry of elements Book. 2. Main r-elements. Moscow: Nedra, p 303 (in Russian)

[2] WHO 2002 Fluorides. Environmental Health Criteria **227**. World Health Organization, Geneva

[3] Lazarev N V 1977 Harmful elements in industry: Reference book. Leningrad: Khimiya. V.III

[4] Agency for Toxic Substances and Disease Registry (ATSDR) 2003 Toxicological profile for Fluorides, Hydrogen Fluoride, and Fluorine. Atlanta, GA: U.S. Department of Health and Human Services, Public Health Service

[5] Yanin Ye P 1996 Fluoride in drinking water of Saransk town and its hygienic significance. Moscow, IMGRE 51

[6] Yanchenko N I, Baranov A N and Chebykin E P 2013 Peculiarities and factors influencing distribution of metal, rare-earth elements, hydrogen and fluoride in snow filtrate and solid deposits of Bratsk city *Vestnik of ISTU* **10 (81)** 141–9 (in Russian)

[7] Bartels-Rausch T et al. 2014 A review of air–ice chemical and physical interactions (AICI): liquids, quasi-liquids, and solids in snow *Atmos. Chem. Phys.* **14** 1587–633

[8] Zarina L M, Lebedev S V and Nesterov E M 2011 Ecological Geochemical Investigations of the Contents of Heavy Metals in the Snow Cover in the Saint-Petersburg Region with Application of GIS Technologies *Int. J. Chem. Eng. Appl.* **2** 117–20

[9] Ouellet M 1987 Reduction of airborne fluoride emissions from Canadian aluminium smelters as revealed by snow chemistry *Sci. Total Environ.* **66** 65–72

[10] Ando M, et al. 2001 Health effects of fluoride pollution caused by coal burning *Sci. Total Environ.* **271** 107–16

[11] Nikolayeva L A and Turchinova D A 2010 The problem of fluoride pollution of soil in vicinity aluminum production *Yestestvoznanie i gumanism* **6** 1 (in Russian)

[12] Rogovenko E S, Blinnikova N V, Shubin A A and Bondarev LG 2010 Environmental Monitoring of Anthropogenic Pollution of Snow Cover of One of the Industrial Districts of Krasnoyarsk *Journal of Siberian Federal University. Chemistry* **4 (3)** 387–94
[13] Khlebopros R G, Taseyko O V, Ivanova Yu D and Mikhailuta S V 2012 *Krasnoyarsk. Ecological essays* (Krasnoyarsk: SFU) p 130 (in Russian)
[14] Davydova N V 2014 Problems of environmental pollution in Siberia *Uspekhi sovremennogo yestestvoznaniya* **5(1)** 186–9 (in Russian)
[15] Znamenskaya T I 2015 Migration and differentiation of pollutants in the steppe landscape in the South of Minusinsk depression Diss. of PhD in Geograph. Sciences Irkutsk p 175
[16] Tandelov Yu P 2015 Fluoride in soil-plant system Krasnoyarsk 146
[17] Ecological atlas of Krasnoyarsk city 2012 http://gis.krasn.ru/blog/ecoatlas/
[18] Manual on monitoring of atmosphere pollution 1991 RD 52.04.186 № 2932-83. Moscow Goskomhydromet p 693
[19] Talovskaya A V, Filimonenko E A, Osipova N A, Lyapina E E, and Yazikov E G 2014 Toxic elements (As, Se, Cd, Hg, Pb) and their mineral and technogenic formations in the snow cover in the vicinity of the industrial enterprises of Tomsk *IOP C. Ser. Earth. Env.* **21** 012042
[20] Mass concentration of fluoride in water. Methods of measurements using potentiometric technique with ion-selective electrode 2008 RD 54.24.360-2008 Goskomhydromet (in Russian)
[21] Predelno dopustimye kontsentratsii khimicheskikh veshchestv v vode vodnykh obektov khozyaystvenno-pitevogo i kulturno-bytovogo vodopolzovaniya 2003 GN 2.1.5.1315–03 [Hygienic standards: Maximum permissible concentration of chemicals in water bodies of drinking and domestic cultural water use] (in Russian)
[22] Arnesen A K M, Abrahamsen G, Sandvik G and Krogstad T 1995 Aluminium-smelters and fluoride pollution of soil and soil solution in Norway *Sci. Total Environ.* **163** 39–53

The influence of dispersing additive on the paraffin crystallization in model systems

A M Gorshkov[1,3], Pham Tien Thang[2,4], L V Shishmina[2,5] and L V Chekantseva[2]

[1] Petroleum Learning Centre, National Research Tomsk Polytechnic University, 30 Lenin Ave., Tomsk, 634050, Russia
[2] Geology and Oil Field Exploitation Department, National Research Tomsk Polytechnic University, 30 Lenin Ave., Tomsk, 634050, Russia

E-mail: [3] gorshkovam89@mail.ru, [4] thanghuongduong1@yahoo.com.vn, [5] shishmina@tpu.ru

Abstract. The work is dedicated to investigation of the influence of dispersing additive on the paraffin crystallization in model systems. A new method to determine the paraffin saturation point of transparent solutions based on the phenomenon of light scattering has been proposed. The linear relationship between the values of critical micelle concentrations of the additive and the quantity of paraffin in solution has been obtained. The influence of the model system composition on the paraffin crystallization has been studied.

1. Introduction
One way to improve rheology properties of high-paraffin oils is application of additives. They inhibit deposit formation or decrease its intensity. Mechanism of dispersing additives action has been researched in diesel fuel and oils [1, 2]. However, mechanism of dispersing additives action in terms of hydrocarbon paraffin solution has not been studied yet. High-molecular paraffin hydrocarbons were the reasons for paraffin build-up in oil pipelines. Therefore, the important task is to study the influence of additive on high-molecular paraffin hydrocarbons when ambient temperature decreases. The aim of the work is to study the influence of dispersing additives C-5A on the process of paraffin crystallization in model systems.

The effectiveness of dispersing additives C-5 action on the process of paraffin crystallization was evaluated by two parameters: the paraffin saturation point of model system and the paraffin pour point.

2. Experimental procedure

2.1. Materials
The process of paraffin crystallization was investigated using model paraffin–heptane system, which is one of the most common solutions in research of paraffin crystallization. Solid paraffins correspond to GOST – 23683-89 (the Russian Federation). Chromatographic grade *n*-heptane was used as a solvent. All experiments were performed on freshly prepared paraffin solutions in *n*-heptane. Paraffin concentration in *n*-heptane was 4%, 6% and 10% of the total mass.

Alkenylsusinimide C-5A was used as dispersing additive, which is 40% concentrate of alkenylsussinimide group in mineral oil and unreacted polybutylene. The additive was a viscous liquid, which was dissolved in toluene with mass ratio 1:1. To study the influence of dispersing additive on the paraffin crystallization, additive concentrations in paraffin solution were 0.03; 0.06; 0.08 and 0.1% of mass. At the beginning of every experiment, the cuvette with the studied solution was dispersed 15 seconds by ultrasound wave to mix all components. Every cuvette was used for one model system. Cuvettes had a cylinder form with diameter 15 mm.

2.2. Method and equipment

Paraffin saturation point is a main parameter which characterizes phase state «liquid phase – solid phase». However, it should be noted that standard method to determine the paraffin saturation point has not been found. Some methods to determine this characteristic can be listed: visual, refractometric, photometric, ultrasonic and viscous method.

The "PhotoCor Complex" equipment was used to determine the paraffin saturation point by the method of photon correlation spectroscopy [3, 4]. To determine temperature models more accurately, the equipment was modified by a temperature sensor DS18B20, which was placed in the thermostat (interval measurement is from -10 °C to 85 °C, accuracy ±0,5 °C). The use of temperature sensor is caused by the cooling process. The cooling process is a dynamic process, in which cooling speed is faster than the speed of temperature distribution in the thermostat. As the result, temperature of the thermostat and the temperature in the cuvette were different. Temperature difference increases in growth of cooling speed. In this paper model system was cooled from 50 °C to 5 °C with the speed 0,05 °C per minute. All experiments were conducted at atmospheric pressure.

A semiconductor laser beam with a wavelength λ = 654 nm was used as a radiation source. The main measuring tool of "PhotoCor Complex" device is photoelectron multiplier tube (Photocor-PC1), which operates in the photon counting mode. The present element of the equipment allows tracking a light scattering intensity change and consequently to determine the beginning of the phase transition in the disperse system. Photicor-PC1 was placed at an angle 90^{0}C in all experiments.

The main element of the control and data analysis unit in "PhotoCor Complex" equipment is a programmable correlator which provides automatic control of the instrument operation according to the algorithm. It can also measure integrated light scattering intensity. In our research, autocorrelated function fluctuation of light scattering was recorded repetitively in cycles. The time of the cycle was 3 minutes. The cycle interval is from 1 to 1000 (figure 1a). Thermosensor data was represented in log. By the input data the dependence of the light scattering intensity on the temperature was drawn (figure 1b).

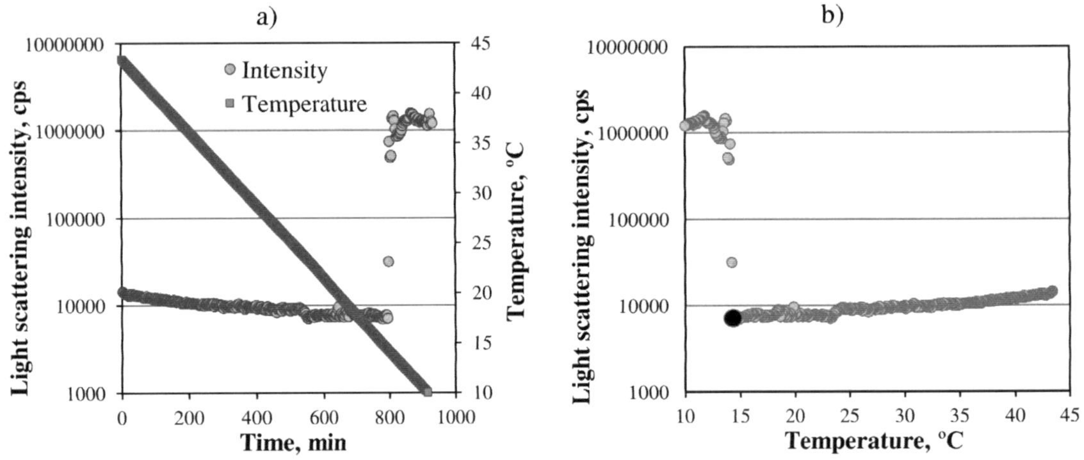

Figure 1. Procedure to determine the paraffin saturation point model system in "PhotoCor Complex" equipment: a) – input data; b) – dependence of the light scattering intensity on the temperature.

In this dependence, the transition change light scattering intensity was shown in the cooling process of the model system. According to the Rayleigh theory, light scattering intensity depends on particle radius (all other conditions being equal) by equation $I \sim R^6$ [5]. The intensity growth for two orders of magnitude indicated increasing radius of paraffin particles and consequently their crystallization. Therefore, the observed break corresponds to initiation of the new coexistent phase (paraffin crystals). We can assume that temperature at the first point of the break is the paraffin saturation point of the model system (figure 1b).

The paraffin pour point of solution was determined by the "CRYSTALL" device which has been created at the Institute of Chemistry of oil Siberian Branch Russian Academy Science (accuracy $\pm 0{,}5^0$C). The main element of the device is a head sensor. In the cooling process, the head sensor rotates and contacts the model system. The rotation speed of the head sensor decreases in the cooling process. The pour point was measured by the speed rotation of the head sensor.

3. Experiments and results

3.1. Critical micelle concentration of the additive in paraffin–heptanes solution

During the work phase, behavior of the additive in paraffin–heptane system was investigated. The model system contains two components which were dissolved in *n*-heptane. Therefore, it is necessary to determine the behavior of each component and the joint impact they have on model systems.

The experiment was performed as follows: the cuvette was placed in the thermostat; the temperature rose up to 50^0C; then the temperature was stabilized for 30 minutes and the light scattering intensity was studied when investigated parameters change (paraffin concentration and additive concentration). It should be noted that to study the phase behavior of dispersing additives, model solutions with additive concentrations of C-5A from 0.03% to 2% of total mass were prepared. Figure 2 shows the dependence of the light scattering intensity on the system composition.

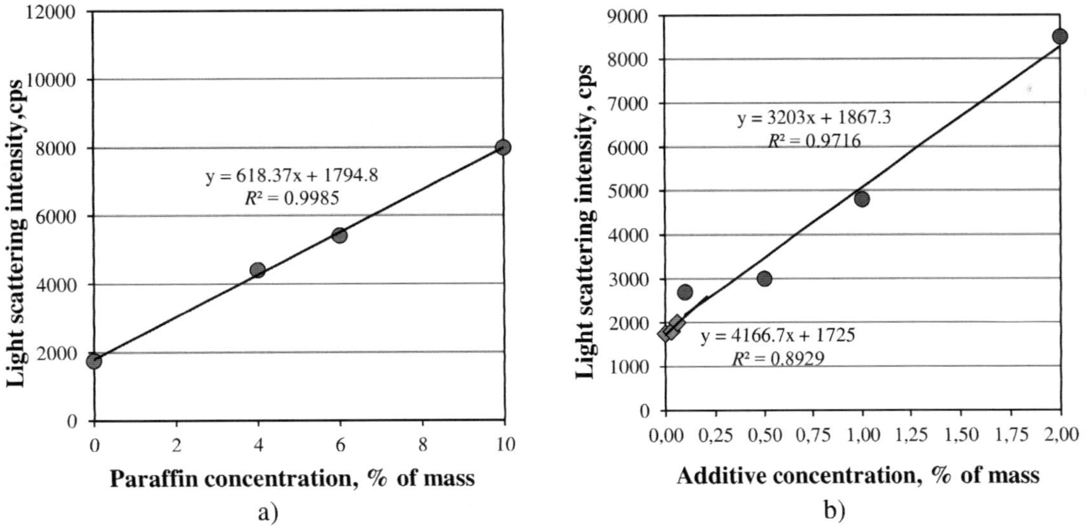

Figure 2. Dependence of the light scattering intensity on the model system composition: a) paraffin–heptane system; b) additive–heptane system.

Figure 2a demonstrates that the light scattering intensity was linearly dependent on paraffin concentration of the system. The reliability of approximation (R^2) was close to unity. This dependence indicates the paraffin presence in *n*-heptane in molecular state. The assumption corresponds to the Rayleigh theory, in which light scattering intensity is directly proportional to concentration of particles at constant value of their radius [5]. A small break of the curve, approximating the dependence of light scattering intensity on additive C-5A concentration, has been revealed (figure 2b). The break indicates the phase transition. When the threshold concentration of additive was exceeded, the additive entered colloid-dispersed state and formed micelles. The point of intersection between two approximating straight lines is the critical micelle concentration (CMC) of additive C-5A in *n*-heptane. Thus, the value of the additive CMC in *n*-heptane was 0.15% of total mass.

Then, the phase behavior of model paraffin–additive–heptane system was considered (figure 3).

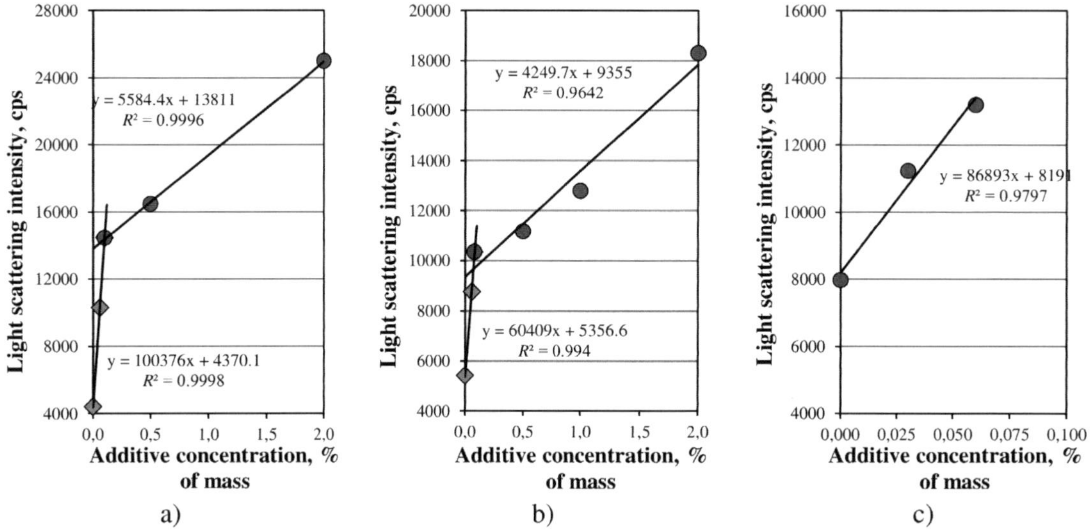

Figure 3. Dependence of the light scattering intensity on the additive concentration system with changes in the paraffin quantity: a) – paraffin concentration 4% of mass; b) – paraffin concentration 6% of mass; c) – paraffin concentration 10% of mass.

As a result (figure 3a and figure 3b), dependences indicate the same phase behavior of the additive in the investigated paraffin–heptane solutions, which is characterized by a clear break in the approximating curve. This break indicates the existence of dispersing additive CMC in paraffin–heptane solution. The high reliability of approximation of the dependences obtained by the linear functions (R^2 was close to unity) confirmed this assumption. Figure 3b shows a linear dependence of light scattering intensity on additive concentration, which indicates the absence of the CMC in this model system composition. This fact can be explained if the dependence of the critical micelle concentration of C-5A additive on the paraffin quantity in *n*-heptane is drawn (figure 4). It should be noted that obtained values of additive CMC are approximated by the linear function ($R^2 > 0.998$). The high reliability of approximation indicates that there is a significant relationship between these two parameters for different paraffin concentrations. Based on figure 4, the additive CMC should be about 0.02% of mass in the solution, which paraffin content is 10% of mass. This value is less than the investigated additive concentrations, which ultimately resulted in the absence of CMC in figure 3b.

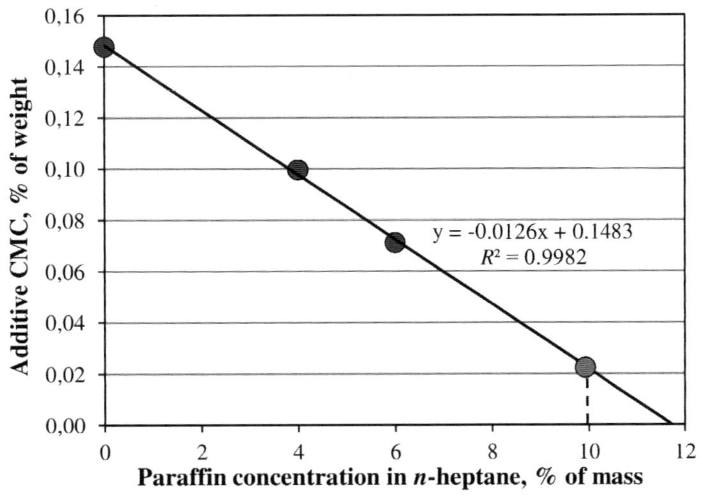

Figure 4. Dependence of the additive C-5A CMC on the paraffin concentration in *n*-heptane.

Thus, it has been established that the paraffin concentration in the model system affects the CMC value of additive C-5A: the higher the paraffin concentration, the lower the CMC value. Hence, additive C-5A may be either in molecular state or in colloid-dispersed state with certain concentrations in different paraffin–heptane solutions. This fact needs to be considered in future with interpretation of data on the additive effect on the saturation point and the paraffin pour point.

3.2. The influence of the additive on the saturation point and the paraffin pour point in model system

The paraffin crystallization occurs due to cooling or changing composition of solution in isobaric process. Ashmyan and co-authors [6] indicate that the paraffin saturation point of oil depends on paraffin hydrocarbons content in oil. When oil paraffin concentration increases, its saturation point will be higher.

The obtained results are represented in figure 5.

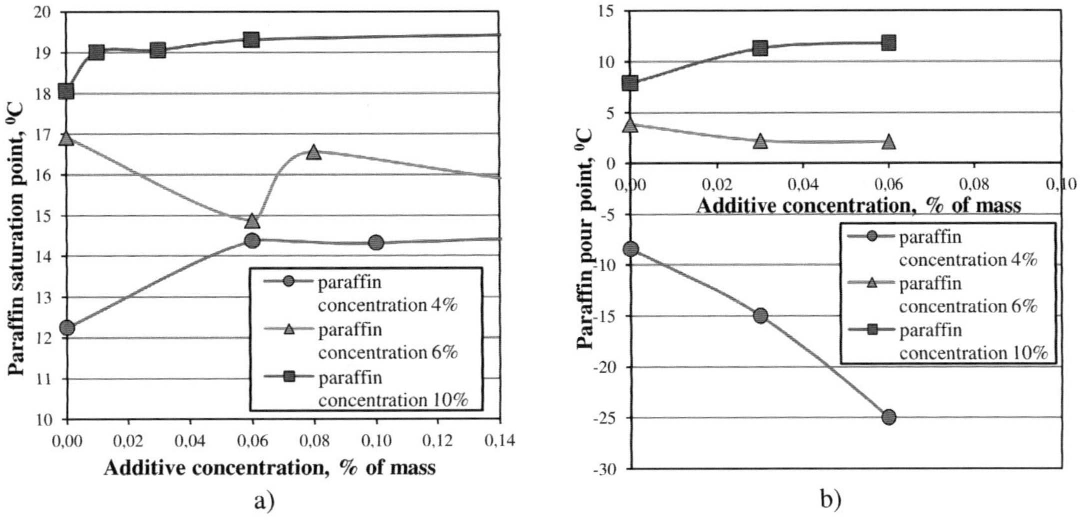

Figure 5. Influence of the additive concentration in a model system on the paraffin saturation point of solution (a) and the pour point of solution (b).

Figure 5a shows that by increasing the paraffin concentration in the model system (all other conditions being equal) the paraffin saturation point increases. Its dependence coincides with sequencing the conclusion of the work [6]. It should be noted that the C-5A additive does not generate the conclusive results on the paraffin saturation point of model system. A positive additive effect is observed in paraffin solution with concentration 6% of mass. It means that the saturation point decreases when the additive concentration less than 0.06% of mass. The negative C-5A additive effect is observed in paraffin solutions with concentrations 4% and 10% of mass. It means that the saturation point increases in all additive concentrations. Having analyzed figure 4 and figure 5a, it can be concluded that the excess of additive concentration higher than the CMC leads to the increase in the paraffin saturation point of model paraffin–heptane system. It might indicate the change in the mechanism of additive action.

Figure 5b shows that the effect of additive on the pour point decreases with the growth of paraffin concentration in model system. When the paraffin concentration grows up to 10% of mass the pour point increases. This fact also confirms the assumption on the change of additive action mechanism in excess of the CMC, as all investigated additive concentrations exceed the CMC value in solution with paraffin concentration 10% of mass (figure 4). It should also be noted that by increasing the paraffin content in the model system (all other conditions being equal) the pour point increases as well as the paraffin saturation point does in the model solution.

4. Conclusion

This paper evaluates the effectiveness of the dispersing additive action in model paraffin solutions with two parameters. They are the paraffin saturation point and the pour point.

It is the first time when the method to determine the paraffin saturation point of transparent solution by means of the method of photon correlation spectroscopy has been proposed. The optical methods are the most informative for transparent solutions, since they allow detecting small changes in the system at the initial moment of paraffin crystallization with high accuracy.

According to the results of experimental data, the following conclusions are made:

1. The CMC of C-5A additive in *n*-heptanes has been established.
2. The influence of the paraffin concentration of model system on the value of additive C-5A CMC has been revealed.
3. It is shown that the phase state of C-5A additive has a significant influence on the paraffin crystallization in the investigated model systems.

References

[1] Hlavaty O L 1989 *Physic chemistry dispersing oil additives* (Kiev: Naukova Dumka) p 184

[2] Rudnick L R 2009 *Lubricant additives: chemistry and applications* ed L R Rudnick (Chemical industries; 124)

[3] Gorshkov A M, Shishmina L V and Roslyak A T 2014 New method for the onset point determination of the petroleum asphaltene aggregation *IOP Conf. Ser.: Earth Environ. Sci.* **21** 012027

[4] Yudin I K, Nikolaenko G L, Gorodetskii E E, Markhashov E L, Frot D, Briolant Y, Agayan V A and Anisimov M A 1998 Universal behavior of asphaltene aggregation in hydrocarbon solution *Petroleum Science and Technology* **3** 401–02

[5] Fridrihsberg D A 1984 *Course of colloid chemistry. Textbook for institute of higher education* vol 2 (Leningrad: Chemistry) p 368

[6] Asmyan K D Experimental methods of investigation of paraffin's oils ed K D Ashmyan (Moscow: OJSC «VNIIneft») p 108

Synthesis of ZSM-5 galloalumosilicate and investigation of their physicochemical and catalytic properties in the course of conversion of propane into aromatic hydrocarbons

L N Vosmerikova[1,2], A N Volynkina[1,3] and A V Vosmerikov[1,4]

[1] Institute of Petroleum Chemistry, Siberian Branch, Russian Academy of Sciences, Akademichesky Avenue 4, Tomsk, 634055, Russia

Email: [2] lkplu@ipc.tsc.ru, [3] a.volynkina@inbox.ru, [4] pika@ipc.tsc.ru

Abstract. Galloalumosilicates having ZSM-5 structure are manufactured from the alkali alumina silica gels via the method of hydrothermal crystallization using different organic templating agents. Their physico-chemical and acidic properties are investigated and their catalytic reactivity is determined in the course of propane conversion into aromatic hydrocarbons. The highest catalytic reactivity and stability are exhibited by the Ga-containing zeolite synthesized using hexamethylenediamine as a structure-forming additive.

1. Introduction

Zeolite catalysts modified with various metals are widely used in different industries. A promising method for modification of zeolites aimed to improve their properties is the introduction of metal cations into their framework at the stage of hydrothermal synthesis.

The elemento-aluminosilicates prepared by this method differ from the unmodified zeolites in their physicochemical properties directly controlled by the structure of the framework aluminosilicate [1-3]. The peculiarity of the synthesis of zeolites having ZSM-5 structure consists in the fact that the hydrothermal crystallization of alumino silicagel is conducted in the presence of different organic compounds (templates). During the synthesis as the crystals grow the molecules of organic templates penetrate into the solid phase providing a stabilization of zeolite structure. The morphology and composition of the zeolite crystals of a certain structure could be modified using a number of structure-forming agents [4-5].

The aim of this work is to study the mechanisms of conversion of propane into aromatic hydrocarbons over ZSM-5 galloalumosilicates prepared using a variety of organic templates.

2. Experimental

Galloalumosilicates were prepared by isomorphous substitution of Si (IV) with Ga (III) in the zeolite crystal lattice at stage of hydrothermal synthesis via partial substitution of aluminum with gallium in the initial aluminum silica gel ($SiO_2/Al_2O_3+Ga_2O_3 = 40$). The liquid glass composed of 9% Na_2O, 29% SiO_2, and 62% H_2O was the source of silicon, while solutions of alumina and gallium nitrates were used as sources of their oxides. Hexamethylenediamine (HMDA), diethylenetriamine (DETA) and urea (carbamide) were used as templates. The content of gallium oxide in the catalysts was 1.85 wt%. The resulting reaction mass was crystallized in steel autoclaves at 170 °C for 4 days. The crystallization completed, the solid phase was separated from the liquid phase by filtration, washed with distilled water until the wash water pH ≤ 9 was reached, and then air-dried at 110 °C and calcined at 540 °C for 16 hrs.

To convert the zeolites to the active H-form, the samples were treated with a 25% aqueous NH_4Cl solution at 90 °C for 2 hours. After removal of the sodium cations, the zeolite samples in the NH_4-form were dried at 110 °C and air-calcined at 540 °C for 6 hrs.

The quality of the resulting samples was controlled by IR spectroscopy and XRD analysis. IR spectra for the samples under study were recorded from the zeolite pellets using a FT-IR Nicolet 5700 spectrometer within the range of 2000-400 cm^{-1} in the presence of KBr (1.2 mg of high silica zeolites per 300 mg of KBr) in air. An XRD analysis of samples was performed in a DRON-3M X-ray apparatus (Cu-anode, Ni-filter).

The morphology and size of the zeolite crystals were determined using the method of scanning electron microscopy (SEM). SEM images were obtained on a LEO EVO 50 apparatus (Zeiss, Germany) at the Nanotech shared use center of the Institute of Strength Physics and Materials Science SB RAS.

The acidic properties of the catalysts were tested by the method of temperature-programmed desorption (TPD) of ammonia. The TPD experimental procedure was described in [6]. The concentration of the acid sites in the investigated samples was determined by the amount of ammonia desorbed at the moment of identification of desorption peaks and expressed as micromoles per 1 g of the catalyst.

The parameters of porous structure and a specific surface area of the samples were determined by the Brunauer, Emmett, Teller (BET) method using a TriStar II (3020) automatic gas adsorption analyzer (Micromeritics, USA). The specific surface area was calculated from the adsorption isotherms for nitrogen at 77 K. The volume and pore size of samples were determined by BJH (Barett-Joyner-Halenda) model from the data on adsorption and desorption isotherms at a relative pressure $P/P_0 = 0.99$.

The experiments for studying the process of propane conversion into aromatic hydrocarbons were carried out in a test flow-type reactor at atmospheric pressure, at the reaction temperature of 400-600 °C and the volumetric feed rate of feedstock 500 hr^{-1}.

The reaction products were analyzed by GLC using a Khromatek-Kristall 5000.2 chromatograph. The volume of the catalyst fed into the reactor was 3.0 cm^3 and its particle size was within 0.5-1.0 mm. The catalytic activity of the samples was estimated from the propane conversion data. The yield and formation selectivity of the gaseous and liquid products were also calculated.

3. Results and Discussion

The results of X-ray studies showed that all samples synthesized correspond to the structure of ZSM-5 zeolite and belong to the orthorhombic syngony system. Figure 1 shows the IR spectra for the resulting galloalumosilicates, indicating the presence of absorption bands characteristic of ZSM-5 zeolites [7]. The samples exhibit absorption bands in the region 550-560 cm^{-1} attributed to oscillations of the tetrahedra, [SiO_4] and [AlO_4], over the external constraints of the skeleton due to the presence of four-, five- and six-member double rings, which determine their structure. The ratio of the intensities of the absorption bands at 550 and 450 cm^{-1} allows us to estimate the purity of the samples synthesized and the degree of their crystallinity.

For the Ga-containing zeolite synthesized with urea, this ratio is less than 0.6, which indicates a low degree of its crystallinity. The crystallinities of the Ga-containing zeolites prepared using HMDA and DETA are found to be 89 and 100%, respectively.

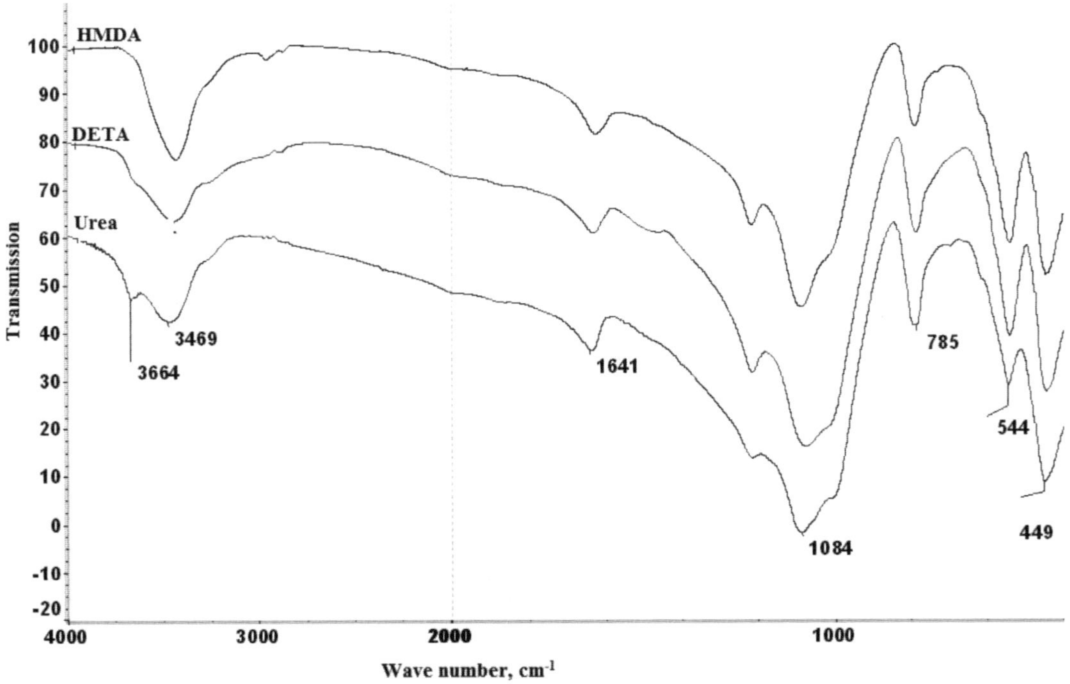

Figure 1. IR spectra of galloalumosilicates.

According to data of microscopic examination the morphology and size of the particles of zeolites synthesized with various structurants are significantly different (figure 2). The particles of zeolite prepared with HMDA have a sphere-like shape (single crystals) and are characterized by a highly-developed surface area. The sizes of the spheroids are within 3-8 microns. The use of urea as a structure-forming additive in the zeolite synthesis the urea results in the formation of finely-faceted isolated crystals in the shape of hexagonal prisms with the faces measuring 3-5 microns. The particles of zeolite synthesized with DETA have the shape similar to that of zeolite particles prepared with HMDA, but their grain sizes are quite different: there are both small and large crystals. An addition of HMDA in the stage of hydrothermal synthesis results in the formation of multiple nuclei, while the synthesis of zeolite using urea gives rise to the formation of a small number of zeolite nuclei during aging of alumosilicagel. This suggests two mechanisms of synthesis of zeolites differing in their induction periods and leading to the formation of crystals with different morphologies.

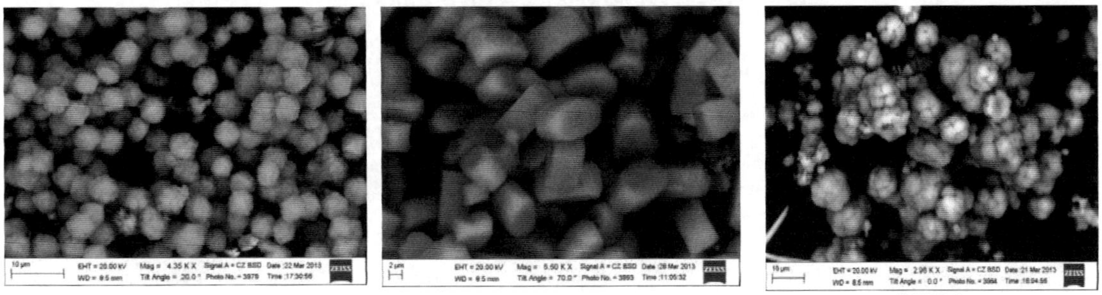

Figure 2. SEM images of galloalumosilicates prepared with different structure-forming agents.

An important characteristic of zeolite catalysts, determining their activity in various chemical reactions are their acid-base properties. To investigate the acidic properties of gallium-containing zeolites use was made of an ammonia probe due to a high basicity of ammonia, small size of its molecule, and the ease of its dosage, which allowed us among other things to determine not only strong but also weak acidic sites. In order to avoid the physical adsorption of ammonia, it was adsorbed over the samples at a temperature of 100 °C. The data on studies of acidic properties of prepared gallium-containing zeolites are presented in table 1.

Table 1. Acidic properties of gallium-containing zeolites.

Template	T_{max}, °C		Concentration, μmol/g		
	T_I	T_{II}	C_I	C_{II}	C_Σ
HMDA	200	425	649	298	947
DETA	210	415	621	267	888
Urea	215	420	605	314	919

Note. T_I, T_{II} are the temperatures of maxima of low- and high-temperature peaks on the thermal desorption curves; C_I, C_{II} *and* C_Σ are the concentrations of weak and strong acid sites and their sums, respectively.

All prepared samples possess acid sites of two types as evidenced by the presence of two ammonia desorption types in the thermal desorption spectrum: weak and strong acid sites with the maximum peak temperatures, T_{max} = 200-215 °C (I) and T_{max} = 415-425 °C (II), respectively. The use of urea as a template in the synthesis of galloalumosilicate results in the formation of a zeolite with the lowest concentration of weak acid sites that are characterized by a higher strength. The galloalumosilicate synthesized using HMDA contains the maximum number of acid sites, whose total concentration is 947 μmol/g, while the galloalumosilicate synthesized using DETA has the minimal number of high-temperature acid sites characterized by a lower strength. Thus, the investigation of acidic properties of the galloalumosilicates has shown that they are characterized by different distributions and ratios of acid sites of different nature, which affects their catalytic properties in the conversion of propane into aromatic hydrocarbons.

An investigation of the structural properties of galloalumosilicates has shown that the samples synthesized using HMDA and DETA have nearly the same specific surface areas and the total pore volume of the second sample is 0.024 cm^3/g only, which is 3.75 times smaller than that of the galloalumosilicate sample synthesized with HMDA. The galloalumosilicate prepared with urea has the largest average pore size but the lowest value of the specific surface area (table 2). This in turn has an influence on the activity and selectivity of the samples during the propane aromatization.

Table 2. Structural properties of galloalumosilicates.

Template	S_{spec}(BET), m^2/g	Total pore volume, cm^3/g	Average pore size, nm
DETA	303	0.024	3.4
HMDA	313	0.09	5.0
Urea	240	0.04	7.1

The investigation of catalytic properties of resulted galloalumosilicates has shown that they differ from each other in the total (measured by propane conversion degree) and aromatizing reactivities. The highest reactivity in the reaction of aromatization of propane is exhibited by the galloalumosilicates synthesized using HMDA and DETA (table 3). The sample synthesized with DETA exhibits a higher activity at the reaction temperatures 400-500 °C, while at 600 °C there is a decrease both in the conversion of propane and in the yield of aromatic hydrocarbons, which is probably due to the low stability of the catalyst performance. The sample synthesized using HMDA exhibits a high catalytic activity at the temperatures 550 and 600 °C. The lowest aromatization activity is exhibited by the catalyst sample prepared using urea. A 76% propane conversion over this catalyst results in a 43.2% yield of aromatic hydrocarbons at the process temperature 550 °C. An increase in the temperature to 600 °C leads to a significant decrease in the catalyst activity, which is apparently due to the presence of a large number of strong acid sites on its surface, which are directly involved into the reactions resulting in the formation of carbon deposits and rapid catalyst deactivation. The reaction product formed over this catalyst at a reaction temperature of 600 °C contains a large number of lower C_2-C_4 olefins.

Table 3. Characteristic of the products of propane conversion over gallium-containing zeolites.

Template	$T_{reaction}$, °C	X, %	A, %	S_{ar}, %	S_{cr}, %	$S_{dehydrogen}$, %	S_{H2}, %
HMDA	400	11	1.6	14.5	74.9	5.8	4.8
	450	26	4.3	16.5	68.0	7.6	7.9
	500	66	36.7	54.3	36.6	4.6	4.6
	550	92	51.9	56.7	35.1	4.0	4.3
	600	97	51.0	52.5	37.7	5.4	4.4
DETA	400	15	2.2	14.2	77.0	3.0	5.8
	450	38	6.7	17.9	71.9	3.3	6.9
	500	81	49.1	60.3	34.1	1.9	3.6
	550	92	50.8	55.1	37.9	2.8	4.3
	600	70	37.0	52.9	26.6	15.9	4.5
Urea	400	11	1.7	15.9	73.6	4.8	5.7
	450	26	5.1	19.3	67.5	5.8	7.5
	500	61	31.9	52.0	39.2	4.2	4.5
	550	76	43.2	56.7	32.5	5.6	4.2
	600	26	7.6	29.8	19.2	47.8	3.3

Note. X is the propane conversion; S_{ar}, S_{cr}, $S_{degydrogen}$ is the selectivity by aromatisation, cracking, and dehydrogenization, respectively; S_{H2} is the selectivity towards a formation of hydrogen; A is the yield of aromatic hydrocarbons.

The major liquid products of propane conversion over Ga-containing zeolite catalysts are aromatic hydrocarbons consisting mainly of benzene, toluene, alkylbenzenes, and aromatic compounds with higher molecular weight including naphthalene, methyl- and dimethylnaphthalenes. Byproducts are represented by gaseous hydrocarbons – methane and ethane. Hydrogen and C_2-C_4 olefins are present in minor amounts. Minor content of olefins in the reaction products suggests that they are intermediates that are actively involved into further interactions, resulting in the formation of aromatic compounds. Fig. 3 presents data on the influence of the template nature on the composition of the liquid products of propane conversion over Ga-containing zeolites. It should be noted that a sufficiently large amount of monoaromatic hydrocarbons (benzene, toluene, xylene etc.) is formed over all catalysts under study. Their fraction in a liquid product is higher than 60%. The lowest amounts of naphthalene and its derivatives were produced on the galloalumosilicate synthesized with an addition of urea. Thus, the use of different templates for the synthesis yields the samples differing in their catalytic properties.

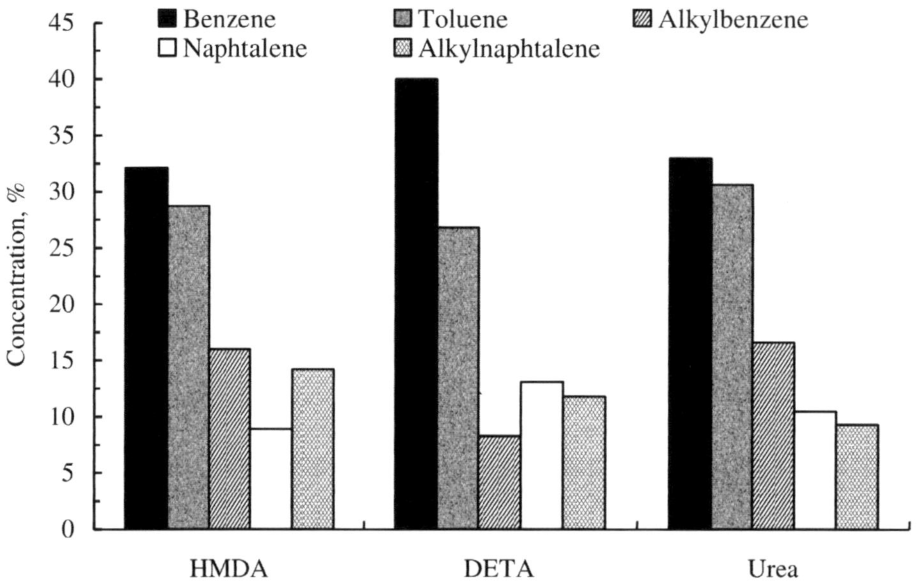

Figure 3. Composition of liquid products of propane conversion over Ga-containing zeolites synthesized using HMDA, DETA, and urea as templates (T = 600 °C).

4. Conclusion

The investigation of activity of Ga-containing ZSM-5 zeolites in the course of conversion of propane into aromatic hydrocarbons has shown that the nature of a structure-forming organic agent used in the synthesis is very important. The most active catalytic reactivity has been exhibited by the Ga-containing zeolite prepared using hexamethylenediamine as an additive. The conversion of propane over this catalyst at 550 °C and 500 h^{-1} has been found to be 92%, while the yield and selectivity towards the formation of aromatic hydrocarbons – 51.9% and 56.7%, respectively.

Referencies

[1] Vosmerikova L N, Ryabova N V and Vosmerikov A V 2008 Synthesis, acid and catalytic properties of element-alumosilicates in the course of lower alkane aromatization *Neftepererabotka i neftekhimia* **2** 35–39 (in Russian)

[2] Dergachyov A A and Lapidus A L 2008 Catalytic aromatization of light alkanes *Russian Chemical Journal. Journal of D.I. Mendeleyev Russian society* **LII** 15–21 (in Russian)

[3] Ione K G and Vostrikova A A 1987 The isomorphism and catalytic properties of silicates with the zeolite structure *Uspekhi khimii* **56** 393–427 (in Russian)

[4] Barrer R M 1985 Hydrothermal chemistry of zeolites (*Moscow: Mir)* p 209

[5] Minachev H M and Dergachev A A 1990 The conversion of low molecular weight hydrocarbons over zeolites. The results of science and technology *Kinetics and Catalysis* **23** 3–90

[6] Vosmerikov A V and Erofeev V I 1995 Effect of mechanical treatment on the catalytic properties of zeolite catalysts for aromatization of lower alkanes *Journal of Physical Chemistry*. Moscow. **69** 787–790 (in Russian)

[7] Weitkamp J 2000 Zeolites and catalysis *Solid State Ionics* **131** 175–188

IOP Conf. Series: Earth and Environmental Science **27** (2015) 012046 doi:10.1088/1755-1315/27/1/012046

Mathematical modeling of diesel fuel hydrotreating

A Tataurshikov[1,3], E Ivanchina[1,4], N Krivtcova[1,5], E Krivtsov[2,6] and A Syskina[1,7]

[1] National Research Tomsk Polytechnic University, Lenin Avenue 30, Tomsk, 634050, Russia
[2] Institute of petroleum chemistry SB RAS, Akademichesky Avenue 4, Tomsk, 634021, Russia

E-mail: [3] tataurshikov@yandex.ru, [4] ied@tpu.ru, [5] krivtcova@mail.ru, [6] john@ipc.tsc.ru, [7] annesyskina@yahoo.com

Abstract. Hydrotreating of the diesel fraction with the high initial sulfur content of 1,4 mass% is carried out in the flow-through laboratory setup with the industrial GKD-202 catalyst at various process temperature. On the basis of the experimental data the regularities of the hydrogenation reactions are revealed, and the formalized scheme of sulfur-containing components (sulfides, benzothiophenes, and dibenzothiophenes) transformations is made. The mathematical model of hydrotreating process is developed, the constant values for the reaction rate of hydrodesulfurization of the specified components are calculated.

1.Introduction

Hydrotreating is one of the main processes of secondary processing of oil raw materials; it is directed on considerable reduction of the content of heteroatomic compounds in oil products. Hydrotreating process is applied for production and upgrading of various oil products: gasoline, kerosene, diesel fuel, vacuum gasoil, and oil fractions. The wide use of hydrotreating process is caused, first of all, by introduction of increasing volume of sulphur and high-sulphur oil to the production cycle. The important aspect making introduction of hydrotreating process relevant is sensitivity of expensive catalysts to heteroatomic components, which are catalytic poisons. Toughening of ecological requirements to sulfur content in fuel, need of increase in oil refining depth necessitate improvement of hydrotreating process [1,2]. Application of mathematical modeling methods allow predicting conditions of hydrotreating process and quality of products that will considerably increase the depth of oil refining and will provide observance of all existing environmental standards [3].

The purpose of this work is to develop the mathematical model of diesel fuel hydrotreating and to calculate the constant values for the reaction rate of hydrodesulfurization of sulfides, benzo- and dibenzothiophenes.

2. Mathematical modeling of diesel fuel hydrotreating

2.1. Thermodynamic analysis and formalized scheme of transformations

Sulphur compounds of diesel fractions are mainly presented by sulfides, homologs and benzologs of thiophenes and, to a lesser extent, mercaptans, and disulfides. Thermodynamic calculation is always an initial stage of mathematical modeling based on physical and chemical laws and pretending to be highly adequate to the real process.

In this work the thermodynamic parameters of hydrotreating process including enthalpy, entropy, Gibbs energy were obtained with the computer program GaussianView and Gaussian 03; the method of calculation is DFT (Density Functional Study). As theoretical basis the B3LYP model, density functional theory of Becke (B3) and electronic correlation of Lee, Yang, and Parr are chosen. The basis is a 6-311G set. The conditions of hydrotreating process for calculation are the temperature of 340 °C and pressure of 3 MPa. The results of thermodynamic calculations of hydrogenation and hydrocracking reactions of the main sulfur compounds are presented in table 1.

Table 1. Values of enthalpy and Gibbs energy for hydrogenation reactions of sulfur compounds.

Sulphur compounds	Reaction	ΔH, kJ / mole	ΔG, kJ / mole
Sulphides	$C_5H_{11}SC_5H_{11} + H_2 \rightarrow C_5H_{11}SH + C_5H_{12}$	–65.60	–81.32
	$C_5H_{11}SH + H_2 \rightarrow C_5H_{15} + H_2S$	–67.66	–77.45
	$C_5H_{12}SC_5H_{12} + 2H_2 \rightarrow 2C_5H_{12} + H_2S$	–133.26	–158.77
Benzothiophenes	$C_8H_6S + H_2 \rightarrow C_8H_8S$	–78.97	–17.96
	$C_8H_8S + H_2 \rightarrow C_8H_{10}S$	–50.61	–13.21
	$C_8H_{10}S + H_2 \rightarrow C_8H_{10} + H_2S$	–72.00	–84.16
	$C_8H_6S + 3H_2 \rightarrow C_8H_{10} + H_2S$	–201.58	–115.33
Dibenzothiophenes	$C_{12}H_8S + H_2 \rightarrow C_{12}H_8 + H_2S$	–94.79	–64.99

On the basis of the data presented in table 1 the hydrogenation rate of sulfur compounds decreases: sulfides > benzothiophenes > dibenzothiophenes that is also confirmed by the literature data [3,4].

As any secondary process of oil processing diesel fuel hydrotreating includes a large number of parallel chemical reactions. Their aggregation in groups, or so-called pseudo-components, is made to reduce the number of individual component reactions. The mathematical model constructed on the basis of the reduced kinetic scheme is to keep the predicting ability, and also remain sensitive to the changes of raw material structure [5]. All groups of sulfur-containing organic compounds (mercaptans, sulfides, disulfides, thiophenes, benzothiophenes, and dibenzothiophenes) were formalized in separate groups of pseudo-components according to the value of isobaric and isothermal potential of their hydrogenation by hydrogen to the corresponding hydrocarbons and hydrogen sulfide. On the basis of the calculations of thermodynamic parameters for the desulphurization reactions of diesel fuel the formalized kinetic scheme of substance transformations (figure 1) is created.

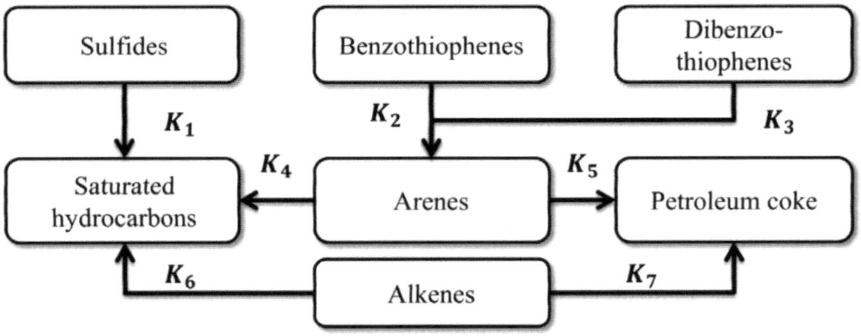

Figure 1. Formalized scheme of diesel fuel hydrotreating transformations.

According to the early results [6,7] it is established that grouping of sulfides, benzothiophenes, and dibenzothiophenes is caused by considerable distinction of chemical reaction rates for different groups of sulfur compounds. Depending on the structure of heteroatomic compounds in the raw materials of hydrotreating, conditions of process and catalyst the total sulfur content can be different at the exit from the reactor.

Except transformation of sulfur compounds such reactions as hydrogenation of aromatic and olefin hydrocarbons to saturated hydrocarbons and formation of coke structures are reflected in the formalized scheme of transformations.

Development of mathematical model based on the created kinetic scheme will allow predicting the quantitative sulfur content in the hydrogenate and selecting the optimum temperature of the process [8,9].

The mathematical model is realized in the form of the computer modeling system (CMS) [10], which is based on the experimental data obtained on the laboratory catalytic setup. When developing the mathematical model (table 2) the following assumptions are made: all reactions of hydrogenolysis are the reactions of the first order in quasihomogeneous conditions; reactions occur in the kinetic area; the order of reactions is the first [11,12]. The mathematical model according to the formalized scheme of compound transformations (figure 1) is presented in the form of a number of the differential equations of substance concentration change depending on time.

Table 2. Mathematic model (C_i – molar concentration; τ – contact time; k_i – reaction rate constant).

Component	Equation
Sulfides	$\frac{dC_S}{d\tau} = -k_1 C_S C_{H_2}$
Benzothiophenes	$\frac{dC_{BT}}{d\tau} = -k_2 C_{BT} C_{H_2}$
Dibenzothiophenes	$\frac{dC_{DBT}}{d\tau} = -k_3 C_{DBT} C_{H_2}$
Alkenes	$\frac{dC_{Alk}}{d\tau} = -k_6 C_{Alk} C_{H_2} - k_7 C_{Alk}$
Saturated hydrocarbons	$\frac{dC_{Sat.h.}}{d\tau} = k_1 C_S C_{H_2} + k_4 C_{Ar} C_{H_2} + k_6 C_{ol} C_{H_2}$
Aromatic hydrocarbons	$\frac{dC_{Ar}}{d\tau} = k_2 C_{BT} C_{H_2} - k_4 C_{Ar} C_{H_2} - k_5 C_{Ar}$
Petroleum coke	$\frac{dC_{coke}}{d\tau} = k_7 C_{ol} + k_5 C_{Ar}$

2.2. Hydrotreating of diesel fuel on the laboratory setup

Hydrotreating of diesel fuel with the total sulfur content of 1,4 mass% was carried out on the laboratory catalytic setup which is the flow-through system with the evaporator, the reactor on the special metal grid with the catalyst, the condenser, and the receiver of the hydrogenate. In the work the alumino-nickel-molybdenum catalyst GKD-202 was used, the volume of the catalyst loaded into the reactor was 10 ml, the size of the granules was 1-2 mm.

Sulphidation of the catalyst was carried out directly in the reactor with the straight-run diesel fraction containing 0,4 mass% of sulphidic sulfur. The catalyst was preliminarily dried up in the nitrogen stream at the temperature of 120 °C. Then hydrogen was supplied to the reactor with increase in pressure to 3.5 MPa. The diesel sulfiding chemical was fed with simultaneous temperature increase to 240 °C. The low-temperature stage of sulphidation was carried out at 240 °C during 2 h. The high-temperature stage was at 340 °C during 2 h. The sulfiding mixture was fed at the volume rate of 2 h-1 with the H2/raw materials = 300/1 ratio.

Hydrotreating of diesel fuel was carried out under the following conditions: the volume rate of raw material supply is 2 h-1, the pressure is 3,5 MPa, the ratio hydrogen/raw materials is 300/1, the temperature is 340, 360, and 380 °C.

The spectrophotometer SPECTROSKAN-S was used for quantitative determination of sulphur content in the initial fraction and hydrogenates.

The method of gas-liquid chromatography (GLC) was applied for identification of sulphur compound groups. GLC was carried out on the chromatograph Crystal-2000M with the quartz capillary column of 25 м×0.22 mm, the stationary phase is SE-54; the carrier gas is helium. When using the flame photometric detector (FPD) for analysis of sulphur compounds the temperature increased linearly from 50 to 290 °C, the rate of column heating was 4 deg/min. The temperature of the detector was 180 °C. The qualitative structure of sulphur compounds was determined by comparison of the retention times of the individual compounds and the analyzed components, and the literature data was also used [5]. The quantitative content of the sulphur compound groups and unattached isomers was determined by the areas of the chromatographic peaks; and the obtained peak areas (in rel. %) were recalculated for the total sulfur content in the sample (in mass%). The change of the sulphur compound structure is presented in table 3.

Table 3. Sulfur content in sulphur compounds of the initial fraction and hydrogenates.

Component	Content, % wt.			
	Feed	340 °C	360 °C	380 °C
S_o	1.4040	0.0980	1.0061	0.0939
S_{sul}	0.2180	0.0175	0.0175	0.0147
C_1BT	0.0341	0.0019	0.0017	0.0021
C_2BT	0.2020	0.0125	0.0128	0.0152
C_3BT	0.3510	0.0220	0.0222	0.0228
$(C_4+C_5)BT$	0.3742	0.0220	0.0236	0.0187
DBT	0.0360	0.0030	0.0036	0.0028
C_1DBT	0.0879	0.0071	0.0095	0.0065
C_2DBT	0.0801	0.0078	0.0104	0.0073
C_3DBT	0.0171	0.0047	0.0047	0.0039

2.3. Calculation results of reaction rate constants of hydrodesulfurization

The reaction rate constants for hydrodesulfurization of the identified groups of sulphur compounds are calculated on the basis of the data on their content change in the initial diesel fuel and hydrogenates (table 3). The basis of the program was formed by the kinetic model developed according to the law of mass action [13,14]. The calculation results of the final concentration of sulphur compounds by means of the developed program showed high adequacy of the obtained data to the experimental one. The relative error doesn't exceed 0,007 mass% for all components. The values of effective reaction rate constants of sulphur compound transformation obtained by the reverse kinetic task solution are presented in table 4.

It is established that with the increase in temperature of hydrotreating the reaction rate constants of all sulphur compound groups hydrogenolysis rise. The reaction rate constants of benzothiophene hydrogenolysis have the greatest values. It is possibly caused by their initially high content in the initial diesel fuel and also easy hydrogenation of C2-C3 bonds. The smallest values of reaction rate constants are noted at the most substituted C3- dibenzothiophene homologs. The obtained results correspond to the literature data [15–17].

The removal of the identified sulphur compound groups depending on the hydrotreating process temperature is shown in figure 4.

Table 4. Effective reaction rate constants of sulphur compound hydrodesulfurization, h^{-1}.

Component	Temperature, °C		
	340	360	380
S_{sul}	1.907	1.971	2.036
C_1BT	2.191	2.153	2.114
C_2BT	2.104	2.030	1.956
C3BT	2.094	2.080	2.065
$(C_4+C_5)BT$	2.140	2.201	2.262
DBT	1.877	1.907	1.937
C_1DBT	1.907	1.938	1.969
C_2DBT	1.761	1.786	1.810
C_3DBT	0.968	1.043	1.118

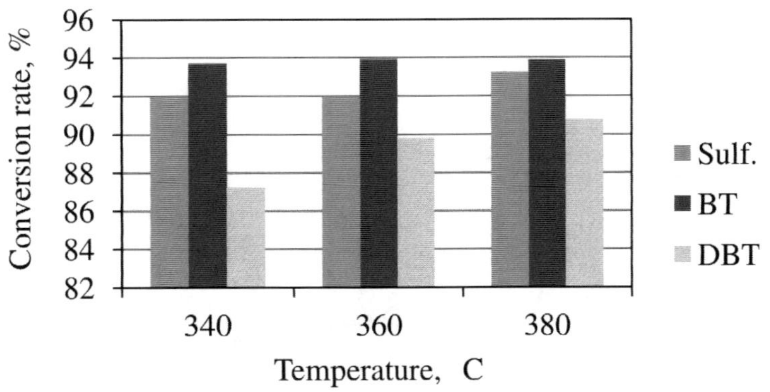

Figure 4. The removal degree of the sulphur compounds depending on the hydrotreating process temperature.

The greatest extent of removal from the raw materials is noted at the benzothiophene homologs (in all experiments). Sulfides are the second in terms of conversion (93,2 mass% is removed at 380 °C). The difference in number of remote sulfides is insignificant at temperatures of 340 °C and 360 °C. The hydrogenolysis depth of the dibenzothiophene homologs very strongly depends on the temperature of the process and reaches 90,8 mass% at 380 °C.

3. Conclusion

In this work the mathematical model of diesel fuel hydrotreating is developed on the basis of the physical and chemical regularities of the process. The thermodynamic and kinetic parameters of the desulphurization reactions of sulfur compounds are estimated in the hydrotreating process. On the basis of the presented data the rate of sulfur compound hydrogenation decreases: sulfides > benzothiophenes > dibenzothiophenes that is also confirmed by the literature data. The developed formalized scheme of transformations and kinetic model consider transformations of sulfur compounds, hydrogenation of aromatic and olefin hydrocarbons, and formation of coke structures. The model was based on the experimental data obtained on the laboratory setup with the GKD-202 catalyst.

The results of calculation of the final concentration by means of the developed program showed high adequacy to the experimental data. The relative error doesn't exceed 0.007 mass% for all components. This fact speaks for high adequacy of the mathematical model to the real process and accuracy of algorithm for effective constant search.

The obtained results can be applied for forecasting the total sulfur content in the hydrogenate depending on the changing structure of raw materials and technological conditions on the hydrotreating setup for process optimization.

Acknowledgements
The reported study has been carried out in the frame of the State Task "Science" 1.1310.2014.

References

[1] Kabe T, Ishihara A and Tajima H 1992 Hydrodesulfurization of Sulfur-Containing Polyaromatic Compound in Light Oil *Ind. Eng. Chem. Res.* **31** 1577–80
[2] Ghosh P, Andrews A T, Quann R J and Halbert Th R 2009 Detailed Kinetic Model for the Hydro-desulfurization of FCC Naphtha *Energy Fuels* **23** 5743–59
[3] Palmer E, Polcar S and Wong A 2009 Clean diesel hydrotreating. Design considerations for clean diesel hydrotreating *PTQ* 91–100
[4] Belinskaya N S, Ivanchina E D, Ivashkina E N, Silko G Yu and Frantsyna E V 2014 Optimization of the process setting mode of diesel fuels hydrodeparaffinization by mathematical modeling *Izvestiya vuzov. Chemistry and Chemical Technology* 57 **11** 90–92
[5] Dolganova I O, Ivashkina E N and Ivanchina E D 2011 Mathematical modeling in problems of increasing the efficiency of the plant production of linear alkyl benzenes *Bulletin of the Tomsk Polytechnic University* **319** 109–12
[6] Krivtsova N I, Ivanchina E D, Tataurshchikov A A and Zanin I K 2014 Study of activity of industrial catalysts of hydropurification of diesel fuel on mathematical model *Izvestiya vuzov. Chemistry and Chemical Technology* **57** 81–83
[7] Krivtsova N I, Ivanchina E D, Zanin I V, Landl Yu I and Tataurshchikov A A 2013 Kinetic laws of turning sulfur-containing compounds at hydrorefining of oil diesel fraction *Izvestia Tomskogo politekhnicheskogo universiteta* **322** 83–86
[8] Reshetnikov S I and Ivanov E A 2003 Mathematical modelling of catalytic processes with non-stationary condition of the catalyst: two-reactor system *Sib. Zh. Ind. Mat.* **6:1** 108–17
[9] Schuit G C A and Gates B C 1973 Chemistry and Engineering of Catalytic Hydrodesulphurization. Department of Chemical Engineering University of Delaware *AlChE Journal* **19** 417–38
[10] Kim S F, Usheva N V, Moyzes O E, Kuzmenko E A, Samborskaya M A and Novoseltseva E A 2014 Modelling of dewatering and desalting processes for large-capacity oil treatment technology *Procedia Chemistry* **10** 448–53
[11] Usheva N V and Kravtsov A V 2006 *Macrokinetics of chemical processes and reactors design: Laboratory practices* (Tomsk: Tomsk Polytechnic University) p 98
[12] Ivanchina E D, Ivashkina (Mikhailova) E N, Frantsyna E V and Dolganov I M 2014 *Optimization of the higher paraffins dehydrogenation reactor unit using the method of mathematical modeling* (Tomsk: Publishing house of TPU) p 180
[13] Boyarinov A I 1975 Optimization methods in chemical technology (Moscow: Chemistry) p 575
[14] Krivtsova N I, Krivtsov E B, Ivanchina E D and Golovko A K 2013 Kinetics of hydrodesulfurization of diesel fraction *Basic Researches* **8-3** 640–44
[15] Michael T Klein et al. 2005 *Molecular modeling in heavy hydrocarbon conversions* (Boca Raton, Florida: CRC Press) p 264
[16] Al-Zeghayer Y S, Sunderland P, Al-Masry W, Al-Mubaddel F, Ibrahim A A, Bhartiya B K and Jibril B Y 2005 Activity of CoMo/g-Al_2O_3 as a catalyst in hydrodesulfurization: effects of Co/Mo ratio and drying condition *Applied Catalysis A: General* **282** 163–71
[17] Satterfield C N Modell M and Wilkens J A 1980 Simultaneous Catalytic Hydrodenitrogenation of Pyridine and Hydrodesulfurization of Thiophene *Ind. Eng. Chem. Process Des. Dev.* **19** 154–60

Analysis of technological conditions influence on efficiency of oilfield treatment

N V Usheva[1], O E Moizes[1,2], E A Kuzmenko[1], S F Kim[1], E S Khlebnikova[1], S N Dyalilova[1] and T V Filippova[1]

[1] National Research Tomsk Polytechnic University, 30 Lenin Ave., Tomsk, 634050, Russia

E-mail: [2]moe@tpu.ru

Abstract. The results of influence of process parameters on oil quality and recommended effective technological modes of oilfield treatment processes are presented in this paper. It is shown that the parameters that significantly affect the efficiency of oil processes are temperature and water-oil emulsion flow rate with a given number of working process units and the structure of flowsheet flows.

1. Introduction

Currently, oilfield treatment requires large-capacity industrial installations, which are characterized by a great variety of equipment used in a complex process flow structure [1-5]. The increase of plant efficiency can be achieved by optimization of technological modes both for existing and designed installations [6-8]. At present, it is becoming possible to achieve by using mathematical modeling methods, which determine the timeliness of this work and allow analyzing the technology of large-capacity industrial oil treatment.

To solve the assigned problems effectively, MS (Modeling Systems) of all kinds can be used. A simulation system of oilfield treatment has been developed. This system based on a modular principle of model formation of process flowsheet units allows defining dehydration and desalting processes [8-10].

The purpose of this paper is to analyze the influence of process parameters on oilfield treatment and to estimate technological modes for enhancing efficiency of dehydration and desalting using mathematical modeling.

2. Objects and research techniques

The object of this study is an oil treatment unit of the oilfield located in Eastern Siberia (figure 1). It is designed to process a large volume of oil, so the flowsheet of the unit has a complicated structure of flows and includes three parallel lines. Since the used equipment is designed for a certain capacity, the process flowsheet includes the required number of concurrent units.

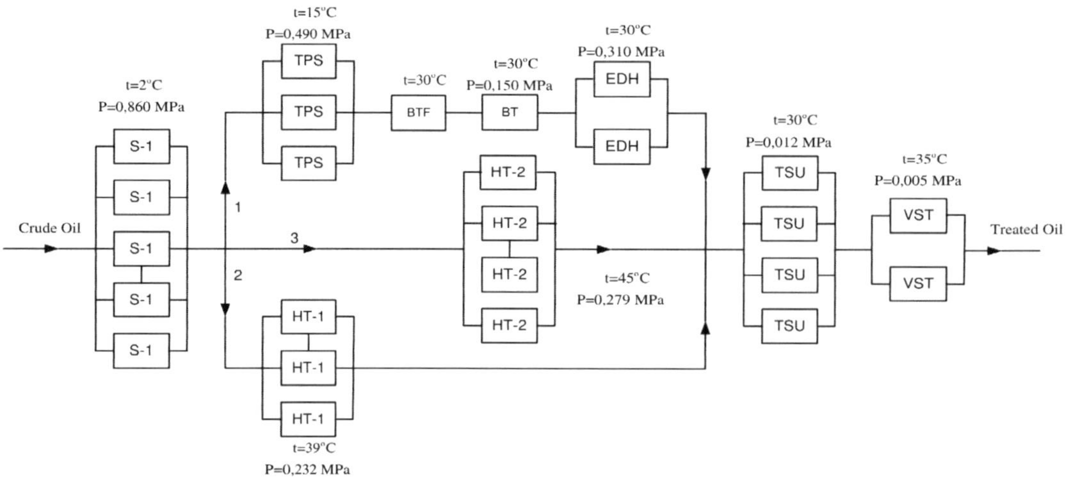

Figure 1. Block diagram of oilfield treatment unit:
S-1 – first stage separator; TPS – three phase separator; BTF - block tube furnace; BT - buffer tank; EDH – electric dehydrator; HT-1 – "Heater Treater" unit of type I; HT-2 – "Heater Treater" unit of type II; TSU – terminal separation unit; VST – vertical steel tank.

The oilfield treatment unit includes three production lines. The first production line consists of three parallel working TPS, in which oil is degassed and dehydrated. Then, the oil flow is heated in the furnace, passes through the buffer tank and enters the electrical dehydrator, in which dehydration and desalting occur by electric field.

The second production line consists of modern modular HT-1 units in which crude oil is heated, degassed and dehydrated.

The third production line includes HT-2 units, equipped with electric dehydrogenating grids, which increase the efficiency of water separation.

Table 1. Physical and chemical oil properties.

Physical and chemical oil properties and technological parameters	Values
Density, kg / m^3	864
Viscosity at 20°C, mm^2 / sec	29.5
Molecular weight , g / mole	292
Water cut, % wt.	20

3. Discussion of results

The influence of process parameters on the oilfield preparation was studied by means of MS. Figure 2 shows the study results of oil-water emulsion separation in TPS.

It was determined (figure 2) that with the increase of temperature the output water cut of TPS decreases and with the increase of emulsion flow rate and water cut at the inlet of TPS increases.

In other units of the process flowsheet, the studies showed similar regularities of the dehydration and desalting processes.

Table 2 shows the options for process parameter variations of oilfield treatment units.

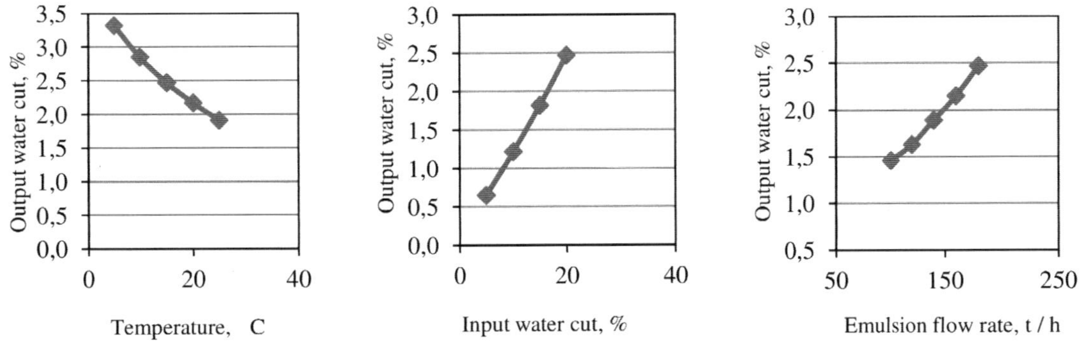

Figure 2. The results of research in TPS.

Table 2. Options of process parameters in calculating oilfield treatment unit.

	Flow rate, kg/h						
	1000238.1						1012142.9
	Input water cut, % wt.						
	20.93					26.00	29.00
Options	1	2	3	4	5	6	7
Unit	Temperature, °C						
TPS	15°C	**25°C**	15°C	15°C	25°C	25°C	25°C
EDH	30°C	30°C	30°C	30°C	30°C	30°C	30°C
HT (I)	Input temperature at oilfield treatment unit: 15°C						**25°C**
	39°C	39°C	**45°C**	39°C	**45°C**	**45°C**	**45°C**
HT (II)	45°C	45°C	45°C	**40°C**	**45°C**	**45°C**	45°C
VST	30°C	30°C	30°C	30°C	30°C	30°C	30°C

For different variants of process parameters (table 2) the studies were conducted using a simulation system, the results of which are shown in table 3 and figure 3.

Table 3. The research results.

	Base Option			Option 2		
Flows	Flow rate, t/h	Water cut, % wt.	Salt content, mg/l	Flow rate, t/h	Water cut, % wt.	Salt content, mg/l
Oilfield Treatment Unit	1000.24	20.93		1000.24	20.93	
TPS	549.91	2.24	1076.38	549.91	1.39	670.17
EDH	453.95	1.15	191.33	444.07	0.62	102.20
HT-1	157.12	3.62	1766.36	157.12	3.62	1766.36
HT-2	209.49	1.11	540.91	209.49	1.11	540.91
	Option 3			Option 4		
Flows	Flow rate, t/h	Water cut, % wt.	Salt content, mg/l	Flow rate, t/h	Water cut, % wt.	Salt content, mg/l
Oilfield Treatment Unit	1000.24	20.93		1000.24	20.93	
TPS	549.91	2.24	1076.38	549.91	2.24	1076.38
EDH	453.95	1.15	191.33	453.95	1.15	191.33
HT-1	157.12	3.13	1527.42	157.12	3.62	1766.36
HT-2	209.49	1.11	540.91	209.49	1.23	597.10

Flows	Option 5			Option 6			Option 7		
	Flow rate, t/h	Water cut, % wt.	Salt content, mg/l	Flow rate, t/h	Water cut, % wt.	Salt content, mg/l	Flow rate, t/h	Water cut, % wt.	Salt content, mg/l
Oilfield Treatment Unit	1000.24	20.93		1000.24	26.00		1012.14	29.00	
TPS	549.91	1.39	670.17	549.91	1.75	840.39	556.46	2.16	1040.93
EDH	444.07	0.62	102.20	420.08	0.53	88.31	412.95	0.59	96.97
HT-1	157.12	3.13	1527.42	157.12	4.03	1976.19	158.99	3.91	1914.52
HT-2	209.49	1.11	540.91	209.49	1.39	675.94	211.98	1.69	821.16

The results showed that the parameters significantly affecting the efficiency of oil treatment processes are temperature and flow rate of oil-water emulsion at a given number of working units and the flow pattern of the process flowsheet.

Figure 3 shows the values of residual water cut and salt concentration at the outlet of oilfield treatment unit under different operating conditions. The study results showed that the most effective option is the temperature condition 5 (table 4), where the residual water cut of oil is about 0.10 % wt. and salt content is 39.75 mg/l.

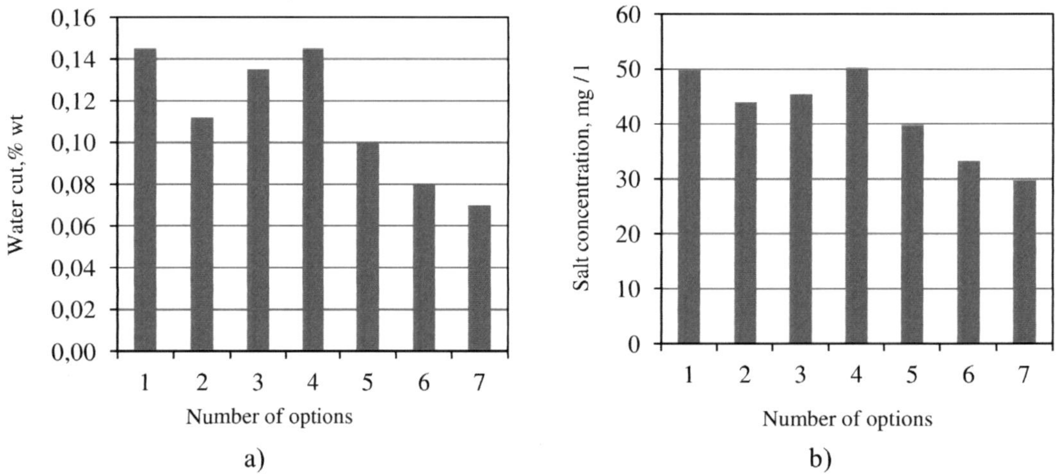

Figure 3. Quality characteristics of treated oil of calculated process options: a) water cut; b) concentration of chlorine salts.

Analysis of technological modes of the operating production unit (table 3) showed that temperature increase up to 25 °C in a three phase separator and up to 45 °C in a heather-treater reduces the residual water cut of oil compared to the base option of 0.10 % wt. water cut.

Having examined the operation of the unit by increasing flow rate and water cut of the formation fluid (options 6 and 7), technological conditions were selected to define the oil quality indicators required by all-Union State Standard: water content and chlorine salts at the outlet from the oilfield treatment unit are 0.07 % wt. and 29.65 mg/l, respectively.

4. Conclusion

The study results based on the simulation system showed that deeper water separation is observed when TPS and EDH are connected in sequence (line 1), even at high load (60% of the total oilfield treatment unit expenditure).

It was shown that temperature and flow rate of emulsion are parameters that have the most significant effect on the residual water cut and salt content in oil.

Thus, the required quality can be achieved by oil treatment both by residual water cut and by salt concentration by varying the process parameters in the units of technological lines. With the application of modeling system the material balance in various process conditions can be determined to give operational forecasts and recommendations on operation modes of oilfield units in the process of oilfield development.

References

[1] Tronov V P 2000 Oilfield treatment *Promyslovaya podgotovka nefti* (Kazan: FEN)
[2] Lutoshkin G S and Dunyushkin M I 2007 *Problems in collecting and preparation of oil, gas and water industries (Sbornik zadach po sboru i podgotovke nefti, gaza i vody na promyslakh)* Textbook for high schools (Moscow: Alyans)
[3] Lutoshkin G S 2005 *Collecting and preparation of oil, gas and water (Sbor i podgotovka nefti, gaza i vody)* Textbook for high schools (Moscow: Alyans)
[4] Howard M, Gioffre P and Swackhammer T. 2013 Spill Prevention, Control and Countermeasure (SPCC) guidance for regional inspectors pp 16–18
[5] Guluyev G, Rzayev A and Pasayev F 2008 Development of methods of determination of disperse composition of oil emulsion *The Second International Conference "Problems of cybernetics and information"* pp 80–82
[6] Frantsina E V, Ivashkina E N, Ivanchina E D and Romanovskii R V 2014 Developing of the mathematical model for controlling the operation of alkane dehydrogenation catalyst in production of linear alkyl benzene *Chemical Engineering Journal* **238** 129–39
[7] Krivtsova N I, Gaga S G, Desiatnichenco A A, Popok E V and Zaitceva E V 2014 Synthetic liquid fuels obtained by thermolysis of animal waste *Procedia Chemistry* **10** 441–47
[8] Kim S F, Usheva N V, Moyzes O E, Kuzmenko E A and Samborskaya M A 2013 Modular design of mathematical models of oilfield treatment devices and technological schemes (Modul'nyy printsip postroyeniya matematicheskikh modeley apparatov i tekhnologicheskikh skhem promyslovoy podgotovki nefti) *Neftepererabotka i neftekhimiya* **10** 41–44
[9] Kim S F, Usheva N V, Moyzes O E, Kuzmenko E A, Samborskaya M A and Novoseltseva E A 2014 Modelling of dewatering and desalting processes for large-capacity oil treatment technology *Procedia Chemistry* **10** 448–53
[10] Kim S F, Usheva N V, Samborskaya M A, Moyzes O E and Kuzmenko E A 2013 Modelling of oil-water emulsion destruction process for large-capacity oil treatment technologies *Fundamental'nye issledovaniya* **8** 626–29

IOP Conf. Series: Earth and Environmental Science **27** (2015) 012048 doi:10.1088/1755-1315/27/1/012048

Electrochemical synthesis of nickel-aluminium oxide system from metals obtained by ore processing

V V Korobochkin[1,2], N V Usoltseva[1,3], K G Shorokhov[1,4] and E V Popova[1]

[1] Department of Chemical Engineering, National Research Tomsk Polytechnic University, 30 Lenin Ave., Tomsk, 634050, Russia

E-mail: [2] vkorobochkin@tpu.ru, [3] usolceva@tpu.ru, [4]shorohof@hotmail.com

Abstract. Separate and combined electrochemical oxidation of aluminium and nickel has been conducted by alternating current of industrial frequency. Concentration increase of electrolyte solution (sodium chloride) in the range from 3 to 25 wt. % and current density from 0.5 to 1.5 A/cm^2 was found to result in the increasing metal oxidation rate, excluding aluminium oxidation which oxidation rate is independent of the electrolyte solution concentration. At the current density of 1.5 A/cm^2 the products of separate oxidation of nickel and aluminium are nickel oxyhydroxides, nickel hydroxides and aluminium oxyhydroxide (boehmite), respectively. In addition to these compounds, the nickel-aluminium oxide hydrate is included in the products of nickel and aluminium co-oxidation. Its content grows with the increasing electrolyte solution concentration. Varying the concentration and current density within the limits indicated, the nickel-aluminium oxide system with nickel oxide content from 3 to 10 wt. % is produced.

1. Introduction

Metal oxides are widely used in industry. Industrial use of natural metal oxides (quartz, corundum, hematite, rutile, cassiterite, magnetite, spinel and others) is limited due to their phase heterogeneity and the presence of impurities and other compounds, as well as varying ore composition of different deposits. Some metals are found in nature mostly as compounds, not containing oxygen. For example, nickel and copper ores consist of metal sulphides. Based on above-mentioned facts, non-ferrous metal ores could be considered as raw materials in producing metals and metal compounds recycled as metal oxides suitable for industrial use. The formation of highly dispersed oxides having improved operating characteristics is promoted by their synthesis in non-equilibrium conditions [1, 2]. Products of electrochemical metal oxidation by alternating current (AC) are characterized by both high specific surface area and high purity because of extremely low content of impurities. This method can be used in producing metal oxides and metal oxide systems. Previously, the AC electrochemical metal oxidation was investigated for the synthesis of copper oxides [3], tin oxides [4] and double copper-containing oxide systems: copper-titanium, copper-cadmium [5], copper-aluminium [6]. Highly-developed aluminium oxide pore structure and its wide application in catalytic and sorption processes determine perspective investigation of the synthesis of alumina-based oxide systems. Nickel-based catalysts are effective in the catalytic partial oxidation of methane in synthesis gas (syngas) [7]. NiO/Al_2O_3 can be used as an oxygen carrier during chemical looping combustion (CLC) of fuels [8]. The absence of direct contact between fuel and air during combustion prevents CO_2 emission into the atmosphere with unreacted oxygen, therefore, avoiding the negative impact on the environment.

Mesoporous nickel-aluminium oxide system can be obtained in several ways. It was established that the bulk $NiO-Al_2O_3$ catalyst is characterized by higher activity and selectivity in the partial methane oxidation to synthesis gas than supported catalyst due to the fact that nickel is distributed not only on the carrier surface, but also within its framework [9]. In this case, Ni content in the sample affects both the conversion of methane and the amount of generated carbon [10].

The research target is to study the possibility of producing nickel-aluminium oxide system during combined electrochemical oxidation of nickel and aluminium under the action of alternating current.

2. Experimental part

AC electrochemical metal oxidation with soluble electrodes was conducted in solutions of sodium chloride of analytical grade with concentrations of 3, 5, 15 and 25 wt. % at the temperature of 100 °C and current densities of 0.5, 1.0 and 1.5 A/cm^2. Sodium chloride is used as the electrolyte because this compound meets most requirements for electrolytes. It has high electrical conductivity and does not chemically react with metals and their oxides. Furthermore, chloride ions contribute to the formation of oxide film on the electrode surface that does not create significant difficulties in the electrolysis process. It is due to both the formation of metal hydroxochloride complexes transforming into a liquid phase and the decomposition of loose oxides that are easily removed from the metal surface. The lower concentration limit is due to high energy consumption and instability of the process, while the upper one – solubility of sodium chloride in water. This process at current density of less than 0.5 A/cm^2 is not of practical interest because, in this case, metal oxidation rate is not high. At current density of 1.5 A/cm^2 the oxidation rate is so high that there is an intensive heating of the electrolyte, its boil-off, gas-filling of interelectrode space, resulting in unstable unit performance. The electrodes were nickel plates (mark H-0) and aluminium plates (mark A-0).

Electrolysis products were washed with distilled water and air-dried at a temperature of 110 °C for 6 hours. The remove of electrolyte ions is conditioned by high mobility of sodium ions that significantly accelerate sintering, as well as by the formation of different sodium-containing compounds of aluminium oxide. At the same time, the formation of nickel-aluminium spinel is prevented [11].

X-ray diffraction (XRD) measurements were performed using DRON-3M diffractometer under the following conditions: CuK_α-radiation (λ=1.5418 Å) at 25 mA and 35 kV, counting rate of 4θ/min and angle scan region (2θ) from 10 to 70°. PDF 2 database was used to identify the phase composition.

3. Results and discussion

The combined electrochemical oxidation of copper and aluminium intensifies the copper oxidation and insignificantly affects the aluminium oxidation rate [3], while above-mentioned nickel reaction is visa versa: rate of combined nickel oxidation is preferably lower than the rate of separate one. On the contrary, combined oxidation of metals accelerates the aluminium oxidation. The dependence of the aluminium oxidation rate from the second metal property substantiates the assumption of interprocess in the combined oxidation of metals with different properties.

Irrespective of synthesis conditions, aluminium oxidation rate (q_{Al}) significantly exceeds nickel oxidation rate (q_{Ni}) (figure 1): 5-20 times in combined oxidation of nickel and aluminium, and 3-7 times in separate oxidation of metals. This can be explained by the hydrogen depolarization at electrochemical oxidation of metals with negative electrode potentials (in this case, aluminium), promoting higher current efficiency, therefore, the predominance of aluminium compounds in electrolysis products [12]. Nickel oxides have significant true density, surface tension and solubility. Moreover, such processes as coagulation, recondensation and coalescence are more typical for nickel resulting in the degradation of the porous structure [1]. As a result, well-developed boehmite porous structure and prevailing of this compound in the electrolysis product predetermine its primary role in the formation of the material porous structure.

Figure 1 shows that the dependence of nickel oxidation rate from electrolyte solution concentration is linear at all current densities. Nickel oxidation rate increases to 2.1-2.3 times at electrolyte solution concentration growth from 3 to 25 wt. %. The concentration influence on the oxidation rate increases when current density grows. At all solution concentrations the increase of current density from 0.5 to 1.5 A/cm^2 accelerates metal oxidation in 3.5-6.0 times.

Comparable to nickel, aluminium oxidation rate does not depend on the sodium chloride solution concentration and increases 3.8 times when current density increase from 0.5 to 1.5 A/cm^2.

Irrespective of current density at combined metal oxidation, the oxidation rates of both metals increase linearly with the growth of electrolyte solution concentration from 3 to 25 wt. %. Depending on the current density the oxidation rate of nickel and aluminium increase to 1.6-5.7 and 1.5-2 times, respectively. Increasing current density from 0.5 to 1.5 A/cm^2 accelerates electrochemical oxidation of nickel to 1.7-6.0 times and aluminum to 1.4-1.7 times.

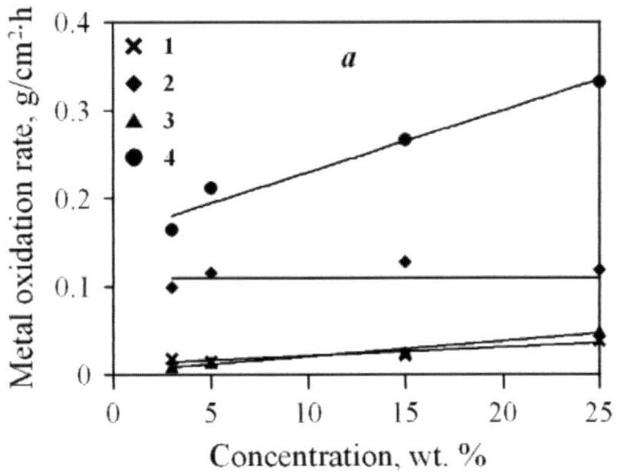

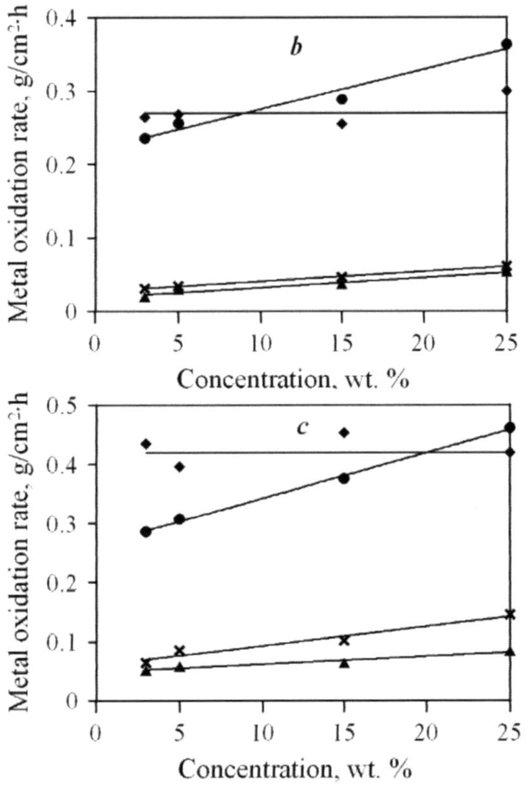

Figure 1. Rate dependence of separate (1, 2) and combined (3, 4) electrochemical oxidation of nickel (1, 3) and aluminum (2, 4) on the sodium chloride solution concentration at current density of 0.5 (a), 1.0 (b) and 1.5 (c) A/cm^2.

The cell voltage is greatly determined by the resistance of electrolyte solution and also depends on oxide film resistance on the electrode surface. Voltage stability in time during electrochemical nickel oxidation at all values of solution concentrations and current densities indicates that formed oxide film does not show significant resistance. Increasing voltage in time during aluminum oxidation results in the formation of a dense oxidation product layer, the thickness of which increases in time and requires additional voltage to overcome the resistance of this layer.

The increase of sodium chloride solution concentration increases electrical conductivity, thus, the voltage required to overcome solution resistance decreases. Measured cell voltage decreases with increasing concentration of the electrolyte solution in both separate and combined metal oxidation (figure 2).

In nickel oxidation the cell voltage reduction is 20 wt. % at current density of 0.5 A/cm^2. At 1.0 and 1.5 A/cm^2 the cell voltage is reduced to ~ 3 times, the most significant decline is observed when the solution concentration is higher than 15 wt. %. In aluminium oxidation the cell voltage behavior is somewhat different. The cell voltage decreases with increasing concentration of 3 to 5 wt. %; the reduction is greater the higher the current density is. At current density of 1.5 A/cm^2, an increase in concentration of more than 5 wt. % does not cause a significant change in the cell voltage, whereas at current densities of 0.5 and 1.0 A/cm^2 the voltage is reduced to 1.5 to 2 times, during electrolysis in solution concentration increase to 15 wt. %, respectively. Further increase of electrolyte solution concentration does not significantly change the cell voltage. Cell voltage during co-oxidation of metals decreases 4-10 times by increasing the concentration of the electrolyte solution from 3 to 25 wt. %, but rises 5-6 times with increasing current density from 0.5 to 1.5 A/cm^2.

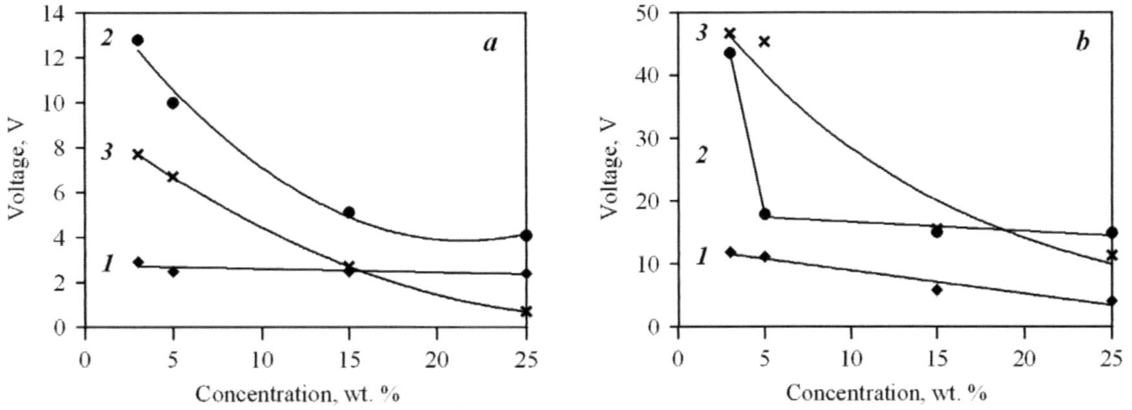

Figure 2. Cell voltage dependence on sodium chloride concentration in separate oxidation of nickel(1), aluminium (2) and nickel and aluminium co-oxidation (3) at current densities of 0.5 (a) and 1.5 (b) A/cm^2.

Comparing oxidation rates and cell voltage revealed that the highest rate of electrochemical oxidation of nickel and aluminium are achieved by carrying out the process in solutions with high concentration and lowest cell voltage. Increasing current density provides higher oxidation rate, but considerably increases the voltage that must be applied to perform this process.

Figures 1 and 2 show that there is a relationship between the correlation of the metal oxidation rates and the cell voltages in combined and separate metal oxidation. It was established that the rate of combined metal oxidation exceeds the rate of separate metal oxidation when the voltage in combined oxidation is lower than in separate oxidation. This could be associated with the structure of the oxide film on the electrode surface. Intensification in combined oxidation of metals is a result of the formation of oxidation product layer on the electrode surface, which does not create significant resistance.

According to the X-ray diffraction patterns (figure 3), the products of separate nickel and aluminium oxidation in the sodium chloride solution with the concentration of 15 wt. % and current density of 1.5 A/cm^2 (provides the highest metal oxidation rate) consist of the same nickel-containing compounds (anhydrous $Ni(OH)_2$ (JSPDS card No. 14-0117) and hydrated $Ni(OH)_2 \cdot H_2O$ (JSPDS card No. 22-0444) nickel hydroxides, nickel oxyhydroxides NiOOH (JSPDS card No. 06-0075), Ni_2O_3H (JSPDS card No. 40-1179)) and aluminum oxyhydroxide (boehmite AlOOH, JSPDS card No. 17-0940), respectively. Wide indistinct peaks of boehmite indicate that slightly crystallized compound with defective structure is formed in ac-oxidation of aluminium.

Since the aluminium oxidation rate is significantly higher than the nickel oxidation rate, the samples predominately include aluminium oxidation products. Products of combined electrochemical oxidation of metals contain the same nickel- and aluminium-containing phase as the products of the separate metal oxidation. Besides, nickel aluminium oxide hydrate ($Ni_5Al_4O_{11} \cdot 18H_2O$, JCPDS card No.22-0452) is formed. The contents of $Ni_5Al_4O_{11} \cdot 18H_2O$ and nickel oxyhydroxide increases as the concentration of the electrolyte solution, used for the electrochemical oxidation of metals, growth (figure 4).

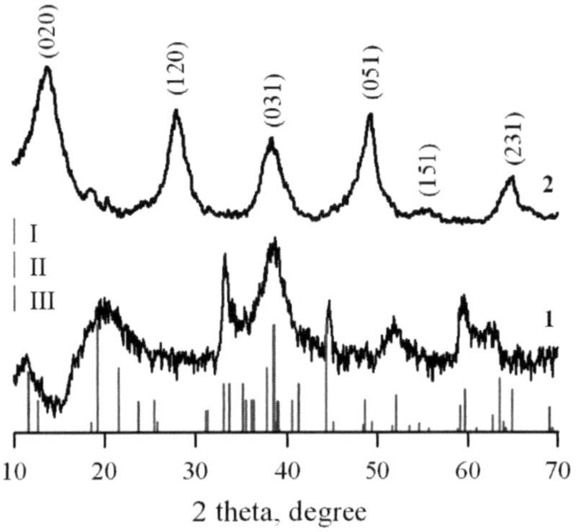

Figure 3. XRD patterns of separate electrochemical oxidation products of (1) nickel and (2) aluminum in sodium chloride solution with concentration of 15 wt. % at current density of 1.5 A/cm^2 (I – $Ni(OH)_2$, II – Ni_2O_3H, III – $2Ni(OH)_2{\cdot}2H_2O$).

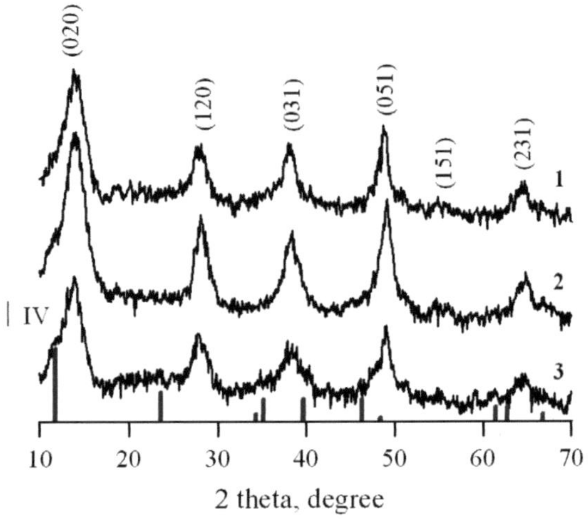

Figure 4. XRD patterns of combined electrochemical oxidation products of nickel and aluminum in sodium chloride solution with concentrations of 3 (1), 15 (2) and 15 (3)wt. % at current density of 1.5 A/cm^2 (IV – $Ni_5Al_4O_{11}{\cdot}18H_2O$).

Regardless of the phase composition of nickel- and aluminium-containing compounds, the nickel oxide NiO and alumina Al_2O_3 are formed during heat treatment of the material [11]. The mass changes of the electrodes were recalculated based on these oxides. The content of nickel oxide in the oxide system increases 2-4 times when the concentration grows from 3 to 25 wt. % and 2-6 times at the changing of current density from 0.5 to 1.5 A/cm^2 (figure 5). Thus, variation of the process conditions of combined electrochemical oxidation of nickel and aluminium results in oxide system formation with nickel oxide content of 3 to 10 wt. %.

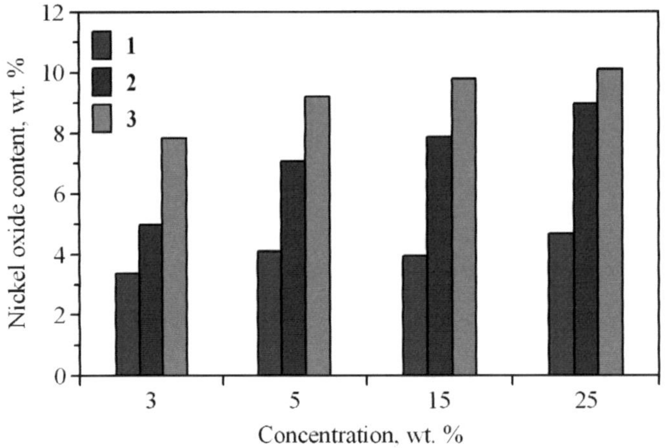

Figure 5. Dependence of nickel oxide content in nickel-aluminum oxide system on concentration of sodium chloride solution at current density of 0.5 (1), 1.0 (2) and 1.5 A/cm^2.

IOP Conf. Series: Earth and Environmental Science **27** (2015) 012048 doi:10.1088/1755-1315/27/1/012048

4. Conclusion

Applying alternating current both individual and combined electrochemical oxidation of nickel and aluminium were performed. It was established that the rate of aluminium oxidation is much higher than nickel oxidation. The rates in combined metal oxidation exceed the rates in separate metal oxidation under conditions when the cell voltage of co-oxidation is less than the voltage of separate oxidation. In electrochemical nickel oxidation, hydrated and unhydrated nickel hydroxides and nickel oxyhydroxides are formed. Electrochemical aluminium oxidation results in aluminium oxyhydroxide (boehmite) formation. Along with indicated compounds of nickel and aluminium the products of combined oxidation of metals include nickel aluminium oxide hydrate. The content of this compound and nickel oxyhydroxide increases with the growth of concentration of sodium chloride solution. Varying the concentration of electrolyte solution from 3 to 25 wt. % and current density from 0.5 to 1.5 A/cm^2 results in the formation of nickel-aluminium oxide with the content of nickel oxide from 3 to 10 wt. %.

References

[1] Zyryanov V V 2005 *Inorganic Materials* **41** 378–92

[2] Merzhanov A G 2012 *40 Years Of SHS: A lucky star of a scientific discovery. A presentation with elements of a scientific lecture* (Bentham Science Publishers) p 104

[3] Korobochkin V V, Gorlushko D A, Balmashnov M A and Usoltseva N V 2010 *Bulletin Томsk polytech.univer. Chem.* **317** 13–16

[4] Korobochkin V V, Balmashnov M A, Gorlushko D A, Usoltseva N V and Bochkareva V V 2013 *Inorganic Materials* **49** 993–99

[5] Dolinina A S, Korobochkin V V, Usoltseva N V, Balmashnov M and Botyanova I V 2014 *Procedia Chemistry* **10** 369–72

[6] Usoltseva N V, Korobochkin V V, Balmashnov M A and Dolinina A S 2014 *Procedia Chemistry* **10** 320–25

[7] Enger B Ch, Lødeng R and Holmen A 2008 *Appl. Catal. A: Gen.* **346** 1–27

[8] Gayбn P, Dueso C, Abad A, Adanez Ju, Diego LF and Garcна-Labiano F 2009 *Fuel* **88** 1016–23

[9] Horiguchia Ju, Kobayashia Ya, Kobayashia S, Yamazakia Yu, Omataa K, Nagaob D, Konnob M and Yamadac M 2011 *Appl. Catal. A: Gen.* **392** 86–92

[10] Huang X, Sun N, Xue G, Wang Ch, Zhan H, Zhao N, Xiao F, Wei W and Sunc Yu 2015 *RSC Adv.* **5** 21090–98

[11] Lazić MM, Hadnađev MS, Bošković GC, Obadović DŽ and Kiss EE *Science of Sintering* **40** 175–84

[12] Kuksina O Yu, Kondrashin V Yu and Marshakov IK 2004 *Protection of Metals* **40** 581–86

Development of engineering solutions for air drilling at Dulisminsk oilfield, Irkutsk oblast

Yu L Boyarko[1], O S Ulyanova[2] and K V Buzanov[1,3]

[1] Department of Well Drilling, Institute of Natural Resources, National Research Tomsk Polytechnic University, 30 Lenin Ave., Tomsk, 634050, Russia
[2] Department of Foreign Languages for Specialists in Natural Resources, National Research Tomsk Polytechnic University, 30 Lenin Ave., Tomsk, 634050, Russia

e-mail: [3] byzanovaltay@mail.ru

Abstract. The article presents the analysis of disastrous mud loss achieved during casing at Dulisminsk oil and gas condensate field (DOGCF) and comparative evaluation of geological structure of DOGCF and Narauksko-Ostashkinsk area of Kemerovo oblast, where the same problem was solved successfully. On the basis of the analysis, an engineering solution is proposed to prevent mud loss in the conditions of DOGCF.

1. Introduction

In recent years, oil and gas sector of the Eastern Siberia has been significantly developed in Russian industry, due to large hydrocarbon resources accumulation. Recoverable oil reserves at Verkhnechonsk field are estimated in 196 million tons, at Tolokansk and Yurubchen-Tokhomsk fields – in more than 170 million tons; at Dulisminsk oil and gas condensate field (DOGCF) – in 81.5 million tons of oil and condensate, and in 80 million m^3 of gas [1]. However, all hydrocarbon reserves in East Siberia are under difficult drilling conditions due to the complex subsurface geological structure of the region.

2. Material and Methods

The paper describes the experience of drilling of the intervals for casing ("advance" drilling) with the application of air drilling and downhole hammer on the example of Dulisminsk oil and gas condensate field of Irkutsk oblast as a possible solution of a mud loss problem. The practice of drilling has shown that drilling of the intervals for casing (0-300 m) at Dulisminsk field of Irkutsk oblast is systematically performed under incompatible drilling conditions, which result from exogenic rock fracturing in the interval of Litventsevsk and Verkholensk suites [2].

The drilled rocks of Verkholensk suite consist of interbedded fractured siltstone of IV category (according to drillability), the 7th abrasiveness category with the density of 2650 kg/m^3; as well as claystone and highly fractured dolomites of III, V category (according to drillability). Also there are marls of V category (according to drillability), the 4th abrasiveness category and the density of 2670 kg/m^3. The density of the above mentioned rocks varies from 20 to 142 kgs/mm^2 according to L.A. Shreyner's method.

IOP Conf. Series: Earth and Environmental Science **27** (2015) 012049 doi:10.1088/1755-1315/27/1/012049

The rocks of Litventsevsk suite consist of fractured limestones, dolomites and gypsum. The mineral density of these rocks of the suite is 2200 – 2850 kg/m^3; hardness varies from 25 to 571 kgs/mm^2. The average abrasiveness of the rocks of this suite is 3.5.

Due to the specific geological structure and, in particular, the interbedded rocks with different permeability and mechanical properties, the drilling in the interval of 0–300 meters for casing is performed with incredible losses of drilling mud at an intensity of 30 cubic meters per hour, which inevitably results in enormous additional economic costs. For example, in well № 1106 incredible losses of drilling mud at an intensity of 20 cubic meters per hour started at the depth of 50 meters, in well № 713 from the depth of 110 meters up to the depth of 300 meters drilling was performed without drilling mud circulation, in well № 202 during the air drilling in the interval of 125 to 282 meters incredible losses of drilling mud at various intensity took place, and as a result, the total mud loss per the interval equal 580 m^3. It is practically impossible to prevent mud loss at such a depth by mud lightening. Clogging material pumping into fractured formations also failed. As a result, the time of section installation for casing of Dulisminsk field is up to 20 days.

The important factor in solution of mud losses problem at Dulisminsk field is considered to be lack of real operational situation in the drilling design. For example, according to pressure scheme, mud loss in the interval of 0–300 meters must occur in case of exceeding pressure gradient of $1{,}34 \cdot 10^{-2}$ MPa/m, however, mud loss of various intensity occurs even when technical water is used to sweep well. Real pressure gradient in this case is much lower than the indicator in pressure scheme. Thus, the information given in the design does not find its practical application in drilling, and in this situation it is necessary to use other sources of data.

3. Results and discussion

To solve the given problem, the analysis of special technical and geological information has been performed. [3–7]. In particular, it has been established that the drilling practice and the results of air drilling with blowdown in Naryksko-Ostashkinsk area of Kemerovo oblast is very important [5]. The drilling in this area was performed to produce gas from coal formations.

The interval in the air drilling technique was made in the following rocks: clay, touchstone, sandstone, siltstone, mudstone and coal. The density of rocks in this interval ranges from 1300 kg/m^3 to 2730 kg/m^3. Maximum compression breaking strength of the rocks in this interval ranges from 60 kgs/cm^2 to 620 kgs/cm^2. Average abrasiveness of the rocks is 1.5. According to drillability, the rocks belong to the following categories; clays – III, touchstones – IX, sandstones – V, siltstones, mudstones and coal – IV. The elastic modulus of the rocks of Litventsevsk suite ranges from 400 kgs/mm^2 to 3200 kgs/mm^2.

Figure 1 shows the pressure scheme of formation and fracture pressures for wells of Naryksko-Ostashkinsk area and well design in compliance with formation pressure and fracture pressure data.

Based on systematic mud loss indicators, as well as the rock structure in Naryksko-Ostashkinsk area the technical specialists suggested using air drilling and blowdown in the interval from 0 to 150 meters, justifying this choice in terms of accidents-free well drilling [2].

In the process of drilling, BHA with downhole percussive drills was used in mud loss interval of 0-150 m. BHA for rotary-percussive drilling is given in table 1.

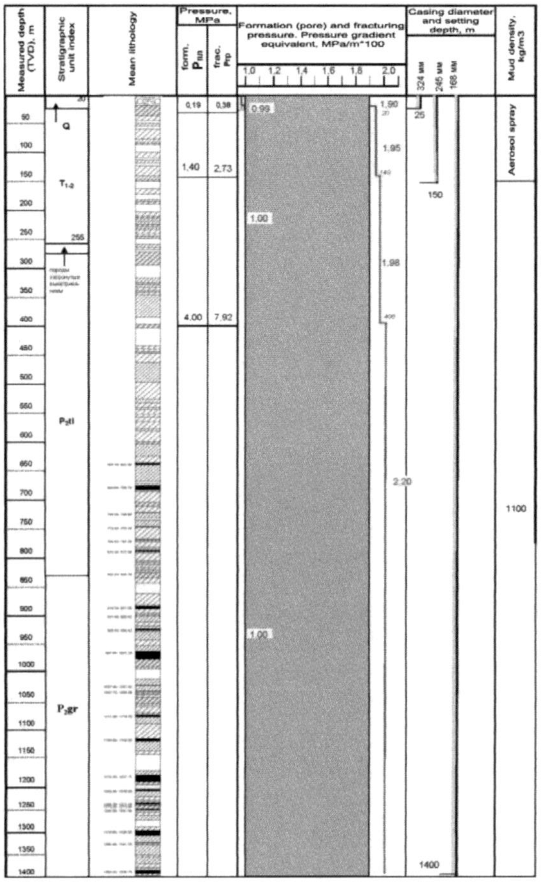

Figure 1. The pressure scheme of formation and fracture pressures for wells of Narykско-Ostashkinsk area.

Table 1. BHA, used for rotary-percussive drilling in Naryksko-Ostashkinsk area.

Series №	BHA components	External diameter, mm	Length (height), m	Mass, kg	Function
	DTH Drill Bit 311 mm	311	0.33	105.3	Bit
1	Downhole percussive drill DTH HAMMER TD 90	196.8	1.6	238.6	Percussive drill
2	Stabilizer sub	139.7	0.38	31	Sub
3	Stabilizer	301.6	3.66	885	Stabilizer
4	HWDP sub	196.8-139.7	0.45	72	Sub
5	HWDP 139.7	139.7	9.1	893	HWDP
6	Sub	196.8-139.7	0.45	72	Sub
7	Stabilizer	301.6	3.66	885	Stabilizer
8	HWDP sub	196.8-139.7	0.45	72	Sub
9	HWDP 139.7	139.7	9.1	893	HWDP
10	Sub	196.8-139.7	0.45	72	Sub
11	Stabilizer	301.6	3.66	885	Stabilizer
12	HWDP sub	196.8-139.7	0.45	72	Sub
13	HWDP 139.7	139.7	rest	11500	HWDP

Downhole percussive drill DTH HAMMER TD 90 shown in figure 2 is used for drilling interval of 0-150 m. in Narykskо-Ostashkinsk area according to its possibility to destroy any rocks. Depending on downhole percussive drill features, it can be used for water supply well drilling, blast holes, open-pit mining, and oil-field exploration, as well as for construction operations in which a great number of rocks need to be removed. The features of downhole percussive drill are the following: diameter is 196.8 mm, length equals 1.6 m; a bottom valve pipe offset, which determines its important characteristics, is 52.3 mm.

The main feature of the downhole percussive drill DTH HAMMER TD 90 by Atlas Copco is controllable supply of compressed air in the pressure vessels. This advantage is achieved due to the presence of a poppet valve in the design of this percussive drill. The control over the position of air channels enables the supply of pressure to the rotochamber when the piston is in its top of stroke, and cut-off of pressure supply, when the piston is near the hit point, thus ensuring maximum efficiency factor [6].

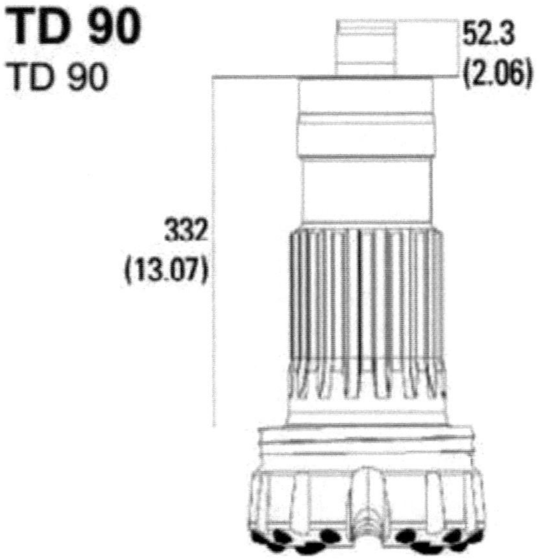

Figure 2. Downhole percussive drill DTH HAMMER TD 90.

In this case a universal flat-shaped drilling bit having a high-capacity gage side with large spherical bit inserts was used. This design of the bore provides the highest efficiency in drilling through hard and abrasive rocks, present in the cross-section of Narykskо-Ostashkinsk area.

According to the operational data, air generation and pumping into wells were done with the following compressors:

- Ingersoll Rand deck-based compressor with an output of 35.4 m^3/min and the operating pressure of 2.5 MPa;
- XRVS606 compressor with the capacity of 36 m^3/min and the operating pressure of 2.5 MPa;
- XRVS336 compressor with the capacity of 19.8 m^3/min and the operating pressure of 2.5 MPa.

It was proved that while drilling, in order to provide a good quality clearing of the wellbore, speed of air-water mixture flow rising should be 15 – 30 m/s. Based on these indicators; air supply rate from compressors was calculated. Maximum output of the three compressors is 91 m^3/min. The speed of flow rising with such an output, taking into account possible water inflow (100 m^3/day) is 19 m/s. In case of minimum water inflow, the speed increases to 23 m/s.

It was also proved by field tests, while the air drilling with a blowdown, the following three types of pressure loss are taken into account [5]:

- the fluid column pressure – 1 MPa,

- the bit pressure – 0.4 MPa,
- In the annulus and feeding lines pressure – 0.4 MPa.

The total pressure loss in air circulation drilling is 1.8 MPa.

Special drilling parameters were designed for the interval: Bit load – not more 20 kN, bit rotating speed – 1 rotation per 0.1 meters of depth, maximum supply of compressed air – 91 m^3/min. Surface casing drilling in Narycksko-Ostashkinsk area with these parameters allows achieving accident-free drilling, and mechanical drilling speed up to 18 m/hour.

Finally, the time for air drilling for wells № 13, № 15, № 17, № 19, № 25, № 27 and № 29 was on average 27 hours.

This result indicates the efficiency of fractured rocks destruction, including high drillability category rocks, with air drilling and blowdown.

The mentioned above facts show the possibility of using this technique at Dulisminsk oil field with mobile drilling rigs by Sramm. We suggest using mobile drilling rig and then the standard drilling equipment. This method was informally called “advance” or “keep ahead drilling”.

The novelty and practical effect of the proposed solution is not only in the justification of the use of downhole percussive drill and air for drilling of the intervals for casing at Dulisminsk field, but also in the development of technique, as well as optimal drilling modes, and also in the practical implementation of the “keep ahead” drilling.

As any other engineering solution, the implementation of blowdown requires a detailed analysis and research. An important component of the research is the results evaluation and the experience analysis of air drilling and blowdown. All these are planned to solve mud loss problems at Dulisminsk field in the nearest future.

To show other aspects of air drilling at Dulisminsk field, it should be noted that there is a disadvantage of actual G&G data for drilling wells. Thus, for example, due to various reasons, there are no data on the geophysical survey of Verkholensk and Litventsevsk suites. It prevents to identify the reasons for mud loss in the interval of 0–300 m for casing and make the development of drilling technique involving downhole percussive drill difficult.

Nevertheless, it can be assumed that the analysis of rock, their genesis, and sedimentation conditions will allow identifying the possibilities for advance drilling.

4. Conclusion

The mentioned analytical data of the tectonic framework in the region, lithological rocks characteristics in the well cross-section, as well as rock composition, the process of geological area formation, and the similarity of geological conditions at Dulisminske and Narycksko-Ostashkinsk fields in terms of rock fracturing, fossilization, and the successful experience of blowdown in Narycksko-Ostashkinsk area are considered as positive aspect for advance drilling at Dulisminsk field.

References

[1] Kudriashov S 2014 ROGTEC looks at Rosneft′s strategic development plans for eastern Siberia and the Far East *ROGTEC Oil and Gas Magazine* **7** 12–25
[2] Kudryashov B B and Kirsanov A I 1990 The drilling of exploration wells using air (Moscow: Nedra) p 263
[3] Pratt C A 1989 Modifications to and experience with air-percussion drilling *SPE Drilling Engineering* **4** 315–20
[4] Whiteley M C and William P 1985 Air drilling operations improved by percussion-bit/hammer-tool tandem *SPE Drilling Engineering* **1**377–82
[5] Drilling program for well №19 in Narycksko-Ostashkinsk area 2011 (Kemerovo) 12-6
[6] Lind Yu B, Mulukov R A, Minikeeva L R and Zairullina E I 2012 Software on automation for design of oil and gas wells construction *Oil industry* **4** 43–54
[7] Brilin V I and Ulyanova O S 2015 Well extension in fractured and broken rocks by wire-line tool (SSK-59) *IOP Conf. Ser.: Earth Environ. Sci.* **24** 012006

The diagram development for Computer Added Control and Monitoring system of drilling

A V Epikhin[1,3], R S Mikhalev[1], A V Anisimov[1] and O S Ulyanova[2]

[1] Department of Well Drilling, National Research Tomsk Polytechnic University, 30 Lenin Ave., Tomsk, 634050, Russia
[2] Department of Foreign Languages for Specialists in Natural Resources, National Research Tomsk Polytechnic University, 30 Lenin Ave., Tomsk, 634050, Russia

E-mail: [3] epikhinav@mail.ru

Abstract. The paper is concerned with the first stage of the extensive research aimed at developing design-automation system and well drilling process control. The proposed system is going to have some advantages over modern analogues, such as economic analysis at all levels, active engineering staff feedback, precedent-related principle for recommendations, etc. It will essentially reduce the risk of human errors and also optimize the well construction process from design to commissioning. The paper considers the results of the first design stage in a form of flow diagrams.

1. Introduction

Due to growing drilling volumes and development of new fields with hard-to-recover reserves, drilling reliable and less-consuming wells in shorten time are considered to be an urgent problem. However, many modern wells constructed with technological failures and characterized by reduced rate, require the use of expensive inflow intensification methods. The problem of reducing the time costs on service and insulation operations, well fishing services and unscheduled repair of drilling equipment is of particular importance. The well design stage errors or non-compliance with rules are the reasons for these problems [1].

The well design stage errors can be caused by various reasons. For example, they could be made by a design engineer while designing the draft template because of a lot of design documents (5 – 12 volumes) to be developed. Although the state expert review increases well design quality, it does not take into account technical and technological errors, since it only assesses the design compliance with the legal requirements. If such errors are made during the well drilling process, it can cause a severe accident.

As mentioned above, well design is made according to a draft template of existing wells for similar drilling conditions. This approach demonstrates its value, but it does not consider the field feedback (how "well-template" was drilled). Therefore, the design drilling well process is still not optimized.

These problems can be partly solved by using modeling methods and automatic control drilling system. The first attempts to develop such models were made in the 1960s, and the first successful results of their implementation were received in the 1970-1980's. Nowadays, the demand for these automatic control systems is constantly increasing which is expressed in great number of researches

[2-7] and continuous introduction and improving applied developments by most international companies. The successful experience of automatic drilling system is available in different areas all over the world, including the Russian Federation (Western Siberia, Povolzhye), Kazakhstan; in offshore drilling. The latter is of particular demand due to high cost of well construction and high level of responsibility of offshore drilling. Drilling design process can be performed theoretical research, laboratory experiments and investigation of field data, as well as combination of theory and field data with further forecasting the model operation, etc. [2-7]

Presently, positive results of automatic system drilling have been achieved by foreign companies specialized in high-technology well construction, such as Sclumberger, Backer Hughes, Halliburton, Shell, etc. In Russia, this problem is being tackled by a number of service companies monitoring drilling, including telemetry. According to the field investigations results, the possibility to decrease the time of a standard well construction up to 15% and its cost estimate up to 5 – 10% has been proved.

Automatic well design systems, which are widely spread among design organizations, are also very important. This significance is confirmed by the fact that a well drilling project is the main document for the drilling crew and engineering staff. A wide range of well design systems with different levels of automation and application have been developed in national practice. Among them there are well design systems for cluster sites [8-10], including complicated geological factors and fields with reservoir management difficulties [11-13], automatic mining design systems (open pit and pit) [14-15], etc.

The given systems have a common disadvantage: that is the design of a separate stage of well construction, for example, well profile calculation and evaluation, or mine opening order, etc. Whereas oil industry requires the system which will measure the cycle of well construction and give valid engineering project out due to all the requirements and normative constraints, responding the given initial data by the operator.

On the other hand, drilling monitoring systems also have a serious disadvantage – the inactivity. In some cases, the systems only register the current process without analysis and management solutions. Therefore, the development of complex system for automatic or semiautomatic design, control and monitoring of drilling is of particular importance.

2. Development the diagram of well automatic design

Critical analysis of existing systems of automatic well design has been carried out. The list of requirements to be met by "the ideal" system has been formed:

- to provide an active feedback with an operator (the system must be in a constant connection with the operator (engineer); it is he who approves the final result and is responsible for the design quality;
- to use modern normative documents for system measurements, and to monitor constantly the legal and normative documents for updating the system data;
- to ensure active system memory: any completed project is saved in the system database and, in case of the next project, the database monitoring system proposes ready-made solutions for specific situations and conditions. Besides, the information about main indicators of the wells drilled efficiency with the option to evaluate the efficiency of adopted project solutions is added to the database;
- module-base organization – the option to do separate measurement;
- to make the module of cost-effectiveness analysis, where two parallel operations are carried out: the measure of specification time costs and well drilling construction cost, as well as the reference to database and making solutions on reducing costs and thereby increasing economic effect.

On the basis of the given list of requirements the diagram of automatic well design was made (figure 1). The system consists of 6 units and 2 databases which ensures the precedent-related principle of its operation. The largest component of the system is an engineering measurement unit, so it was divided into modules for the convenient development and subsequent use.

For convenience, schemes, units and functional relations are shown in different colors. Design units are colored in green, purple means the output unit, while black – the input unit, facilities unit is colored in pink. Functional relations are indicated by solid lines, the color of which is determined by the relation source. For example, the information can be given to other units only from the input unit, while the output unit can both receive the information (green lines) and give the information out (purple lines). A model development of the given system with further integration and debugging by experimental measurements for the existing well is scheduled.

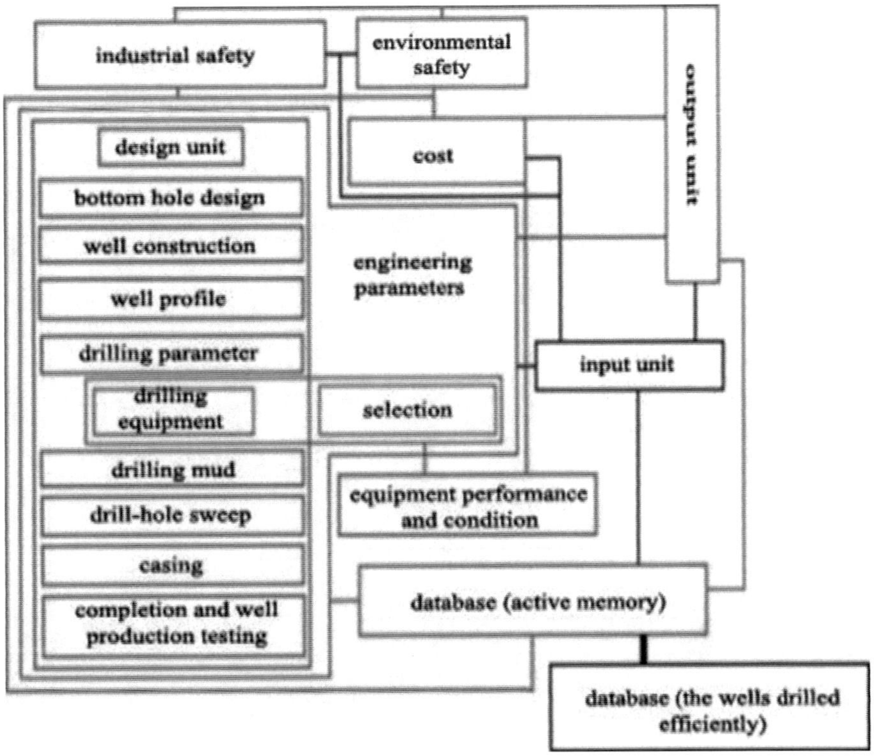

Figure 1. Automatic well design diagram.

3. Development of Computer Added control and monitoring system of drilling

The selection of national control and monitoring systems was made; KUB-2, Leuza-2, APK "Volga", KIBR – M1, SKUB-M1, MCU were included in this selection. It was analyzed according to GOST 14169-93 requirements "The systems of ground-control monitoring of drilling process of oil and gas drill-holes. General technical requirements and test methods" determining the minimum list of the parameters for the systems [17]. According to the analysis, GOST 14169-93 is proved to have lost its importance, because none of the systems met the standards, but they were actively used in drilling practice.

Therefore, the importance of modern Computer Added Control and Monitoring system of drilling is obvious. The diagram of the system has been decided to develop for modern drilling conditions satisfying the following requirements:

- database with option of "self-learning",
- precedent-related principle of operation,
- preparation of monitoring solutions for an operator in accordance with design and real drilling parameters;

- data recording system on media and server in Internet;
- monitoring technological parameters of drilling;
- drilling cost estimation;
- considering the feedback from drilling crew and engineers;
- control of drilling crew efficiency
- video surveillance of main components of drilling rig.

Therewith, the system itself should meet the standards: reliability, minimality (i.e. minimum of additional equipment and interference into a technological process), usability, quickness of analysis and solutions.

To identify the research problems correctly the diagram of a standard system of Computer Added Control and Monitoring system of drilling has been developed (figure 2). The prototype is made on a basis of the analytical comparison of national systems according to GOST 14169-93.

The diagram of the system includes mostly one-way communication. Measurement instruments fix technological parameters and transmit them to database which can either be stored on the media or used by an operator. In some cases, for example, in APK "Volga" diagram is complicated due to the analysis of obtained information by the system and suggestion of solution and recommendations (in figure 2 such version is shown by dashed lines). The given diagram emphasizes lack of using the information obtained in drilling, experience of a drilling crew and engineers.

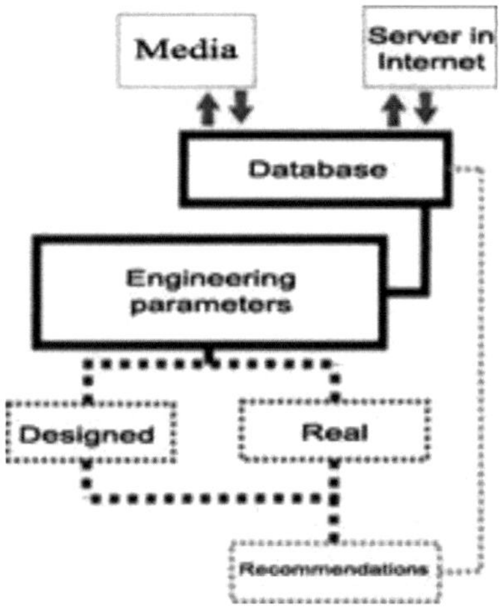

Figure 2. The diagram of a standard system of Computer Added Control and Monitoring system of drilling.

The alternative diagram of Computer Added Control and Monitoring system was further developed (figure 3). This system includes 4 functional units, such as: unit of engineering parameters registration, unit of technical conditions of drilling equipment evaluation, unit of feedback collection and estimation of drilling crew efficiency, unit of drilling cost assessment.

The unit of engineering parameters registration like its modern analogues gathers the information from measurement instruments in drilling operation. Then the data are processed and compared with those of the design data. After their comparison, taking into account the precedent-related operation principle of existing database (about already drilled wells), the system gives some option of monitoring solutions to an operator who, in turn, gives them to the engineers and drilling crew.

The unit of equipment performance and conditions gathers information for the analysis from the following sources: changes of technological parameters, with the other conditions being equal, operational video surveillance of main drilling rig components, the drilling crew feedback. The obtained information is analyzed according to the wear standards of drilling equipment, which are given as initial data. On the basis of the previous drilling experience the system also develops its recommendation which an operator transmits to the drilling crew and engineers in the form of possible solutions for the final approval.

The unit of collection feedback and estimation of drilling crew efficiency implies performing two tasks. The drilling crew can give useful information about the drilling equipment performance and drilling conditions to the system and an operator. The system operator, in turn, registers the information and interprets correctly.

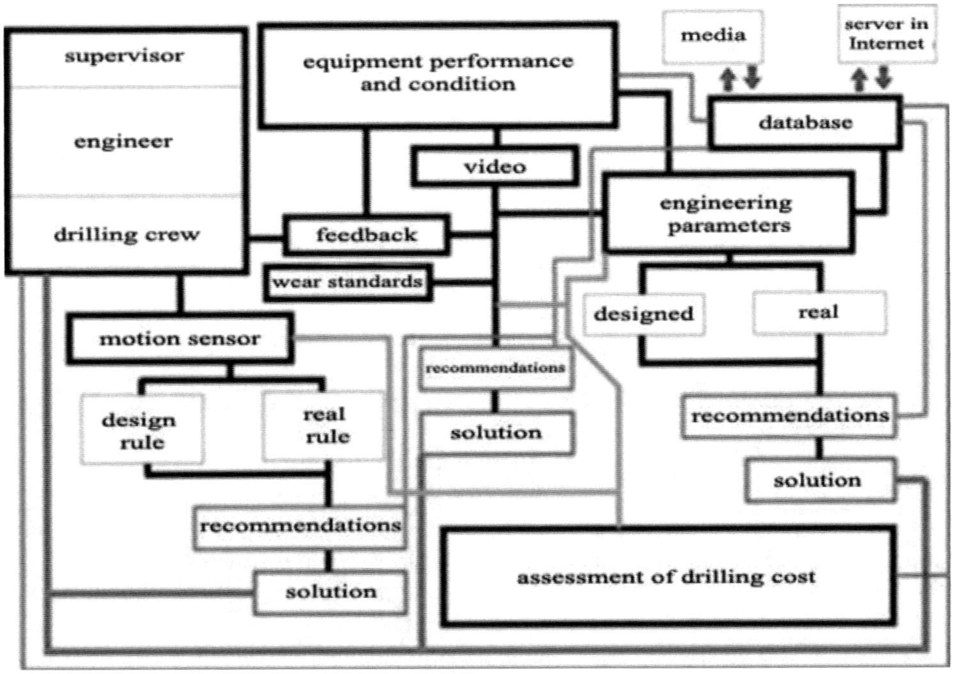

Figure 3. The developed diagram of Computer Added Control and Monitoring system of drilling.

Estimation of the drilling crew efficiency can be organized with the help of a motion sensor, for example, in each helmet. It will allow the system to analyze the efficiency of time management and improve working conditions. The unit of drilling cost assessment is to generalize and analyze the data from described above components of the system. The result of its operation is a real-time dynamic pattern of efficiency. The possibility of administrative decisions in the course of the given well project allows evaluating their influence on the well construction cost, and adjusting the drilling crew work, as well as control well project of the drilling process.

All the information is recorded in the database that operates according to the precedent-related principle and develops the recommendations for solutions. The data are stored on the server in the Internet with the possibility of receiving the report of the customers and additionally saved on the media within the range of the drilling site.

4. Conclusion

It should be noted that the suggested approach to well design is an alternative one. It allows optimizing the work of engineers of the design organization and decreasing the number of human errors. To be sure, it does not reduce the importance of engineer in design and drilling a well, but the system is going to improve the quality and the quantity of operations. The peculiarity of the system is that there is an active memory which allows the engineers to develop ideal technical solutions for the chosen areas and formulate the concept of "the model" for a well.

The suggested model of Computer Added Control and Monitoring system for drilling can increase the speed and quality of the well construction due to the recommendation system and application of successful experience of the drilled well constructions. This system is complicated at the design stage

IOP Conf. Series: Earth and Environmental Science **27** (2015) 012050 doi:10.1088/1755-1315/27/1/012050

and building the relations, revealing the trends in drilling. The next stage of the research is to specify the operating procedure for each unit and describe the character and the level of relations.

References

[1] Gutorov Yu A and Nikiforov A A 2012 Problems of well construction quality management based on performance assessment of wells as oil production objects *Oil and Gas Business* **5** 141–52

[2] Sitnikov N B 2000 Modelirovanie i optimizaciya processa bureniya geologorazvedochnyh skvazhin: avtoref. dis. ... doktora tehnicheskih nauk (Ekaterinburg) p 43

[3] Litvinov M A 2005 Sistema kompleksnogo modelirovaniya processov pri burenii neftyanyh i gazovyh skvazhin na osnove nechetkih mnozhestv: dis. … kand. tehn. nauk (Orenburg) p 156

[4] Cuprikov L A 2008 Razrabotka sistemy adaptivnogo upravleniya processom rotornogo bureniya neftyanyh i gazovyh skvazhin: dis. … kand. tehn. nauk (Krasnodar) p 193

[5] Primera A, Perez-Damas C, Kumar S and Rodriguez J E 2006 Simulation while drilling: Utopia or reality? *SPE Intelligent Energy Conference and Exhibition (Amsterdam, 6 April 2006)* **2** 460–70

[6] Zhou J and Nygaard G 2011 Automatic model-based control scheme for stabilizing pressure during dual-gradient drilling *Journal of Process Control* **21** 1138–47

[7] Breyholtz O and Nikolaou M 2012 Drilling automation: Presenting a framework for automated operations *SPE Drilling and Completion* **27** 118–26

[8] Bastrikov S N 2006 Proektirovanie i stroitelstvo skvazhin s kustovyh ploshadok na neftaynyh mestorozhdeniyah Zapadnoj Sibiri: avtoref. dis. … doktora. tehn. nauk (Tumen) p50

[9] Itkin V O 2004 Matematicheskie modeli prostranstvennyh traektorij pri proektirovanii kustovyh skvazhin: dis. … kand. tehn. nauk (Moscow) p 148

[10] Breyholtz Q and Nygaard G. 2009 Deep water drilling: Full pressure profile control in open hole section using model predictive control *Society of Petroleum Engineers – International Petroleum Technology Conference (Doha, 7-9 December 2009)* **1** 600–08

[11] Breyholtz Q, Nygaard G and Nikolaou M 2010 Automatic control of managed pressure drilling *Proceedings of the 2010 American Control Conference (Baltimore, 30 June – 2 July 2010)* Article number 5531008 442–7

[12] Harlamov A K 2007 Sovershenstvovanie metodiki proektirovaniya kustov i profilej skvazhin na mestorozhdeniyah so slozhnymi shemami razrabotki: dis. … doktora. tehn. nauk (Tumen) p 138

[13] Oppelt J and Reinicke K M 2015 Modeling, simulation and automation to optimize wellbore delivery *Oil Gas European Magazine* **41** 10–1

[14] Müller J, Jung S, Kruspe T, Herbel A and Jörns S H 2015 Measurement while drilling for directional drilling in deep geothermal reservoirs *Oil Gas European Magazine* **41** 34–5

[15] Korol G G and Lomako L S 2012 MINEGRAME System 5.0 *Credo-Dialogue* **1** 40-3

[16] Hauge E, Aamo O M, Godhavn J-M and Nygaard G 2013 A novel model-based scheme for kick and loss mitigation during drilling *Journal of Process Control* **23** 463–72

[17] GOST 14169-93 requirements "Systems ground-control monitoring of boring process oil and gas drill-holes. General technical requirements and test methods" *Support for GOST* URL: http://www.gosthelp.ru/text/GOST1416993Sistemynazemno.html

IOP Conf. Series: Earth and Environmental Science **27** (2015) 012051 doi:10.1088/1755-1315/27/1/012051

Experimental research of drilling mud influence on mud motor mechanical rubber components

A V Epikhin[1,3], A V Ushakov[1], V V Barztaikin[1], V V Melnikov[1] and O S Ulyanova[2]

[1] Department of Well Drilling, National Research Tomsk Polytechnic University, 30 Lenin Ave., Tomsk, 634050, Russia
[2] Department of Foreign Languages for Specialists in Natural Resources, National Research Tomsk Polytechnic University, 30 Lenin Ave., Tomsk, 634050, Russia

E-mail: [3] epikhinav@mail.ru

Abstract. The paper describes the experimental research of drilling mud influence on engineering parameters of mud motor mechanical rubber components. It is believed to be urgent due to increase in using mud motors in oil and gas well construction now, and, consequently, the issue of increasing their exploitation is becoming current. The development test results of elastomer IRP-1226 dependent on the mud type (alkaline, hydrocarbon or salt-saturated ones) and the temperature are shown in the paper. It is proved that the type of drilling mud and the temperature in bottom-hole zone have an influence on wear of mud motors elastomers.

1. Introduction

In modern mechanical engineering the quality of products, especially their durability, is one of the major parameters. It is important for mud motors used for drilling of oil and gas wells [1]. The main difference of a mud motor from other downhole drilling motors is using a multithread screw rotor gear as an operative part which presents internal cylindrical epicyclic gear including stator and rotor with teeth number equal 1 [1].

However, now well drilling technology is characterized by the problems connected with unstable mud motors operation, their failures, low durability of operative components, and accidents, such as backed off and drilling string failures. According to the data of drilling companies' reports, 5-12 accident happened per year caused by mud motors failure that lead to long workover or abandonment of wells. The using of mud motors shows that reasons for approximately 50% failures involves wear of operative components because of rubber-metal friction (steel rotor – rubber-metal stator). Engine operational life can vary from 90 to 235 hours depending on dimension type and operating conditions [2].

The issue of elastomer wear and destruction has been urgent by now. For example, D.F. Baldenko, F.D. Baldenko, M.V. Dvoinikov and V.P. Ovchinnikov have confirmed that wear of operative component per 100 hours is equal to 45%, and 33% among them (more intensive wear of stator rubber elastomer) is for the first 60 – 80 hours of engine operation in a well. The causes lie in strength capacity of interacted surfaces of operative components (rubber – steel), increased initial tension and growth of hydraulic and mechanical resistance in operative components in the process of running-in under radial forces [3–7].

In most cases, to evaluate the high wear problem of operative components (rotor-stator), researchers pay attention to mud motors constructional irregularity which is clearly revealed in operation under extreme conditions of cyclic load, high bottom hole pressures and temperatures, as well as aggressive drilling mud. So, the main methods to increase durability of mud motors components are thought to be connected with improvement of engine design and its kinematic scheme by size and component exchange, as well as rigidity and stator armoring [6 – 12]. It allows increasing engine operational life significantly, for example, operating time of some modernized models in development testing is up to 585 hours [10-12], and also, decreasing negative thermal and mechanical influence on elastomers. It cannot be denied that elastomer durability is significantly impacted by drilling mud in which an engine works. Therefore, one of the urgent issues to increase mud motor operation life is to improve drilling mud composition or special additives development.

2. Methods and materials

The influence of different mud types on rubber samples IRP-1226, which is oil-resistant rubber with improved wear-resistance and range of temperature from - 20 to +100^0 C. has been studied. Its nominal strength is equal to 9.8 MPa, failure elongation is 125% and Shore hardness number is 65 – 95.

At the first stage of the experiments the influence of mud type (saturated salt solution, alkaline solution (pH=14), oil, diesel fuel) on sample size in long interaction was analyzed. The decision of mud type for analysis is determined by the necessity to reveal aggressiveness of drilling mud to rubber IRP-1226. It will allow the researchers to evaluate additives efficiency for drilling mud. The evaluation was made according to one parameter: geometric sample size, because in research works elastomers instability in regard to aggressive ambient as swelling or decrease of a size [11-12] is often mentioned.

At the second stage of the experiments the influence of temperature on sample parameters in different types of drilling mud was studied. According to the given data, the most intensive wear of mud motor occurred using hydrocarbon drilling mud. Therefore, the following types of drilling mud were taken for the experiment: oil, diesel fuel, biodegradable drilling mud. The evaluation results were made according to geometric sample size too. While processing the experimental results, the change of sample relative size per time in mud was analyzed.

The samples were put into plastic containers during the experiment and completely plunged into mud under atmospheric pressure. To verify the experimental results, 3-5 samples were analyzed for each type of drilling mud. At the first stage of the experiment, one batch of samples was studied under normal temperature, while at the second stage two batches of samples were used: the first one was studied at room temperature, and the second one was analyzed under temperature 50^{0}C reached in cabinet dryer. The time of experiment was 15 days for each stage. This was because mud motors operation time equals to 200 – 600 hours with an average values of 240-300 hours [12, 14].

3. Results and Discussion

The results of the first stage of the experiments are shown in figure 1. The sample size was taken as a standard of comparison because it helps to evaluate the sample changes (swelling or decrease of a size). During the analysis it was revealed that all types of drilling mud had an influence on samples with the similar dynamics of size change: at the first stage (50-100 hours) it was noticed that the stable swelling (saturated salt solution) or decrease of a size (oil, diesel fuel, or alkaline solution). After it, the size dynamics changes oppositely and the next 100-150 hours of the experiment the sample size becomes stable. Besides, the final sample size change was fixed for all types of drilling mud reaching 0.5 – 1 cm^3 (increase of size in saturated salt solution, diesel fuel and or alkaline solution and decrease of size in oil). The largest change of the sample size between the initial and the final measurements was registered in alkaline solution. The experiment was carried out without additional conditions, such as mechanical load, high temperature and pressure. Therefore, it was concluded that in bottom hole the changes of parameters (shape, size) and elastomer wear would be faster.

The results of the second experimental stage are given in figure 2. In all cases, like in the first stage, the active changes of geometric sample size were observed. The following dependences were noted:

- at room temperature decrease of a size in diesel fuel and oil is more intensive than under temperature equal 50°C that is caused by high evaporation degree of their volatile compound;
- significant increasing sample size is registered for biodegradable drilling mud under higher temperature that is proved by the predictions and stipulated by its low volatility;
- the most intensive changes of geometric sample size occurs in the period of 0 – 50 hours;
- in the period of 50 – 300 hours the intensity of sample changes tends to subside, and the trend line tends to a horizontal line;
- in general terms, the obtained dependences for different temperatures show the similar dynamics;
- the geometric sample size change is insignificant that is determined by static nature of the experiment.

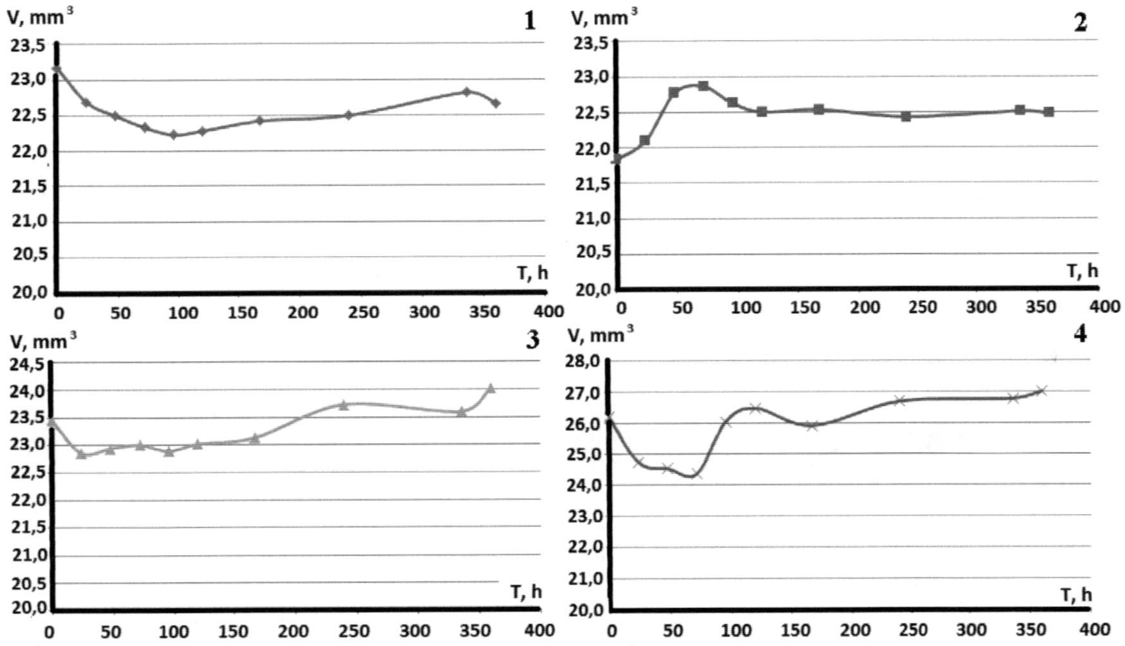

Figure 1. The dependence of size of sample IRP-1226 on the period of its putting in mud: 1-oil, 2 – saturated salt solution, 3 – diesel fuel, 4 - alkaline solution.

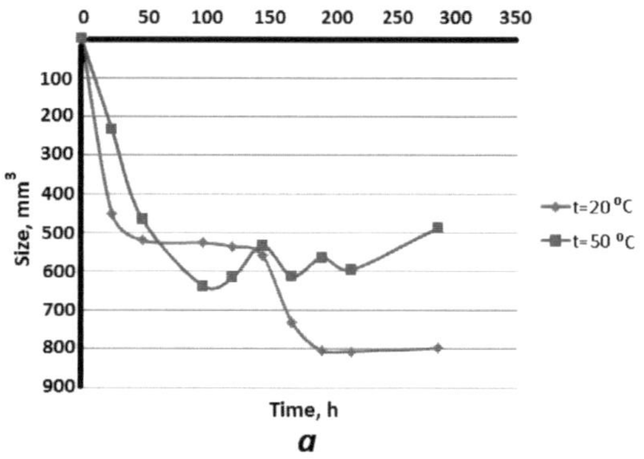

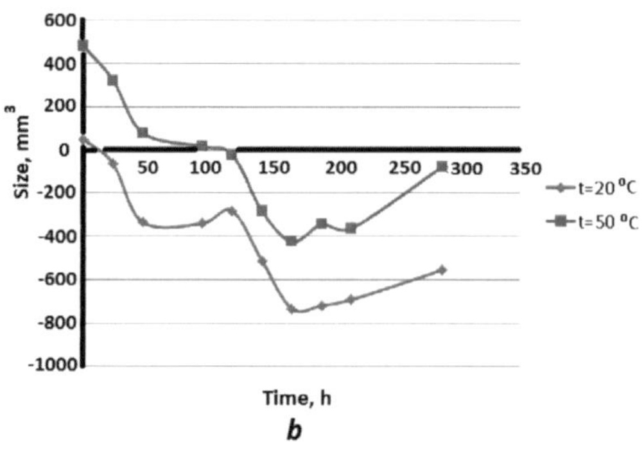

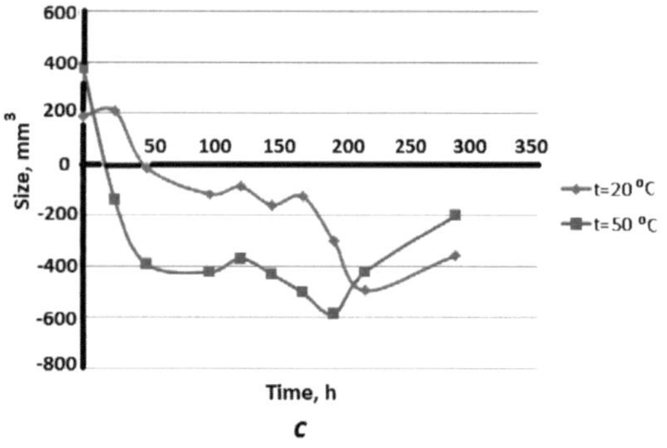

Figure 2. Relative change of sample size IRP-1226 in different types of drilling mud: a – oil, b – diesel fuel, c - biodegradable drilling mud.

During the experiment (to 50 hours) intensive splitting of samples was observed to be plunged into biodegradable drilling mud and diesel fuel at temperature 50°C (figure 3). As can be observed in the photos, the mud became dark and at the end of the experiment the detached pieces of samples could be found in it. It also confirms the high intensity of elastomer wear in hydrocarbon drilling mud even if the experiment is static.

a b

Figure 3. View of sample and types of drilling mud: a – diesel fuel, b – biodegradable drilling mud, 1 – t = 20° C; 2 – t = 50° C.

4. Conclusion

During the experiment the influence of different types of drilling mud on size of rubber samples IRP-1226 used as elastomer of mud motors has been analyzed. It has been proved that under long-term exposure, intensive change of rubber sample size takes place that is gradually stabilized for 200 – 250 hours. Total change of the sample size amounts 2 – 3% relative to the initial size.

Besides, the influence of temperature on sample size IRP-1226, which was put in hydrocarbon drilling mud has been analyzed. It has been noticed, that the most intensive wear occurs in such types of drilling mud as diesel fuel and biodegradable drilling mud that was revealed due to intensive splitting of samples. The dependence of the temperature on the degree of sample size change is more typical for biodegradable drilling mud, while it was not observed for oil and diesel fuel. As a conclusion, it can be said that the reason for this is high evaporation degree of hydrocarbon volatile compounds and decrease in mud aggressiveness to the samples. The following issues have been defined as possible research problems:

- to analyze the temperature influence from -30°C to +90°C;
- to evaluate the mass of a sample and its durability;
- to study abrasive wear of drilling mud to elastomer;
- to assess abrasive influence of mud motor on samples IRP-1226 in different types of mud and at different exposure.

References

[1] Korotaev Yu A 2003 *Issledovanie i razrabotka tehnologii izgotovleniya mnogozahodnyh vintovyh gerotornyh mehanizmov gidravlicheskih zabojnyh dvigatelej*: dissertaciya ... doktora tehnicheskih nauk (Perm) p 386

[2] Goldobin D A 2011 *Razrabotka i issledovanie vintovyh zabojnyh dvigatelej s oblegchennymi rotorami i armirovannymi statorami*: avtoref. dis. ... kandidata tehnicheskih nauk (Perm) p 22

[3] Anyanwu O N, Klotz C, Labrecque D and Ulrich C 2012 Optimized downhole mud motor delivers outstanding performance improvement in Alaska coiled tubing drilling *Society of Petroleum Engineers – Coiled Tubing and Well Intervention Conference and Exhibition* 88–100

[4] Ismakov R A and Al-Suhili M H 2015 Research of the effects of different agents on the work of power section downhole drilling motors *Oil and Gas Business* **1** 64–78

[5] Thompson J and Rossing M 2011 High-temperature downhole motor facilitates cleanout of obstructed thermal well, saving production downtime *SPE Automotive and Composites Divisions - Coiled Tubing and Well Intervention Conference and Exhibition* 455–61

[6] Baldenko D and Baldenko F Design solution to improve the working of downhole PDM *Journal "Burenie i neft"* URL: http://burneft.ru/archive/issues/2013-02/10

[7] Anyanwu O N, Klotz C and Ulrich C 2012 Rib-steered motor technology extends CT drilling applications *World Oil* **233** 135–9

[8] Dvoynikov M 2009 Research of wear resistance of engine bodies of the positive displacement motors *Journal "Burenie i neft"* **5** 15–9

[9] Ross M, Anyanwu O N, Klotz C and Ulrich C 2012 Rib-steered motor technology: The revolutionary approach extends the coiled tubing drilling application scope *Society of Petroleum Engineers – Coiled Tubing and Well Intervention Conference and Exhibition* 101–15

[10] Herbig C, Hohl A, Oueslati H, Reckmann H, Heuermann-Kühn L, Baumann J and Heinisch D 2014 Data-driven drilling optimization: Using historical data to improve future performance *SPE/IADC Drilling Conference, Proceedings (Madrid, 8-9 April 2014)* **1** 83–94

[11] Varela R, Guzman F, Cruz D, Atencio N, Iturrizaga F, May R, Solano R and Perez R 2015 First application of turbodrill and hybrid bit to optimize drilling times in cretaceous formations with high chert content in Mexico South Region *Oil Gas European Magazine* **41** 32–3

[12] Ignatov D N 2012 Reliability indicators of downhole mud motors *Molodyezh i nauka: Sbornik materialov VIII Vserossijskoj nauchno-tehnicheskoj konferencii studentov, aspirantov i molodyh uchyonyh* (Krasnoyarsk: Siberian Federal University) URL: http://conf.sfu-kras.ru/sites/mn2012/section18.html

[13] Fufachev O I 2011 *Issledovanie i razrabotka novyh konstrukcij rabochih organov vintovyh zabojnyh dvigatelej dlya povysheniya ih energeticheskih i ekspluatacionnyh harakteristik*: avtoref. dis. ... kandidata tehnicheskih nauk (Moscow) p 138

[14] Yadav A and Ali A 2014 Drilling technology taking drilling operations to new frontiers *Society of Petroleum Engineers - SPE Bergen One Day Seminar* 210–7

Efficiency increase of hard rock destruction with the use of eccentric pulses

P Pushmin[1,2] and G Romanov[1,3]

[1] Institute of Mineral Resources Management and Processing, Irkutsk national research technical university, Lermontov str., 83, Irkutsk, 664074, Russia

E-mail: [2] dog-eat-dog-79@mail.ru, [3] GRom123456@mail.ru

Abstract. The increase of efficiency of expendable well-drilling directly depends on the mechanism of rock destruction. The depth of hole coverage should be at the maximum possible value and equal along the whole contact area. The design features of performing rock-destruction tool in rotary drilling prevent the cutters, located on different distance from rotation axis, from working with equal intensity. The application of the shock pulses to a boring tool increases the mechanical drilling speed and sinking per revolution. One of the techniques for the efficiency increase is the application of impactor machines. On the other hand, the high values of mechanical drilling speed require the high values of impact energy which results in high power inputs energy consumption. The use of eccentric impact loading instead of central one can provide the enhancement of rock destruction efficiency, lower the power inputs, and decrease the risk of borehole deviation.

1. Introduction

One of the relevant issues of modern drilling engineering is the enhancement of rock destruction process. The basic measure in this area is mechanical drilling speed which depends on the combination of basic parameters – axial thrust, rotary speed of the tool, and expense of cleaning agent. The increase of the values of mentioned parameters does not always result in mechanical drilling speed increase as exceeding of these values can lead to different operating problems and breakdowns. The increase of the depth of bottom-hole coverage with the cutters of rock-destruction tool without exceeding of optimal parameters is a key to the enhancement of rock destruction efficiency.

2. Theoretical Background

The feature of embodiment of boring tool, based on rotation as a breaking force, consists in unequal performance level of cutters located on different concentric circles of the butt [6]. The linear speed of movement is higher than that of the cutters located further from the rotation axis. As a result, one can observe the decrease of sinking per revolution and mechanical drilling speed. It is caused by nonoptimal ratio of unit axial load and cutter load values.

This difference in performance is appreciable particularly in non-core drilling where the difference in linear speed (cutting rate *Vp*) of close and distant cutters can reach high values. Thus, the cutters which are close to rotation axis perform rock destruction with higher intensity (figure 1).

Another reason of efficiency decrease is overgrinding of drill cuttings which is carried out by peripheral cutters to a greater extent. It is due to the increase of hardening of isolated rock debris with its size reduction [1]. The multiple overgrinding also leads to the rise of power inputs and high wear rate of rock-destruction tool.

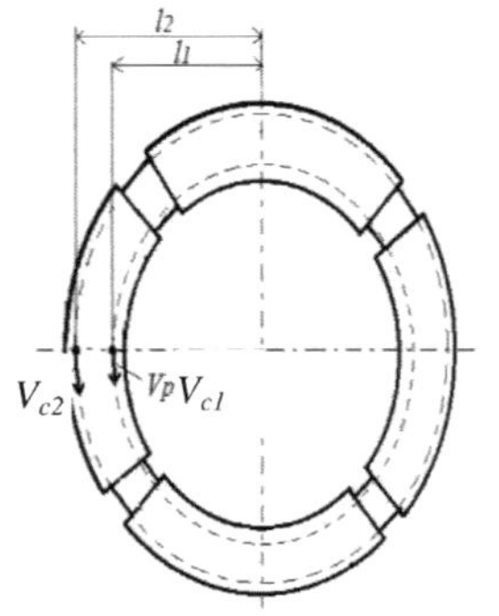

Figure 1. Scheme of diamond cutters performance analysis: V_{ci} – cutting rate; l_i – distance between rotation axis and cutter.

It allows making a conclusion that frequent contact of peripheral cutters with abrasive particles of drill cuttings creates conditions for their polishing. The farther is the location cutter from the rotation axis the higher is the value of its wear degree which occurs due to a greater amount of abrasive particles to contact (figure 2).

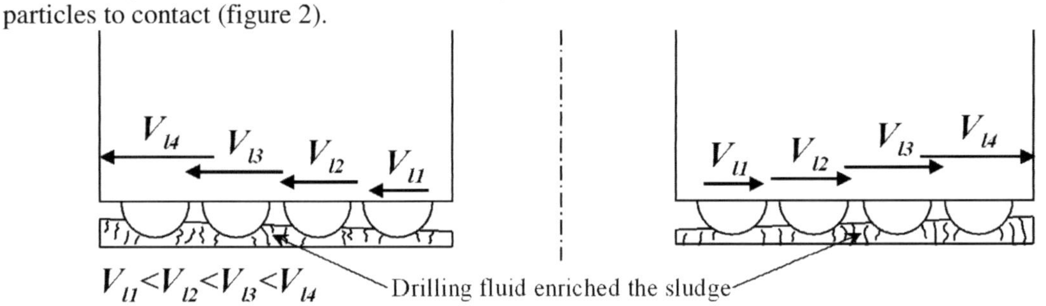

Figure 2. Scheme of unequal wear of rock-destruction tool cutters analysis: V_{li} – linear speed of movement of cutter *i*.

The quality of bottom-hole cleaning should be highly effective and, at the same time, the amount of drill cuttings contacting rock cutting elements should be minimal. It requires flow regime under the butt of boring tool to be turbulent and radially directed [5].

Considering foregoing, the search for the methods of intensification of rock destruction by the peripheral cutters is required. It should be noted that, according to the theory of rock failure, the ensuring of uniformity of cutters wear would enable the values of sinking per revolution and mechanical drilling speed to increase and therefore, to achieve the efficiency enhancement of hard rock destruction as a whole.

The use of impactor machines such as hydraulic hammers and pneumatic impact tools enables increasing sinking per revolution due to a greater depth of penetration of cutters to a rock and improvement of energy intensity of rock failure at borehole bottom as well as lowering of wastes of axial load communicated to a boring tool through the system of drill pipes. The considerable experience of Atlas Copco (Sweden) company in the development and operation of pneumatic impact tools showed that the rise of mechanical drilling speed virtually is directly proportional to pneumatic air increase. The mechanical drilling speed of pneumatic impact tools like Cop performing solid rock failure at high air pressure reaches 15-25 meters per hour. Headway per bit also increases due to the improvement of energy intensity of rock failure at borehole bottom.

Efficiency indexes of hydropercussion drilling rise with the increase of the specific energy of single shock pulse. According to the authors of the research [3], the mechanical drilling speed rose three and a half times when energy of the shock pulse increased from 75 to 200 J. The lifetime of rock-destruction tool also increased in 7.4 times.

The tendency of development of modern engineering in the area of exploration drilling for solid raw materials is oriented towards rotary-percussion drilling (particularly in our country) with the application of high pressure pneumatic impact tools [3]. This is one of the most productive methods of shaft sinking. Some of the countries implement the combined method. Before the ore zone the drilling with reverse circulation of compressed air and sludge sampling is applied while in ore zone the core drilling is applied. The quality of deposit sampling is maintained at a considerable reduction of time and resources for execution of boring.

At the same time, the rise of air pressure, required for increasing of impact energy, consumes greater amounts of compressor energy which is quite expensive. Besides, application of axial shock pulses does not solve the main challenge of rotary boring – unequal performance of cutters located on different concentric circles of rock- destruction tool butt as the percussion energy is dispensed throughout borehole bottom (figure 3, *a*). As a result, it is necessary to develop method for enhancement of failing percussion action efficiency which is not connected with the increasing of expenses on cleaning agent and, as a sequence, on energy consumption. Different mechanic of rock failure should be applied instead.

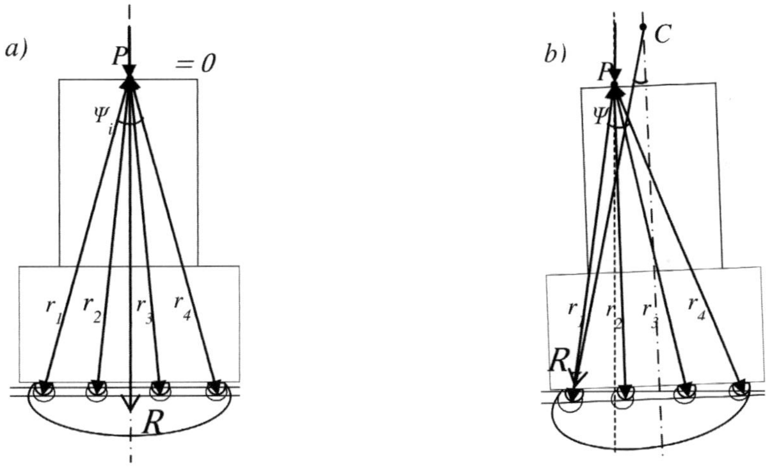

Figure 3. Scheme of the percussion rock failure process study:

a – central position of shock pulse; *b* – eccentric application of shock pulse.
P – impact force; r_i – the distance between center of percussion and rock cutting element;
Ψ_i – the angle between vertical line and direction from center of percussion to rock cutting element *i*; – angle between borehole axis and shock pulse vector R_y; R_y – shock pulse vector; *C* – center of gravity of hammer of pneumatic drill.

One of such methods can be the application of eccentric shock pulse [6] the vector of which is directed towards location of peripheral cutters (figure 3, *b*). Theoretical research has shown that maximum values of tensions are observed under the cutters located on the side of application of shock pulse [2]. Maximum values of tensions, herewith, exceed and minimum tension values are considerably lower than corresponding values of central percussion which are all equal.

This modification of the character of shock loading can provide the matching of wear rate and performance of the cutters throughout butt of rock-destruction tool.

Thus, eccentric application of shock pulses can provide the efficiency enhancement of boring in whole due to the improved mechanics of the process of rock failure, rise of mechanical drilling speed, and stabilization of boring tool performance.

3. Experimental research

Experimental work on the boring of rock slabs (marble and dolerite) at different regime parameters showed the efficiency of rotary-percussion drilling method with eccentric application of shock pulses. The inequality of eccentricity *E* for different rocks was demonstrated. Thus, the destruction of dolerite

– elastobrittle rock – is performed with isolated diamond cutter causing lateral chipping with low value of *E*. For marble, as a more plastic rock, higher values of *E* are observed as higher amplitude of lateral movement of the cutter is required [5].

It is known [3] that pneumatic pressure *P* is proportionate to unit weight of hammer (*Q/F*) and shock pulse angle *Д* and, therefore, eccentricity of shock load application *E*. Thus, decreasing of *E* can lead to decreasing of air pressure *P*, required for efficient percussion hard rock failure. This tendency was confirmed by experimental research results.

It is also known that rotary-percussion drilling efficiency is dependent on distance *r* between center of percussion and the butt of rock-destruction tool. Both factors are interdependent and their ratio should be tuned empirically.

To estimate the influence of combination of these factors (*E*, *r*, energy of single percussion) on the efficiency of rock failure, a number of additional experiments were carried out. The experiments were conducted with the planning procedure of complete factorial experiment which can reduce the number of experiments without loss of reliability of research.

Based on experimental results, the mathematical models of mechanical speed and sinking per revolution–vs–regime parameters at different values of *E* and *r* curves were drawn. For convenient results presentation, the curve was drawn with maximum values of mechanical drilling speed obtained from mathematical models (figure 4). At minimum value of *E* the successive distance decrease leads to the gradual rise of mechanical speed.

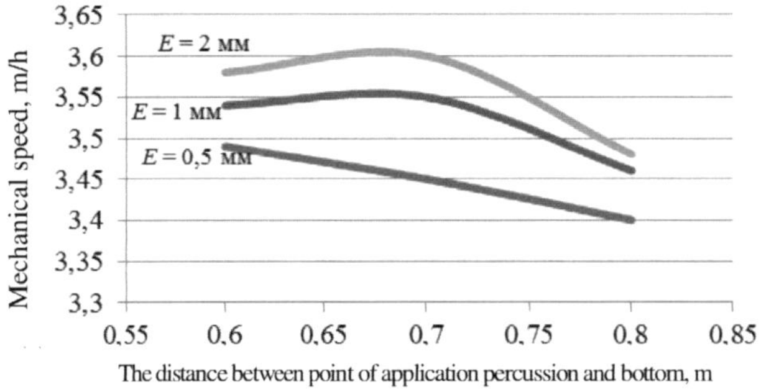

Figure 4. Mechanical speed dependence on *E* and *r*.

With increasing of *E* the ambiguous influence of *r* on rock failure process was noted. Thus, at value $E = 1$ mm the maximum mechanical speed was reached at $r = 0.7$ m and its subsequent decrease promoted the speed reduction but not less than the value at maximum distance from borehole bottom.

The similar tendency is observed with increasing of E up to 2 mm (figure 3). It indicates that the distance value 0.7 m is optimal for current *E* values, and for $E = 2$ mm to a greater extent.

The reduction of efficiency indexes of rock failure with maximum drawing close to the borehole bottom (at all other parameters being equal) can be explained that with *r* reduction the shock load vector is directed closer to borehole axis which prevents the shock pulse energy from being implemented in full.

4. Research with Autodesk software

For the highest performance the tensions under cutters should exceed the compression rock strength. The axial stress under each cutter is equal and their linear motion speed is different. Therefore, the attainment of high values of mechanical speed requires greater expenditure of energy.

The shifting of shock pulse application results in redistribution of tensions in boring tool and, as a sequence, on borehole bottom. Maximum tension values are located on the side of center of shock load application and the minimum values are on opposite. The difference between these tensions rises with increasing of *E*.

To investigate the influence of eccentricity values on tensions arising in wellbore zone at the moment of eccentric impact, the computer models were developed by AutoCAD Mechanical and

Autodesk Inventor software. The example of such model is present on figure 5, *a*. The different tension values are colored accordingly to special scale to the left.

A number of curves (figure 5, *b*) based on computer models were drawn reflecting the change of tension values for different values *r*. The analysis of diagrams allowed us to highlight the most significant moments which indicate the positive influence of eccentricity of shock load on the drilling performance.

Axial stress values are proportionate to penetration of cutters into rock. With the increasing of *E* the tension values on the side of shock load center rise which indicates the higher depth of bottom-hole coverage. As a result, the mechanical speed increases but excessive penetration of cutters, however, can negatively affect the boring process.

Lateral stress, aroused with the shifting of shock pulse application center, thus promoting the level increase of the destructed rock due to the lateral movement of cutters. The vibrations created in the rock face increase the chip loading. However, with the increasing of *E* the transverse vibrations can negatively affect the core-forming and wellbore wall durability.

Equivalent tensile stress values allowed us to find the energy losses depending on *E* and *r* values. The highest energy losses were observed at maximum values of dot E and r which can be due to partial energy conversion into torque loads.

In general, all these conclusions confirm the experimental results mentioned above.

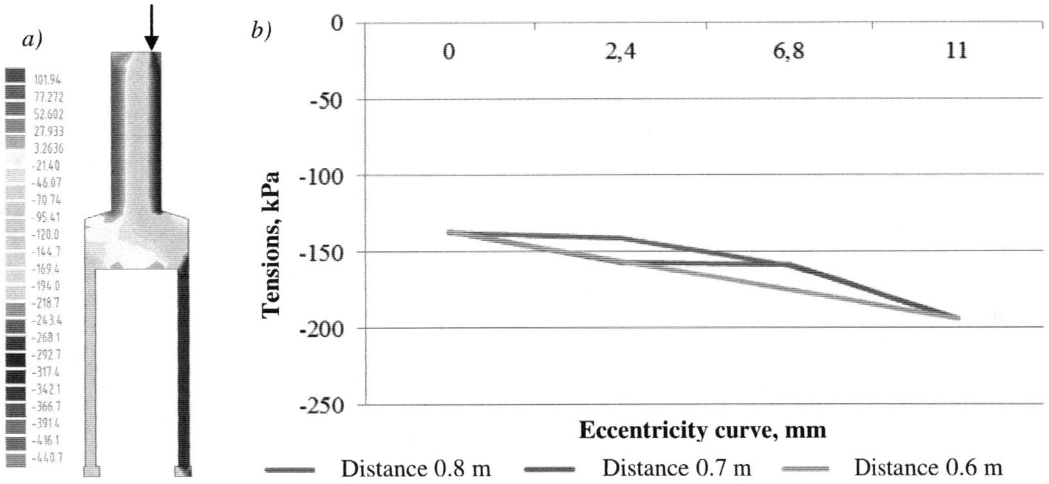

Figure 5. *a* – Graphic model of tensions, arising in core tube; *b* – Tensions-vs-eccentricity curve.

5. Possible ways for implementation of suggested technique

The results of the research have shown that the maximum performance of rock failure requires the decreasing of *r* and at the same time the tuning of optimal value of eccentricity of shock load. The implementation of such scheme is possible in non-core drilling by constructive change of rock-destruction tool.

Figures 6 and 8 show bit design examples, which are able to transform the central impact into eccentric one. The bit on figure 6 is applied for rotary-percussion drilling which is characterized by low and medium energy values of single percussion and high shock pulse frequency. The principle of operation is in changing of shock pulse vector direction directly in bottom hole area. Joint 2, established in the center of bit case 1, changes the direction of shock pulse by horizontal cranking due to virtue of elastic insert 3 which also serves as a shock absorber. Vector changes its direction so that it can get to peripheral part of the butt of rock-destruction tool 4.

The bit on figure 7 is applied for rotary-percussion drilling which is characterized by the high energy of single percussion. The design feature of the bit is movable anvil 2 which is hammered by anvil block 3 and can alleviate the influence of warp on the bit case 4 and therefore to prevent the tool from damage. Metal insert 3 changes the direction of shock pulse vector from rotation axis to the peripheral side of the butt of rock destruction-tool 6. The spring 7 in combination with the elastic insert 8 promotes the lateral movement of cutters with regard to hinged joint 5.

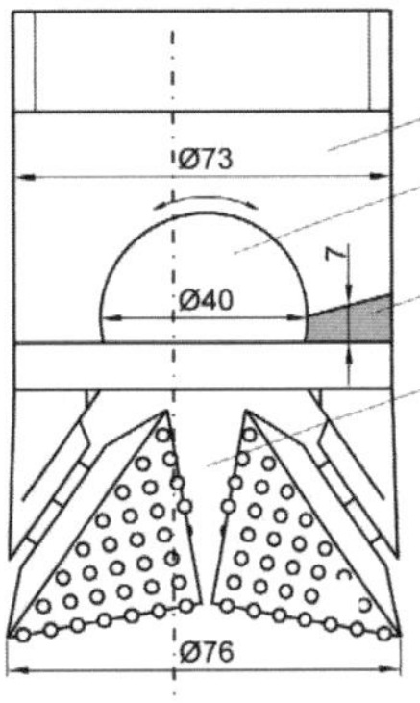

Figure 6. The bit for rotary-percussion drilling.

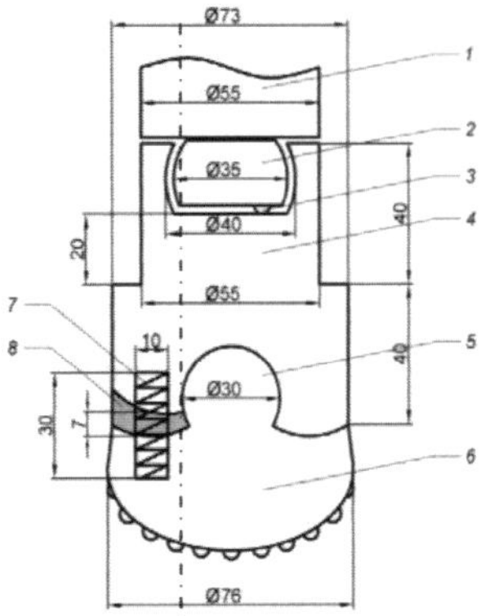

Figure 7. The bit for percussion-rotary drilling.

6. Conclusion

The results of analytical and experimental research demonstrated the rise of the mechanical speed and sinking per revolution. In our opinion, this can occur due to the deflection of shock pulse vector from rotation axis. In combination with certain ratio of E and r it directs the percussion energy to peripheral (working) part of the butt of diamond crown bit and increases the penetration of cutters into the rock. As a sequence, the performance of cutters located on different concentric circles of rock-destruction tool equalizes.

Thus, the boring with the application of eccentric shock pulses:

- can be applied in both coring and non-core drilling
- enables the performance to rise without increasing of power inputs, including sampling of core and drill cuttings
- lowers the risk of borehole crooking
- lowers energy intensity power consumption of rock failure process
- can be implemented by presented constructions of rock-destruction tool.

7. References

[1] Evseev V D 2004 *Physics of rock fracture in drilling oil and gas wells* (Tomsk: Publishing house TPU)

[2] Neskoromnyh V V and Fakhrutdinov A A 1995 Study the possibility of forming a stem of different curvature in wells off-center dot blows *Izvestiya vuzov. Geology and exploration* **2** 83–89

[3] Neskoromnyh V V 2008 *The Destruction of rocks during exploration* (Irkutsk: Publishing house of ISTU)

[4] Neskoromnyh V V 1999 The results of experimental studies of rock failure of off-center shock pulses *Izvestiya vuzov. Geology and exploration* **6** 115–20

[5] Pushmin P S, Neskoromnyh V V and Romanov G R 2014 Study on the effect of off-center application of shock pulses on the efficiency of diamond drilling *Bulletin of the Siberian branch of the section of Earth Sciences the Academy of natural Sciences. Geology, prospecting and exploration of ore deposits* **6(49)** pp 54–61

[6] Pushmin P S and Romanov G R 2014 Features of the mechanism of destruction of the hard rock diamond rock cutting tools *Bulletin of the Siberian branch of the section of Earth Sciences the Academy of natural Sciences. Geology, prospecting and exploration of ore deposits* **5(48)** pp 59–64

IOP Conf. Series: Earth and Environmental Science **27** (2015) 012053 doi:10.1088/1755-1315/27/1/012053

The research of sapropels as the drilling fluids in dispersed phase (Lake Kirek)

R R Sagitov[1,2] and K M Minaev[1,3]

[1] Department of Well Drilling, Institute of Natural Resources, National Research Tomsk Polytechnic University, 30 Lenin Ave., Tomsk, 634050, Russia

E-mail: [2] sagitov_r@bk.ru, [3] minaevkm@bk.ru

Abstract This research describes the application of Kirek Lake sapropel as a drilling fluid in dispersed phase which could replace traditionally used clay powders in drilling fluids. Sapropel is century-old bed silt of freshwater lakes of more than 12 000 years, i.e. Holocene. It consists of natural organic and inorganic substances and chemically is a complex multicomponent biogenic genesis system. Humic complexes and wulfonic acids, polysaccharides, carbonic and protein polymers comprise sapropel suspension texture. This article introduces formulations and laboratory research of sapropel suspensions and thermal activation.

1. Introduction

Oil and gas drilling development has a century and a half history. This was a time of significant changes- automated technological operations and processes and increasing well-bore depths and lengths. Further, drilling flushing fluids have become more complicated and are multicomponent systems, fine dispersed emulsions and suspensions, aerated fluids. Clay powders are applied as gelling agents. High-quality and yield bentonitic powders containing clay mineral montmorillonite of 70% are applied [1]. Hydromicaceous and palygorskite clay powders are used in the cases of drilling intervals of combined or artesian horizons with salt-saturated fluids [2]. Low-quality and yield kaolinic clay powder is used as weighting up in drilling high formation pressure intervals.

Due to drilling complication conditions and resource potential exhaustion more and more investigations are conducted in clay raw materials activation, quality improvement, increasing drilling fluid yield [3]. Platform drilling requires salinity resistant drilling fluids [4]. Clay powders should be substituted due to the absence of high-quality productive deposits, especially, in the Russian Federation. Such a potential substituting material could be sapropel [5]. Sapropels are freshwater bottom sediments(sea / lake) accumulating under anaerobic conditions as a result of physico-chemical and biological decomposition of decayed aquatic vegetation, debris, planktons and oil humus particles. Mineral and organic components of terrigenous discharge influence deposit composition. Another factor in deposit formation is flowage [6].

Near sapropel as dispersed - phase drilling fluid possesses good colloidal properties. Sapropels include all necessary minerals and organic elements, biopolymers and natural surfactants providing high aggregative stability of sapropel suspensions [7].

Both in Russia and abroad sapropel is being investigated. The application domain of various sapropel – products is diverse: fertilizers, sorbents and raw materials for chemical industry, production of forage additives and biostimulants, medical industry. In the Tomsk Oblast the geological resources

IOP Conf. Series: Earth and Environmental Science **27** (2015) 012053 doi:10.1088/1755-1315/27/1/012053

(comparable highly mineralized difference) were estimated to be 3,98 billion tons [8]. Tomsk Oblast is situated in quite a favourable climatic zone furthering the formation of large sapropel deposits. Ho wever, today sapropels are poorly studied and are used in limited spheres as in the medical industry and as a fertilizers. Putting on production unutilized subsurface resources is promising, which, in its turn, predetermines the necessity of overall geological research of sapropel deposits for further licensing and development [8].

One of the most significant estimated deposits situated in Tomsk Oblast is Kirek Lake deposit. The lake area is approximately 0.49 km^2, length -1.45 km, width – 0.4 km, average depth – 2.7 m. elongating N-E southward. The lake water is alkaline, salt composition includes hydrocarbonates with mineralization of 0.2 g/dm^3. Sapropel deposit type is calcareous, organic-ferrigenous, peaty, thickness of 0.2-8.5 m [8]. Sapropel resources in null contour is 2242.6 ths. m^3. Kirek Lake sapropel is finely-ground and weakly alkaline medium of pH 7-7.6. The deposits are potential for geological survey and commercial deposits for future development and application.

2. Sapropel suspension formulation

Sapropel suspension based on Kirek Lake raw materials as drilling fluid has not been analyzed. Sapropel formulation has been developed and sapropal, an alternative aqua-gel, as dispersed-phase drilling fluid was analyzed in Wash &Cement Drilling Slurries Lab., Drilling Department, Tomsk Polytechnic University. Preparation technique of sapropel suspensions involves the following: a wet sapropel sample of a particular mass and of various concentrations of sodium hydroxide (NaOH) wAS placed into 0.6 dm^3 water and mixed at rotation frequency of 10,000 revolutions per minute for 10 minutes. All experiments were based on standard API tests. Rheological properties were determined by a viscometer OFITE 800. Filtration quality properties were determined by filter press OFITE. Density was measured by a beam balance OFITE.

In the first experiment stage different sapropel concentrations were used (table 1). The experiment aim was to determine the optimal sapropel concentrations which would correspond to the industrial applied clay powder, according to TS 39-0147001-105-93 "Clay powder for drilling fluids." Besides, the chosen formulation should have minimum solid phase content.

Table 1. Technological properties of experimental suspensions.

Sapropel concentration g/cm^3	ρ, g/cm^3	Relative viscosity, s	Filtration, sm^3 per 30 min	Plastic viscosity, sP	Dynamic sheer stress, $lb/100ft^2$
400	1.07	20	39.4	13	2
500	1.07	22	38	14	5
600	1.07	25	36	17	6
650	1.08	28	32	18	9
750	1.08	31	28	19	12

Technological properties of prepared suspensions were measured during the period of 7 days. The maximal alteration is observed for relative viscosity (figure 1). The increase of relative viscosity is due to flocculation of sapropel particles. As a result, effective area of dispersed phase is reduced, negatively affecting drilling fluid quality.

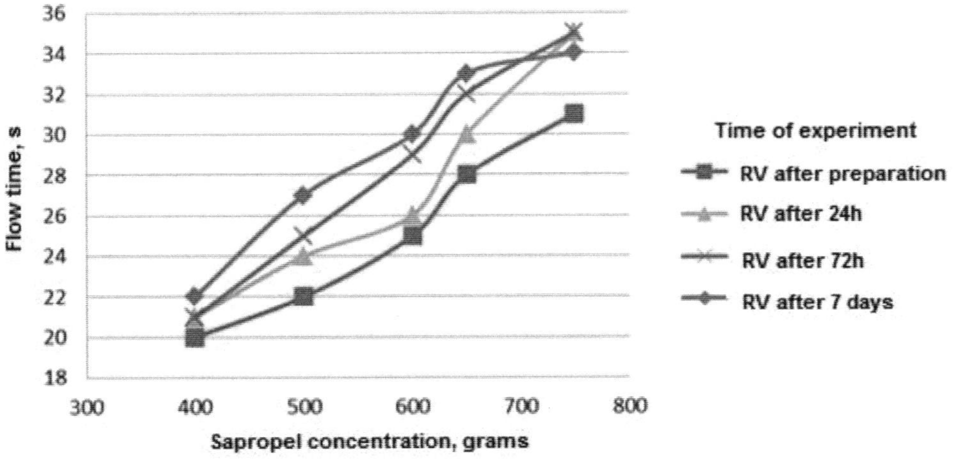

Figure 1. Relative viscosity (RV) of suspension with different sapropel concentrations.

To produce high quality drilling fluid from sapropel, the effect of alkali on suspension was examined. Experimental suspension with sapropel concentration of 500g/dm^3 was chosen, sapropel moisture was 64%, respectively, drilling fluid yield was up to 8m^3, which corresponds to clay powder "PBMG", according to TS 39-0147001-105-93 "Clay powder for drilling fluids".

The results of the experiment are represented in figure 2. The alkali concentration increase does not result in relative viscosity increase. Moreover, in alkaline medium organic and inorganic elements interact which promotes drilling fluid quality and its yield.

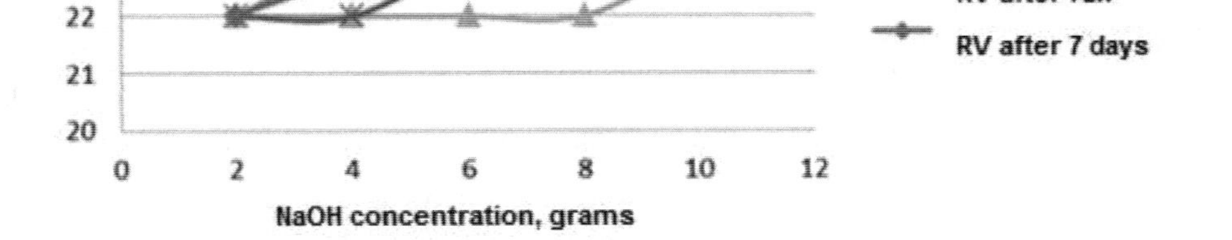

Figure 2. Relative viscosity (RV) of suspension with different concentrations of sodium hydroxide (NaOH).

3. Thermal activation of sapropel suspension

Produced suspension possess high filtration indexes. To improve the coefficient thermal activation of suspension was carried out (table 2). Suspension with 300g of sapropel, 6g content of NaOH and 600g water was heated to 90°C and mixed at this temperature for 60 minutes. After the thermal processing the filtration index decreases twofold (figure 3).

Table 2. Technological properties of experimental suspensions.

Properties	Before thermal processing	After thermal processing
Relative viscosity, s	22	22
Plastic viscosity, sP	16	17
Dynamic sheer stress, lb/100ft^2	3	1
Filtration, sm^3 per 30 min	40	15
PH	12.5	13.5

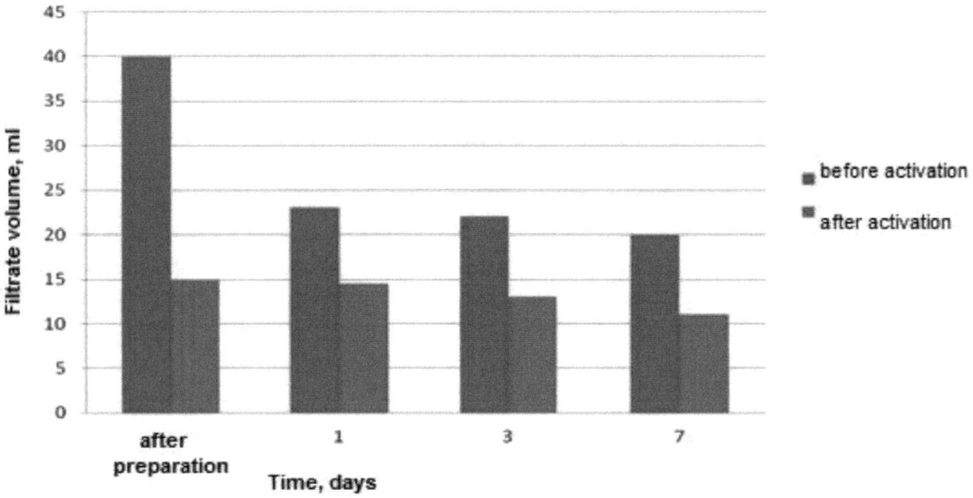

Figure 3. Filtrate volume after thermal activation.

4. Conclusion

The research showed the potential application of sapropel as dispersed phase drilling fluid. In the future the research of suspensions based on sapropel will be continued, as well as their modification and activation, as well as the development of formulations of drilling fluids based on low-moisture sapropels. It would be necessary to study the features of dispersion structure formation, depending on their dryness, solid phase concentration, chemical additives and other factors.

References

[1] Garrett Rodney L 1987 *Mining Engineering* **11** 1011–15
[2] Galán E 1996 *Clay Minerals* **4** 443–53
[3] Sarrafi A and Schaffie M 2015 *Research on Chemical Intermediates* **5** 2823–31
[4] Uti L O and Joel Ogbonna F 2013 *37th Nigeria Annual International Conference and Exhibition – To Grow Africa's Oil and Gas Production: Required Policy, Funding, Technology, Techniques and Capabilities, NAICE 2013: Proceedings of a meeting held (Victoria Island, Lagos; Nigeria, 30 July – 1 August 2013)* **2** 988–97
[5] Lishtvan I I, Lozhenitsyna V I, Lerman, A S, Lobov A I and Artamonov V Yu 1989 *Journal of Engineering Physics* **4** 1123–27
[6] Topachevsky I V 2011 *Geology and Mineral Resources of the World Ocean (Ukraine)* **1** 66–72
[7] Belyaeva L A 2010 *Management of Natural Resources: Ecology, Economics, Technology. Materials of International Scientific Conference: Proceedings of a meeting held (Minsk, Belarus, 6–8 October 2010)* 26–30
[8] Kudashev I G 2004 *Sapropel of Tomsk Region: Geology, Genesis, Resources and Prospects of Use:* avtoref. dis. ... kandidata geologo-mineralogicheskikh nauk (Tomsk) p 23

Arresting and supplying apparatus for increasing pellet impact drilling speed per run

A V Kovalyov[1], Ye D Isaev[1], A V Veryevkin[1], V M Gorbenko[1] and O S Ulyanova[2]

[1] Department of Well Drilling, National Research Tomsk Polytechnic University, 30 Lenin Ave., Tomsk, 634050, Russia
[2] Department of Foreign Languages for Specialists in Natural Resources, National Research Tomsk Polytechnic University, 30 Lenin Ave., Tomsk, 634050, Russia

E-mail: arteka011287@mail.ru

Abstract. The paper describes pellet impact drilling which might be used to increase the drilling rate and the penetration rate of hard and tough rock drilling. Pellet impact drilling implies rock destruction by metal pellets having high kinetic energy in the immediate vicinity of the earth formation encountered. The pellets are recirculated in the bottom of the bore hole by a high velocity fluid jet, which is the principle component of the ejector pellet impact drill bit. The arresting and supplying apparatus is supposed to increase speed per run in pellet impact drilling, as it not only replenishes the pellets but also supplies and then picks up the pellets from the bottom hole. The paper presents the design of the pellet-supplying component which ensures a portion of pellets supply to the bottom hole.

1. Introduction

Pellet impact drilling is currently investigated from both practical and theoretical perspectives by the Department of Well Drilling, Tomsk Polytechnic University. This drilling method has a great potential for enhancing the penetration rate of hard and tough rock drilling. Besides, pellet impact drilling can be easily incorporated into the existing mechanical drilling technology without significant reconstruction of the drilling rig.

Pellet impact drilling implies rock destruction by metal pellets having high kinetic energy in the immediate vicinity of the earth formation encountered. The pellets are recirculated in the bottom of the bore hole by a high velocity fluid jet, which is the principle component of the ejector pellet impact drill bit.

We have proved that the most effective way to drill hard and tough rock is t use ejector pellet impact drill bit with the nozzle and the tubular mixing chamber arranged consequently in line [1]. Furthermore, the tubular mixing chamber should end in the diffuser and the pellet arrestor should be installed for the pellets to be sent to the tubular mixing chamber.

2. Designing the arresting and supplying apparatus

Wear pellets is inevitable in pellet impact drilling. In the descriptions of field work results, the initial 63.5 kilo of pellet decrease by 1.13 kilo for 3.5 hours in quartzite drilling [2]. A.B. Uvakov in his work [3] pointed out that average pellet wear for 100 hours is equal to 4-5 mm in diameter.

Pellet impact drilling is provided by several consecutive stages:
1) to supply the pellets to the bottom hole;
2) to pull down ejector pellet impact drill bit;
3) to drill a well;
4) to hoist a drill bit;
5) to pick up the pellets.

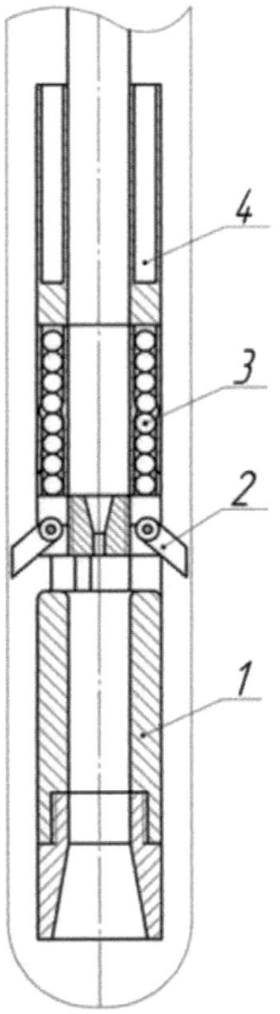

Figure 1. Bottom hole assembly (BHA) with the arresting and supplying apparatus.

1 – drill bit;
2 – arrestor;
3 – pellet supplier;
4 – inlet chamber

According to different investigations [2 – 5] the diameter of used pellets is bigger than the diameter of primary nozzle that prevents pellets to be supplied through a drill string. It is necessary to avoid supplying pellets to the annular space between the drill bit and the wellbore because of its possible pinching or needless space between them and the bottom hole. Two ways for pellets to be supplied to the bottom hole were tested. The first way involved using paper boxes for pellets which after the start of the drilling mud circulation softened and were lifted to the well mouth. The second way used a special container (a core barrel, tapered in upper part and filled with clay). It was pulled down by drill bit, where the pumping fluid under pressure drew pellets. After this, the container was lifted up.

The pellet arrestors of different design were used to pick up pellets after drill bit to be lifted up.

High drilling cost, low reliability and time waste of round-trip operations when is required for the pellet to be replenished are the major factors to design new tool for bit run speed increasing.

In this way, it could be possible to design a new tool to increase the penetration rate, to replenish the pellets, supply and pick them up from the bottom hole. According to its function, it can be called the arresting and supplying apparatus.

Design bottom hole assembly (BHA) with the arresting and supplying apparatus is shown in figure 1. The arresting and supplying apparatus consists of a petal-shaped arrestor (2), pellet supplier (3), inlet chamber (4).

The arresting and supplying apparatus provides the following operations:

1) to pull down BHA with the arrestor in off-position to bottom hole;
2) to influence the pellet suppler in order the pellets to be supplied;
3) to turn the arrestor on;
4) to drill a well while pellets to be worn significantly;
5) to put the arrestor in off-position so that the petals close space between the wellbore and drill bit;
6) to switch a pump on and direct the pellets to the inlet chamber;
7) to act on the pellet supplier so that the pellets to be supplied to the bottom hole;
8) to turn the arrestor on again;
9) to switch a pump on again and continue to drill.

3. Designing the pellet supplier

At the first stage of design the pellet supplier is developed [6], and it is shown in figure 2.

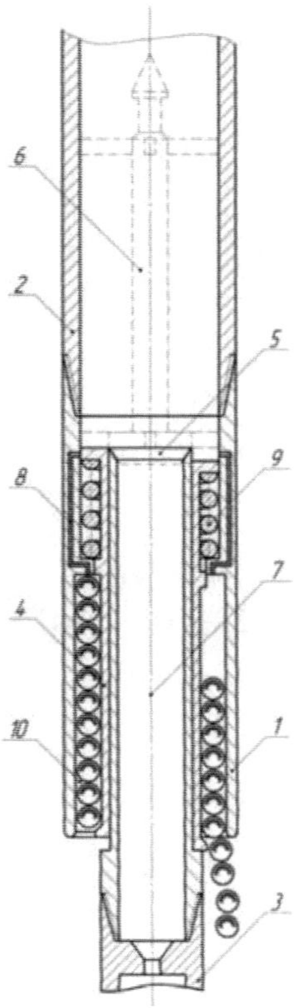

Figure 2. Pellet supplier.

1 – body;
2 – drill string;
3 – ejector pellet impact drill bit;
4 – traveling element;
5 – saddle;
6 – block valve;
7 – central valve;
8 – puff port;
9 – spring;
10 – pellet sector

The pellet supplier includes body (1), connected with a drill string (2) in the upper part and a pellet drill tool in the lower part (3), as well as a traveling element (4) with a saddle (5) to provide a leak-proof fit of the block valve (6). There are two valves in the body (1): a central valve (7) and a puff port (8). The construction is supplied with a spring (9) and a pellet section (10) between the body (1) and the travelling element (4).

The operation principles of the pellet supplier are the following: the pellet section (10) is filled with pellets before the pellet supplier is pulled down to the well. The drilling mud through the central valve (7) is pumped in drilling while the traveling element (4) is kept by the spring (9) in a maximum upward position. To supply drilling mud circulation by pellets the block valve (6) is moved from the well mouth to the drill string (2) which fits the saddle (5) and blocks the central valve (7).

Due to increasing the pressure of drilling mud on the saddle (5), the spring (9) shrinks and drives the traveling element (4) into downwardmost position to open the puff port (8). Being subjected to gravitation and drilling mud pressure, the pellets fall to the annulus space. After this, drilling mud circulation stops putting the traveling element (4) back in the initial position. The block valve (6) is retrieved from the drill string (2) with a help of a back end to be pulled down by a work line. Then the pump is turned on again and drilling continues.

4. Conclusion

Design of the arresting and supplying apparatus is considered to replenish pellets in connection with a drill bit. The design of the pellet supplier which allows pellets to be supplied the bottom hole is suggested.

The possible investigating problems are the following:

- to develop an arrestor and the way of its turning on/off;
- to design a pellet supplier to supply pellets to the bottom hole by portion;
- to develop a low-priced arrestor;
- to find alternative ways influencing the components of the arresting and supplying apparatus.

References

[1] Kovalyov A V, Ryabchikov S Ya, Isaev Ye D, Aliev F R, Gorbenko M V and Strelnikova A B 2015 Designing the ejector pellet impact drill bit for hard and tough rock drilling *IOP Conf. Ser.: Earth Environ. Sci.* **24** 1–5 URL: http://iopscience.iop.org/1755-1315/24/1/012016

[2] Eckel I E and Deily F H 1956 Development and testing of jet pump pellet impact drill bits *Transaction AIME* **207** 15

[3] Uvakov A B 1969 *Sharosruynoyue bureniye* (Moscow: Nedra) p 207

[4] Zaurbekov S A 1995 Povyshenie effektivnosti prizaboynykh gidrodinamicheskikh protsessov pri sharostruynom burenii skvazhin: avtoref. dis. … kand. tehn. nauk (Almaty) p 18

[5] Kovalyov A V, Ryabchikov S Ya, Isaev Ye D, Aliev F R, Gorbenko M V, Baranova A V 2015 Pellet impact drilling operational parameters: experimental research *IOP Conf. Ser.: Earth Environ. Sci.* **24** 1–5 URL: http://iopscience.iop.org/1755-1315/24/1/012015

[6] Kovalyov A V, Isaev E D, et al. *Saboiniy sharopitatel* Patent RU № 2014148550, E21B7/16 Filed 02.12.2014

Energy-saving compression valve of the rock drill

A N Glazov[1,2], A A Efanov[1,3] and T Yu Aikina[1,4]

[1] Institute of High Technology Physics, National Research Tomsk Polytechnic University, 30 Lenin Ave., Tomsk, 634050, Russia

E-mail: [2] glazov@tpu.ru, [3] arter@sibmail.ru, [4] aikina@tpu.ru

Abstract. The relevance of the research is due to the necessity to create pneumatic rock drills with low air consumption. The article analyzes the reasons for low efficiency of percussive machines. The authors state that applying a single distribution body in the percussive mechanism does not allow carrying out a low-energy operating cycle of the mechanism. Using the studied device as an example, it is substantiated that applying a compression valve with two distribution bodies separately operating the working chambers makes it possible to significantly reduce the airflow. The authors describe the construction of a core drill percussive mechanism and the operation of a compression valve. It is shown that in the new percussive mechanism working chambers are cut off the circuit by the time when exhaust windows are opened by the piston and air is not supplied into the cylinder up to 20% of the cycle time. The air flow rate of the new mechanism was 3.8 m^3/min. In comparison with the drill PK-75, the overall noise level of the new machine is lower by 8-10 dB, while the percussive mechanism efficiency is 2.3 times higher.

1. Introduction

Percussive mechanisms of pneumatic rock drills have high compressed air flow, low efficiency and high exhaust noise level [1], which leads to significant energy costs, decrease in productivity and restrictions concerning the usage of powerful machines if there is central supply of devices with compressed air, as well as deterioration of hygiene and sanitary working conditions. Therefore, creating highly efficient and energy-saving percussive mechanisms is an urgent problem when drilling equipment is designed.

The main reasons for high air consumption are: considerable leakage of compressed air into the atmosphere from pneumatic circuits through the cylinder when an exhaust window is opened by a piston, and faulty process operating cycle in chambers. The longest airflow takes place through a reverse chamber. These leaks are specific characteristics of a valve and a slide-valve; their duration can reach up to 32% of the operating cycle time. For instance, in the core drill PK-75, the direct airflow from a circuit into the atmosphere takes 23% of the cycle duration. The proportion of such leakages is significantly larger since the speed of the air supplied into the cylinder increases during the exhaust which intensifies the noise level of air exhaust to a greater extent.

Properties of machines are considerably determined by the structure of a compression valve. A single distribution body (DB) [2-7] is used in most types of percussive machines. Its functions are accomplished by a slide-valve, a piston and, most commonly, a valve. As DB has two reversals within an operation cycle, it is impossible to fully cut off the working chambers from the circuit before an exhaust window is opened and to realize an energy-saving operating cycle. The search for chambers cutoff from the circuit leads to premature air inlet into working chambers and, as a result, there is a loss in energy parameters and other indicators.

The increase in the percussive mechanism efficiency is associated with the working chambers cutoff from the circuit before opening an exhaust window and the improvement of process operating cycle. Applying a compression valve with two DBs separately controlling the air inlet into working chambers opens the greatest opportunities in achieving maximum efficiency.

This paper describes the construction and the working principle of a new compression valve with two self-sustainable valve DBs which separately operate working chambers. It also outlines the main results of the prototype test of the core drill percussive mechanism which was performed in the drilling laboratory of the Research and Design Institute (NIPIrudmash) and under industrial conditions at the mine.

2. Materials and Methods

The study of percussive mechanism was carried out in the laboratory of Tomsk Polytechnic University at the stand with a hydraulic feeder by a short stem with a cross bit on a concrete block and granite. Air pressure in the distribution box and working chambers, command channels and the inlet channel of the reverse chamber was registered by strain gauge pressure sensors. Immediate air temperature in working chambers was measured by resistance sensors. An inductive sensor was used to record the piston motions. Contact sensors were applied to record valve reversal. Strain-measuring amplifiers TA-5, the loop oscillograph N-115, resistance and shunt bridge R-155 were used to record signals from the sensors. Airflow rate was measured by the differential pressure flowmeter which consisted of a restriction orifice and the differential manometer DT-50. Determining the characteristics of the impact unit operation was fulfilled according to pressure diagrams on the computer or a track curve of the piston.

The mechanism was tested in the drilling laboratory of the Research and Design Institute (NIPIrudmash) at the stand with a screw feeder while drilling granite with a hardness of 10–12 (according to the scale of the professor M.M. Protodiyakonov) by a cross bit 65 mm in diameter.

The tests of the new and the commercial impact units applied in the drill PK-75 were performed at the mine on a rod-piston feeder with the use of the same rotary head. The drill was operated by remote control. Water-flush drilling was performed in rocks with a hardness coefficient of 14 16 at the same feeding force. The phonometer Sh-63-1 was used.

3. Results and Discussion

In Tomsk Polytechnic University a percussive mechanism construction (figure 1) with a new compression valve has been designed as applied to the core drill PK-75 with independent rotation of the drill and a piston 120mm in diameter, and its samples have been investigated. Some results of research and development of the operating processes in this percussive mechanism are outlined in the works [8-10].

Let us describe the mechanism operating principle.

The chamber ***o*** communicates with the atmosphere through the command channel ***c*** and the piston 2 groove through the exhaust window ***g***. The valve 4 and the pushrod 5 are in the upper position. Compressed air inflows through the channel ***d*** into the cylinder reverse chamber 1, the piston 2 moves upwards. The valve 6 and the pushrod 7 are in the down position. During further motion, the piston opens the command channel ***c*** compressed air is supplied from the reverse chamber into the chamber ***o***. The sum of forces acting on the upper faces of the pushrod 5 and the valve 4 is greater than the force under the valve. The valve 4 and the pusher 5 are reversed to the down position and the air inlet into the cylinder stops until the exhaust window ***в*** starts opening by the piston. Then the piston closes the exhaust window ***e*** and compresses air in the direct chamber. The air supplied through the command channel ***f*** into the chamber ***p*** enables a steady down position of the valve 6. Then the chamber ***p*** communicates with the atmosphere through the channel ***f***, through the piston groove and the exhaust window ***e***. The valve 6 and the pushrod 7 move upwards, and compressed air flows into the direct chamber. After that, the piston holds the upper operating position and starts direct motion (in figure 1 – down).

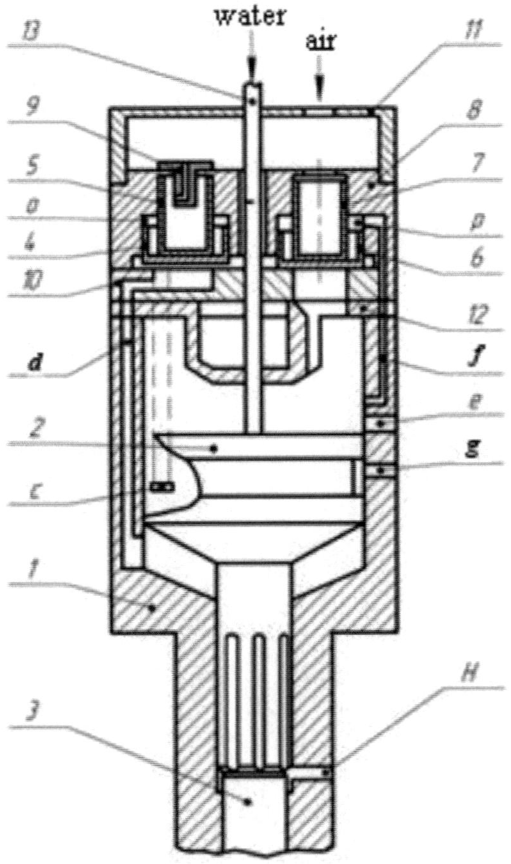

Figure 1. The scheme of the percussive mechanism.

A percussive mechanism comprises:
1 – cylinder with exhaust windows ***e*** and ***g***;
2 – piston; 3 – drill rod;
4, 5 – valve and pushrod of the reverse chamber;
6, 7 – valve and pushrod of the direct chamber;
8 – distribution box; 9 – throttle; 10 – valve body;
11 – cover; 12 – cylinder cover;
13 –flushing out supply tube;
o, ***p*** – chambers above valves;
c, ***f*** – command channels of the direct and reverse chambers, respectively; ***d*** – inlet channel of the reverse chamber;
H – outlet.

Under air pressure, the piston moves forward and after a while it opens the command channel ***f***, compressed air flows from the direct chamber into the chamber ***p***. The valve 6 and the pushrod 7 move as far as they can go to the valve body 10, i.e. direct chamber is cut off from the circuit. Remaining air flows out into the atmosphere when the exhaust window ***e*** is opened by the piston. The piston compresses air in the reverse chamber, air flows through the channel ***d*** under the piston and into the chamber ***o*** above the valve 4. The valve 4 and the pushrod 5 are in the down position. At further motion of the piston, the chamber ***o*** and the command channel ***c*** are connected with the atmosphere through the piston groove and the exhaust window. The valve 4 and the pushrod 5 move upwards, then the piston strokes the jackrod 3. Next operating cycle of the mechanism begins under rebound impulse and compressed air pressure.

Let us consider typically occurring moments during the compression valve operating (figure 2).

The indicative points on the diagram of air pressure in the reverse chamber correspond to: **a** – piston forward position; **c** – start of opening the command channel ***c***; **e** – opening the exhaust window ***g*** (figure 1). Points **d** and **l** on the line 1 correspond to the time of valve 4 seating and lifting from the valve body seat 10. On the diagram of the air pressure in the direct chamber, points **i** and **j** stand for the time of opening the command channel ***f*** and the exhaust window ***e*** (figure 1) during the piston direct motion, respectively. The start of the valve 6 lifting and seating on the valve body 10 (figure 1) is observed on the line 2 at the moments referring to points **g** and **k**. The air pressure diagram P_1 up to point **b** shows intensive air inlet into the chamber after the valve lifting from the seat. Point **f** corresponds to air cutoff in the cylinder. Up to that moment, drop in pressure is observed temporarily, i.e. increase in air consumption due to opening the command channel. On the whole, the air pressure curve $\boldsymbol{P_l}$ reflects the change in immediate airflow and the percussive mechanism operating process.

The valve operating diagrams 1 and 2 demonstrate that compressed air is not supplied into the cylinder in periods of time from point **d** to point **g** and from point **k** to point **l**. The moments of air inlet and cutoff in working chambers do not coincide. This all enables to accomplish the necessary process cycle in working chambers and reduce inefficient air consumption.

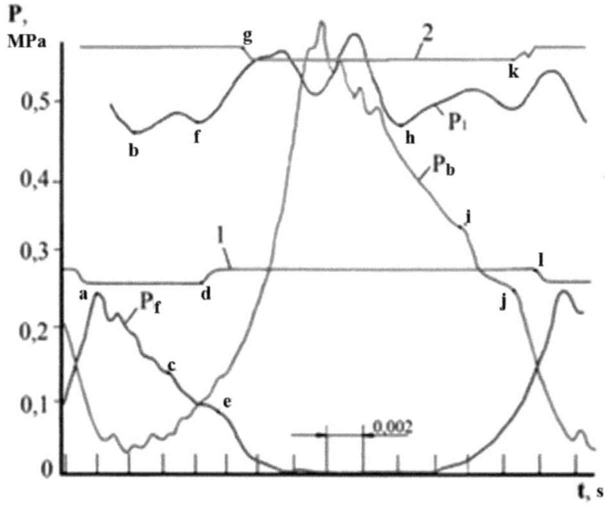

Figure 2. Oscillograph of the percussive mechanism operating processes: 1 and 2 – valve reversals of reverse and direct chambers, respectively; P_b, P_f и P_l – air pressure in the direct and reverse chambers, pre-valve chamber of the percussive mechanism, respectively.

4. Conclusions

High efficiency of the mechanism is achieved owing to implementing two autonomous DBs with inverse connection with the piston through the command channels and coordinated characteristics of elements in distribution and operation units. Unlike the extensively used machines with two DB reversals where the air inlet into the cylinder is continuous, the new device has four DB reversals within an operating cycle which allows performing an energy-saving operating cycle.

The features of the energy-saving operating cycle in the designed mechanism are: the working chambers are cut off from the circuit before the exhaust window is opened by the piston, air is not supplied into the cylinder up to 20% of the operating cycle time, the values of thermodynamic processes are less than 1.4, the internal energy of compressed air is used, maximum air pressure in the direct chamber is higher than that of the circuit, distribution phases in chambers are asymmetrical.

The laboratory and industrial tests of the drill have shown that it possesses higher technical and economic indicators in comparison with the core drill PK-75. The comparative tests of drills have shown that the new machine develops drilling speed 20-25% higher and has overall noise level lower by 8-10 dB. Its percussive mechanism has the efficiency 2,3 times higher, airflow rate being 3,8m^3/min.

The results of the conducted research have proved a possibility to create energy-saving constructions of impact machines that significantly surpass their known industrial counterparts in efficiency.

Implementing drills with the new percussive mechanism will enable to reduce energy consumption, increase productivity and improve hygiene and sanitary working conditions.

IOP Conf. Series: Earth and Environmental Science **27** (2015) 012055 doi:10.1088/1755-1315/27/1/012055

References

[1] Kivade S B, Murthy S N and Vardhan H 2011 Study of Noise Sources in Pneumatic Rock Drills – its effect and control – A Critical Review *J. the Indian Mining and Engineering* **50** 12–23

[2] Esin N N 1978 *Pneumatic Percussive Machines for Drilling Blastholes* (Novosibirsk: Nauka) p 104 (in Russian)

[3] Esin N N, Kostylev A D, Gurkov K S and Smolyanitskiy B N 1987 *Air Hammer Machines for Borehole and Blasthole Making* (Novosibirsk: Nedra), p 214 (in Russian)

[4] Vasiliev V M 1989 *Drills: Reference Book* (Moscow: Nedra) p 215

[5] Sudnishnikov B V and Yesin N N 1965 *Compression Valves of Pneumatic Percussive Machines* (Novosibirsk: Institute of Mining Affairs) p 46

[6] Ivanov K I, Latyshev V A and Andreev V D 1987 *Drilling Technique at Exploitation of Mineral Deposits* (Moscow: Nedra) p 272

[7] Jimeno E, Jimino C and Carcedo A 1995 *Drilling and Blasting of Rocks* (Rotterdam; Brookfield: A.A. Balkema) p 391

[8] Glazov A N 2005 Operating Processes of the Drill Pneumatic Mechanism *Bulletin of the Tomsk Polytechnic University* **308** 132–36

[9] Glazov A N 2007 Energy-saving Constructions of Pneumatic Rock-drills *Gornyi Zhurnal* **1** 76–78

[10] Glazov A N 2006 Estimated Dependences of Static Model of Pneumatic Mechanism Operating Processes *Bulletin of the Tomsk Polytechnic University* **309** 150–54

Drill pipe threaded nipple connection design development

A L Saruev[1,2], L A Saruev[1,3] and S S Vasenin[1,4]
[1] National Research Tomsk Polytechnic University, 30 Lenin Ave, Tomsk, 634050, Russia

E-mail: [2] saruev@tpu.ru, [3] saruevla@tpu.ru, [4] vasenin8@gmail.com

Abstract. The paper presents the analysis of the behavior of the drill pipe nipple connection under the additional load generated by power pulses. The strain wave propagation through the nipple thread connection of drill pipes to the bottomhole is studied in this paper. The improved design of the nipple thread connection is suggested using the obtained experimental and theoretical data. The suggested connection design allows not only the efficient transmission of strain wave energy to a drill bit but also the automation of making-up and breaking-out drill pipes.

1. Introduction

The analysis of the behavior of the up-to-date thread connections used in drill pipes shows that the total loads applied to them are represented by the coupling torque, axial load, and impacts produced by percussive mechanism with the definite frequency.

A combination of these factors has a significant effect on the drilling rate which is proportional to the strain wave energy [2]. Moreover, the impact loads produced with the definite percussion frequency, generate the additional energy which allows increasing the drilling rate and has a wave-like behavior. The aim of this study is to analyze the operation of threaded nipple connection while passing of a strain wave through it and improve its design.

2. Experimental part

During the power pulse propagation over the drill string from the percussive mechanism to a drill bit, the energy losses are observed in drill pipe connections that lead to the decrease in the strain wave energy over the drill string.

The drilling tool presented in figure 1, operates in the following way. The percussive mechanism induces and maintains the in-and-out movement of the hammer. The hammer produces impacts on the steel shank in which the strain wave is formed at each impact. The strain wave is characterized by the shape (elevation), amplitude, length, and energy. These properties depend on the geometry of colliding elements, material properties, and speed of collision. A common operating mode of the percussive mechanism is such that the collision can be regarded as elastic, and the effect from the strain wave can be neglected.

While the strain wave propagates along the drill pipe and through thread connections, its shape is changed, and the energy dissipation is observed. The intensity of these processes depends upon the design and the size of the drilling tool members, their material properties, contact stiffness, etc.

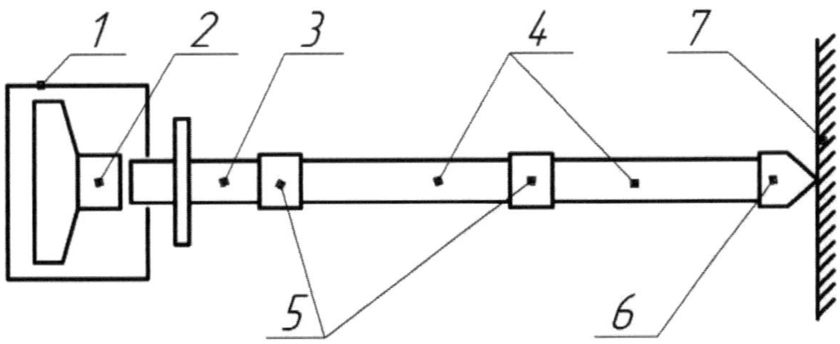

Figure 1. Schematic view of the drilling tool: 1 – percussive mechanism; 2 – hammer; 3 – steel shank; 4 – drilling pipe; 5 – thread connections; 6 – drill bit; 7 – rock.

While propagating along the drill pipe and affecting the drill bit, the strain wave makes the drilling tool to penetrate in the rock that leads to its subsequent failure. In order to maintain the drilling tool near the rock, the feed thrust should be applied to it. Usually, the value of the feed thrust is several tens less than that of the strain wave.

Except for impact loads, in many types of percussion drilling machines the coupling torque is applied to the drilling tool. Depending on the feed thrust value, the coupling torque provides merely either an angular rotation of the drilling tool between two subsequent impacts or the additional rock failure with a drill bit.

The section of the drilling tool covered with the strain wave, the stress state is generally considered to be spatial. Axial, flexural, radial, and shear stresses occur in this tool section.

Having achieved the running end of the drill string, the power pulse is reflected. Its waveform depends on conditions of the running end of the drill string and can be of three types:

a) compression pulse at a total absence of the drilling tool penetration;
b) tensile pulse at a total absence of resistance to the drilling tool penetration;
c) tension-compression pulse at drilling tool penetration in the rock possessing a certain resistance.

Figure 2 presents the schematic movement of the running end of the drill string at different interactions between the drilling tool and the rock.

The reflected pulse propagation is accompanied by phenomena similar to that of the direct pulse. The difference is that tensile pulse causes the loss of contact between the drill string joints, and the further energy transmission through the nipple body increases the loss of the reflected strain wave energy. While propagating over the drill string, the reflected strain waves interfere with each other. Stresses within the interference area are determined by the vector summation of stresses produced by individual strain waves. Therefore, the stress peak in the connection can exceed the direct pulse stress by not over than two times.

Experimental studies carried out by Fisher [4], show approx. 50% increase of the surface stress that occurs due to the flexural pulse. These additional stresses turn to be the strongest at the beginning of the drill string, and then completely vanish. This is because the flexural pulse propagates more slowly than the compression pulse.

The strain wave energy is changed while propagating along the drill string and can be obtained from [3]

$$W = W_0 \cdot e^{-2\,\alpha_1 \cdot L + \alpha_2 \cdot n}\,, \tag{1}$$

where is the strain wave energy with distance L from the end of the drill string; is the strain wave energy at the end of the drill string; is the damping decrement of elastic vibrations in the drill string material per unit length; is the energy loss rate in one connection of the drill string; is the number of connections over length *L*.

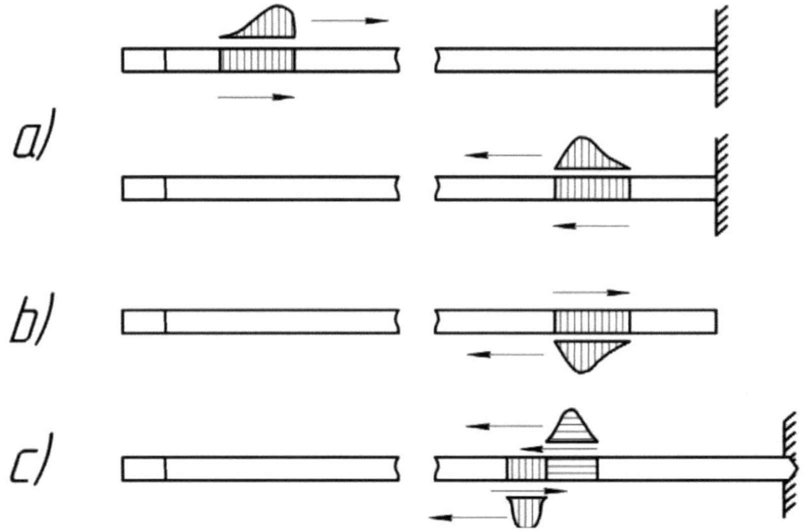

Figure 2. Schematic view of pulse propagation and the drill string section movement: a) no drilling tool penetration; b) no resistance to drilling tool; c) partial drilling tool penetration.

Research results obtained by Shadrina *et al.* in [5] show that the damping decrement is $\alpha_1=4\cdot10^{-3}$ (0.4 %), while the energy loss rate ranges between 0,8 - 18% depending on the design and the quality of the pipe connection.

The strain wave energy loss in one connection of the drill string is mostly caused by two factors, namely: reflection of strain waves in places of cross-sectional differences and joints of the drill string; frictions in the drill string or connecting element at pulse propagation over the connection.

During the strain wave propagation over the places of cross-sectional differences, the reflected pulse is observed, amplitude of which is defined by the direct pulse amplitude [2]:

$$F_r = \frac{S_2 - S_1}{S_2 + S_1} \cdot F_a \tag{2}$$

The strain wave energy is proportional to the squared peak value; the rate of impact energy loss due to reflection can be obtained from the ratio between the strain wave energy approaching to the connection and the reflected strain wave energy

$$a_r = \frac{W_r}{W_a} = \frac{F_r^2}{F_a^2} = \left(\frac{S_2 - S_1}{S_2 + S_1}\right)^2 \tag{3}$$

The strain waves reflection from the drill string joint depends on the coupling torque affecting the connection. The higher coupling torque, the tighter joint and the lesser the movement of drill string and connecting element at strain wave propagation. Hence, the lower are its energy losses. The increase in the coupling torque decreases the drill string relative displacement in the connection owing to which the inelastic resistance force is reduced during the loading cycle in the connection (figure 3). This behavior is explained by the transverse strain occurred in thread loading which is directed away from the nipple. At the certain value of the coupling torque, both the pressure in thread contacts and inelastic resistance forces in thread connection decrease. The increase of the coupling torque results in the increase of the contact stiffness between threads and the nipple.

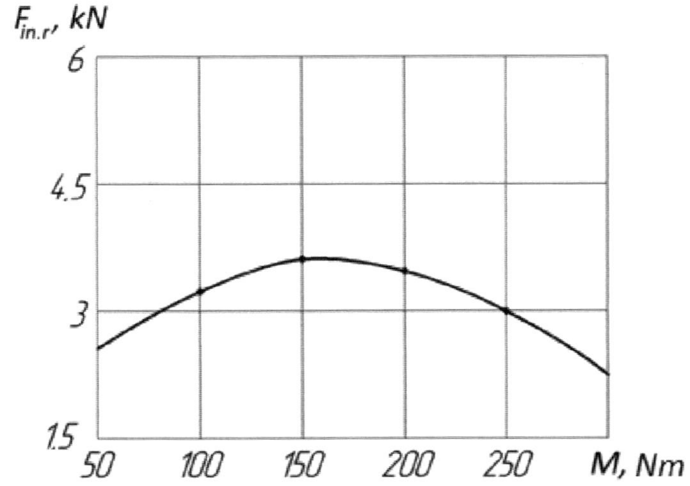

Figure 3. The dependence between inelastic resistance force and coupling torque.

Strain wave energy losses are mainly caused by the inelastic resistance forces occurred at the pulse loading and displacement of the nipple relative to the connected drill strings. Thus, it is advisable to design the thread connection that will reduce energy losses and increase the drill string operating capacity and the efficiency of well drilling [1].

The suggested design of the nipple thread connection allows the improvement of the drill string reliability, cutting transport and the automation of making-up and breaking-out drill pipes. The nipple connection shown in Figure 4, comprises two drill pipes having the female straight thread and the nipple having the continuous male thread. The nipple has a collar with two annular grooves on its both sides. Collar grooves are cut on the internal side of the drill pipe ends. Non-threaded segment of one of the nipple ends is rigidly fixed in the drill pipe and has longitudinal cuts that divide the non-threaded segment into elastic plates. Elastic plates have collars with a bevel front face which are fixed in the annular groove after the achievement of strain tolerance in the tapered groove. Wrench flats are cut on another non-threaded segment of the nipple end.

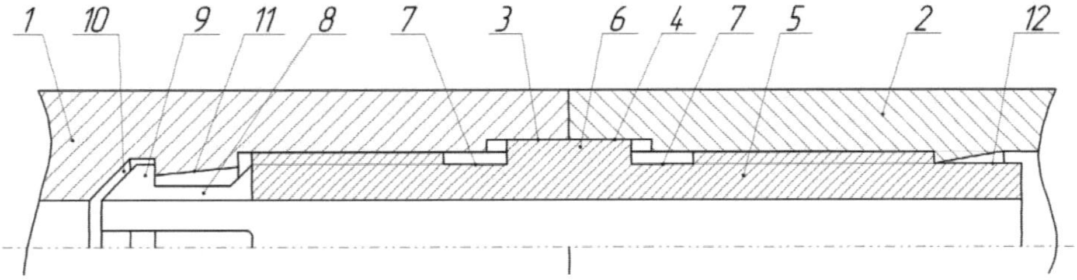

Figure 4. Cross-sectional view of nipple thread connection: 1, 2 – drill pipes; 3, 4 – grooves; 5 – nipple thread; 6 – collar; 7, 10 - annular groove; 8 – elastic plates; 9 – collar with bevel front face; 11 – tapered groove; 12 – wrench flat.

In percussion drilling, the drill string and connecting elements experience tensile, compressive and flexural loads and the coupling torque effect.

The suggested design of the nipple connection has the following advantages:

- Nipple locates inside two butt-coupling pipes providing the minimum energy loss at strain wave transmission.
- In strain wave propagating through the connection, pipe sections are compressed, while the nipple is unloaded from normal tensile and deformation stresses produced by the pre-tightening force.
- The collar and annular grooves cut in the nipple improve the connection durability at bending strain.
- The automation of making-up and breaking-out drill pipes is possible due to the nipple fixation in one of the drill pipes.
- The nipple connection hidden inside the drill pipe provides the consistency of the outer diameter of the drill string that improves the cutting transport and eliminates the possible drill string jamming in the well.

3. Conclusion

The suggested nipple thread connection can be used in percussion drilling of medium-hard and harder rocks (f = 6…14). The transmission rate of strain wave energy over the drill string increases due to the increase of the contact stiffness between thread connections.

References

[1] Zhukov I A 2005 *Elastic wave formation in waveguides at impacts with semi-cathenoidal hammers: PhD thesis* (Tomsk: Tomsk Polytechnic University Press) p 132
[2] Ivanov K I, Varich M S, Dusev V I and Andreev VD 1987 *Drilling technique for exploitation of mineral deposits* (Moscow: Nedra Publ.) p 272
[3] Slistin A P 1990 *Parameter determination for longitudinal impact transfer by compound bars: PhD thesis* (Tomsk: Tomsk Polytechnic University Press) p 184
[4] Fisher G 1962 *Rock failure and mechanics* (Moscow: Gosgortekhizdat) p 300
[5] Shadrina AV, Saruev LA and Saruev AL 2009 *Dynamic processes in drill string at percussive drilling small holes in underground digging* (Tomsk: TPU Publ.) p 175

The behavior of enclosed-type connection of drill pipes during percussive drilling

A Shadrina[1,2] **and L Saruev**[1,3]
[1] National Research Tomsk Polytechnic University, 30 Lenin Ave., Tomsk, 634050, Russia

E-mail: [2] avshadrina@rambler.ru, [3] saruevla@tpu.ru

Abstract. Percussion drilling is the efficient method to drill small holes (≥ 70 mm) in medium-hard and harder rocks. The existing types of drill strings for geological explorations are not intended for strain wave energy transfer. The description of the improved design of the drill string having enclosed-type nipple connections is given in this paper presents. This nipple connection is designed to be used in drilling small exploration wells with formation sampling. Experimental findings prove the effectiveness of the enclosed nipple connection in relation to the load distribution in operation. The paper presents research results of the connection behavior under quasistatic loading (compression-tension). Loop diagrams are constructed and analyzed in force-displacement coordinates. Research results are obtained for shear stresses occurred in the nipple connection. A mechanism of shear stress distribution is described for the wave strain propagation over the connecting element. It is shown that in the course of operation the drill pipe tightening reduces the shear stress three times.

1. Introduction

During the percussive drilling, the pipe connection is exposed to the axial load, impacts produced by percussive mechanisms, and coupling torque [1]. The study of the nature of these loads, their estimation and consideration in the drill string design allow improving the durability of thread connections.

Such types of connections as pipe-in-pipe, coupling, coupling-and-tool refer to inefficient design methods of the percussive drilling because cyclic tensile stresses and a low rate of power pulse propagation are observed.

A safe operation of the thread connection is provided, first of all, by the reduction of the strain wave energy loss. The strain wave energy dispersed in thread connections is determined by the multiple reversed strains, thread frictions and, as consequence, heating of thread connections rather than by longitudinal movement of the drill string [3-8].

The nipple connection hidden inside the drill pipe is the alternative to the thread connection which lack most of the above stated disadvantages [9].

2. Experiment technique

The behavior of the suggested nipple connection presented in figure 1 was studied on the test bench under the quasistatic loading (compression-tension), the schematic view of which is shown in figure 2. The properties of dissipative forces were obtained.

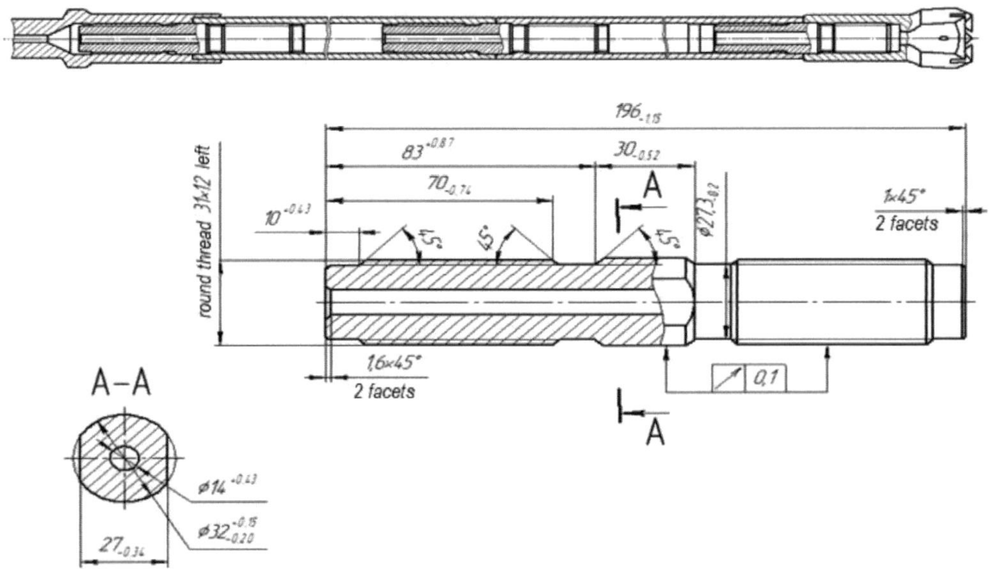

Figure 1. Cross-sectional view of enclosed-type nipple connection.

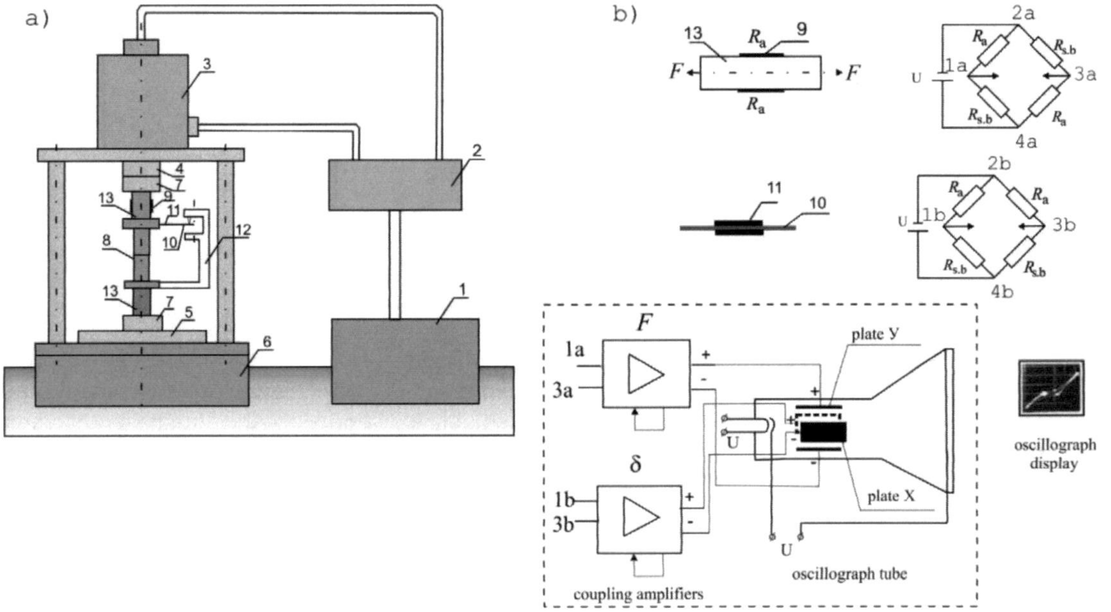

Figure 2. Schematic view of the test bench for hysteresis diagram construction at thread connection loading: a) test bench components: *1* – oil-pumping station; *2* – distribution device; *3* – hydraulic press; *4* – hydraulic press shaft; *5* – cellar floor; *6* – foundation; *7* – clamper; *8* – thread connection (coupler); *9* – strain gauges; *10* – displacement measuring elastic plate; *11* – displacement transducers; *12* – clamp; *13* – pony drill collar; b) circuitry: R_a, $R_{s.b}$ – active and self-balancing strain gauges respectively; *Y* and *X* – vertical and horizontal oscilloscope beam deviation plates, respectively; δ – drill pipe displacement relative to coupling element; *F* – longitudinal force.

Loop diagrams were obtained by the following way. A segment of the drill pipe was fixed by clampers, a clamp being mounted.

The coupling torque was applied to the thread connection. Balancing was performed for the bridge joint and oscilloscope S1–117.

The pulse and relative displacement signals were displayed on the oscilloscope by the *x* and *y* axes. The beam scanning was switched-off. The loop diagram obtained on the oscilloscope display was then photoed.

Shear stresses were studied on the drop-hammer test bench. The axial load was created by the air-feed mechanism and ranged from 3.4 to 11.3 kN. The rate of load application was controlled by the compressed air pressure supplied to the percussive mechanism. The coupling torque of the drill string was obtained using the weight lever and ranged from 49 to 245 Nm. Pulses obtained on the oscilloscope display were photoed. Resistance strain gauges were stuck in the centre of the nipple as shown in figure 3, since shear stresses achieve their maximum in the given section under static and dynamic loads.

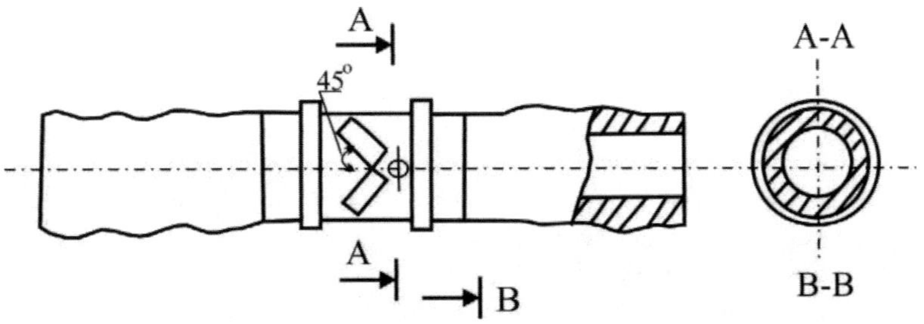

Figure 3. Schematic view of sticking resistance strain gauges for recording shear stresses.

The obtained data were then processed using the mathematical statistics techniques.

3. Results and discussion

The analysis of the curves shown in figure 4, shows the deformation properties of the nipple thread connection.

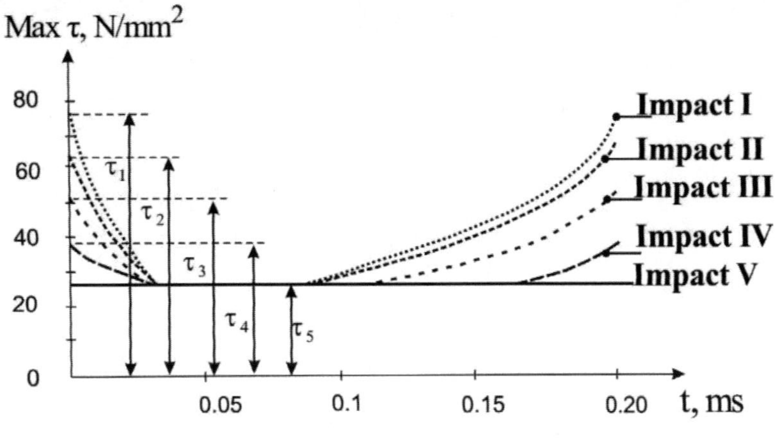

Figure 4. Shear stress distribution in the nipple at combined action of coupling torque and power pulses.

The area of the hysteresis loop represents the behavior of inelastic resistance force or the dissipated strain wave energy at the applied cyclic compressive and tensile loads.

A mechanism of deformation of the enclosed-type nipple connection during the quasistatic loading is given in the table below.

Table. Deformation mechanism of the enclosed-type nipple connection during the quasistatic loading.

Diagram sectors	Description
I–II	Thread connection is pre-tightened: the value of the applied coupling torque defines deformation of the drill pipe and nipple threads. The tightening force in drill pipes and nipple is larger than the compressive force. Further compression of the connection increases the compressive force in the drill pipe joint and unloads the nipple. Static friction is transformed to sliding friction. In sector *II* friction is absent, and the nipple is completely unloaded; a gap between the adjacent threads is observed both in the drill pipe and nipple. This sector corresponds to the pipe stiffness.
II–I	At the compressive load removal (down to *I* point), the nipple returns to its original position. This sector corresponds to the connection stiffness.
I–III	Tensile force applied to the thread connection causes a larger nipple tension and increases friction over the contact area between threads in drill pile and nipple. The total thread stiffness increases.
III-IV	Further decrease of tensile force results in insignificant relative displacements in drill pipe and nipple threads due to the high stiffness contact between these threads.
IV–V	The decrease of tensile force results in the decrease of friction in drill pipe and nipple threads.
Interval V–I	Tensile force is absent in *V* point. The significant friction is observed between the adjacent threads of drill pipe and nipple. This sector corresponds to inelastic deformation in the nipple connection.
V–VI	The total stiffness of the nipple connection is considerably lower than that in sector *I–II,* since the reversed friction is observed in the drill pipe thread.

The external coupling torque affecting the drill string evokes shear stresses in the connecting elements. At the same time, the coupling torque can be transmitted through the nipple threads and the drill pipe joints.

The modification of shear stresses is observed in the nipple as a result of a series of percussions generated by the hammer at constant pre-impact velocity of 5.1 m/c (figure 4).

The behavior of these dependencies can be explained by the coupling torque fractions redistributed between the nipple thread and the drill pipe joint. This complex wave process occurring in the pipe connection can be described as follows.

1. The tightening force acting on the thread connection at the beginning (some 3 t) results in the drill pipe compression and the nipple tension.
2. During the strain wave propagation, the compressive force increases in drill pipes up to the its peak values (15-18 t), while the tensile force in the nipple decreases. The contact deformation of adjacent threads in connecting elements is considerably reduced resulting in the nipple threading into the pipe due to the external coupling torque.
3. In the ending of strain wave propagation, the drill pipe acts on the nipple thread as a compressed spring, thereby causing its larger tension than before the strain wave propagation. Contact forces, friction, and shear stresses in threads are increased.
4. By the direct pulse propagation, shear stresses turn to be lower than the preceding ones. This is because the increase of the external coupling torque fraction in the pipe joint, and the decrease of that one in the nipple.

This wave process is repeated until about 70-75% of the coupling torque in the connection is transmitted through the drill pipe joint. The transmission of the coupling torque is then stabilized. In this experiment, the transmission of 70-75% of the coupling torque was observed after 5 impacts.

4. Conclusions

1. Experimental research is connected with the understanding of the process dynamics occurred in the drilling tool.
2. The new design of the coupling element allowed minimization of strain wave energy losses at its propagation over the drill string, providing the sufficient strength of coupling elements.
3. The combined effect from the coupling torque and strain waves generated by the longitudinal impact, the drill pipe tightening occurred while in operation that reduced the shear stress 2-3 times. Being located inside the drill pipe, the nipple was unloaded from the most of transverse and longitudinal loads, while 70% of coupling torque was transmitted through the drill pipe joints.

References

[1] Beccu R and Lundberg B 1987 *International Journal of Impact Engineering* **6** 157–73
[2] Ivanov K I, Latyshev V A and Andreev V D 1987 *Drilling technique for exploitation of mineral deposits* (Moscow: Nedra Publ.) p 272
[3] Li X B, Summers D A, Rupert G and Santi P 2001 *Tunnelling and Undegraund Space Technology* **16** 107–14
[4] Luiz F P Franca 2011 *International Journal of Rock Mechanics and Mining Sciences* **48** 827–35
[5] Lundberg B and Collet P 2010 *International Journal of Impact Engineering* **37** 901–6
[6] Lundberg B and Okrouhlik M 2006 *International Journal of Impact Engineering* **32** 1573–83
[7] Thuro K. 1997 *International Journal of Earth Sciences* **86** 426–38
[8] Keskinen E, Karvinen T, Montonen J and Heinonen M 2011 Dynamics of stress wave during percussive drilling process *Proc. 13th World Congress in Mechanism and Machine Science. Guanajuato (Mexico, 19–25 June 2011)* pp 1–5
[9] Shadrina A V and Saruev L A 2014 *International Journal of Mining Science and Technology* **24** 245–49

IOP Conf. Series: Earth and Environmental Science **27** (2015) 012058 doi:10.1088/1755-1315/27/1/012058

Safety enhancement of oil trunk pipeline crossing active faults on Sakhalin Island

E Tishkina[1,2], N Antropova[1,3] and T Korotchenko[1]

[1] Department of Petroleum Asset Engineering, National Research Tomsk Polytechnic University, 30 Lenin Ave., Tomsk, 634050, Russia

E-mail: [2] tishkina_ekaterina@mail.ru, [3] antropova@tpu.ru

Abstract. The article explores the issues concerning safety enhancement of pipeline active fault crossing on Sakhalin Island. Based on the complexity and analysis results, all the faults crossed by pipeline system are classified into five categories – from very simple faults to extremely complex ones. The pipeline fault crossing design is developed in accordance with the fault category. To enhance pipeline safety at fault crossing, a set of methods should be applied: use of pipes of different safety classes and special trench design in accordance with soil permeability characteristics.

1. Introduction

The TransSakhalin pipeline system passes through the areas characterized by complex geological setting and high natural risks. It stretches the areas of high seismic activity – up to 9-10 earthquake intensity (MSK). Besides vertical and horizontal surface displacements, the following adverse effects of seismic activity should be considered: soil liquefaction, landslide, seismic shaking and wave propagation [1]. When selecting pipeline route, it is not always possible to avoid active fault crossing. This presents a significant hazard, as earthquakes are more likely to occur along the faults, and displacements can be up to several meters. Therefore, it is of particular importance to develop a set of methods aimed at enhancing oil trunk pipeline safety, especially in those areas where it crosses the active faults [2].

Basically, active faults are those faults which have moved at least once over the past 10-11 thousand years. In addition, there are faults that can be evaluated as potentially active even if the movement has occurred over the past 500 thousand years. Thus, it can be stated that an active fault is a seismogenic fault that regularly rejuvenates, which, in its turn, constitutes a danger for the society [2, 3]. It is worth noting that the faults on Sakhalin Island occupy a zone with a width of a few kilometers [1].

Sakhalin Island is located along a diffuse plate boundary zone between the Eurasian tectonic plate and the Okhotsk microplate (North American plate). The Kuril trench that stretches northeastward from eastern Hokkaido is regarded as an eastern boundary of this zone. Three large N-S faults stretch along the whole Sakhalin Island: Western-Sakhalin, Central-Sakhalin (Tym-Poronaysk), and Hokkaido-Sakhalin (Northern Sakhalin).

The Sakhalin-II pipeline system crosses seismic zones of Central-Sakhalin and Hokkaido-Sakhalin faults. It passes through active faults that can lead to surface fault rupture. The pipeline system crosses a total of 10.5 km of faults [3]. There is one crossing of the Goromay fault and 18 crossings of the

Kliuchevskoy fault which stretches mainly along the pipeline route. In the course of pipeline design, several fault crossings were eliminated (faults №2 and №19). However, due to populated areas, roads and transport corridors, there is no further possibility to reduce the number of pipeline fault crossings. Active faults constitute a threat for the structural integrity of buried pipelines: in case of surface fault rupture during a seismic event, the pipeline is to accommodate significant differential ground displacement across the ruptured fault. The conventional buried pipelines are preferable for active fault crossings as there is possibility to avoid technical issues concerning a long run of unrestrained pipe; in addition, it reduces the risk of pipeline third-party damage. Crossing of active faults on Sakhalin II onshore pipeline route requires more complex engineering solutions [4].

2. Fault categories

The conventional engineering solution for active fault crossing is pipeline construction in a shallow, sloped-wall trench with loose backfill in order to provide pipeline flexibility within the trench. However, seasonal frost penetration on Sakhalin necessitates considering additional engineering solutions which are directly dependent on the category of the active fault [4].

The active faults that are crossed by the Sakhalin II onshore pipeline are divided into the following categories:

Category № 1 – very simple faults (4 faults). This category includes strike-slip faults with vertical and/or along the strike displacement up to 0.5 m [4].

Category № 2 – simple faults (4 faults). This category includes strike-slip faults with vertical and/or along the strike displacement exceeding 0.5 m.

Category № 3 – complex faults (1 fault). The category involves thrust faults with vertical and horizontal movements up to 0.7 m., where the orientation of the fault trace is such that the intersection angle of pipeline and fault can be easily adjusted perpendicularly (along - 20^0 N) to the horizontal fault displacement (along + 70^0 N) in order to control compression load on the pipe itself.

Category № 4 – very complex faults (3 faults). The category involves thrust faults with vertical and horizontal movements exceeding 0.7 m., where the orientation of the fault trace is such that the intersection angle of pipeline and fault can be easily adjusted perpendicularly (along - 20^0 N) to the horizontal fault displacement (along + 70^0 N) in order to control compression load on the pipe itself. In this case, the pipeline is subjected to relatively low compression loads. Pipeline expansion loops (bend offsets) are believed to reduce the risk of upheaval buckling.

Category № 5 – extremely complex faults (7 faults). The category involves thrust faults with vertical and horizontal movements exceeding 0.7 m., where the orientation of the fault trace is such that the intersection angle of pipeline and fault can be easily adjusted perpendicularly (along - 20^0 N) to the horizontal fault displacement (along + 70^0 N) in order to control compression load on the pipe itself. In this case, the pipeline is subjected to relatively high compression loads. Pipeline expansion loops (bend offsets) are believed to reduce the risk of upheaval buckling.

The above-described fault classification is based on the idea that pipeline should be oriented at fault crossing so that it is subjected to tension stress instead of compression loads. This is due to the fact that the stress which causes local bending and pipeline buckling is always less than tension stress that can result in pipeline rupture [5].

The calculations for fault 1 ALT are given as an example (table 1).

Table 1. 1 ALT fault parameters [6].

Parameters	Values
Pipeline band angle north upwards (degree)	-93.17
Fault line north upwards (degree)	-10
Vertical displacement (m)	1.00
Overthrust faulting (m)	0.13
Strike-slip faults (m)	5.40
Direction of fault movement	Westward

Table 2 compares the calculated strain values with the strain values allowable for earthquake-resistant pipes that are used at fault crossing 1 ALT.

Table 2. 1 ALT fault crossing [6].

Pipeline	Max. compression strain	Max. tensile stress
Oil pipeline	0.18% (<4.80%)	1.08% (<3.00%)
Gas pipeline	0.32% (<4.80%)	1.22% (<3.00%)

As evident from table 2, the calculated maximum values of compression strain and tensile stress both for oil and gas pipelines are less than the allowable values.

3. Measures aimed at enhancing pipeline safety

Let us consider the measures which are intended to ensure pipeline safety at active fault crossing on Sakhalin Island [4].

- Use of earthquake-resistant pipes in the fault trace zone taking into account the fault error.
- Perpendicular pipeline fault crossing (regarding fault horizontal movements) provides the control over the excessive compression loads, decreases the risk of buckling and excessive bending stresses and deformations.
- Use of bend offsets, i.e. combination of straight sections and pipe (lateral) bends, to accommodate compression loads.
- To provide additional safety, pipeline shutdown valves are installed on both sides of fault crossing.
- Implementation of seismic monitoring system that is intended for digital recording of ground shaking at sites and real-time processing of accelerograms.
- Establishment of pipeline emergency response and control center to make adequate decisions in accordance with the accidents (Pipeline Maintenance Depots).
- At pipeline fault crossing, the trench is dug along the entire width of the fault taking into account the fault errors. This decreases local bending stress caused by soil-pipe interaction forces (axial and transverse, lateral and vertical).
- Trench depth, insulation material at trench, backfill material (natural loose soil or artificial material, for example, expanded clay that can be put under pipeline or close to the pipeline at the fault) should be modified in accordance with pipeline operation conditions, soil characteristics, frost penetration depth, etc.

4. Special trench design concept

The concept of special trench design is based on the following principle: to facilitate displacement of the pipeline inside the trench without being damaged in response to earthquake activity. This means that pipeline accommodates the fault displacement by moving inside the trench without being subjected to significant stresses and deformations, which, in its turn, ensures structural integrity of the pipeline. To achieve this, backfill must be easily compressed and, by doing so, minimizes soil-pipe friction [6]. To guarantee that the pipeline behavior inside the trench corresponds to the stress analysis results, special trenches must not freeze, as ice formation alters the mechanical properties of the backfill. Besides, freeze-thaw cycles might result in undesired pipe deformations or local stresses. To avoid soil freezing, two essential factors must be considered: water entry and thermal equilibrium.

The first factor, i.e. water entry is controlled by constructing waterproof trenches, while the second – by installing insulation slabs above the pipe and inside the trench. To avoid water entry to the trench, the following three engineering solutions have been proposed considering hydrogeological and morphological conditions at each fault crossing:

- draining trenches (at sites with high permeable soil);
- waterproof trenches (at sites with low permeable soil);

- waterproof trenches on embankments (when water drain is impossible due to landscape peculiarities).

The main objective of special drainage system (draining trenches) is to discharge water collected along the special trenches during the winter months when soil is frozen.

The special trench design is intended to ensure pipeline safety during the design earthquake. As an example, let us consider enlarged waterproof trench at fault crossing №1 Alt [7]. Sand or light material was chosen as a backfill.

The trench is sealed with the geomembrane (1) made from high-density polyethylene, with edges being welded (figure 1) [4]. Prior to geomembrane installation and welding, heavy nonwoven geotextile is placed to prevent geomembrane puncture. The sealed waterproof trench is fitted with inspection pits (2) to monitor water accumulation inside the trench. The inspection pits are connected with the trench bottom by means of non-perforated pipes (3) made of high-density polyethylene. This provides additional safety to the entire system since it is possible to control water ingress.

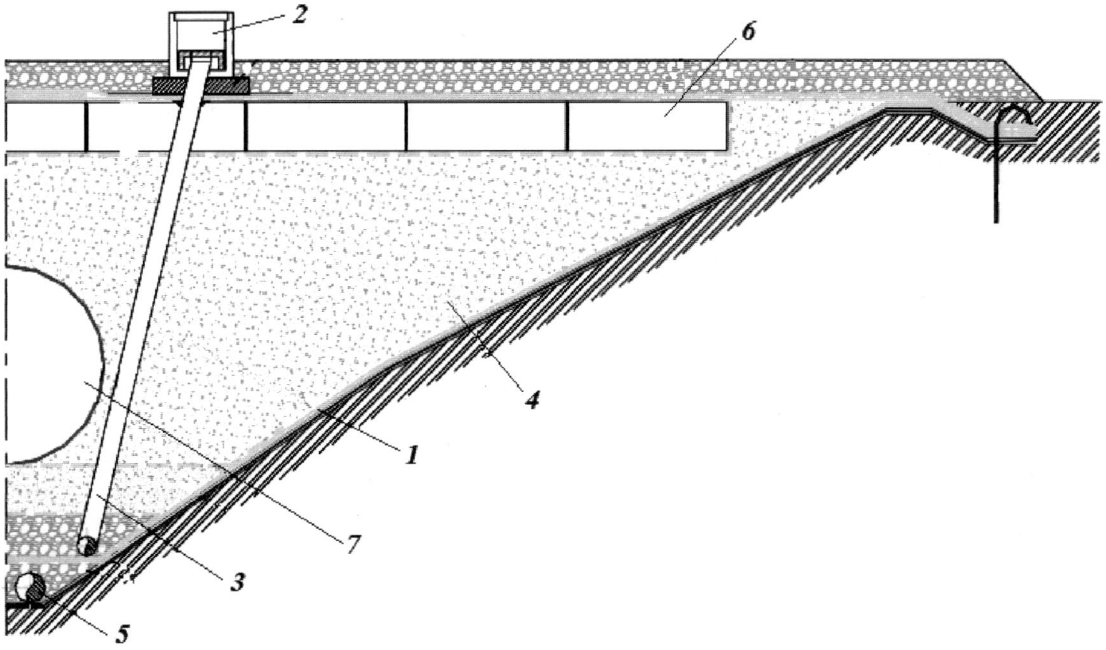

Figure 1. Waterproof trench design 1 – geomembrane; 2 – inspection pits; 3 – nonperforated pipe; 4 – light backfill material; 5 – perforated pipe; 6 – insulation slabs; 7 – pipeline.

Due to the possible sharp water level rise during the summer months, the drainage system for removing water through the high-density polyethylene perforated pipes (5) is installed at the trench bottom filled with the light backfill material (4). This decreases water stress on the trench walls and bottom. The length of waterproof trench ranges from 80 up to 100 m, which minimizes potential for pipeline damage and enables to localize the damaged section in case of waterproof barrier failure.

5. Conclusion

Thus, to provide the safety of pipeline fault crossing, the engineering solutions must be taken on the basis of the analysis results and, accordingly with the following principles:

- pipeline shall withstand a design strength level earthquake without any damages or with minimum damages. In addition, minimal interruption of normal operation can be required for insignificant pipeline repair;

IOP Conf. Series: Earth and Environmental Science **27** (2015) 012058 doi:10.1088/1755-1315/27/1/012058

- pipeline shall withstand the effect of ductility level earthquake without damages. In this case pipeline can be seriously damaged leading to pipeline shutdown for repair.

References

[1] Resolution of the Problem-Oriented Scientific Engineering Council: "Innovative Pipeline Construction Technologies in Sakhalin I and Sakhalin II projects" URL: http://pandiaweb.ru/text/77/473/7402.php. Reference date: 08.04.2015.

[2] Suschev T S 2010 *Povyishenie bezopasnosti magistralnyih nefteprovodov na uchastkah peresecheniy s aktivnyimi tektonicheskimi razlomami:* avtoref. dis. ... kandidata tehnicheskih nauk (Ufa) p 26 (in Russian)

[3] Nadein V A, Ivantsov O M 2009 Sahalinskiy truboprovodnyiy meridian *Truboprovodnyiy transport* **3(15)** 10–17

[4] Project Sakhalin II: Stage 2. Fundamentals of fault crossing design Document № 5600-Z-90-42-T-9006-00 2008 p 116 (in Russian)

[5] Vazouras P, Dakoulas P, Karamanos S 2014 Structural performance of buried steel pipelines crossing strike-slip faults *Proc. Biennial Int. Pipeline Conf. (Calgary, Canada, 29 September – 3 October 2014)* vol 4

[6] Project Sakhalin II: Stage 2. Analysis of oil and gas pipeline fault crossing 20 – fault 1Alt Document № 5600-Z-90-42-T-9002-00 2008 p 51

[7] Project Sakhalin II: Stage 2. Trench design in accordance with fault category – soil-pipe interaction analysis Document № 5600-Z-90-42-T-9003-00-P2 2007 p 127

The dependence of cylindrical resonator natural frequencies on the fluid density

V Rudachenko[1,2], V Filushin[1,3] and T Korotchenko[1]

[1] National Research Tomsk Polytechnic University, 30 Lenin Ave., Tomsk, 634050, Russia

E-mail: [2] RudachenkoVA@tpu.ru, [3] filushinvv@mail.ru

Abstract. The article examines the dependence of cylindrical resonator natural frequencies (sensitive element) on the density (mass) of different fluids flowing through it. The cylindrical resonators are being widely applied in automatic control systems of technological processes as oscillating transducer density meter. The article presents the experimental results that prove the dependence of natural frequencies and vibration amplitude on the fluid density.

1. Introduction

The natural frequency of mechanical resonator under ideal conditions and without considering vibration damping is defined by the following formula [1, 2]:

$$f = \frac{1}{2 \cdot \pi} \cdot \sqrt{\frac{K}{m}}, \tag{1}$$

where: K – resonator rigidity;

m – resonator mass.

Under real conditions, resonator frequency depends on the parameters of a given system such as mass, rigidity, and energy loss. As these parameters are of the system itself, this frequency is termed as natural one [2]. The natural frequency of cylindrical resonator with the fluid flowing through is defined by the formula [3, 4]:

$$f = \frac{\lambda^2}{2 \cdot \pi \cdot l^2} \cdot \sqrt{\frac{K}{(m_{st} + m_m)}}, \tag{2}$$

where: λ – root of frequency equation of cylinder bending vibration, which is defined depending on the restraint type [1, 3];

l – equivalent length of resonator branch;

m_{st} – mass per unit length of resonator walls;

m_m – mass per unit length of the medium inside resonator.

The dependence of resonator vibration frequency that has close values to natural frequencies on the fluid density is widely applied, for example, in density sensors [4–6]. Based on the above equations and experimental measurements, it is possible to create a calibration curve for density sensor. For example, this experiment can be carried out using sensor resonance frequency in air and distilled water [4, 5, 7].

Natural frequency of resonator depends on its parameters, precisely, its shape, size, mass, and elasticity of the material it is made of. An illustration for oscillating element is given in figure 1.

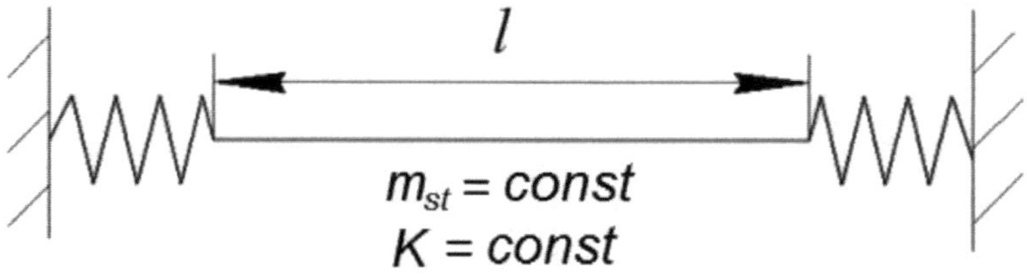

Figure 1. Resonator oscillating element.

To exclude the effect of temperature on resonant frequency calculation, ambient temperature remained constant throughout the course of the experiment. All the experiments were carried out at atmospheric pressure.

Table 1. Oscillating element design parameters.

Tube length, mm	External diameter, mm	Wall thickness, mm	Material	Bellows
1000	32	2.5	precision steel 36NHTYU	38-4-0.21x3 GOST R 55019-2012

The value of resonant vibration frequency of the oscillating element directly depends on the mass of the system itself. It should be emphasized that the design parameters of cylindrical resonator remained constant (table1) during the experiments. Thus, the current value of natural frequency altered in response to fluid density change. The present work is aimed at clarifying this dependence.

Based on the number of degrees of freedom, oscillatory systems are divided into two basic types: system with concentrated or lumped parameters and system with distributed parameters. The former presents a set of elements which is usually reduced to one of the basic parameters, i.e. elasticity and time response. These parameters vary independently from each other. Having a single degree of freedom, these systems are characterized by a finite number of natural frequencies. The latter is one in which all the elements have the same elasticity and time response. These two parameters are distributed within the system so that a change in parameter values causes alteration of the other parameter. It is obvious that the quality of mechanical system with distributed parameters is significantly higher. The quality of the system is referred to the rate of mechanical energy dispersion per oscillation period. Therefore, the oscillatory system with distributed parameters guarantees more precise measurements [4].

The fix-ends of the resonator are conditioned by a high rate of mechanical energy dispersion that passes through the substrates to the resonator body [5, 9]. To decrease energy loss, the resonator is fixed by means of bellows (figure 2).

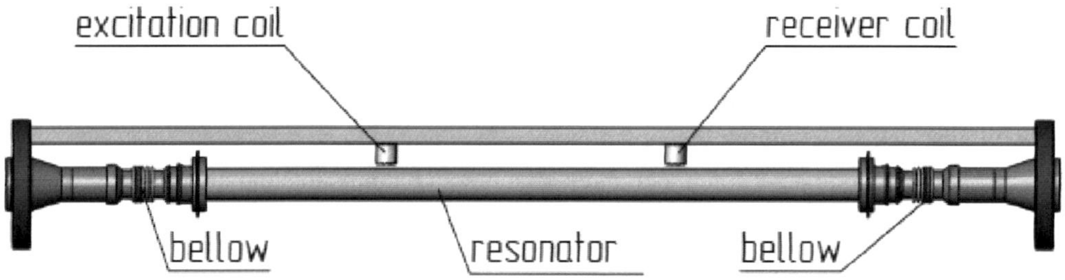

Figure 2. Oscillating system.

2. Materials and methods

Resonator is a cylindrical tube fixed by bellows. It was filled in sequence with fluids (media) of different density (table 2). With an electromagnetic sensor, the resonator oscillated vibrations at a definite frequency. The parameters of the generated signal are listed in table 3.

Table 2. Density of investigated medium.

Medium	Density, kg/m^3
Air in laboratory	1.21
Isopropyl alcohol	774.3
Distilled water	989.7
Solvent 646	831.6
Orange juice	1038.8
Non-freezing solution	957.6

Table 3. Generated signal parameters.

Initial frequency, hertz	End frequency, hertz	Amplitude, V	The number of periods generating the signal at intermediate frequency	The number of points	Signal type
300	2500	10	40	2000	continuous, sinusoidal

The experiment has revealed the difference between the received signals in terms of amplitude and frequency. Figure 3 illustrates the received signals. The presence of several resonance frequencies indicates the fact that the system has more than one degree of freedom.

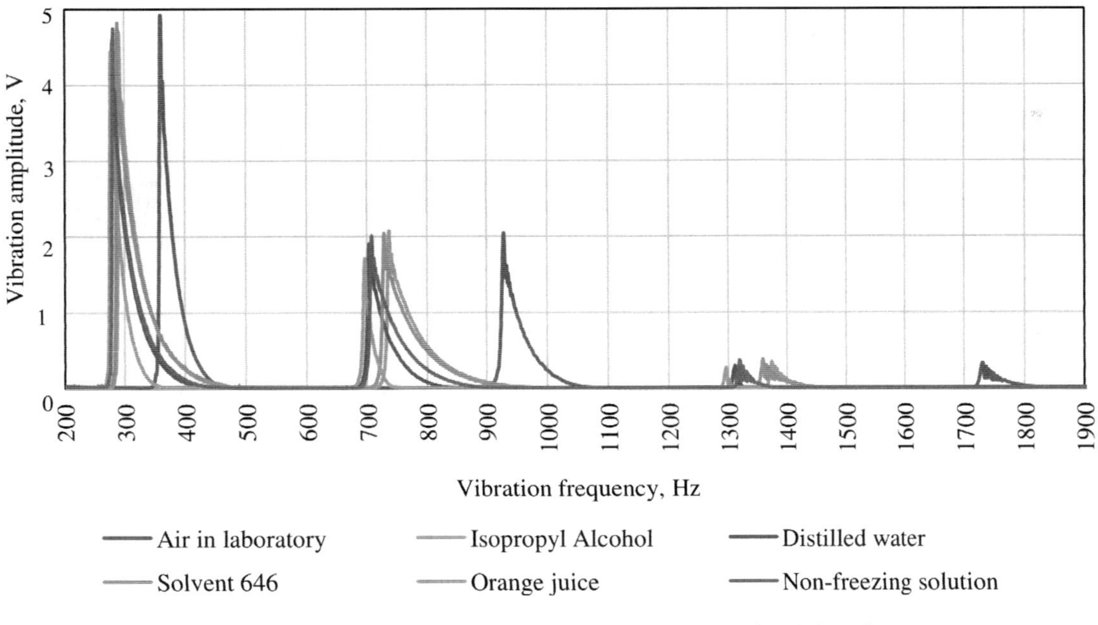

Figure 3. Amplitude-frequency curve of the received signals.

It is possible to specify the value of resonant vibrations for each fluid (medium) and for each harmonic by reducing the frequency range, i.e. generating new signals within narrower frequency range. The parameters of generated signals for determining resonant frequencies for each harmonic are given in table 4.

Table 4. Parameters of the second generated signal.

Generated signal №	Initial frequency, hertz	End frequency, hertz	Amplitude, V	The number of periods generating the signal at intermediate frequency	The number of points	Signal type
1	200	500	10	40	2000	continuous, sinusoidal
2	630	980	10	40	2000	continuous, sinusoidal
3	1200	1750	10	40	2000	continuous, sinusoidal

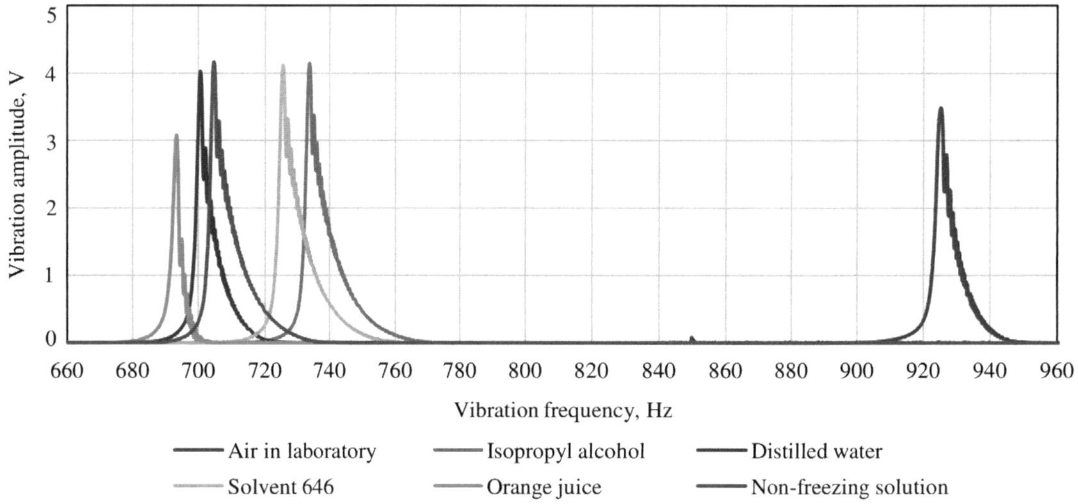

Figure 4. Amplitude-frequency curve of the received signal that corresponds to the vibrations within secondary resonant frequency (harmonic).

Figure 4 shows the measurements within the second resonant frequency range. Due to the differences in density and, consequently, in mass, resonant vibration range of each fluid (medium) is different from that of the others. The analysis of the diagrams has revealed that resonant vibration frequency value increases with increasing density. Therefore, air demonstrates the highest value of resonant frequency.

3. Conclusion

To sum up, the conducted experiments enable to define the resonant vibration value of the cylindrical resonator filled with different fluids (media). The resonant frequency values for each investigated fluid (medium) and for each harmonic are listed in table 5.

Table 5. Density of investigated fluids (media).

Investigated fluid (medium)	Resonant frequency for 1st harmonic	Resonant frequency for 2nd harmonic	Resonant frequency for 3rd harmonic
Unit of measurement	Hertz	Hertz	Hertz
Air in laboratory	357.82	923.46	1724.38
Isopropyl alcohol	287.7	732.996	1368.231
Distilled water	275.98	699.157	1306.364
Solvent 646	284.773	723.84	1352.947
Orange juice	273.443	691.95	1291.594
Non-freezing solution	277.389	703.19	1314.228

IOP Conf. Series: Earth and Environmental Science **27** (2015) 012059 doi:10.1088/1755-1315/27/1/012059

It is worth noting that the dependence of vibration resonant frequency on the density of fluid inside the resonator is approximately linear (figure 5).

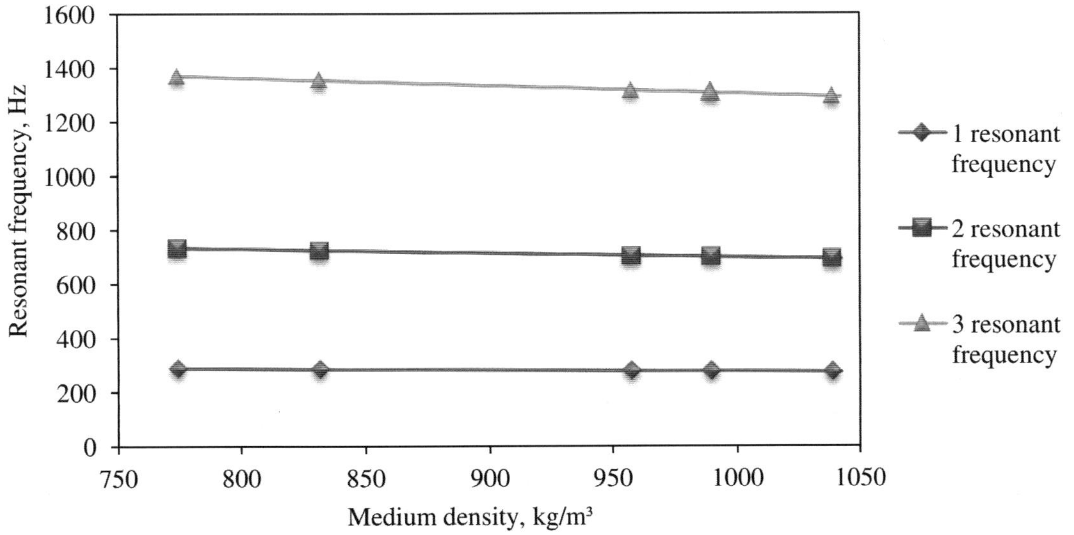

Figure 5. Dependence of vibration resonant frequency on the density of fluid inside the resonator.

Fluid (medium) mass obviously increases with increasing fluid density, which in its turn, contributes to increasing the mass of the whole oscillatory system. The vibration frequency is in inverse proportion to mass. Therefore, density increase results in resonant frequency decrease. Consequently, natural frequencies also decrease.

References

[1] Babakov I M 1968 *Teoriya kolebaniy* (Moscow: Nauka) (in Russian)

[2] Birbraer A N and Roleder A J 2009 *Extreme actions on structures* (St. Petersburg: Publishing House of Polytechnic University) p 594

[3] Shikina V E 2014 Calculation of frequency of oscillations of a piezoceramic primary converter for mass flowmeter of liquids *University proceedings. Volga region* **1 (29)** 54–63

[4] Zhukov V E 1991 *Vibratsionnye plotnomery* (Moscow: Energoavtomizdat) p 144 (in Russian)

[5] Guseynov Ya Y, Guseynov T K and Abdulova 2012 N A Odnotrubnye rezonatory povyshennoy dobrotnosti dlya vibratsionnykh plotnomerov *Proc. of Gubkin Russian State University of Oil and Gas* **4(269)** pp 48–53

[6] Najmzadeh M, Haasl S and Enoksson P 2007 A silicon straight tube fluid density sensor *J. Micromech. Microeng.* **17** 1657–63

[7] Fatkhudinov A Sh 1999 *Avtomatizirovannyy uchet nefti i nefteproduktov pri sbore, transporte i pererabotke* (Ufa: Nafteavtomatika) (in Russian)

[8] Kvilis S S 1980 *Plotnomery* (Moscow: Energiya) p 278 (in Russian)

[9] Guseynov T K 2007 Issledovanie nizkochastotnogo trubchatogo rezonatora s sosredotochennoy neodnorodnost'yu dlya vibratsionno-chastotnogo plotnomera zhidkikh sred *The electronic scientific journal "Oil and Gas Business". 2007. №1. URL: http://ogbus.ru/authors/Guseynov/Guseynov_1.pdf*

Design of automated oil sludge treatment unit

N Chukhareva[1,2], T Korotchenko[1,3] and A Yurkin[1,4]

[1] National Research Tomsk Polytechnic University, 30 Lenin Ave., Tomsk, 634050, Russia

E-mail: [2] natasha@tpu.ru, [3] tatyana1003@mail.ru, [4] yurkin0660@mail.ru

Abstract. The article provides the feasibility study of contemporary oil sludge treatment methods. The basic parameters of a new resource-efficient oil sludge treatment unit that allows extracting as much oil as possible and disposing other components in efficient way have been outlined. Based on the calculation results, it has been revealed that in order to reduce the cost of the treatment unit and the expenses related to sludge disposal, it is essential to apply various combinations of the existing treatment methods.

1. Introduction

The continued progress of oil industry advances the application of resource-efficient technologies based on technical, economic, and environmental principles. The violation of the above principles generally causes the loss of irreplaceable natural resources and, as a result, environmental damage. Therefore, the international ecological standards ISO 14000 aim to persuade companies to take more responsibility for violation of these principles and regulations [1]. However, financial costs annually incurred by petroleum companies because of oil and gas transportation accidents do not decline [2]. This obviously necessitates improvements in the existing technologies and prompts development of new ones which would be effective in oil spill detection, elimination, and further treatment of oil sludge.

The annual volume of spilled hydrocarbons to be utilized urges to develop efficient, reasonably priced, and environmentally friendly technologies based on the various physical and chemical methods of separation. In addition, treatment of oil sludge has been one of the most important aspects of environmental protection in petroleum industry [3].

2. Materials and methods

Today, there are both foreign and Russian companies which are focused on oil spill elimination and further utilization and treatment of oil and oil sludge. The literature review reveals the most widely applied technologies which have been developed both by Russian (LLC «Avantage», LLC "Spetzautocom", ZAO "RusEcoProject") and foreign companies (LLC «Alfa Laval», AG «MOG») [4-6].

The oil sludge treatment unit produced by Swedish company «Alfa Laval» has been chosen as one of the most successful decisions in terms of operational parameters and cost efficiency. Its basic parameters are listed in table 1. All calculations were made per one month of machine operation given that it was a five-day working week and 8-hour work day. The calculations involved the current price for oil sludge treatment (1250 rub/m^3).

Table 1. Technical characteristics of *"Alfa Laval"* oil sludge treatment unit.

Company	Power	Price	Treatment capacity	Treatment methods
LLC «Alfa Laval»	45 kW	13 000 000	9 m^3/h	Physical, chemical, biological
Price of a block	**Operational cost (mains)**	**Operational cost (on-site)**	**Total profit**	**Payback period**
	36 956		1 800 000	7.2 months

The advantages of Swedish project are as follows: high treatment quality of any oil products, compliance with modern environmental standards, high reliability and technical support guaranteed by the manufacturer. However, the discussed oil sludge treatment unit has the following limitations: impossibility to regenerate the initial state of spilled oil or petroleum product and need for consumables (filter cartridges and chemical reagents).

3. Results and discussion

Having considered all the advantages and limitations of "Alfa Laval" project, the authors have made an attempt to develop a new facility within import substitution framework with due regard to the listed characteristic features.

Based on the calculations which involve the costs related to the proposed unit operated from different power sources, it has been revealed that one of the most effective methods to cut sludge treatment cost is the possibility to carry out all works stationary. Depending on the operating conditions, autonomous field work will be 3-4 times more expensive (table 2).

Table 2. Technical characteristics of the proposed treatment unit.

Company	Power	Price	Treatment capacity	Treatment methods
proposed unit	15 kW	1 745 200	2 m^3/h	Physical methods
Price of a block	**Operational cost (mains)**	**Operational cost (on-site)**	**Total profit**	**Payback period**
300 000	12 318.75	47 232	400 000	4.4 months

Another main feature of the proposed oil sludge treatment unit is the possibility to apply physical treatment methods. Unlike biological and chemical methods, physical treatment method allows achieving the required quality level of the treated sludge as its structure remains constant [7-11]. Therefore, it can be stated that the new treatment unit is resource efficient and it allows the second use of oil sludge and its further utilization (for example, for road pavement or building materials production, etc.)

It should be noted that the change of the Russian Federation towards industry development implies application of domestic details and components because of financial benefits. As a result, the project cost at the stage of facility assembly, as well as the cost of the final product could be significantly reduced with the little changes in quality level as opposed to Swedish and British analogues.

At the same time, the possibility to combine several treatment methods is also considered. This would contribute to the competitiveness of the proposed unit as the possibility to apply several treatment methods could significantly expand areas of usage. To be more precise, such a sludge treatment unit can be applied when the treatment capacity is more important than preservation of oil initial characteristics. As it is stated in [8,12], when it is impossible to preserve the initial characteristics of spilled oil, it is more reasonable in terms of financial losses and treatment capacity (m^3/h) to apply biological treatment method. In addition, this method should be considered when there is higher risk of expanding the contaminated area. The main advantage of bioremediation is that it allows treating large volumes of oil sludge within a relatively short period in comparison with physical treatment methods. However, it is worth noting that biological method also has a number of shortcomings: difficulty in regenerating oil characteristics, limitations related to environmental conditions such as temperature, moisture, pH, and contaminant composition.

Thus, the combination of several oil sludge treatment methods would make the proposed oil sludge treatment unit more multi-applied, which in its turn, means that it could be used in solving various engineering tasks.

It should be noted that the change of the Russian Federation towards industry development implies application of domestic details and components because of financial benefits. As a result, the project cost at the stage of treatment unit assembly, as well as the cost of the final product could be significantly reduced with the little changes in quality as opposed to Swedish and British analogues.

To verify the advantages of the new oil sludge treatment unit, the comparison of the proposed treatment unit and Swedish one, as well as, the feasibility study in terms of short-term perspective (3 years) and long-term perspective (10 years) has been carried out. The calculations have been made for physical treatment methods only. The cost of one cubic meter of the sludge was 500 rubles. Among other possible expenses, energy consumption and price of purchased equipment were considered. Unlike "Alfa Laval" facility, the proposed treatment unit has obvious advantages in terms of short-term perspective. This is explained by lower price of domestic assembly components used in new treatment unit. However, long-term perspective does not evidence any significant difference in profits (table 3).

Table 3. Profit and work volume comparison for short- and long- term perspectives.

Parameters	**Alfa laval (1 unit)**	**Proposed treatment unit (4 units)**
Processing capacity (m^3/hour)	9	8
Expenses (unit assembly and energy, 3 years)	14 304 160 rubles	8 754 700 rubles
Total profit (3 years)	25 920 000 rubles	23 040 000 rubles
Volume of treated sludge (3 years)	51 840 m^3	46 080 m^3
Expenses (unit assembly and energy consumption, 10 years)	17 434 750 rubles	12 893 800 rubles
Total profit (10 years)	86 400 000 rubles	76 800 000 rubles
Volume of treated sludge (10 years)	172 800 m^3	153 600 m^3

In addition, it should be noted that in searching for new solutions in oil sludge treatment the main task was to develop not a huge stationary facility, but a compact mobile treatment unit. It means that the feasibility study for short-term perspective should be more accurate under planned operation conditions.

4. Conclusion

The most significant findings to emerge from this study are as follows:

- the proposed oil sludge treatment unit is of low cost in comparison with European analogues;
- due to high mobility, insensitivity to the sludge composition, small number of staff required to operate the unit, it is possible to provide more cost-effective in-situ oil spill response as compared with the ex-situ methods;
- having a proper financing, the further development and serial production of the proposed unit will provide qualified specialists with new workplaces and contribute to dealing with small local spills far from big refineries;
- the proposed treatment capacity is enough to treat oil sludge of a small refinery.

References

[1] Russian Business Consulting (RBC) [Electronic resource]: Russian Business Consulting. Electronic journal, 2000. URL:http://t.rbc.ru/tyumen_freenews/19/11/2014/956527.shtml (reference date 27.01.15)

[2] Independent Newspaper [Nezavisimaya gazeta] 2000 [Electronic resource]: URL: http://www.ng.ru/ng_energiya/2014-12-09/11_vred.html (reference date 27.01.15).

[3] Li H, Song S F, Qu C T, Xie Q and Yang B 2013 Study of treatment process and development of oily sludge *Advanced Materials Research* **838-841** 2667–72

[4] Storm-15 Machine [Electronic resource]: official web site "Man oil group". URL: http://www.manoilgroup.com/media/storm-15-ru.pdf (reference date 17.09.14).

[5] Patent 94012433 Russian Federation, Int. Cl. H 04 B 1/38, H 04 J 13/00 Method for Treatment of Oil Sludge and Soil Decontamination / Zorkin V A, Bushueva N N, Pobedinskiy N A, Beznosov V N, Chevardova N P, Aisin E Kh, Moiseev P A and Chalchenko V P Proprietors: Zorkin V A, Bushueva N N, Pobedinskiy N A, Beznosov V N, Chevardova N P, Aisin E Kh, Moiseev P A and Chalchenko V P № 94012433/26; date of filing. 08.04.94; date of publication 20.08.96, Bull. № 36 (Part II) p 3

[6] Patent 2434051 Russian Federation, Int. Cl. H 04 B 1/38, H 04 J 13/00 Mobile Module Unit for Utilization of Oil-Slime and Rejects of Products of Oil and Gas Processing / Il'in R J, Luk'janov A S, Seregin S N, Zakhar'ev G G, Magzanov S I and Sidorenko V N Proprietors: ZAO "Rusehkoproekt" (RU) № 2000131736/09; date of filing 11.06.10; date of publication 20.11.11, Bull. № 32 (Part II) 3 p.

[7] Pleshakova E V 2010 *Ekologo-funktsional'nye aspekty mikrobnoy remediatsii neftezagryaznennykh pochv* (Saratov: Saratov State University) p 47

[8] JSOC Bashneft [Electronic resource]: official web site, Alfa-laval oil sludge treatment unit maintenance. URL: http://zakupki.bashneft.ru/purchase/11028/ (reference date 24.10.14).

[9] Kaminsky E F and Khavkin V A 2001 *Glubokaya pererabotka nefti: tekhnologicheskiy i ekologicheskiy aspekty* (Moscow: Technics) p 383

[10] Pikovsky Y.I. 2003 Problema diagnostiki i normirovaniya zagryazneniya pochv neft'yu i nefteproduktami *Pochvovedenie* **9** p 1132–40

[11] Khavizov A R 1996 *Utilizatsiya otrabotannykh masel* (Ufa: Reaktiv) p 260

[12] Akhmetov S A, Ishmiyarov M Kh and Kaufman A A 2009 *Tekhnologiya pererabotki nefti, gaza i tverdykh goryuchikh iskopaemykh (*Saint-Petersburg: Nedra) p 827

IOP Conf. Series: Earth and Environmental Science **27** (2015) 012061 doi:10.1088/1755-1315/27/1/012061

Suppression of flow pulsation activity by relaxation process of additive effect on viscous media transport

S Kharlamov[1,4], P Dedeyev[1,5], L Meucci[2,6], I Shenderova[1,7], A Manastirniy[3,8] and M Usenko[3,9]

[1] National Research Tomsk Polytechnic University, 30 Lenin Ave., Tomsk, 634050, Russia
[2] Engineering school, University of Florence, 4 P.zza S.Marco, Florence, 50121, Italy
[3] Engineering and technical center, 13/3 Mira Ave., Tomsk, 634057, Russia

E-mail: [4] kharsn@mail.ru; [5] dedeyev@mail.ru; [6] laurina.meucci@gmail.com, [7] inna-shenderova@yandex.ru, [8] manastirniy@tomskiy-itc.ru; [9] usenkomu@tomskiy-itc.ru

Abstract. The article presents the analysis of the processes occurring together with the turbulent transfer of impulse in mixture of hydrocarbon fluid and polymer solutions (anti-turbulent additives). The study evaluates complex shear flows by popular theoretical and practical methods. Understanding of hydrodynamic and dissipative effects of laminar-turbulent transition tightening and turbulence suppression is provided. The peculiarities of "thin" flow structure in pipeline zones with complex shape walls are evaluated. Recommendations to forecast the local flow parameters, calculation of hydraulic resistance are given.

1. Introduction

Currently, oil and gas transporting companies face the challenges of effective development and application of technologies, which increase flow capability of pipeline systems [1, 2]. The application of antiturbulent additives (ATA) is valuable due to the effective raw hydrocarbons transportation management without influencing pipeline system structure.

Thus application of ATA looks promising, mostly because it enlarges pipeline transmission capacity. But numerous instabilities in technological processes, as well as changes in heat and vortical transfer significantly influence flow regimes. These range of questions, connected with the understanding of viscous media and polymer additive flow features, defines the relevance of the article and requires detailed analysis within physical, mathematical and chemical modeling. The study of heat, mass and impulse transfer in the above mentioned systems will allow to minimize energy loss and to make positive forecast of laminarized flow dynamics. The study will enlarge the analysis database of transition flows and create the basis for making integral linkage of hydraulic resistance.

The article aims to systematize the data concerning hydrodynamics of laminarized flows with ATA in nature and mechanics; to develop criteria linkages for monitoring and control of hydraulic resistance in complex shear flows with polymer additives; to develop mathematical model of steady and transition hydrocarbon flow regimes in pipelines.

2. Questions of inner flow modeling with inhomogeneous turbulence

It is known [8-10], that pipeline flow could be laminarized within approaches, which consider the specificity of viscous media flow: in the areas of variable cross-sections (the acceleration in joints of confusser and diffuser types) or by an inlet of special high-molecular polymer additives. Particularly, the Toms' effect [3], discovered at 1946, has found a broad application in various fields, such as pipeline oil transportation, mud injection into wells, fire safety. In spite of significant number of supplements, physical and mathematical essence of the effect is still having some questions to be considered. There is a need in detailed research of flows with inhomogeneous anisotropic turbulence in inner systems. The physical modeling is based on L. Prandtl and T. Karman's, studies about multiband structure of turbulent flow [4].

The analysis of data, which have been obtained in [5], makes it possible to state, that the injection of such additive polymer solutions leads to considerable energy economy, mostly caused by flow laminarization in viscous sublayer and buffer zone, at that, the core flow does not undergo any changes. Considering the physical description of mechanisms, it should be mentioned that the most widespread theory [1, 2, 6, 7] of such processes description, states that the long-length polymer molecule prevents the development of vortex by creating the "cocoon, coil" structure or "polymer solution – media". It has been discovered [1, 6, 7], that the efficiency of ATA depends on its concentration. A large number of ATA function in minor concentration zone ($10^{-3}\div10^{-4}\%$), but the growth of concentration leads to the decrease of laminarization rate. The bibliographic data shows there are no unified laws and mechanisms of turbulent flow generation in these areas. Thus, the detailed theoretical and experimental analysis of convection-diffusion processes in solutions at laminarization process within the models with anisotropic turbulence. In particular, the study of the hydrodynamic questions of disturbances relaxation in laminarazing flow, gives the results, which allow estimating Toms' effect [8] time initiation by criterion:

$$\frac{u_*^2\theta}{\nu} \geq 1, \qquad (1)$$

where u_* – dynamic velocity,
ν – kinematic viscosity,
θ – perturbation relaxation time in polymer solution.

However the criterion (1) is inefficient for the description of complex shear flow evolution and imposed external forces. Certain problems have been studied in [2, 9], where conditions of process limits (flow laminarization) were formulated with the help of Lumley criterion and description of turbulent flow by Blasius's friction law. This approach is practically valuable for industrial testing and producing of ATA, but from the point of rigorous flow modeling, these criterions do not explain ATA interaction with pulsating flow parameters.

3. Integral parameters forecast

Among the number of mentioned questions, special attention should be paid to hydraulic friction coefficient modeling. There are many formulas that have been successfully applied for the calculation and construction of pipelines, as well as for the determination of the required number of pumping stations and their distribution. It is considered, that the equation (2) should be applied for calculating smooth pipes parameters, while equation (3) is applied for calculating rough inner surface parameters:

$$\frac{1}{\sqrt{\lambda}} = 2\lg(\mathrm{Re}\sqrt{\lambda}) - 0{,}8 \qquad (2)$$

$$3{,}75 + \frac{2\sqrt{2}}{\sqrt{\lambda}} - 5{,}75\lg\frac{a}{k} = \Phi\left(\frac{k\cdot\upsilon_*}{\nu}\right) \qquad (3)$$

It should be mentioned that the application of (2) and (3) relations for the description of viscous hydrocarbon flow (media and ATA) motion is inappropriate because these equations do not consider the effect of ATA on the wall friction. The article attempts to derive an appropriate formula for λ in flows with additives.

In particular, when describing the friction loss in flow media, it should be considered that ATA are unable to create chemical reaction of liquid and polymer solution and that the concentration of ATA is not enough to change the rheology of transported media. This allows basing only on concentration and ATA parameters during the modification of equation (2), but due to the lack of data of FLX-O additive testing, the concentration influence becomes the single parameter of modification. Modification of equation (2) is realized by introducing $\psi(C)$ dependence, where C is the concentration of the additive:

$$\frac{1}{\sqrt{\lambda}} = 2(\lg e)\cdot\psi(C)\ln(\mathrm{Re}\sqrt{\lambda}) - 0,8 \tag{4}$$

Industrial test data of FLO-X additive [10] allows calculating $\psi(C)$ (table 1).

Table 1. ψ(C) values with different concentrations of FLO-X.

***C*, ml/m³**	0	5,2	11,3	19	28,8	37,5	**50**
***Ψ* (C)**	**0.979**	**1.018**	**1.140**	**1.32**	**1.49**	**1.64**	**1.84**

Due to the fact that the equation (4) is given implicitly, the calculation of function dependencies is performed with numerical methods. The mathematical processing of the calculation was performed by Mathcad and Maple 17 software programs. Obtained dependencies, calculated values of approximation accuracy and correlation coefficients are presented in table 2.

Table 2. Types of **Ψ (C)** function and evaluation of the results reliability.

Function type	***Formula description***	***Approximation accuracy R^2***	***Correlation coefficient***
Linear	$\Psi(C) = 0.0184\cdot C + 0.9429$	0.997	0.998
Exponential, 1st type	$\Psi(C) = 0.989\cdot\exp(0.0131\cdot C)$	0.978	0.989
Exponential, 2nd type	$\Psi(C) = 1 - \exp(-18.525\cdot C^{-1})$	0.972	0.979
Logarithmic	$\Psi(C) = 0.357\cdot\ln(C) + 0.343$	0.935	0.967
Degree	$\Psi(C) = 0.631\cdot C^{0.262}$	0.969	0.985

It should be noted that the presented dependences of ψ (C) are only practical particular equations and there is a need in further evaluations for the real pipeline transportation conditions.

4. Prospects of physical and mathematical modeling of complex shear flows

In case of complex shear flows with inhomogeneous turbulence in transition regimes there is a clear in creating connections between parameters of turbulent additives and "thin" parameters of pumped medium with its rheological properties. However the question about the types of such connections remains open. Modeling of such complex spatial flows is impossible without PC and the whole spectrum analysis of viscous media flows. It is well-known [11], that conservation laws of mass, momentum and energy are the ground of the researched phenomenon. These laws can be described by equations:

$$\frac{\partial\rho}{\partial t} + div(\rho\bar{v}) = 0 \tag{5}$$

$$\frac{D(\rho\vec{v})}{Dt} = DIV\,\vec{\vec{\tau}} + \vec{F} \tag{6}$$

$$\frac{D(\rho c_p T)}{Dt} = div(\lambda gradT) + S_T \quad (7)$$

$$\Phi(P, R, T) \equiv 0 \quad (8)$$

Symbols in the equations (5) – (8) are commonly accepted.

Bibliographic data analysis of complex shear stress flow modeling shows [11, 12] that applied engineer calculations are based on RANS method. Besides, interpretation of results requires validation and verification of chosen turbulence model. It should be noted that at "flexible" semi-empiric turbulent model application, the following aspects should be considered:

- ATA inlet does not cause chemical reaction of liquid and polymer solution;
- Use rate of ATA does not change the rheology of transported media;
- The temperatures of media and ATA are the same; there is no processes of interphase heat exchange.

On the ground of bibliographic data analysis [1 – 12] it is possible to make a conclusion that there are benefits of changes prediction in local structure of complex flow by two-parameter models ((k-ε), (k-ω), (k-L)). These models have the parameters which show changes in local properties of pulsating flow: here (k) is kinetic energy, (ε)-rate of dissipation, (ω)-pseudovorticity, (L)-integral scale of energy-vortex. However, it should be noted that attempts to make the analysis of flows with strong anisotropic structure on the basis of these models are unlikely to be correct. The reason is that all these models are based on the assumption that molar diffusion coefficients are scalar (viscosity, thermal conductivity). In such cases it will be particularly valuable to apply RSS (Reynolds Shear Stresses) models with transport equations for one-point correlation moments of second order velocity fields fluctuations with the support base of two-parameter turbulence models. For example, (RSS-L) [11, 12].

5. Discussion of the results

Despite the relative limits of presented relations of λ forecast, it is possible to draw the conclusion about ATA interaction with fluid. The highest rate of accuracy is obtained by linear and exponential (I type) dependencies application. This is proved by calculated coefficients of correlation (0.998 and 0.989) and approximation accuracy (0.997 and 0.978).

The obtained dependencies (figures 1 – 2) show satisfactory dynamics in smooth pipe, which was originally built into foundation as the result of choosing the equation (2) as a basic one. In the range of Re numbers higher than 30,000 it is necessary to take into account the relative roughness of pipe walls in order to obtain the correct flow description.

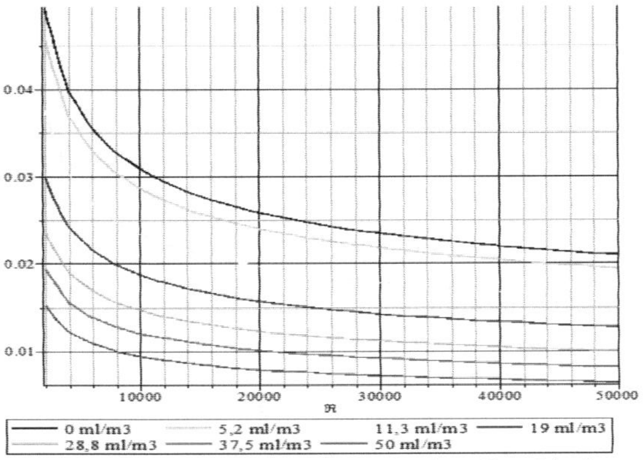

Figure 1. The dependence of hydraulic friction coefficient λ on Re number and concentrations of ATA (FLO-X) in case when $\psi(C)$ is described by linear dependence.

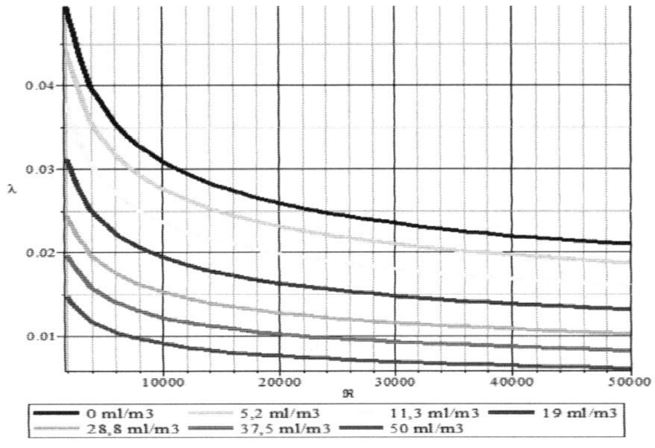

Figure 2. The dependence of hydraulic friction coefficient λ on Re number and concentrations of ATA (FLO-X) in case when $\psi(C)$ is described by exponential dependence.

In figures 3, 4 turbulent viscosity field changes (figure 3) and speed (figure 4) are given to show the dynamic parts of the averaged and pulsating flow structure, as well as the properties evolution of space vortex formations in pipelines with U-joints. Calculations of isothermal flow of homogeneous hydrocarbon droplet (low viscosity oil) are made with the introduction of RSS-kL-model. Features of the solution method and formulation of the model is described in details in [12]. It is evident that in areas with a curved boundary background. There is a field that results in both turbulence and laminarization, due to the acceleration and deceleration of flow. Localization of these areas is only possible on the basis of models with high-level circuits of constitutive equations. Moreover, the suppression of turbulence intensity by introducing polymer additives in these zones will effectively manage the process of natural resources transportation.

Figure 3. Molar dynamic viscosity field of Newtonian dropping fluid with low viscosity. Properties are similar to the oil in pipelines of complex shape (Re_{in}=8000, U- shaped junction of the constant diameter, calculation - RSS-*kL*- model [12])

Figure 4. Velocity field distribution in pipelines with U-sections (D=0.5, m, Re_{in}=10000)

6. Conclusion

Despite significant progress in modeling of relaminarazing flows in complex pipelines there is still no clear understanding of processes, occurring in mixtures of viscous media with polymer solution of random concentration. The majority of experimental research is concentrated on integral parameters of flow without considering such characteristics as turbulent viscosity, frequencies of turbulent vortex pulsations and etc. It is obvious that experiments with fixation of abovementioned parameters should be held. On the other hand, the description of hydraulic resistance coefficient should be studied due to the prospect to get a distribution of velocity profile. Systematic investigations of flow dynamic structure should be carried out depending on ATA parameters changing: molecular weight of the polymer, the effective length of the polymer chain, elasticity of the molecule and its ability to dampen.

Calculations of the "thin" flow structure shows that the closure of the defining equation system (5)–(8) with RSS turbulence model, allows to formulate valuable practical conditions for pipeline optimal functioning with complex pipeline wall structure and connections. The turbulence model is justified by the class of laminirazing flows in pipes. Obtained results enable to enlarge the experimental database to derive new semi-empirical models which are necessary to study transportation of gas hydrates and liquefied natural gas. These issues could become subjects of further investigations.

References

[1] Tarasov M Yu and Yuzhakov I S 2011 Field tests of anti-turbulence additive agents for increasingflow capacity of heavy oils pipelines *Oil Industry Journal* **10** 117–19

[2] Rakhmatulin Sh I 2005 O turbulentnom techeniyi slaboconcentrirovannikh rastvorov polimerov v truboprovodakh *Neftegazovoye delo* – URL: http://www.ogbus.ru/authors/Rahmatullin/Rahmatullin_1.pdf

[3] Toms B A 1949 Some observations on the flow of linear polymer solutions through straight tubes at large Reynolds numbers *In Proceedings of the 1st International Congress on Rheology (North Holland, Scheveningen, 21–24 September 1948)* **2** 135–41

[4] Laufer J 1954 The structure of turbulence in fully developed pipe flow *NACA Technical Report* (Washington, DC, USA) **1174** 1–18

[5] Bogdevich V G et al 1980 Some problems of wall flows control *Journal of applied mechanics and technical physics* **5** 99–109

[6] Voskoroboynic V A, Grinchenkov V T and Makarenko A P 2007 Snizhenie gidrodinamicheskikh shumov rastvorami visokopolimernikh polimerov *Acusticheskiy vestnik* **10** 33–42

[7] Kornilov V I 2005 Problems of turbulent friction with active and passive methods *Thermal physics and aeromechanics* **2** 183–208

[8] Lumley J F 1969 Drag reduction by additives *Fluid Mechanics* 367–84

[9] Zaryankin A E and Baranovskiy B V 1975O svyazi stepeni turbulentnosti s chislom Reynoldtca. *IVUZ, seriya "Energetika"* **5** 144–47

[10] Bogdevičius M et al 2013 Mathematical modeling of oil transportation by pipelines using anti-turbulent additives *Journal of Vibroengineering* **15** 419

[11] Bubenchikov A M and Kharlamov S N 2001 Mathematical models of heterogeneous anisotropic turbulency in inner flows (Tomsk: TGU publishing) p 440

[12] Kharlamov S N and Alginov R A 2014 Modelling of Complex Shear Flow Structure in Pipelines *International Journal of Engineering, Science and Innovative Technology* **3** 500–9

IOP Conf. Series: Earth and Environmental Science **27** (2015) 012062 doi:10.1088/1755-1315/27/1/012062

Formalization of hydrocarbon conversion scheme of catalytic cracking for mathematical model development

G Nazarova[1,2], E Ivashkina[1,3], E Ivanchina[1,4], S Kiseleva[1,5] and V Stebeneva[1,6]

[1] Department of Fuel Engineering and Chemical Cybernetics, National Research Tomsk Polytechnic University, 30 Lenin Ave., Tomsk, 634050, Russia

E-mail: [2] silko@tpu.ru, [3] ivashkinaen@tpu.ru, [4] ied@tpu.ru, [5] svk22077@mail.ru, [6] stebeneva_valeriya@mail.ru

Abstract. The issue of improving the energy and resource efficiency of advanced petroleum processing can be solved by the development of adequate mathematical model based on physical and chemical regularities of process reactions with a high predictive potential in the advanced petroleum refining. In this work, the development of formalized hydrocarbon conversion scheme of catalytic cracking was performed using thermodynamic parameters of reaction defined by the Density Functional Theory. The list of reaction was compiled according to the results of feedstock structural-group composition definition, which was done by the n-d-m-method, the Hazelvuda method, qualitative composition of feedstock defined by gas chromatography-mass spectrometry and individual composition of catalytic cracking gasoline fraction. Formalized hydrocarbon conversion scheme of catalytic cracking will become the basis for the development of the catalytic cracking kinetic model.

1. Introduction

At the present stage of fuel chemical technology development the advanced petroleum refining processes are the most significant for refining, because the industrial sector is focused on the efficient use of natural fuel-power resources and the demand in fuel is increasing every year.

Currently, the construction of modern technologies for advanced petroleum processing such as hydrocracking, catalytic cracking, hydrotreating of heavy hydrocarbon fractions, hydrodewaxing and hydroisomerization is organized at the major refinery plants. Modernization of the refinery industry is principally connected with the aim to create a competitive refining and petrochemical sector, corresponding the high level of modern foreign refineries (oil refining depth – 85-95%).

Advanced petroleum processing is aimed to the production of light hydrocarbon fractions (mainly gasoline and diesel fractions) from the heavy fraction of petroleum feedstock.

Exploitation of advanced petroleum processing units in the energy and resource-efficient modes should be carried out with the application of the mathematical models which are based on the hydrocarbon conversion mechanism reflecting the physical and chemical essence of process [1, 2].

It is important to take into account factors that influence the quality and quantity of production (multi-component of feedstock, catalyst types, process conditions, associativity of flowsheet devices and others.) during the development of mathematical model of the reactor. In addition, in petroleum fractions with boiling point above 200 °C the cyclic hydrocarbons are of mixed (hybrid) nature. Usually cyclic hydrocarbons contain the side paraffinic chains and, simultaneously, a part of naphthenic and aromatic rings.

In general, the mathematical model of reactor represents a set of hydrodynamic and kinetic components supplemented by the material and heat balance equation.

The aim of this work is to develop the formalized hydrocarbon conversion scheme in advanced petroleum processing on the example of the catalytic cracking vacuum distillate on the basis of which the mathematical model of the catalytic cracking reactor-regenerator will be further developed.

Technological aggregation on the certain characteristics is generally applied during the development of mathematical description of the refining process. Reaction system components are aggregated according to their belonging to the specific hydrocarbon series [3,4] and the least probable reaction is excluded from the list in order to simplify schemes. This approach is characterized by the reaction mechanism with the average reactivity of hydrocarbon groups.

Since the difficulty of advanced petroleum processing modeling lies in the determination of group composition of heavy hydrocarbon fraction, the most popular approach to the formalization of such processing scheme will be the technological aggregation on the fraction boiling point [5–10]. However, the chemical composition of the fractions is diverse and the scheme does not take into account the different reactivity within the selected groups.

There are combined models which take into account not only interactions of pseudocomponents on the fractional composition but also the chemical conversion of the main hydrocarbon groups of the catalytic cracking reaction system [11–14] and also deactivation functions, diffusion effects, chemical reaction and etc. [15–22].

The selection of the formalization level of hydrocarbon chemical conversions is an important stage in the development of mathematical description, which is preceded by the stages of studying of the theoretical foundations and experimental data on the hydrocarbon composition of installation streams, analysis of technological modes of reactor operation and the identification of features of industrial units exploitation, laboratory investigation (such as chromatography mass spectrometry, chromatography, n-d-m-method, Hazelvud-method) and thermodynamic analysis of the catalytic cracking reactions.

2. Determination of quality and structural-group composition of catalytic cracking feedstock

Data concerning quantitative and qualitative group composition of catalytic cracking feedstock and individual composition of gasoline fraction is required for the development of the formalized hydrocarbon conversion scheme in the catalytic cracking.

Liquid adsorption chromatographic separation on silica gel ASA with a grain size of 0.2 - 0.3 mm was performed according to VNII NP algorithm with the aim to determine the structural-group composition of catalytic cracking feedstock. Hydrocarbon fractions were extracted sequentially by desorption method with the application of solvents with different polarity (hexane, toluene and hexane, ethyl alcohol and toluene). The laboratory investigation results of catalytic cracking feedstock showed that paraffinic and naphthenic content equal to 61.20% wt, aromatic content of 35.57% wt. and resin content of 3.23% wt.

Hydrocarbons were separated by the refractive index on paraffin-naphthenic and aromatic fraction. Measurement accuracy of refractometer IRF-22 was tested by hexane refractive index. Structural-group composition of paraffin-naphthenic fractions was determined by the n-d-m-method, for aromatic hydrocarbon fraction by the Hazelvud-method (table 1).

Besides, the research in the determination of molecular weight and total sulfur content in catalytic cracking feedstock was performed. The devices CRYETTEWR, determining the freezing point of the samples and X-ray fluorescence energy-dispersive analyzer of sulfur "SPECTROSCAN SL», were applied for this purpose. The sulfur content and molecular weight of the feedstock is 0.0461% wt. and 382.43 g/mol respectively.

Mass-spectra of the catalytic cracking feedstock components were obtained in the laboratory by gas chromatography-mass spectrometry for determining qualitative composition of raw materials. Chromatography-mass spectrometer Hewlett Packard 6890 Gas Chromatograph System with 5973 Mass Selective Detector with GC Chemstation software was applied as the analytical equipment.

Table 1. Structural-group composition of catalytic cracking feedstock.

Parameter	Paraffin-naphthenic fraction	Aromatic fraction
Carbon content in the aromatic structures (Ca)	2.59	27.52
Carbon content in the naphthenic structures (Cn)	31.27	55.73
Carbon content in paraffinic structures (Cp)	66.14	16.76
Average number of aromatic rings (Ka)	0.14	1.40
Average number of naphthenic rings (Kn)	2.05	3.90
Total number of rings (Ko)	2.19	5.30

According to the structural group composition data of paraffin and naphthenic fraction of catalytic cracking feedstock (n-d-m-method), the fraction contains some amount of aromatic structures. It can be explained by the fact that the sorption of this structure is substantially similar with the sorption of naphthenes and that the small amount of this structure is washed out by the solvent. Content of aromatic structure is 2.59 % in catalytic cracking feedstock. Content of carbon quantity in naphthenic structures of feedstock is 31.27 %, the average number of naphthenic rings - 2.05 units.

According to the results of gas chromatography-mass spectrometry, the paraffin hydrocarbons are presented by hydrocarbons with the carbon number C_{13}–C_{40}. Naphthenic hydrocarbons are presented by mono- and polycycloalkanes with long alkyl substituents of normal and iso-structure with carbon atoms in the alkyl substituents alkilnyh C_1–C_{25} (figure 1).

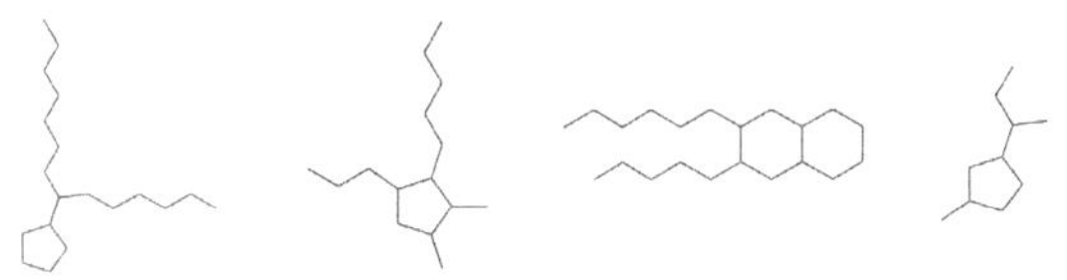

Figure 1. Examples of naphthenic hydrocarbons molecular structures (methylalkylcyclohexanes, bicycloalkanes, dimethylalkylcyclohexanes, etc.)

According to the analysis of aromatic concentrates by the Hazelvude-method the aromatic structure is presented by hybrid structures. In aromatic fractions isolated from catalytic cracking feedstock the average number of naphthenic rings (3.9 units) is more than the number of aromatic rings (1.4 units). The total content of rings in catalytic cracking feedstock is 5.3 units. The carbon content in aromatic and naphtenic structures of feedstock is 27.52 and 55.73 % respectively. Carbon content in paraffinic fragments of aromatic concentrate indicates the presence of alkyl substituted of aromatic hydrocarbons and naphthenes; the content of paraffins fragments – 16.76%. The simplest representatives of aromatic hydrocarbons of catalytic cracking feedstock are monoalkylbenzenes, methylalkylbenzenes, methyl and substituted naphthalenes, monoaromatic steroids, etc.

Laboratory investigation by the gas-liquid chromatography was conducted to determine an individual composition of catalytic cracking gasoline fraction using a gas chromatograph "HROMATEK - CRYSTAL 5000" version 2 with a flame ionization detector and capillary column DV -1 100·0.25·0.5. Software "Chromatec analyst" for the management, collection and processing of chromatographic information was used. The results of determining the individual composition of gasoline fraction was processed using the software developed at the Department of Fuel Engineering and Chemical Cybernetics of Tomsk Polytechnic University (table 2). The program is destined for processing of chromatographic data to determine the allocation of high-octane components of gasoline and aggregation of gasoline hydrocarbons according to the different chemical characteristics (molecular structure, reactivity, octane number etc.)

3. Thermodynamic analysis of the catalytic cracking reactions

The feedstock of advanced petroleum processing is characterized by the high boiling point of fractions (350-570 °C). Thermodynamic parameters of reactions involving the high molecular weight hydrocarbons are virtually absent in directory. Furthermore, it is also important to take into account the real operating conditions of industrial unit at calculation of reactions thermodynamic parameters, as well as the interaction effects of hydrocarbons molecules with the reaction mixture molecules in the

Table 2. Individual and combined components of the catalytic cracking gasoline fraction.

Component	% wt	Component	% wt	Component	% wt
propane	0.082	trimethylhexanes	0.633	tetramethylbenzenes	1.303
n-butane	0.331	isoaraffins C_9	1.214	aromatics C_{10}	6.213
n-pentane	0.504	methylnonanes	0.957	aromatics C_{11+}	5.416
n-hexane	0.586	dimethyloctanes	0.835	butene-1	0.483
n-heptane	0.411	isoaraffins C_{10}	0.438	butene-2	1.422
n-octane	0.382	isoaraffins C_{11+}	1.394	isobutylene	0.468
n-nonane	0.236	cyclopentane	0.114	olefins C_4	0.007
n-decane	0.215	methylcyclopentane	1.939	pentene-2	0.959
paraffins C_{11+}	0.284	ethylcyclopentane	0.120	pentene-1	0.327
i-butane	0.984	dimethylcyclopentanes	0.460	Methylbutenes-1	0.933
i-pentane	3.900	ethylmethylcyclopentanes	0.133	2-methylbutene-2	1.741
2-methylpentane	3.149	trimethylcyclopentanes	0.474	cyclopentene	0.243
3-methylpentane	2.095	n-propylcyclopentane	1.906	olefins C_5	0.031
2,2-dimethylbutane	0.013	cyclohexane	0.706	hexene-1	0.166
2,3-dimethylbutane	0.713	methylcyclohexane	1.428	hexene-2	0.292
2-methylhexane	2.537	1,2-dimethylcyclohexanes	0.226	hexene-3	1.003
3-methylhexane	1.852	1,3-dimethylcyclohexanes	0.302	methylcyclopentenes	0.662
3-ethylpentane	0.128	naphthenes C_8	0.968	methylpentenes-1	0.613
2,3-dimethylpentane	0.073	naphthenes C_9	1.375	methylpentenes-2	2.058
2,4-dimethylpentane	0.518	naphthenes C_{10+}	1.148	cyclohexene	0.077
2,2,3-trimethylbutane	0.019	benzene	0.583	olefins C_6	0.896
2-methylheptane	1.094	toluene	3.295	heptene-1	0.042
3-methylheptane	1.277	ethylbenzene	1.355	heptene-3	0.323
3-ethylhexane	0.299	i-propylbenzene	0.063	methylhexenes	0.647
2,3-dimethylhexane	0.12	n-propylbenzene	0.517	dimethylpentenes-2	0.156
2,4-dimethylhexane	0.352	1,3,5-trimethylbenzene	0.887	olefins C_7	1.632
2,3,4-trimethylpentane	0.011	1,2,4-trimethylbenzene	2.675	octene-2	0.067
methylethylpentanes	0.435	1,2,3-trimethylbenzene	0.682	octene-3	0.098
methyloctanes	1.982	dimethylbenzene= xylols	6.154	olefins C_8	0.771
dimethylheptanes	0.598	methylethylbenzenes	3.251	olefins C_{9+}	1.705

calculation of the thermodynamic parameters of reactions.

The software Gaussian which implements the quantum-chemical methods of calculation of the molecule electronic structure and provides adequacy and sufficient accuracy of calculations for chemical reactions simulation of refinery processing was used to solve these difficulties.The ab initio (non-empirical) method DFT – Density Functional Theory was used as a main calculation method. Theoretical approximation was B3LYP model (Becke's density functional theory (B3) with theoretical approach using Lee, Yang and Parr electron correlation (LYP), 3-21G basis.

The list of anticipated catalytic cracking reactions was made on the basis the laboratory investigations to determine the structural-group and quality composition of feedstock and individual composition of the gasoline fraction.

The calculation results of the average values of the thermodynamic parameters of the most probable catalytic cracking reactions at the process conditions (T = 504 ° C, P = 0.108 MPa).are presented in table 3. The values of Gibbs energy reaction (ΔG) characterizes the probability of reaction and it value is necessary for formalization of the hydrocarbon chemical transformations scheme, which will be the foundation to the catalytic cracking kinetic model.

According to the calculated thermodynamic values, the reactions of high molecular weight paraffins cracking (ΔG = –76.26 kJ/mol), hydrogen transfer (ΔG = –140.24 kJ/mol), dehydrogenation of naphthenes (ΔG= –114.33 kJ/mol), dealkylation of aromatic hydrocarbons (ΔG= –76.39 kJ/mol) and naphthenes (ΔG= –107.65 kJ/mol) as well as coke formation reactions (ΔG= –597.2 kJ/mol) are having the most thermodynamic probability.

Table 3. The average values of the thermodynamic parameters of catalytic cracking reactions

Hydrocarbon groups	ΔH, kJ / mol	ΔG, kJ / mol
Cracking of high molecular weight n-paraffins C_{13}–C_{40}	69.46	–76.26
Cracking of high molecular weight n-paraffins with formation of isoparafins	66.96	– 66.28
Cracking of medium weight n-paraffins and C_5–C_{12}	73.60	– 55.63
Isomerization of medium weight paraffins C_5-C_{12}	–1.78	– 0.56
Cracking of olefins;	92.72	–32.36
Dealkylation of naphthenes;	118.14	–107.65
Hydrogen transfer;	74.26	–140.24
Dealkylation of aromatic hydrocarbons;	109.68	–76.39
Dehydrogenation of naphthenes;	210.68	–114.33
Cyclization of olefins to naphthenes	–53.8	–7.54
Polycondensation and coke formation	104.89	–597.20

4. Development of formalized hydrocarbon conversions scheme in the catalytic cracking

Formalized hydrocarbon conversions scheme in the catalytic cracking was composed according to the results of thermodynamic analysis taking into account the reversibility of reaction. Detailed hydrocarbon conversion scheme with taking into account the feedstock components, light and heavy gasoil are presented in figure 2.

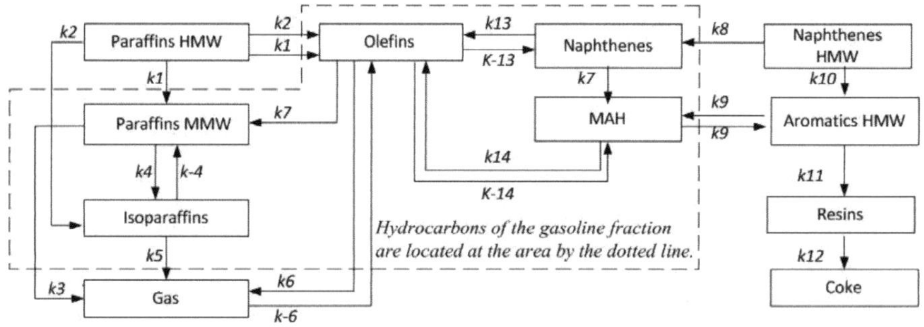

Figure 2. Formalized hydrocarbon conversion scheme in the catalytic cracking: kj and k-j – reaction rate constants in forward and reverse direction.

Paraffinic hydrocarbons are represented by two groups in the formalized hydrocarbon conversion scheme: "Paraffins MMW" group contains the normal paraffins of gasoline fraction C_5-C_{12} and "Paraffins HMW" group contains paraffinic hydrocarbons of vacuum distillate C_{13}–C_{40}. Such distribution of paraffinic hydrocarbons on groups was based on the analysis results of catalytic cracking feedstock by gas chromatography-mass spectrometry, analysis of the gasoline fraction by gas chromatography and also according to the detailed thermodynamic analysis of the cracking reactions.

Separation of the paraffinic hydrocarbon feedstock on iso-paraffins and normal paraffins within the group "Paraffins HMW" is not provided due to the complexity of their quantitative determination. However, formalized hydrocarbon conversion scheme takes into account the formation possibility of isoparaffins of gasoline fractions from the feedstock that is connected with the carbonium ion mechanism of hydrocarbon conversion.

Unsaturated hydrocarbons was combined in "Olefins" group due to the difficulty of olefinic and diene hydrocarbon identification in the heavy fractions of petroleum feedstock. Olefin hydrocarbons of gasoline fraction are represented by C_3–C_{13} hydrocarbons according to the chromatographic analysis results.

Aromatic hydrocarbons according to conversion scheme are divided in two groups: the "MAH" group includes aromatic hydrocarbons of gasoline fraction C_6–C_{13}; the "AH HMW" group is presented by the more complex aromatic hydrocarbon of feedstock, light and heavy gasoil with long alkyl substituents of normal and iso-structure.

According to conversion scheme, naphthenic hydrocarbons are divided in two groups: «Naphthenes» group contains the naphthenic hydrocarbons of gasoline fraction C_5–C_{11}; "Naphthenes HMW" group is presented by the more complex naphthenic hydrocarbon of feedstock, light and heavy gasoil (mono- and polycykloalkanes with long alkyl substituents of normal and iso-structure) with carbon number in alkyl substituents C_1–C_{25}.

According to the developed conversion scheme, reaction of coke formation occurs through successive conversion of complex naphthene-aromatic structures with further condensation of resins.

5. Conclusion

Compiling the formalized hydrocarbon conversion scheme and evaluation of their reactivity are an important stage in mathematical description development as the accuracy of the model calculations depends on the level of detail of the chemical reactions mechanism. Selected level of the catalytic cracking reaction mechanism formalization provides the recording of feedstock hydrocarbon groups to hydrocarbons of gasoline fraction conversions, light and heavy gasoil and coke in the kinetic model. Thus, the kinetic model remains sensitive to the changes of feedstock composition and at the same time allows estimating the yield and group composition of the main cracking product.

References

[1] Frantsina E, Ivashkina E, Ivanchina E and Romanovsky R 2014 *Chemical Engineering Journal* **238** 129–39
[2] Kang X and Guo X 2013 *Energy Sources* Part A **35** 1921–28
[3] Belinskaya N, Ivanchina E, Ivashkina E, Frantsina E and Silko G 2014 *IOP Conf. Ser.: Earth Environ. Sci.* **21** 012030
[4] Fernandes J, Verstraete J, Pinheiro C, Oliveira N and Ribeiro F. 2007 *Chemical Engineering Science* **62** 1184–98
[5] Baudrez E., Heynderickx G.J., Marin G.B. 2010 *Chemical engineering research and design* **88** 290–303
[6] Dagde K and Puyate Y 2012 *IJERA* **2** 557–68
[7] Hua Zhou, Jianxiang Lu, Zhikai Cao, Jia Shi, Ming Pan, Wei Li and Qingyin Jiang 2011 *Fuel* **90** 3521–30
[8] Sildir H, Arkun Y, Cakal B, Gokce D and Kuzu E 2012 *Journal of Process Control* **22** 1956–65
[9] Martinez J and Ancheyta J 2012 *Fuel* **100** 193–99
[10] Varshney P, Kunzru D and Gupta S 2014 *Indian Chemical Engineer* 21
[11] Barbosa A, Lopes G.C., Rosa L.M., Mori M., Martignoni W.P. 2013 *Chemical engineering transactions* **32** 637–42
[12] Randolph A 1988 *Academic Press* 231–54
[13] Lan X, Xu Ch, Wang G, Wu L and Gao J 2009 *Chemical Engineering Science* **64** 3847–58
[14] Behjata Y., Shahhosseini S. and Hashemabadi S.H. 2011*Chemical engineering research and design* **89** 978–89
[15] Sedighi M., Keyvanloo K. and Towfighi J. 2013*Fuel* **109** 432–38
[16] Bartholomew C and Fuentes G 1997 *Elsevier* 695
[17] Lopez-Isunza F 2001 *Studies in Surface Science and Catalysts* **134** 187–99
[18] Froment G 2014 *Current Opinion in Chemical Engineering* **5** 1–6
[19] Fu´nez A, Thybaut J, Marin G, Sa´nchez P, De Lucas A and Valverde J 2008 *Applied Catalysis A: General* **349** 29–39
[20] Choudhury I R, Hayasaka K., Thybaut J., Narasimhan C.S.L., Denayer J.F.M., Martens J.A. and Marin G.B. 2012 *Journal of Catalysis* **290** 165–76
[21] Vandegehuchte B, Thybaut J, Martinez A., Arribas M.A. and Marin G. 2012 *Applied Catalysis A: General* **441–442** 10–20
[22] Deglmann P. V., Schafer A., Lennartz C. 2015 *International Journal of Quantum Chemistry* **115** 107–36

IOP Conf. Series: Earth and Environmental Science **27** (2015) 012063 doi:10.1088/1755-1315/27/1/012063

Experimental evaluation of the effectiveness of water mist automated fire extinguishing systems for oil transportation

D O Glushkov[1], G S Nyashina[1,2], P A Strizhak[1] and R S Volkov[1]

[1] Department of Heat and Power Processes Automation, Institute of Power Engineering, National Research Tomsk Polytechnic University, 30 Lenin Ave., Tomsk, 634050, Russia

E-mail: [2] gsn1@tpu.ru

Abstract. Experimental investigation of regularities of carryover of water mist droplets (radius of 50 – 500 μm) by high temperature (500 – 1800 K) products of combustion of typical petroleum products (oil, gasoline, kerosene, etc.) was carried out. The panoramic optical methods and high-speed hardware and software systems were used. Speeds of droplets after mixing with oncoming high temperature gases were determined. Conditions of continuation of droplets movement through combustion products with preservation of initial trajectory in spite of intensive evaporation and braking were found. The predictive evaluation of effectiveness of water mist use for extinguishing of fires involving oil and typical petroleum products.

1. Introduction

Modern technologies of fire extinguishing using water mist, water spray, water curtain and other variants of applications of vapor-droplet aerosol have sufficiently limited implementation [1–3]. Especially if the fires have high spread rate, high flame temperature and high heat flow. These particularly include fires involving oil and petroleum products [4]. As a rule, foam of different expansion rate is used to suppress such fires. Such foams are not always effective and are expensive. Therefore, it is necessary to study the possibilities of application of vapor-droplet water aerosol with the most simple content for extinguishing such fires.

One of the main reasons and possible difficulties for the use of vapor-droplet water sprays to extinguish fires with high speed distribution and high heat flows is carryover of droplets. In the works [5, 6] the features of turning and next carryover of water droplets by flow of high-temperature products of kerosene combustion were studied. The typical values of velocities of water droplets, which they have after the contact with the flame were determined. The conditions under which the droplets can maintain the initial direction of motion, in spite of braking by the combustion products and intensive evaporation were determined. To determine the conditions of applicability of the water vapor-droplet aerosols for the extinguishing of fires with typical petroleum products (particularly kerosene) it is necessary to have experimental data on the macroscopic regularities of evaporation and carryover of droplets of such aerosols. Therefore, it is advisable to develop the understandings [5, 6] for the high temperature combustion products of oil and several typical liquid petroleum products.

The purpose of the article is experimental investigation of macroscopic regularities of carryover of water mist droplets by high temperature combustion products of several widely used petroleum liquid products.

2. Experimental setup and research methods

During the experiments, we used a setup with elements, which are similar to the used in the works [5, 6]. The basic elements of the setup are: high-speed camera *1* (figure format – 1024 × 1024 pixels, the frame frequency up to $6 \cdot 10^5$ per second); *2* – cross-correlation digital camera (figure format – 2048 × 2048 pixels, the minimum delay between two serias shots – not more than 5 ms); double pulsed solid-state laser *3* with active area "YAG" and neodymium additives (wave-length is 532 nm, minimum energy in impulse is 70 mJ, maximum impulse time is 12 ns, recurrence frequency is 15 Hz); synchronizing processor *4* supporting modes of internal and external launch (signal sampling is not more than 10 ns and supporting modes of internal and external launch).

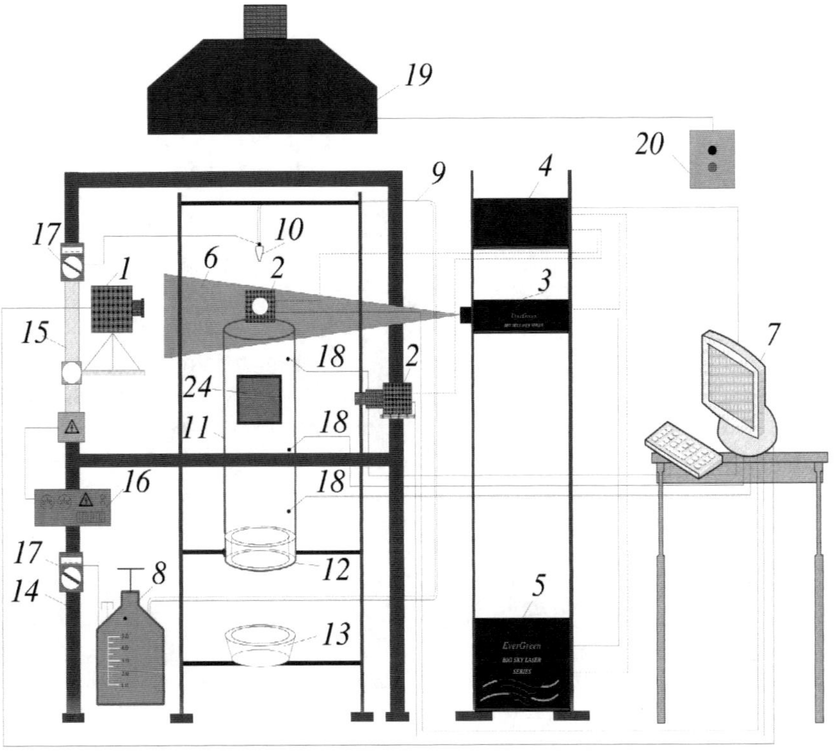

Figure 1. A scheme of experimental setup: *1* – high-speed camera; *2* – cross-correlation digital camera; *3* – double pulsed solid-state laser; *4* – synchronizer of PC, cross-correlation digital camera and laser; *6* – light "pulse"; 7 – PC; *8* – vessel with experimental liquid; *9* –channel of experimental liquid supply; *10* – dosing device; *11* – cylinder of quartz glass; *12* – hollow cylinder with combustible liquid in internal medium; *13* – trap; *14* – mount; *15* – motorized pointing device (MPD) to move the camera; *16* – power supply unit MPD; *17* – digital multimeters; *18* – thermocouples; *19* – replacement exhaust system; *20* – switching on and off board of replacement exhaust system.

The methodology of the experiments included several steps:

- vessel *8* was filled with preliminary prepared water;
- dosing device *10* was connected to the output of the vessel *8* through the channel *9*. The dosing device *10* are configured depending on the required parameters of the flow of liquid (special metal nozzles, generating polydisperse stream of water mist droplets were used in the experiments);
- dosing device *10* was installed on the mount *14* 0.5 m above the top face of the cylinder *11*;

- depending on the type of experiment the height of installation of cross-correlation digital camera *2* and laser *3* was chosen (in the case of application of PIV and IPI methods the height of installation of the laser was chosen so that the optical axis of the camera and plane light «pulse» *6* of laser intersect at an angle of 90 degrees; in the case of SP methods in front the camera the stroboscopic light source was set. It was the screen connected to the laser *3* by means of light guide. The height of installation of the laser is not important in this case);
- calibration of measuring system was carried out (the scale coefficient and the cross-correlation optical zoom of the camera *2* were determined), and also set of the "waist" and the angle of inclination of the light "pulse" *6* of the laser *3* was executed;
- the base of the hollow cylinder *12* was filled with combustible liquid (about 250 ml), which was ignited before the experiment;
- after 5 minutes (the time required for heating of the inner cavity of the cylinder *11* to the required temperature) on the PC *7* the specialized software was run (realizing optical diagnostic methods "Particle Image Velocimetry" (PIV), "Stereoscopic Particle Image Velocimetry" (Stereo PIV), "Particle Tracking Velocimetry" (PTV), Interferometric Particle Imaging (IPI) and "Shadow Photography" (SP)) [7], the dosing device *10* was turned on and the recording of video with droplets of working fluid was carried out;
- the values of velocities and the sizes of water mist droplets were calculated by the PC *7* by means of special software.

The most widely used liquid petroleum products were used for the experiments as combustible liquids: gasoline (RON 92), kerosene (TS-1), acetone (technical, GOST 2768-84), alcohol (ethyl synthetic, GOST 52574-2006), oil (Chayandinskoye condensate field exploration well 75 with ρ=667 g/l). The products combustion temperature in the cylinder channel *11* was controlled in three points of the channel (0.15 m, 0.5 m, 0.85 m) on its symmetry axis using the a chromel-alumel thermocouples (the range of temperature measurement is 273–1373 K, the measurement error is $\pm$ 3.3 K).

The water temperature was controlled in two points (in the vessel *8* and in the inlet of dosing device *10*) using two chromel-copel thermocouples (the range of temperature measurement is 273–473 K, the measurement error is $\pm$ 1.5 K), and was 298–300 K.

Sizes (radiuses) of water droplets R_m, and velocities of droplets U_m and combustion products U_g were accepted as objective function as in [3, 5, 6].

IPI and SP methods were used to measure the sizes of water mist droplets. IPI method based on measuring the droplet diameter using the distance between the strips on the interference pattern, which was formed by light being reflected and once refracted by droplet. Droplets were illuminated by the light "pulse", the interference pattern was observed for all droplets in the measuring area. Next, using the algorithm "IPI Kit" the droplet sizes were determined. SP method is based on registration of the shadow pictures of an object having a refractive index that is different from its environment. Special screen connected to the laser *3* by means of light guide was placed in front of the camera behind the droplet flow. The cross-correlation digital camera with the colour filter recorded the shadow images of water droplets. Post-processing of images using series of filters and algorithms allowed to determine the droplet size.

PIV, Stereo PIV, PTV methods were used to measure the velocities of the droplets. PIV method was used to estimate the velocities of high-temperature gas flow. These methods are based on determining the movement of special particles – "tracers" during a fixed time interval (the time between the laser flashes). Special polyamide fluorescent particles (particle size is 1–5 microns) were put into water as the "tracer" particles immediately before the experiment. Application of PIV and Stereo PIV methods allowed obtaining averaged two- and three-component fields of the gas flow and droplet flow. The use of PTV allowed determining the velocity of each droplet in the flow.

The final decision in the choice of methods of video image processing was taken from the preliminary analysis of a group of captured video. The main attention was paid to parameters such as noise in the image, the apparent size of the droplets in the stream, the droplet concentration, the movement of "tracers" during the time between laser flashes.

A hardware-software complex based on high-speed video camera "Phantom" and software "Tema Automotive" [8] was used for additional assessment of velocities and tracks of moving water mist droplets under their mixing with high-temperature gas flow. The flow of droplets was recorded by high-speed video camera *1* with recording frequency is not less than 10^4 frames per second. Video processing with the use of algorithms "Correlation" and "Circular Symmetry" [8] allowed to visualize track of movement for each droplet in the registration area as well as to determine the velocities of their movement.

Systematic errors of determining of the velocities U_m and U_g were about 0.005 m/s, of the sizes R_m were about 10^{-5} m.

3. Results and discussion

Figure 2 shows a typical videogram at the moment of mixing the droplet flow with the counter gas flow (left) and the corresponding velocity field (right). As can be seen from figure 2, all velocities (of droplets and gases) are close to zero at the interface "water droplets – high temperature gases". Significant deceleration, and the subsequent reversal and entrainment of water mist droplets take place exactly at the interface of two medias. However, as it has been found in [5], the initial dimensions R_m of droplets considerably affect to the braking and subsequent reversal of water droplets in the stream of high-temperature gases. For example, figure 2 shows, that relatively large droplets are lower than a pronounced interface "water droplets – high temperature gases" while above it a significant set of smaller droplets, which were not able to overcome the counter resistance is observed.

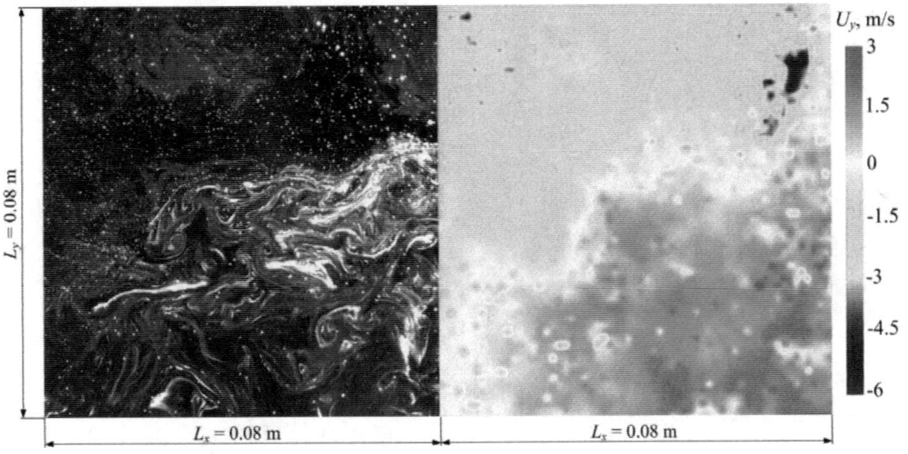

Figure 2. Videogram and velocity field at the moment of mixing the flow of water mist with a counter flow of high-temperature gases.

The dependences of relative velocities of water droplets on the absolute velocities of the counter gas flow were constructed for three typical combustible liquids: gasoline, acetone and oil (figure 3). It was made for analysis of macroscopic regularities of carryover of water mist droplets by high temperature products of combustion of typical combustible products.

To do this, the droplets in flow of water mist (as in [3, 5]) were divided into four groups in accordance with the sizes. After that for each group of droplets the experimental points were found, which later were used for approximation curves construction.

Analysis of figure 3 shows that small droplets with sizes R_m=0.03–0.1 mm completely stop at high velocities U_g of oncoming gas flow from 0.3 m/s to 0.7 m/s. These data are in good agreement with the results of [5], where water droplets of similar size were completely stopped and unfolded to the opposite direction in the combustion products of kerosene at the U_g≈0.3 m/s.

IOP Conf. Series: Earth and Environmental Science **27** (2015) 012063 doi:10.1088/1755-1315/27/1/012063

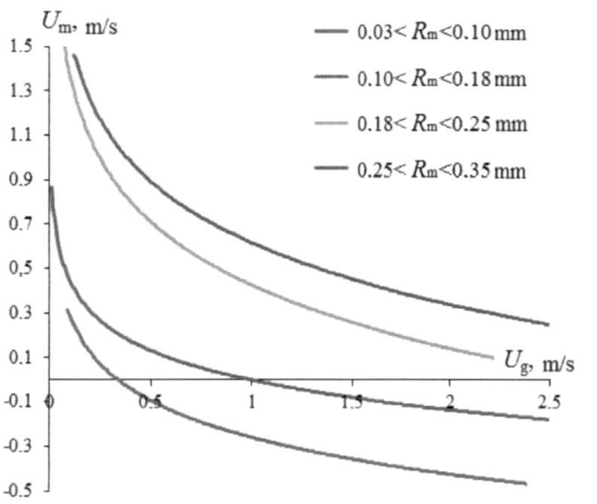

a

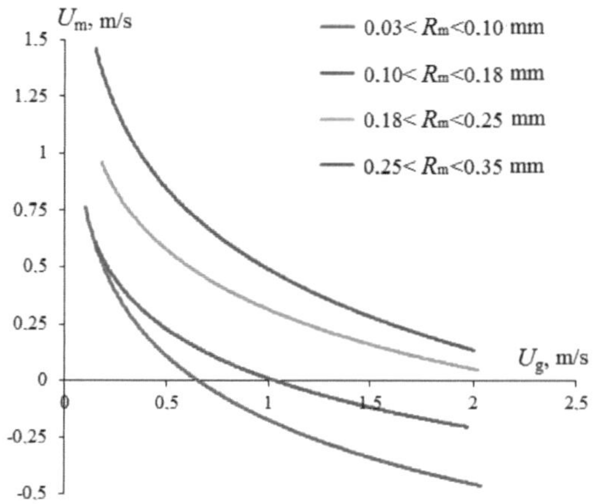

b

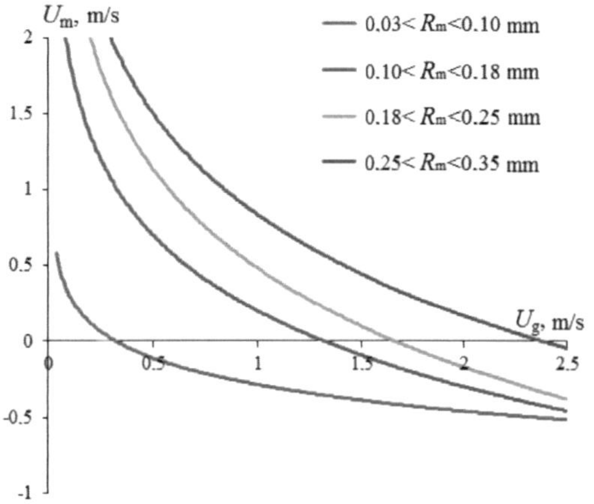

c

Figure 3. The velocities of the droplets in dependence on the velocities of the combustion products of gasoline (*a*), acetone (*b*) and oil (*c*).

Droplets with sizes R_m=0.01–0.18 mm fully turn during their motion through products of combustion of gasoline and acetone even at velocities of the oncoming gas flow $U_g \approx 1$ m/s. The droplets of the same size in the combustion products of oil change their tracks to the opposite at velocities of the oncoming gas flow $U_g \approx 1.3$ m/s. To explain this phenomenon it is necessary to pay attention to the temperature of the combustion products of fuels. Table 1 shows the temperature distribution of the combustion products in a cylindrical channel *11* for the 4 flammable liquids.

Table 1. Temperatures ($T_f \pm$ 15 K) of the combustion products in cylindrical channel.

The position of the thermocouple relative to the base of cylinder *11*	0.1 m	0.3 m	0.5 m	0.7m	1 m	1.5 m
Gasoline	1906	1540	712	615	475	397
Acetone	1744	1140	832	710	412	384
Kerosene	1738	824	514	485	376	348
Oil	1139	634	506	454	372	339
Alcohol	1105	560	428	404	366	324

The experiments have shown that the highest temperature of the combustion products belongs to gasoline. Its maximum temperature reaches to 1906 K. The acetone and kerosene burns with a lower intensity, their temperatures do not exceed 1744 K and 1738 K respectively. These temperatures are less by 150–200 degrees than the temperatures of the gasoline combustion products. Oil and alcohol have the lowest temperature of the combustion products which does not exceed 1150 K.

It follows that the temperature of the oncoming gas flow has a great influence on the degree of braking of water droplets. Droplets significantly slow down during intensive evaporation (due to intensive mass loss during vaporization). The more the temperature of the surrounding gas environment, the more rapidly drops brake. These results coincide well with the results of [6], where the conditions of braking and evaporation of sequentially moving drops were analyzed. The temperature of the combustion products of oil is much lower (about 600–800 K) than for gasoline and acetone, therefore large velocity U_g are required (figure 3) for complete stop and turn of water droplets in high temperature gas flow.

Relatively large droplets of water mist (figure 3) is completely stopped and deployed at gas flow velocities are even more than 2–2.5 m/s. That can be explained by the fact that the drops with R_m>0.2 mm evaporate much more slowly than small droplets [3, 5, 6]. Thereby large velocities of oncoming gas flow are required.

After comparing the received data with the results of [3, 5, 6], it is possible to predict the track of the water mist droplets and share of their evaporation during movement in oncoming high temperature combustion products. Experiments have shown that droplet with R_m<0.2 mm are practically completely deployed further input to the space of high-temperature combustion products, without going a distance even 0.2 m. Larger droplets R_m=0.2–0.5 mm are capable to penetrate into the combustion area and almost completely evaporate.

In works on the theme of fire, it is traditionally emphasized [1–4], that fire extinguishing of liquid combustible substances by water droplet flow is possible and necessary, but the liquid must be properly shredded. Different sources give different data about optimal droplet size for the extinguishing of fires of flammable liquid substances (from 100 to 400 microns). These droplets sizes depend on many factors: the flash point of the substance, the area and height of the fire, and others. But the majority of the sources said that for the effective extinguishing droplets must evaporate completely during movement through the flame. Thus, the results of this study can expand the understanding of the modern technologies of fire-fighting and help in the selection of the optimal parameters of water spraying for fire extinguishing at oil and gas facilities. Also, these results may contribute to the development of the theory of two-phase gas vapor-droplet and multiphase systems.

4. Conclusions

The experiments have shown that for typical oil and all considered liquid petroleum products the conditions of entrainment of water droplets by combustion products are typical. The effectiveness assessment of the use of water mist (droplet radius is 50–500 microns) for suppression of oil and petroleum products burning have shown that the main disadvantage of this approach is quite intense carryover of droplets by high temperature flow of product combustion. The main advantage of this approach is almost complete water evaporation, i.e. full involvement of water in the combustion zone.

Acknowledgement
The investigation was performed by Russian Foundation for Basis Research grant (No. 15-38-20006).

References

[1] Xiao X K, Cong B H, Wang X S, Kuang K Q, Yuen Richard K K and Liao G X 2011 *J. Fire Sci.* **29** 339
[2] Meenakshi Ga, Rajora R, Sahai S, Shankar R, Anjan R and Kale S R 2012 *Fire Safety J.* **54** 130
[3] Vysokomornaya O V, Kuznetsov G V and Strizhak P A 2014 *Fire Safety J.* **70** 61
[4] Miyashita T, Sugawa O, Imamura T, Kamiya K and Kawaguchi Ya 2014 *Fire Safety J.* **63** 1
[5] Kuznetsov G V, Kuibin P A and Strizhak P A 2014 *Thermophys. Aeromech.+.* **21** 609
[6] Volkov R S, Kuznetsov G V and Strizhak P A 2014 *Thermophys. Aeromech.+.* **21** 255
[7] Bilsky A V, Lozhkin Yu A and Markovich D M 2011 *Thermophys. Aeromech.+.* **18** 1
[8] Janiszewski J 2012 *Metrol. and Measur. Sys.* **19** 797

Explosion of heterogeneous water droplet in a high-temperature gaseous region

M V Piskunov[1,2] and A A Shcherbinina[1,3]

[1] Department of Heat and Power Processes Automation, Institute of Power Engineering, National Research Tomsk Polytechnic University, 30 Lenin Ave., Tomsk, 634050, Russia

E-mail: [2] piskunovmv@tpu.ru, [3] nushenka17@gmail.com

Abstract. Using high-speed video recording tools (up to 10^5 frames per second) and «TEMA Automotive» and «Phantom Camera Control» software packages the experimental features of explosive disintegration, boiling and evaporation of water droplets with comparably sized solid inclusions heated in high-temperature (more than 650 K) gaseous region were determined. The necessary and sufficient conditions of explosive vapor formation achievement with the next heterogeneous water droplet disintegration were found.

1. Introduction

The technologies of use of the heterogeneous and multiphase flows in power engineering, chemical and petrochemical applications are applied enough in recent years [1-4]. These technologies primarily mean the presence of solid inclusions (solid phase) in a flow. Most often these solid inclusions are added artificially [1-4]. Thus, the phase change mechanisms at the «solid inclusion – liquid droplet» interfaces are of interest. The research of vapor bubble occurrence, growth and movement processes [5, 6] near solid inclusions in a heterogeneous flow attracts special interest. Uninvestigated heat exchange features of different phases in a flow determine one of the main unsolved fully problems – mechanism determination of liquid droplet phase changes with solid inclusions in gaseous (especially high-temperature) region taking into account the optical and thermal-physical particle material properties.

One of the main applications of multiphase heterogeneous gas-vapor-droplet flows is the forest fire extinguishing by using the water atomization. Both the theoretical [7, 8] and experimental [8-10] investigations in the given branch show that the addition of typical solid metallic and nonmetallic inclusions (particles of different shapes and sizes – from 50 μm to 500 μm) in extinguishing liquid increases significantly the efficiency factor. An experimental investigation of an impact of solid inclusions comparably sized with a liquid droplet on the evaporation and boiling rate in the combustion area and possible explosive vapor formation conditions are of interest.

The pulverization of extinguishing agent droplets by probable explosive vapor formation will increase the flame coverage area by the vapor-droplet cloud. In this case, it seems advisable to determine necessary and sufficient conditions whereby the phase change (evaporation and boiling of liquid, explosive vapor formation) differences of water droplets with solid inclusions can be identified.

The aim of the present work is to determine the explosive vapor formation conditions of a heterogeneous water droplet heated in a high-temperature gaseous region corresponding to combustion products of the typical petroleum products.

2. Experimental setup and methods of investigation

During experimental investigations the setup (figure 1) similar to the one used in the tests [9, 10] was applied.

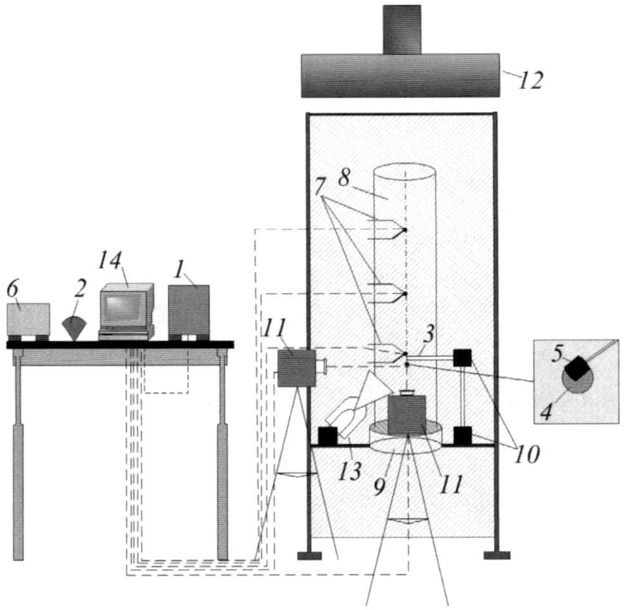

Figure 1. Scheme of the experimental setup:

1 – analytical weighing system,
2 – electronic dosing devise,
3 – rod (holder for inclusion),
4 – droplet,
5 – inclusion,
6 – manufacture of inclusions,
7 – thermocouples,
8 – cylinder of quartz glass,
9 – burner/heat gun,
10 – moving mechanisms,
11 – high-speed cameras,
12 – air flow system,
13 – spotlight,
14 – personal computer (PC)

Droplets moved through a high-temperature region in the experiments [9, 10]. In this work a heterogeneous water droplet was fixed on the rod *3*. Figure 2 illustrates the typical configurations of heterogeneous liquid droplets.

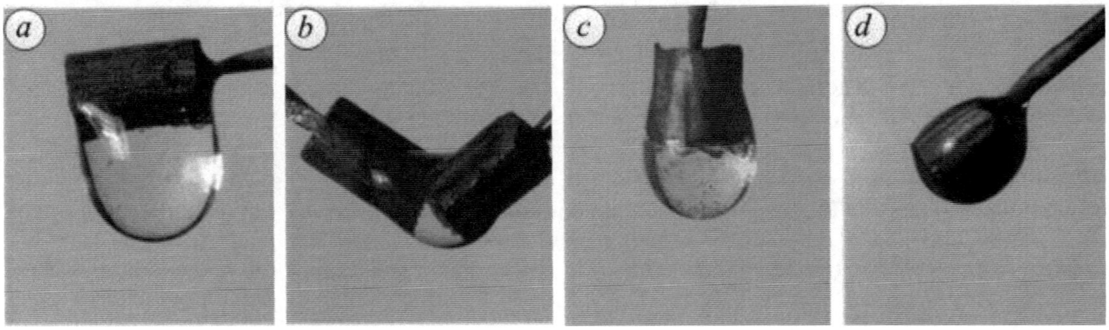

Figure 2. Typical variants of heterogeneous liquid droplets: *a* – water droplet (volume is 15 μl) hangs from the horizontal inclusion in the parallelepiped shape and covers only one inclusion edge; *b* – water droplet (volume is 15 μl) is between two inclusions in the parallelepiped shape, inclusions are in contact; *c* – water droplet (volume is 15 μl) hangs from the vertical inclusion in the cylindrical disk shape and fully covers the inclusion; *d* – water droplet (volume is 5 μl) fully covers the inclusion in the parallelepiped shape with the formation of almost uniform in thickness film.

The decision on the heterogeneous droplet fixation on the rod was accepted according to the results of the primary tests illustrated the regular partial protrusion of inclusions out of the droplet surface during their motion through a high-temperature gaseous region.

High-speed video recording of the heterogeneous droplet evaporation on the ceramic rod *3* was performed. Two high-speed (up to 10^5 frames per second) video cameras («Phantom V411» and «Phantom Miro M310») *11* were used during this process. A heterogeneous water droplet was fixed into the quartz cylinder *8* (height is 1 m, inside and outside diameters – 0.2 m and 0.206 m). This volume contained the temperature-controlled combustion products. The processing of the videograms was provided by «Phantom Camera Control» and «TEMA Automotive» software packages.

The burner *9* was at the bottom of the cylindrical channel *8*. Industrial alcohol used to generate high-temperature gases was in the burner *9*. The chosen industrial alcohol as a combustible liquid produces less smoke than kerosene as used in previous experiments [9-11]. With such configuration, the droplet images had sufficient contrast to be analyzed.

The temperature of the gaseous region in the cylinder *8* was controlled by the chromel-alumel thermocouples *7* (measuring temperature range 273–1373 K, measurement error ± 3.3 K). The temperature was registered at heights 0.3 m, 0.5 m and 0.7 m relative to the bottom of burner *9*. Three holes were drilled the cylindrical channel *8* for these purposes. The hole at height 0.7 m was also used to insert the ceramic rod *3* with a fixed heterogeneous droplet into the channel (after measurement of gas area temperature).

The experiment included several steps. The droplet of fixed mass was lowered on the graphite inclusion *5* by using the dosing device *2*. Then the droplet was inserted into the hole at height 0.7 m. After insertion of the droplet in a high-temperature gaseous region the video recording of droplet evaporation was activated. The obtained video materials were processed on PC *14* by using the «Phantom Camera Control» and «TEMA Automotive» software packages: the sizes of droplet (d), liquid film thicknesses (δ) and the characteristic time of heating (τ_h) until explosive disintegration or complete water evaporation were determined.

The experiment of the radiation energy impact assessment on the heating and phase changes of heterogeneous liquid droplet was carried out by using the same setup, but instead of burner *9* was installed the heat gun of fixed motion speed (3 m/s) and temperature (up to 1000 K) of air flow. Also, the preliminary investigations included the tests using the household multi-fuel burner where the types of fuels were industrial alcohol and gas mixture (butane, propane). The video recording, processing of water droplets and solid inclusions during the preliminary tests correspond above mentioned procedure.

Errors on measurements of inclusion, water droplet sizes (d) and film thickness (δ) did not exceed 0.01 mm. Random (characterize the measurement repeatability) errors of d and δ determination were not more than 7-10 %. The error on time of heating τ_h did not exceed 1 ms.

3. Results of experiments and discussion

It was found that the intensive boiling and vapor formation of a heterogeneous liquid droplet heated in a high-temperature gaseous region are carried out much faster when a heat flux is supplied to an inclusion. Then a heat flux begins to heat a droplet.

To check this phenomenon the control experiments with two droplets and one inclusion were carried out. Figure 3 shows that the phase changes of the second water droplet in relation to the air flow direction were realized moderately until fully evaporating the first droplet. Then the intensive formation of the vapor layer at the «solid inclusion – water droplet» interface and the liquid boiling were observed. It is worth to note that the explosive vapor formation did not record during the tests in the hot air. It proves the fundamental importance of the radiation energy to achieve conditions of the process under investigation.

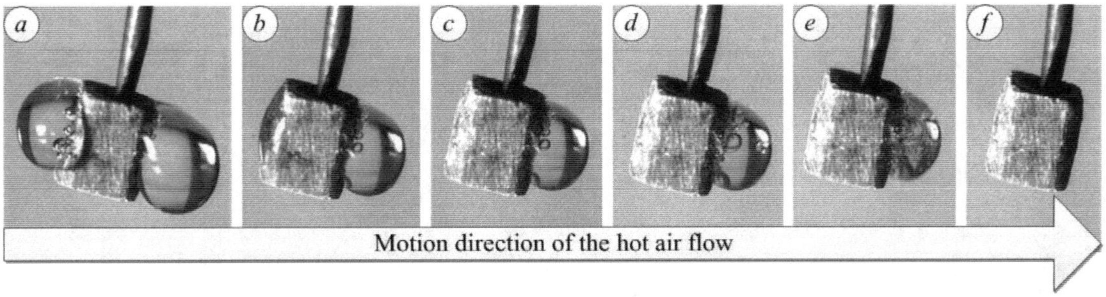

Figure 3. Typical video frames of two water droplets (the initial droplet volume is 7 μl) during evaporation and boiling at the graphite inclusion surface (the characteristic size is 2 mm) in various time points (the temperature of the air is 623 K): *a* – 6 s; *b* – 14 s; *c* – 15 s; *d* – 16 s; *e* – 18 s; *f* – 22 s.

Also, the increase of a heat exchange surface (for instance, a heterogeneous droplet consists of two inclusions) is one of conditions to intensify phase changes. «Water droplet – two inclusions» case (figure 4) showed that such processes as intensive boiling and onset, joining of bubbles take place during the first second of heterogeneous droplet heating in a gaseous region. Figures 4, *b* and 4, *c* show the bubble joining at the interface and single bubble growth with increase of a heterogeneous droplet size in relation to the initial size. When exceeding the vapor pressure of a critical value of the pressure caused by forces of surface tension and outside environment the explosion of a heterogeneous liquid droplet takes place. Figure 4, *d* shows the water droplet after the collapse (explosion) of single vapor bubble. The consequences of explosive vapor formation can be different (formation of a small droplet group or large liquid agglomerates, collapse of vapor bubbles without water droplet disintegration). In this case (figure 4) the water droplet disintegration on the small droplet group or large agglomerates was not observed.

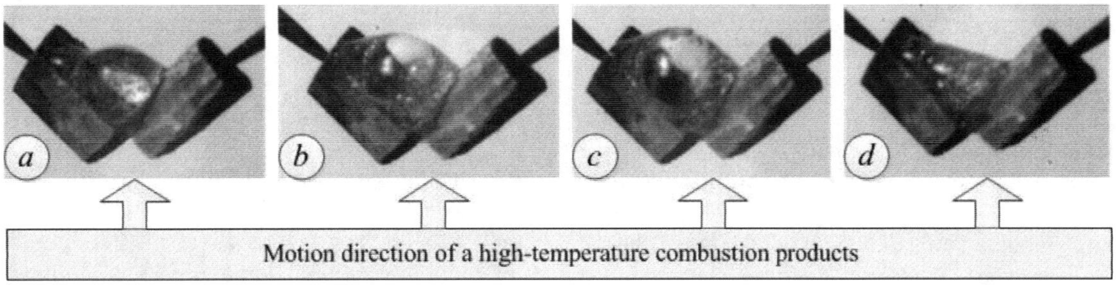

Figure 4. Typical video frames of the heterogeneous droplet consisting of the water droplet (the initial volume is 15 μl) and two graphite inclusions (the characteristic size of each inclusion is 3 mm) during evaporation and intensive vapor formation in various time points (the temperature of industrial alcohol combustion products is about 650 K): *a* – 0.33 s; *b* – 3 s; *c* – 3.45 s; *d* – 4 s.

It was revealed that an explosive vapor formation can continue for a long period (figure 5). The liquid evaporation from free surface of a water droplet takes most of the heating time. Figure 5, *b* illustrates the significant decrease of the water droplet size in relation to the initial size (figure 5, *a*). The significant increase of heterogeneous droplet size was recorded during the onset, growth and joining of the vapor bubbles (figure 5, *c*).

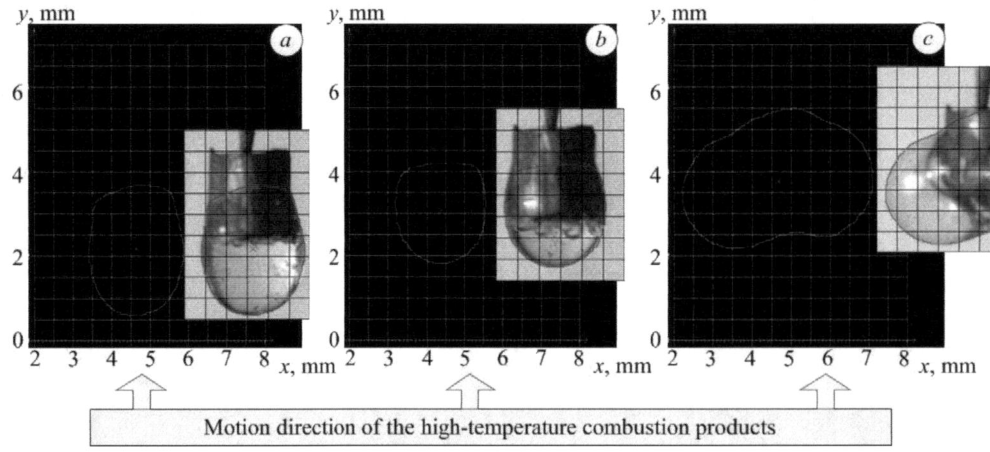

Figure 5. Typical video frames of the heterogeneous droplet heated in the industrial alcohol combustion products with the indication of change in water droplet (the initial volume is 15 μl) sizes during the explosive vapor formation in various time points: *a* – 0.01 s; *b* – 9 s; *c* – 12.175 s.

The obtained results allowed revealing the conditions of explosive vapor formation achievement. A temperature of gaseous region and radiation energy are the most important. The carried out control tests showed that the shape, material, sizes of inclusion (inclusions) and position of a heterogeneous droplet in relation to a motion direction of a gaseous flow.

The carried out experimental investigations showed that the explosion of a heterogeneous water droplet 3-4 mm in size at the outside gaseous region temperatures of not less than 650 K occured during several seconds (1-3 s). It is worth noting that the processes of long duration (up to 12-13 s) are also possible. If we tell about heterogeneous droplets with such sizes [11], then the existence times at the same conditions, as a rule, are not less than 10 s. The intensive evaporation from the outside surface and the bubble onset near the inside interface was observed during all experiments at the temperatures of not less than 650 K. Such phase change mechanism is characterized by large existence times of droplets (30-40 s).

4. Conclusions

It was found that the explosive vapor formation conditions of heterogeneous droplets 3-4 mm in size are possible at the gas temperatures of more than 650 K. So, the existence times till breaking up the droplets, as a rule, were several seconds. The phase change features of heterogeneous water droplets were recorded. It was revealed the significant impact of radiation energy on the considered process. Therefore, the explosive disintegration conditions of heterogeneous water droplets heated in combustion products of petrochemicals are realized during a short time versus hot air at the same temperatures.

5. Acknowledgement

This work was supported by the grant of the President of the Russian Federation (MD-2806.2015.8).

References

[1] Doh D H, Yum J H, Cho G R, Kim M H, Ryu G W and Takei M 2013 *J. Therm. Sci.* **22** 478
[2] Roy G, Palm S J and Nguyen C T 2005 *J. Therm. Sci.* **14** 362
[3] Gherasim I, Roy G, Nguyen C T and Vo-Ngoc D 2011 *Int. J. Therm. Sci.* **50** 369
[4] Yang Y-T and Lai F-H 2011 *Int. J. Therm. Sci.* **50** 61
[5] Wang H, Zhang Z-Y, Yang Y-M, Hu Y and Zhang H-S 2008 *Chin. Phys. B* **17** 3847
[6] Han R, Zhang A-M and Li S 2014 *Chin. Phys. B* **23** 034703
[7] Sazhin S S, Elwardany A E, Krutitskii P A, Castanet G, Lemoine F, Sazhina E M and

Heikal M R 2010 *Int. J. Heat Mass Transfer* **53** 4495
[8] Sazhin S S, Elwardany A E, Krutitskii P A, Depredurand V, Castanet G, Lemoine F, Sazhina E M and Heikal M R 2011 *Int. J. Therm. Sci.* **50** 1164
[9] Volkov R S, Kuznetsov G V and Strizhak P A 2015 *Int. J. Therm. Sci.* **88** 193
[10] Volkov R S, Kuznetsov G V and Strizhak P A 2014 *Tech. Phys.* **59** 1770
[11] Vysokomornaya O V, Kuznetsov G V and Strizhak P A 2014 *Fire Saf. J.* **70** 61

Power autotransformer choice for ensuring power supply of oil and gas fields

I S Tsoy[1,2] **and N M Kosmynina**[1,3]

[1] Department of Power Engineering Institute,
National Research Tomsk Polytechnic University, 30 Lenin Ave., Tomsk, 634050, Russia

E-mail:[2] isc2@tpu.ru, [3] kosm_nm@tpu.ru

Abstract. In this article the algorithm of the power autotransformer choice is offered. According to this algorithm the program for the power autotransformer choice is developed. The program can be used as an educational tool for studying of the theoretical section. Also the program can be used for the analysis of the power autotransformer operation modes for power station and substation after carrying out calculation of the operational modes on the specified objects. Composition checks of the entered data and a correctness of the autotransformer choice are provided in the program, and in all cases of a deviation from the correct decisions the relevant information is given.

1. Introduction

Now the oil and gas industry is one of the major branches in world economy. This branch includes exploration of oil and oil and gas fields, drilling of slits, production, transportation, oil refining and associated gas. At all these stages electrical energy is required.

The power stations generated energy is transferred on power lines to high-voltage switchgear (HV switchgear) of substations. Then energy will be transformed to the required voltage level of low-voltage switchgear (LV switchgear). The medium-voltage switchgear (MV switchgear) is possible on powerful substations.

For electric power transformation from one voltage level to other voltage levels power transformers are used. In case of HV switchgear and LV switchgear on substation double-wound transformers are used; in case of HV switchgear, MV switchgear and LV switchgear - triple-wound. At a voltage of HV switchgear of 220 kV and above and voltage of MV switchgear of 110 kV and above the Russian industry produces power autotransformers – variety of triple-wound transformers (figure 1).

Also power autotransformers are used on power stations for connection between of HV switchgear and MV switchgear (figure 2) and as generator autotransformers in case of connection of the generator in the winding of the low-voltage of the autotransformer (figure 3).

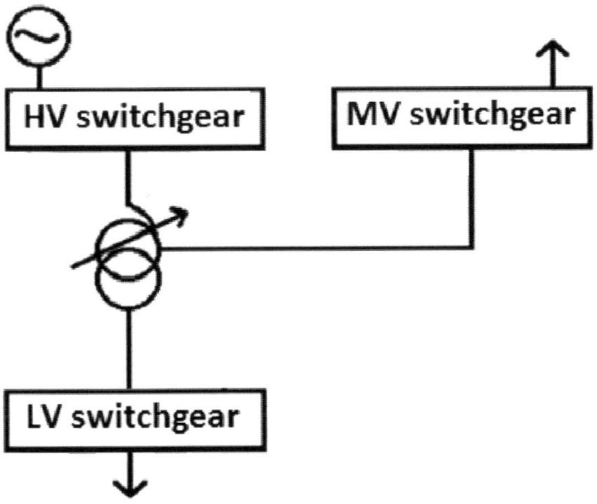

Figure 1. Substation simplified block diagram.

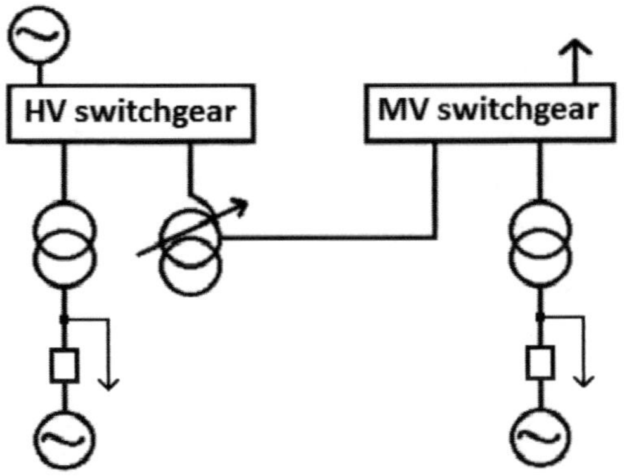

Figure 2. Power plant simplified block diagram with connection power autotransformer.

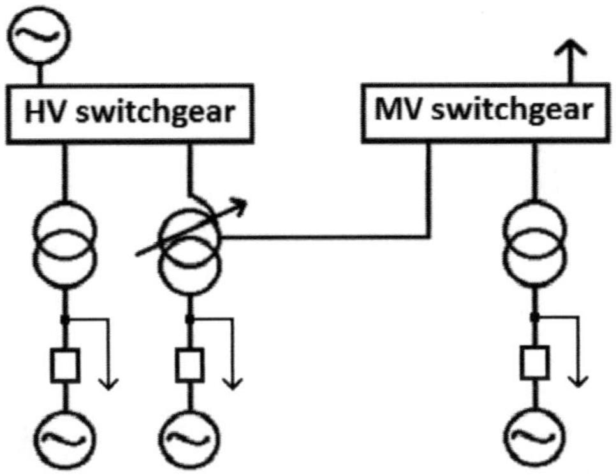

Figure 3. Power plant simplified block diagram with generator power autotransformer.

2. Algorithm of the power autotransformer choice

For a choice of the autotransformer it is necessary to calculate power flows between switchgears; to define an operation mode of the power autotransformer and the most loaded winding: the common, series, low-voltage winding.

The power autotransformer has such operation modes: the autotransformer, the transformer, the combined. In case of the autotransformer modes the side of LV is opened, there is a power flow from HV side on MV side or from MV side on HV side. In case of the transformer modes: power flow from LV side on HV side or from HV side on LV side when opened MV side; from LV side on the MV side or from MV side on LV side when opened HV side; from LV on the sides of MV and HV or from the sides of MV and HV on LV side. In the combined mode all three windings participate in power flow. In the first case power flow goes on the side of MV or from MV, in the second – on the side of HV or from HV. As an example the power flows of the combined modes is shown (figures 4-7).

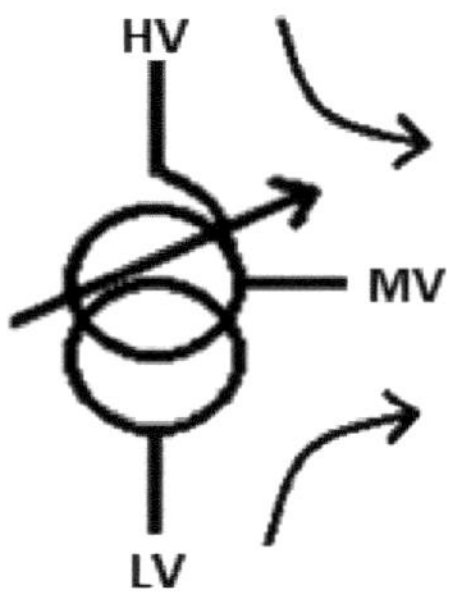

Figure 4. Power flows on the side of MV of the combined mode.

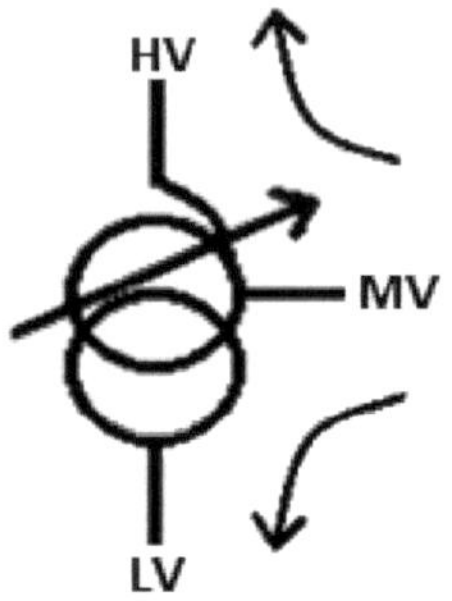

Figure 5. Power flows from the side of MV of the combined mode.

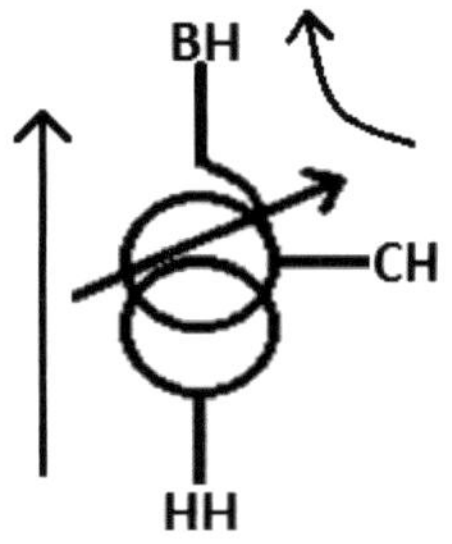

Figure 6. Power flows on the side of HV of the combined mode.

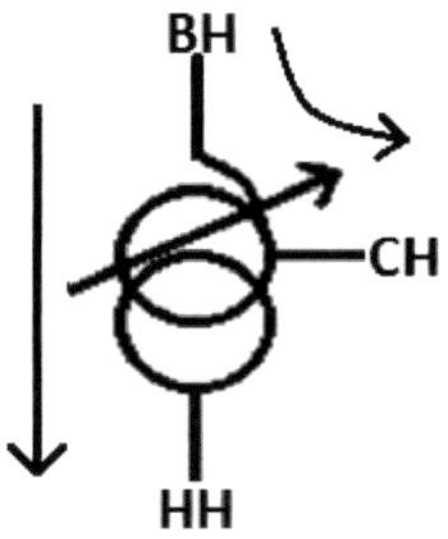

Figure 7. Power flows from the side of HV of the combined mode.

After mode identification it is necessary to make calculation of the demanded rated power for the following formulas [1]:

$$S_{calc} = \max\{S_c, S_S, S_{LV}\},$$

where S_C –power flow on the common winding
S_S – power flow on the series winding
S_{LV} –power flow on the low-voltage winding

$$S_{rate} \geq \frac{S_{calc}}{k_{co\_ratio}},$$

where k_{co_ratio} – co-ratio of an autotransformer coefficient is defined as

$$k_{co\_ratio} = \frac{U_{HV} - U_{MV}}{U_{HV}}$$

Calculation of the required power rating of the power autotransformer will be:

for the autotransformer modes:

$$S_{rate} \geq \frac{S_{C(S)}}{k_{co\_ratio}} \text{ or } S_{rate} \geq S_{MV} = S_{HV}$$

for the transformer modes:

$$S_{rate} \geq \frac{S_{LV}}{k_{co\_ratio}}$$

for combined modes:
in case of a power flow on the side of MV or from the side of MV:

$$S_{rate} \geq \frac{\sqrt{\left(P_{LV} + k_{co\_ratio} \cdot P_{HV}\right)^2 + \left(Q_{LV} + k_{co\_ratio} \cdot Q_{HV}\right)^2}}{k_{co\_ratio}}$$

in case of a power flow on the side of HV or from the side of HV:

$$S_{rate} = \max\left(\sqrt{\left(P_{LV} + P_{MV}\right)^2 + \left(Q_{LV} + Q_{MV}\right)^2} \cdot k_{co\_ratio}, S_{LV}\right).$$

The selected autotransformer needs to be checked on loading of a low-voltage winding [3]:

$$S_{LV} \leq S_{LV\_rate}$$

3. The program development for the power autotransformer choice

The program development for research of operation modes of the power autotransformer is carried on Electric Power Systems' Department of Power Engineering Institute of National Research Tomsk Polytechnic University. This development will allow selecting the autotransformer set both on substation, and on power station.

The program is written in the Delphi programming language. The choice of the Delphi programming language is caused by that this programming language allows to create the convenient interface and mobility, for example, buttons, functions of saving, comparing, the dropdown list were used [2, 4].

The program consists of some windows: the description, input of basic data, results, check and literature.

The description window contains the name of the program, assignment, the executed functions, data on the developer and the research supervisor.

In the input of basic data window (figure 8) lines for input of powers and voltages are created. Lines for voltages have the dropdown list with rated values of networks voltages to which the autotransformer [3] is connected. The user needs to enter values of the active and reactive powers and the side voltages and to press the Calculation button. Check of input of all parameters and powers balance with submission of specific information to the user is created in the program.

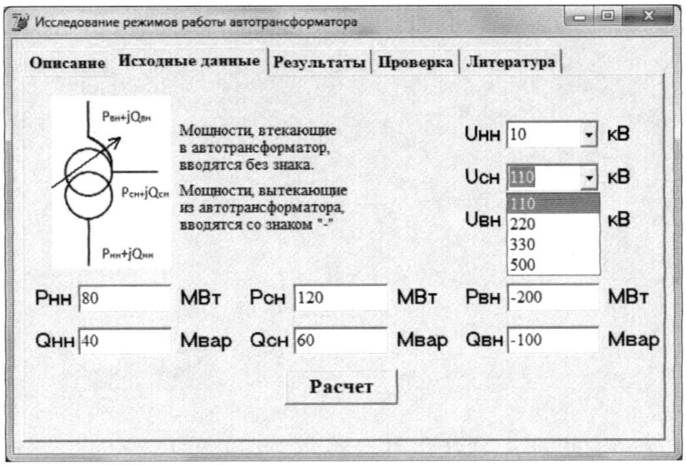

Figure 8. Input of basic data window.

After clicking of the "Calculation" button there is a transition to the results window (figure 9). In this window the next information are shown: the basic data, power autotransformer operation mode (transformer, autotransformer, combined); values of an autotransformer co-ratio coefficient; required power rating and power of the most loaded autotransformer winding. Results can be saved in the txt file form. By the "Back" button it is possible to return to the window of basic data in case of faulty action (input of parameters). When clicking the "Further" button there will be a transition to the check window.

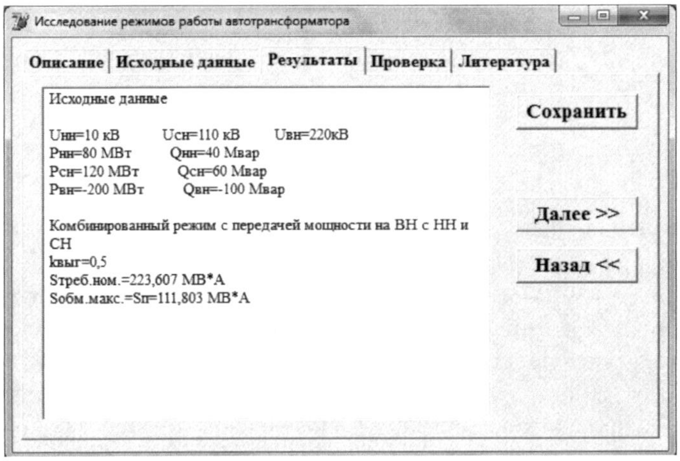

Figure 9. Result window.

The check window of the power autotransformer choice is shown on a figure 10. The check is carried out on voltages, power rating of a low-voltage winging, power rating of the autotransformer. At lines in the left part of a window catalog data of the power autotransformer are entered. When clicking the "Check" button comparing of catalog data with the estimated is made. If any value doesn't correspond to necessary conditions, in the right window the message on inappropriate parameter and on the wrong choice of the autotransformer will be displayed. If all values correspond to conditions, the message on a right choice of the autotransformer will be displayed. In a window of check there is a Reference manual button opening the reference manual with catalog data of autotransformers [3].

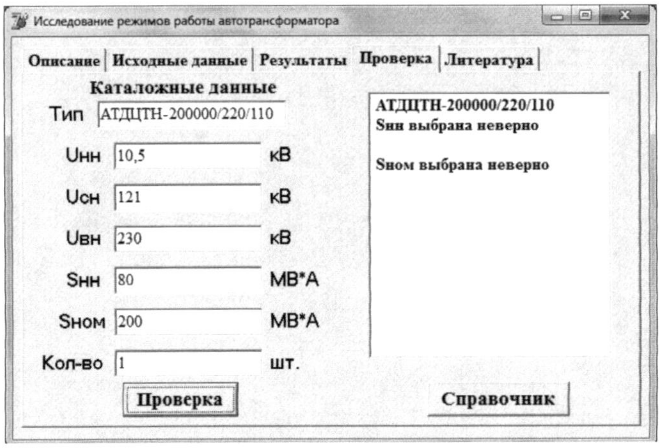

Figure 10. Check window.

In the literature window a list of the used sources is shown. Also there is a "Help" button in case of which clicking the document with methodical material on this subject opens.

4. Conclusion

The provided material contains the concentrated presentation of operation modes and algorithm of the power autotransformer choice, one of the main electric equipment participating in process of electrical power supply of customers including an oil and gas field. On this basis the program allowing on according to the entered data for the autotransformer is developed:

1. to define the most loaded winding;
2. to determine an operation mode by power transmission;
3. to calculate the required power rating;
4. to save calculated data in the file form;
5. to check correctness of the autotransformer choice;
6. to study reference methodical material on this subject.

Now operation on implementation of the developed program for educational and research tasks in TPU is carried.

References

[1] Vasilyev A A, Kruchkov I P, Nayashkova E F and Okolovich M N 1990 *Electric part of stations and substations* (Moscow: Energoatomizdat) p 576
[2] Kultin N B 2007 *Programming bases in Delphi 7* (St. Petersburg: BHV-St. Petersburg)
[3] Neklepayev B N 1989 *Electric part of power plants and substations. Reference materials for course and degree design* (Moscow: Energoatomizdat) p 608
[4] Homonenko A D 2005 *Work with databases in Delphi* (St. Petersburg: BHV-St. Petersburg)

Approaches of Russian oil companies to optimal capital structure

T Ishuk[1,3], O Ulyanova[2] and V Savchitz[1]

[1] Department of Natural Resources Economics, National Research Tomsk Polytechnic University, 30 Lenin Ave., Tomsk, 634050, Russia
[2] Department of Foreign Languages for Specialists in Natural Resources, National Research Tomsk Polytechnic University, 30 Lenin Ave., Tomsk, 634050, Russia

e-mail: [3] tana.itl@mail.ru

Abstract. Oil companies play a vital role in Russian economy. Demand for hydrocarbon products will be increasing for the nearest decades simultaneously with the population growth and social needs. Change of raw-material orientation of Russian economy and the transition to the innovative way of the development do not exclude the development of oil industry in future. Moreover, society believes that this sector must bring the Russian economy on to the road of innovative development due to neo-industrialization. To achieve this, the government power as well as capital management of companies are required. To make their optimal capital structure, it is necessary to minimize the capital cost, decrease definite risks under existing limits, and maximize profitability. The capital structure analysis of Russian and foreign oil companies shows different approaches, reasons, as well as conditions and, consequently, equity capital and debt capital relationship and their cost, which demands the effective capital management strategy.

1. Introduction

Oil companies are known to be the main sector of Russian economy. They manufacture more than a quarter of the industrial production volume in Russia whereas their share of income in the Federal budget comes to about 50%. Oil sector accounts for more than half of all export and one third of total capital investments [1]. Furthermore, at present there is an idea that this sector due to raw material profits will lead the Russian economy onto the way of innovative development like a locomotive. Therefore, the urgent problem now is to provide favorable internal and external conditions to fulfil this idea.

2. Material and Methods

The special feature of Russian oil companies is a large capital output ratio with a great part of equity capital in the source of financing: their debt capital comes to only 15–20 %, while the debt capital share in foreign oil companies can be up to 50 % [2].

The purpose of the article is to find out the reasons for differences in capital structure and to choose the approaches to its optimal structure for Russian oil companies. Therefore, to achieve this purpose, it is necessary to define the concepts and criteria of the optimal capital structure, and to analyze the capital structure by the example of Russian oil companies which are included into the 10-top companies, to find out their main problems in making the optimal capital structure, and to suggest the solutions of the problem.

The capital structure decision is believed to find a compromise between risk and profitability. The growth of debt capital share increases risk and at the same time raises profitability of equity capital. Accordingly, the companies using only equity capital, on the one hand, have maximal financial stability (equity to total assets), on the other hand, considerably decrease the development rates losing additional source of financing assets growth.

The optimal capital structure implies such a combination of equity capital and debt capital (leverage ratio) that provides maximal market-value capital of the company and, consequently, enterprise value as a whole. The fundamental criteria to make optimal capital structure are the following:

1) Financing Policy which is intended to find out the required amount and the share of financing sources by analyzing leverage ratio.

2) Enterprise Value (EV). The capital structure is considered optimal if it provides maximal enterprise value.

3) Cost of Capital. Any financing source has its cost. The capital structure is defined optimal if debt capital cost is minimal.

4) Risks (unpaid interests, loan default, bankruptcy, etc).

5) Profitability. Raising debt capital can increase return on equity due to degree of financial leverage.

There is no doubt that each criterion is important for the optimal capital structure. The analysis of some of them is given below.

Table 1. Debt capital share in oil companies capital structure (2011 – 2013), mln. RUB [3].

Company	Capital, total (RUB)			Dept capital, %		
	2011	2012	2013	2011	2012	2013
OJSC “Gazprom”	10 786 140	11 956 836	12 555 820	29.8	29.1	27.9
OJSC “Surgutneftegas”	1 653 382	1 797 066	1 962 162	6.4	6.0	7.5
OAO LUKOIL	1 182 920	1 187 985	1 253 476	47.7	37.8	31.9
JSOC Bashneft	272 649	339 572	366 406	62.9	54.2	53.7
JSC TATNEFT	462 243	474 563	487 569	28.9	20.2	18.7

According to table 1, the companies seek to decrease the debt capital share. Oil company OAO LUKOIL decreases by 15.8%, JSOC Bashneft – 8.7%, JSC TATNEFT – 10.2%, only company OJSC “Surgutneftegas” shows a growing tendency of 1.1%. The above mentioned data make it possible to draw the conclusion that the debt capital share in oil companies varies from 6% (OJSC “Surgutneftegas”) to 62% (JSOC Bashneft). It makes us think deeply about existing capital structure efficiency of these companies and choose the most efficient structure. For this purpose, the analysis of return on equity and equity to total assets is required.

Table 2. Return on equity (ROE) and equity to total assets [3].

Company	DuPont ROE		Equity to Total Assets		
	2012	2013	2011	2012	2013
OJSC “Gazprom”	6,3	6,6	0.7	0.7	0.7
OJSC “Surgutneftegas”	6,0	8,5	0.9	0.6	0.9
OAO LUKOIL	36,5	18.3	0.5	0.9	0.7
JSOC Bashneft	6.1	8,3	0.4	0.5	0.5
JSC TATNEFT	4.6	7.0	0.7	0.8	0.8

According to the data obtained, the largest return on equity is 36% (OAO LUKOIL), while 15–20% indicator is considered as enough. However, 36% indicator is not pernicious for the company; but it means company OAO LUKOIL is taking more risky business than other companies. It proves the main business law: the higher risk, the higher profitability.

Also, it is shown, profitability decreases up to "safe" indicator (18.3%) over time. Return on equity in other companies is considerably lower (4–8%). That fact proves that the companies are not employing its capital effectively.

Further, equity to total assets is required to be analyzed in order to make a conclusion about capital structure. Maximal indicator with value of 1, shows full financial independence of the company, which is a good sign of business effectiveness. On the other hand, the company is considered financially sound if the indicator is more than 0.5. Paying special attention to OAO LUKOIL, one can conclude that when equity to total assets equals 0.6% (2012), return on equity achieves 36%, with debt capital share being 38%.

The company has higher potential for its economic growth and at the same time it is financially independent.

In this respect, it is important to consider OJSC "Surgutneftegas": equity to total assets equals 0.9, with debt capital share being 6–7%. Being financial independent, the company considerably limits its economic development rates refusing to obtain debt capital.

JSOC Bashneft had a different financial strategy. Having debt capital share 60% in 2011 with low equity to total assets 0.4, it had a risky policy to obtain capitals. However, by 2013 the financial situation had become stable.

As seen, having low dept capital, most large oil companies are financially independent. Any capital structure decision is a choice of the company itself between profitability and risk, because increasing in debt capital leads to increasing both risk and profitability. Each company makes such a decision taking into account the company overall strategy. For a comparative analysis, it necessary to consider foreign oil companies.

Table 3. The debt capital share in the capital structure of foreign oil companies (2011 – 2013) [3].

Company	Capital, total всего, (mln, $.)			Dept capital, %		
	2011	2012	2013	2011	2012	2013
Schlumberger	2 315 670	2 583 980	2 814 830	43.4	43.6	41.2
Total	14 937	19 350	20 199	82.9	85.1	83.3
Royal Dutch Shell	14 155 274	14 694 758	14 997 552	53.0	50.1	49.6
Exxon Mobil	13 887 560	14 002 628	14 548 521	49.8	50.3	53.4
Statoil	4 870 259	4 969 685	5 610 853	63.7	59.3	59.9

According to table 3, debt capital share of foreign oil companies varies from 41% to 85 %. Such high debt capital share gives additional possibilities to foreign companies for economical growth of business and ramp up.

Table 4. Return on equity and equity to total assets of foreign oil companies [3].

Company	DuPont ROE		Equity to Total Assets		
	2012	2013	2011	2012	2013
Schlumberger	15.8	17.1	0.6	0.6	0.6
Total	41.3	40.3	0.2	0.1	0.2
Royal Dutch Shell	15.3	9.1	0.5	0.5	0.5
Exxon Mobil	27.1	18.7	0.5	0.5	0.5
Statoil	21.6	11.2	0.4	0.4	0.4

On average, return on equity of foreign companies is 25%. Reduction of the indicator over time shows that the net income of companies is growing much slower than equity capital. In compliance with Table 4, equity to total assets is 0.5 that is enough for the companies with the debt capital share being more than 50% to be financially independent.

3. Conclusions

Considerable difference between capital structures of Russian and foreign oil companies can be explained by the fact that foreign companies were gradually founded under the influence of objective factors of the market system as well as regulatory actions of the government, while Russian companies were immediately founded on the basis of commercial acts under the condition of centrally planned economy (CPE) [4, 10].

Also, the peculiarity of Russian oil companies is in high level of state participation. It is not only the result of government's block of shares and the state participation in company management respectively, but due to the fundamental principles of Russian companies. Most assets of Russian companies were made due to the Federal Budget, whereas, the state participation in foreign oil companies is characterized as insignificant [5].

The essential difference in the capital structure is the cost of attraction of borrowed funds. The rate percent in Russia is from 18% to 24%, while abroad it does not exceed 8% [5].

The experts opinion from "Deloitte & Touche CIS[4]" confirm that 21% of respondents, i.e. Russian oil companies, are planning to increase the amount of financing of the core activities due to the investments of third party organizations. As for state participation, most companies (76%) believe the influence of state participation on oil industry is going to increase. Simultaneously, according to opinions of oil companies' experts, the following factors played a vital role in developing of oil industry: export tax cut, disaggregation of tax on natural resources production depending on index complexity of petroleum field geology, and reduced rate of tax on natural resources production (oil) for private oil deposits, as well as investment bonus (accelerated amortization) [6].

To sum up, not only internal organizational, financial but macroeconomic factors could have a good influence on increasing company cost as an essential condition for capital raising, making the debt capital and participating in large projects together with world leading companies.

The given analysis of mentioned-above indicators of profitability and financial independence of Russian and foreign oil companies [7, 8, 9] proves the necessity to update the certain principles and concepts of financial management theories taking into account Russian institutional and manufacturing conditions.

References

[1] Grinkevich L S, Sharf I V 2014 Analysis of Russian Federation budget revenues from the implementation of export duties "60-66-90" scheme *IOP Conf. Ser.: Earth Environ. Sci.* **21** 1–5

[2] Voronina E V 2004 Formirovanie ehffektivnoj struktury kapitala korporacij neftegazovogo kompleksa: dis. ... kand. ehkon. nauk (Surgut) 201 RGB OD, 61:04-8/2730

[3] Annual reports of oil industry *(Electronic Materials)* URL: http://www.lukoil.ru/static.asp?id=85; http://www.gazprom.ru/investors/; http://www.surgutneftegas.ru/; http://www.bashneft.ru/; http://www.tatneft.ru/ http://raexpert.ru/ratings/expert400/2011/capitalization/table2/

[4] Bulakh D A Peculiarities and prospects of Russian oil companies. *(Electronic Materials)* URL: http://www.vipstd.ru/gim/content/view/83/247/

[5] Credit book *(Electronic Materials)* URL: www.creditbook.ru/voproscredit/29-public/548-stavki-po-kreditam-v-evrope

4 "Deloitte" is committed to serving state and private companies in different branches of engineering independent audit process, management and financial support.

[6] Oil and gas survey – 2013, Presentation "Deloitte & Touche CIS" 2013 (*Electronic Materials)* URL: http://www.hse.ru/data/2014/01/25/1326319826/Deloitte_oil&gas%20survey.pdf.
[7] Tsibulnikova M R, Pozharnitskaya O V and Strelnikova A B 2015 Designing economic and legal mechanism of land management in oil and gas companies *IOP Conf. Ser.: Earth Environ. Sci.* **24** 012032
[8] Markov V K 2011 Raising efficiency of the country's oil and gas enterprises *Vestnik Saratov State Socioeconomic University* (*Electronic Materials)* **3** URL: http://cyberleninka.ru/article/n/raising-efficiency-of-the-countrys-oil-and-gas-enterprises
[9] Sokolov A Y and Giniatullin Y M 2015 Management accounting and costs controlling in oil producing companies: Historical perspectives **6** 430–4
[10] Clint O and Pyle I 2014 Competing and partnering for resources and profits: Strategic shifts of oil Majors during the past quarter of a century *Energy Strategy Reviews* **3** 72–87

Study of the electrodes length influence on the trajectories of water droplets dispersed in oil and affected by non-uniform electric field

S N Kharlamov[1,2], V V Zaikovskii[1,3] and I V Shenderova[1,4]

[1] Institute of Natural Resources, National Research Tomsk Polytechnic University, 30 Lenin Ave., Tomsk, 634050, Russia

E-mail: [2] kharsn@mail.ru, [3] alkast@tpu.ru, [4] inna-shenderova@yandex.ru

Abstract. The paper presents the results of numerical modelling of the processes accompanying movement of drop viscous media (water) in oil under the influence of exterior forces of the electric and dynamic nature. Systematic calculations of influence on the electric field heterogeneity drops, created by a symmetric and asymmetrical configuration of electrodes are carried out both in inter electrode and behind electrode areas taking into account a complex operation of dielectrophoresis forces, buoyancies and drag, as well as the variability of electrode sizes. The analysis of drop movement trajectories shows that the asymmetrical configuration of electrodes can be applied for an electro-coalescence intensification of water-in-oil emulsion. Correctness of calculations of the mathematical model and numerical methods are confirmed by good results if compared with the available data of the other authors.

1. Introduction

Nowadays there are two basic physical principles, which are applied during electroseparation of water-in-oil emulsions: electrophoresis (the motion of charged particles along the lines of the electric field) and dielectrophoresis (motion of polarized substances toward the greatest increase of electric field).

The analysis of literature has shown [2-4] that in recent years the attention of researchers focuses on the issues of understanding the nature of the influence of "small-scale" dielectrophoresis effect. It is known that this physical mechanism has a number of disadvantages as it functions only at small distances between polarized droplets [5, 6]; in the certain range of parameters of the electric field it may be a reason for secondary droplets of dispersed phase (which are much smaller size) origination [7]; there is a possibility for a short circuit between non-insulated electrodes [8].

In an inhomogeneous electric field, water droplets will tend to move in the direction of the greatest increase of the electric field [6, 10]. Thus, an additional bonding mechanism emerges. Non-uniform electric field can be generated due to the asymmetric, confuser, diffuser or cylindrical configuration of the electrodes [9].

Considering the abovementioned, the aim of this paper is: 1) to perform the development and verification of a mathematical model and numerical algorithm of calculation of dispersed conductive phase movement processes in water-oil emulsions under electric field action; 2) to establish the efficient algorithm application in systematic researches of an intensification problems of electro coalescence process at asymmetrical configurations of electrodes, and in the forecast of the potential changes, electric field intensity and drop movement trajectories of water in emulsions;3) explanation and an assessment of influence character of dielectrophoresis forces on drop movement trajectory.

2. Numerical simulation of the electric field

Currently the different approaches to the calculation of the non-uniform electric field characteristics and its effect on polarized dielectrics are described in detail in the literature. In particular, there is an approach in which the potential of the electric field is modeled by implicit analytic function obtained by the method of conformal mappings [11]. Another approach, which allows obtaining analytical solution on the more general geometry of the electrodes with mixed boundary conditions, is the method of Wiener – Hopf [12]. However, its usage is presented in the literature only by cases of electric field calculation in the interelectrode space.

A much more common method of obtaining values of the electric field potential on arbitrary geometry of the electrodes and arbitrary distance from them - is the numerical solution of the Laplace equation [13, 14].

In the assumption of free charges absence in a simulated two-dimensional region, Laplace equation can be written in the form [15]:

$$\frac{\partial^2 \varphi}{\partial x^2} + \frac{\partial^2 \varphi}{\partial y^2} = 0 \tag{1}$$

This equation must be supplement by boundary conditions of several types. For the boundaries of the computational domain, which should be at a sufficient distance from the plates of the capacitor, the potential is zero. For left plate (index "lp") and right plate (index, "rp") of the condenser the potential should be known and specified by initial conditions of the calculation.

Relation to the stationary electric field, based on the determination of this value, could be formulated for the two-dimensional case, as a vector [15]:

$$\overline{E} = \left\{ -\frac{\partial \varphi}{\partial x}; -\frac{\partial \varphi}{\partial y} \right\} \tag{2}$$

Numerical integration of equation (1) is performed with the involvement of implicit finite-difference schemes and approximation of the derivative with the second order accuracy [$O(\Delta x^2)$, $O(\Delta y^2)$] regarding the steps for axial and longitudinal variables on a "cross" grid-type pattern.

Verification of the numerical algorithm was implemented on a two-dimensional flat vertical condenser, in the ranges of define parameters, that are presented in table 1.

Table 1. Initial data for verification of the algorithm.

Parameter, designation, measurement unit	Value
Length of the left capacitor plate, L_{lp}, 10^{-2} m	1 ÷ 10
Length of the right capacitor plate, L_{rp}, 10^{-2} m	10 ÷ 1
Voltage on the left capacitor plate, U_{lp}, V	0 ÷ 5000
Voltage on the right capacitor plate, U_{rp}, B	0
Distance between capacitor plates, d, 10^{-2} m	1 ÷ 20
Discretization step along the longitudinal coordinate, Δx, 10^{-3} m	1 ÷ 0,05
Discretization step along the transverse coordinate, Δy, 10^{-3} m	1 ÷ 0,05

Individual results of the calculation are shown on figures 1-2, which illustrate combined field potential φ and the electric field strength E in the study area at fixed $\Delta x = \Delta y = 10^{-4}$ m, $U_{lp} = 5000$ V, $U_{pp} = 0$ V, $d = 0{,}5 \cdot 10^{-2}$ m. Variety of the scalar field potential $\varphi(x,y)$ is illustrated by the background color of figures 1-2, the vector electric field E (x, y) is indicated by arrows.

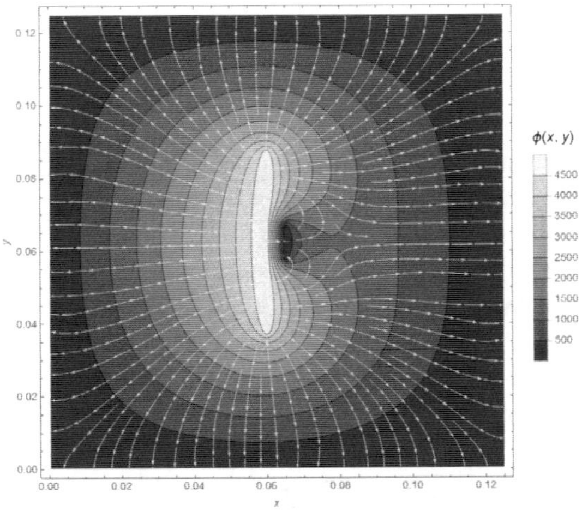

Figure 1. The change in the potential φ(x,y) and electric field E (x,y) at $L_{lp} = 5 \cdot 10^{-2}$ м, $L_{rp} = 1 \cdot 10^{-2}$ m

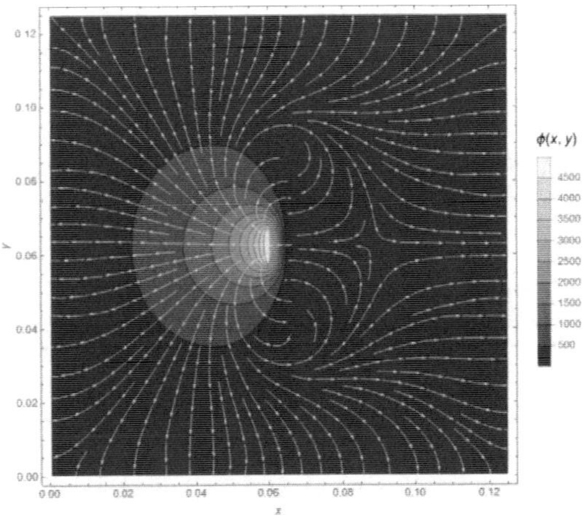

Figure 2. The change in the potential φ(x,y) and electric field E (x,y) at $L_{lp} = 1 \cdot 10^{-2}$ м, $L_{rp} = 5 \cdot 10^{-2}$ m

To validate and assess the accuracy of calculations graphical comparison of power distribution dielectrophoresis (8) obtained using the described algorithm and obtained by other researchers (for example, [13]) was carried out. The comparison of results is shown on figure 3. Graphical field mapping shows that the features associated with the predominant direction of the field vectors to the edges of the capacitor plates are forecasted quite accurately.

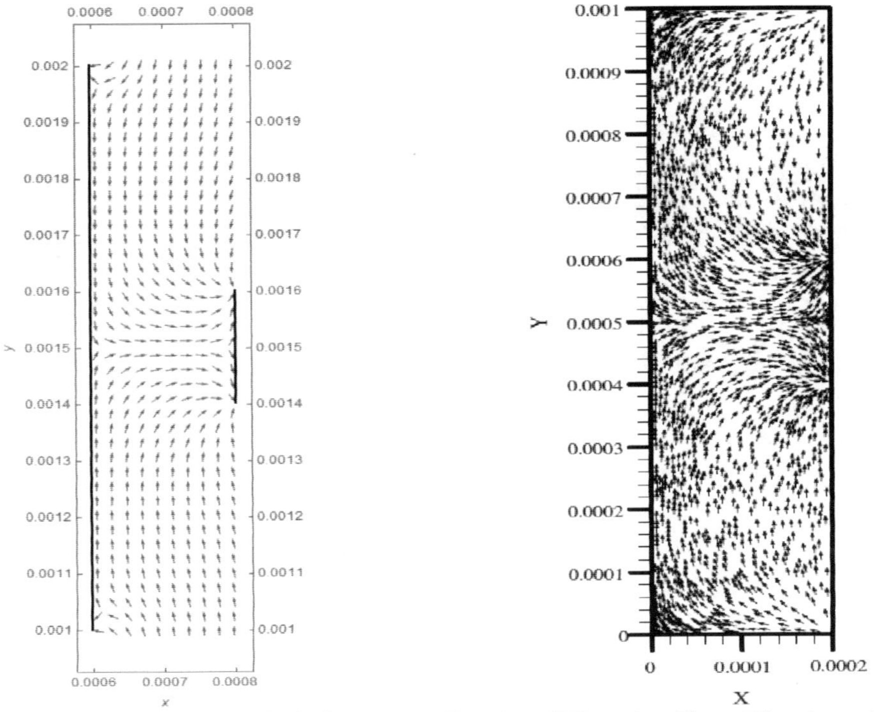

The results of the present calculation Results of Zhenqian Chen, Xiaozhong Shen [13]

Figure 3. Graphical comparison of the dielectrophoretic force spatial distribution at $U_{lp} = 0$ V, $U_{rp} = 70$ V, $d = 200 \cdot 10^{-6}$ m, $L_{лп} = 1000 \cdot 10^{-6}$ m, $L_{пп} = 200 \cdot 10^{-6}$ m.

3. Numerical modeling of water droplets motion

For modeling the motion of single spherical incompressible water droplets in oil under the influence of an inhomogeneous electric field, it is necessary to determine the forces which influence such a drop. They are: gravity force, the force of Archimedes, drag force, the dielectrophoretic force.

The effect of gravity and Archimedes force is modeled by the ratio for buoyancy force $F_{buoyancy}$ [16]:

$$\overline{F}_{buoyancy} = \overline{F}_{gravity} + \overline{F}_{Archiimedes} = -(\rho_{drop} - \rho_{oil}) \cdot g \cdot V_{drop} \quad (3)$$

The effect of drag force F_{drag} for liquid particles, moving with a slight acceleration (taking into account the internal fluid circulation inside the droplet), is modeled by Hadamard – Rybczynski ratio [16]:

$$\overline{F}_{drag} = -3\pi \cdot \frac{\lambda + \frac{2}{3}}{\lambda + 1} \cdot \mu_{oil} \cdot d_{drop} \cdot \overline{\upsilon}_{drop}, \quad where\ \lambda = \frac{\mu_{water}}{\mu_{oil}} \quad (4)$$

The effect of dielectrophoretic force F_{DEP} on conducting drop of water, that dissolved in a non-conductive oil environment and being under the influence of an inhomogeneous electric field is modeled by the ratio [11]:

$$\overline{F}_{DEP} = 2\pi \cdot \varepsilon_{oil} \cdot \frac{d^3}{8} \cdot \frac{\varepsilon_{water} - \varepsilon_{oil}}{\varepsilon_{water} + 2\varepsilon_{oil}} \cdot \nabla E^2 \quad (5)$$

Thus, the equations of motion in projections on coordinate axes for a water drop could be written as follows:

$$m_{drop} \cdot a_y = F_{buoyancy} + F^y{}_{drag} + F^y{}_{DEP}, \quad a_y = \frac{d^2 y}{dt^2} \quad (6)$$

$$m_{drop} \cdot a_x = F^x{}_{drag} + F^x{}_{DEP}, \quad a_x = \frac{d^2 x}{dt^2} \quad (7)$$

4. The discussion of research results

Application of numerical integration methods, in the computer algebra system "Mathematica", the trajectory of single water droplets were simulated in the computational domain H = 0.125 × 0.125 m with initial coordinates: y_0=0.125 m (upper edge of computational domain), $x_0 \in \{0.015;0.045\}$ m with a step. The range of varying physical parameters of the simulated environment is presented in table 2.

Table 2. The values of physical quantities in the simulated water drop movement.

Parameter, designation, measurement unit	Value
Length of the left capacitor plate, L_{lp}, 10^{-2} m	2 ÷ 5
Length of the right capacitor plate, L_{rp}, 10^{-2} m	5 ÷ 2
Voltage on the left capacitor plate, U_{lp}, V	5000
Voltage on the right capacitor plate, U_{rp}, B	0
Distance between capacitor plates, d, 10^{-2} m	1
Discretization step along the longitudinal coordinate, Δx, 10^{-3} m	0,05
Discretization step along the transverse coordinate, Δy, 10^{-3} m	0,05
Droplet diameter, d_{drop}, 10^{-6} m	50
Density of water, ρ_{water}, kg/m^3	1000
Density of oil, ρ_{oil}, kg/m^3	850
Dynamic viscosity of water, μ_{water}, 10^{-3} Pa·sec	1,002
Dynamic viscosity of oil, μ_{oil}, 10^{-3} Pa·sec	7
Dielectric permittivity of water, ε_{water}, 10^{-12} F/m	81
Dielectric permittivity of oil, ε_{oil}, 10^{-12} F/m	2,5
Step of varying the initial droplet position, x_0, 10^{-4} m	2,483

It should be noted that the above model doesn't consider the coalescence of water droplets – the trajectory of each droplet is considered separately, as if it was moving in the absence of other drops. The trajectories of water droplets with different initial position x_0 are placed on a single figure in order to provide a more effective visualization of the non-uniform electric field effect.

To study the dependence of the trajectory of water droplets from the ratio of plates dimensions and, consequently, influence of inhomogeneity of the electric field generated by the asymmetric configuration of the plates, various calculations were carried out with changing of the parameters L_{lp}, $L_{гр}$ in the range $(2 \div 5)\ 10^{-2}$ m. Some results of numerical simulation are presented in figures 4-5.

In particular, these results allow us to conclude that the configuration of the electrodes had a major influence on water droplets trajectories.

In the case of asymmetrical flat capacitor and the prevalence of the length of a charged plate over the length of the grounded plate (figure 4), there was a proportional deviation of one half of the drops in the direction of a charged plate, the other half – towards the grounded plate. This is a consequence of large-scale inhomogeneity of the electric field in the interelectrode space. In this case, due to the large number of horizontal deviations, the length of the path of water droplets in this field increases. The number of drops that slipped between zones of influence, with this configuration of electrodes, visually is about 3%, and the zone of influence of a grounded electrode in the "right" outer space is slightly less than the zone of influence of a charged electrode in the "left" outer space.

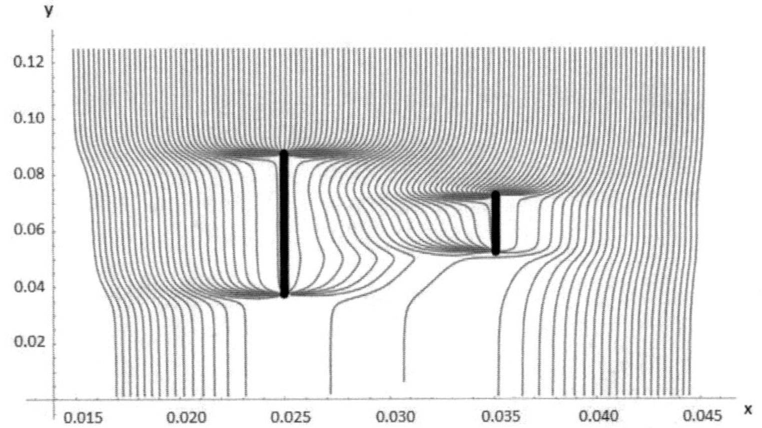

Figure 4. Water droplets trajectories in oil with the asymmetric configuration of the electrodes, $L_{лп} = 5{\cdot}10^{-2}$ m, $L_{пп} = 2{\cdot}10^{-2}$ m, $U_{лп} = 5000$ V, $U_{пп} = 0$ V.

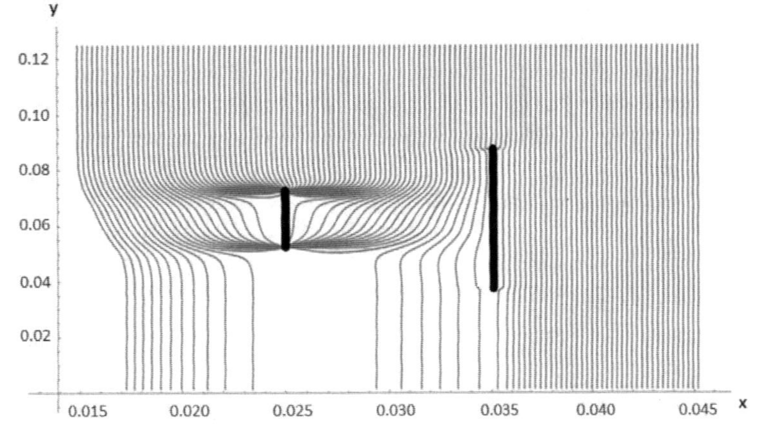

Figure 5. Water droplets trajectories in oil with the asymmetric configuration of the electrodes, $L_{лп} = 2{\cdot}10^{-2}$ m, $L_{пп} = 5{\cdot}10^{-2}$ m, $U_{лп} = 5000$ V, $U_{пп} = 0$ V.

In case when the length of the grounded plate is more than the length of a charged plate (figure 5), almost all particles in the interelectrode space are attracted to a charged plate. However, about 15% of the particles do not reach the boundary of the zone of attraction, which is much greater than for the case when the charged plate length is more than the length of the grounded one. The distance between the "slipped out" drops increases if compared to the original, which creates an even greater negative effect.

The overall effect after treating the plane capacitor by asymmetrical vertical electric field, in both cases, is expressed in reduction in the number of drops in the interelectrode space and some distance behind the electrodes, due to their attraction to the edges of the plates.

5. Conclusion

The paper presents the results of mathematical modeling and numerical calculation of the movement process of water droplets in oil, as well as the efficiency of the algorithm in the range of investigated parameters. Verification of the model and algorithm is performed by comparison with the known results of other authors and with high degree of agreement in the prediction of the effects of power dielectrophoresis on the motion of the dispersed phase. For the single incompressible spherical water drops in oil, the calculation of the trajectories of water droplets in an inhomogeneous electric field, created by different configurations of electrodes, was performed. Detailed calculations have shown, that the capacitors configuration with prevalence of a charged plate length over the grounded plate length provides the least amount of "slipped" drops, greater path length of drops in the interelectrode space and a greater area of outer space impact on the drops trajectories.

References

[1] Levchenko D N, Bergshtein N V, Hudyakova A D and Nikolaeva N M 1967 *Emulsii nefti s vodoi i metody ih razrusheniya* (Moskow: Khimiya)
[2] Eow J S, Ghadiri M, Sharif A O and Williams T J 2001 *Chemical Engineering Journal* **84** 173
[3] Mohammadi M, Shahhosseini S and Bayat M 2014 *Chemical Engineering Research and Design* **92** 2694
[4] Eow J S and Ghadiri M 2002 *Chemical Engineering Journal* **85** 357
[5] Panchenkov G M and Tzabek L K 1969 *Povedenie emulsii vo vneshnem elektricheskom pole* (Moskow: Khimiya)
[6] Vinogradov V M and Vinokurov V A 2007 Obrazovanie, svoistva I metody razrusheniyaneftyanih emulsii (Moskow: RGU nefti i gaza imeni I.M. Gubkina)
[7] Mousavi S H, Ghadiri M and Buckley M 2014 *Chemical Engineering Science* **120** 130
[8] Mhatre S, Vivacqua V, Ghadiri M, Abdullah A M, Al-Marri M J, Hassanpour A, Hewakandamby B, Azzopardi B and Kermani B 2015 *Chemical Engineering Research and Design* **96** 177
[9] Kharlamov S N and Zaykovskiy V V 2013 *Proc. 8th Int. Forum on Strategic Technology IFOST (Ulaanbaatar, Mongolia, 28 June–1 July 2013)* **1** 784
[10] Jones T B 1995 *Electromechanics of particles* (New York: Cambridge University Press)
[11] Khainovskii V I, Kozirev A E and Nikitin P V 2013 *Vestnik APK Stavropoliya* **4** 99
[12] Kua C H, Lam Y C, Yang C, Youcef-Toumi K and Rodriguez I 2008 *Journal of Electrostatics* **66** 514
[13] Chen Z and Shen X 2010 *Building and Environment* **45** 968
[14] Tarantzev K V, Krasnaya E G and Korosteleva A V 2011 *Izvestia Penzenskogo gosudarstvenogo pedagogicheskogo universiteta* **26** 654
[15] Nagel J R 2014 *Antennas and Propagation Magazine* **56** 209
[16] Melheim J A and Chiesa M 2004 *Proc.5th Int. Conf. on MultiphaseFlow (Yokohama, Japan, 30 May–4 June 2004)* **1** 126

Potential of renewable and alternative energy sources

V Konovalov[1,3], O Pogharnitskaya[1,4], A Rostovshchikova[1] and I Matveenko [2,5]

[1] Natural Resource Economics Department, National Research Tomsk Polytechnic University, 30 Lenin Ave., Tomsk, 634050, Russia
[2] Foreign Language Department, National Research Tomsk Polytechnic University, 30 Lenin Ave., Tomsk, 634050, Russia

E-mail: [3] sltan@ngs.ru, [4] pov@tpu.ru, [5] mia2046@yandex.ru

Abstract. The article deals with application potential of clean alternative renewable energy sources. By means of system analysis the forecast for consumption of electrical energy in Tomsk Oblast as well as main energy sources of existing energy system have been studied up to 2018. Engineering potential of renewable and alternative energy sources is evaluated. Besides, ranking in the order of their efficiency descending is performed. It is concluded that Tomsk Oblast has high potential of alternative and renewable energy sources, among which the most promising development perspective is implementation of gasification stations to save fuel consumed by diesel power stations as well as building wind-power plants.

1. Introduction

In recent years increasingly greater attention is paid to the problems of utilizing clean alternative renewable energy sources to meet energy supply demands in different agricultural and industrial spheres. The relevance and perspectiveness of the given trend are conditioned by the two key factors: catastrophic situation in the environment and necessity of searching for new kinds of energy [1].

Due to slow and sure resource depletion of basic energy sources, first of all, oil and gas, as well as limits in development of hydro- and atomic power, many countries carry out the research in extension of alternative energy sources use – peat, oil shale, bitumen, unconventional gases, energy of Earth heat, sun, wind, ocean, biosynthesis etc [2]. The research in alternative energy sources utilization is in progress in Russia as well. The Government Executive Order of the Russian Federation by 28 August 2003 № 1234 approved the energy strategy in Russia for the period up to 2020. The order is focused on necessity of using renewable sources of energy and local types of fuel.

2. Research object and methods

Studying the forecast for energy consumption and capacity within the period 2013-2018 developed by JSC «SO EES» based on facts and forecasts on production and commercial operation of electrical power engineering entities, approved programs of regional social-economic development as well as information on concluded utility connection contracts (table 1), there seems no doubt in the conclusion that in the nearest future the demands for renewable and alternative energy sources will be increasing, therefore, the research problem becomes even more urgent.

Table 1. Forecast of energy consumption in Tomsk Oblast energy system, mln. kWh.

Indicator	2014 (forecast)	2015 (forecast)	2016 (forecast)	2017 (forecast)	2018 (forecast)
Electrical energy consumption, mln. kWh	9344	9421	9531	9609	9688

The energy system was developed in the 1960's, mostly in Tomsk city and neighboring regions. The intensive process of energy system development took place in the 1970-80's. Within this period the railway road "Asino – Beluy Yar" was built, the consumers of the adjacent to railway road areas were supplied with electricity, agricultural consumers of south-eastern regions were provided with electrical energy, and intensive development of the northern oil fields started. In the field area the new towns (Strezhevoy, Kedrovuy) were built, Lyginetskoye, Igol'skoye, Gerasimovskoye, and other oil fields were developed, for their energy supply substations and transmission lines were built.

The primary power sources of Tomsk Oblast energy system are power stations Tomsk state district power plant (SDPP)-2 (Ru–331 MW) and Tomsk thermal power plant (TPP)-3 (Ru–140 MW), Tomsk auxiliary boiler (Ru–14.7 MW) incorporated in Tomsk branch of JSC «TGK–11» and SCC TPP (549 MW) belonging to JSC «Siberian Chemical Combine». The other generating capacities include, as a rule, units of small generation at some industrial enterprises such as gas-turbine and gas-engine power plants (stations of industrial plants) of installed capacity 61.2 MW. The energy supply of Tomsk Oblast is scarce. In 2012 the power exchange with neighboring regions united in energy system of Siberia and Ural amounted 3.635 bln. kWh or 40 % of annual energy consumption.

The scarcity of Tomsk Oblast in electrical energy causes the search for the ways of decreasing its dependence on outside power providers and, hence, increasing its power safety.

Availability of natural gas deposits and its production at the level of 4–4.2 bln.m^3 allows the potential possibility of building new gas power generating units in Tomsk TPP–3, mini–TPP on the basis of gas-turbine and gas-diesel power generating units of low capacity. Besides, the forecasted tendency of outstripping growth of natural gas cost in comparison with coal cost conditions the necessity of using gas at efficient power equipment only.

As an alternative for gas power engineering development one can consider the development of atomic energy and coal power stations.

It should be taken into account that the significant part of generating equipment at the Tomsk Oblast power stations are worn-out in service and require repair or replacement.

One more peculiarity of Tomsk Oblast energy system is the fact that primary energy sources are located in the south of Oblast, whereas significant part of power consumers, most oil production enterprises are situated in the north of the Oblast. Transportation of electrical energy from south to north is performed in long transmission lines of 110–220 kW in voltage, with very high current loads. It results in limitation of consumers' loads in transmission lines in post-emergency and repair conditions. In addition, at many substations of the energy system there is much tension around the issue of shortage in transformer capacity. These circumstances explain the fact that possible connections of new consumers to the power network are essentially used up in some nodes of power systems.

All mentioned above necessitates the improvements in the Tomsk Oblast energy system, among which much attention is paid to utilization of alternative energy sources and renewable energy resources.

Evaluation of technically feasible potential of renewable and alternative energy sources allows determining technical potential of principal kinds of renewable and alternative energy sources, among which wind energy seems to be the most profitable (table 2). In different parts of the world the application of wind energy has achieved the level which permits it to become the primary energy source. In developed countries, especially in Europe, the growth of wing energy generation was conditioned by the problem of global climate change for a long time.

Wind energy is the most preferable solution of international energy problems. It does not pollute environment and does not depend on fuel. Moreover, wind resources are available in any part of the world and they are sufficient to meet growing demands in energy [3].

Table 2. Annual technical potential of renewable and alternative sources of energy for Tomsk Oblast.

Kind of energy	Annual energy production, bln kWh	Annual heat production, mln Gcal
Wind energy	100 – 402	–
Solar energy	15 – 25	36
Small hydrostations	4.1	–
Biomass	4.7	26
Geothermal energy	–	76

The presented values describe total technical opportunities of energy production by means of renewable and alternative energy sources.

Comparison of the table data with the contemporary production level of electrical (5,5 bln KWH) and heat energy (12.3 mln Gcal) in Tomsk Oblast shows that technical potential of renewable and alternative energy sources exceeds significantly the energy demands.

To evaluate the real possibility of utilizing renewable and alternative energy sources, it is necessary, first of all, to determine their economic potential, i.e. that part of technical potential, the usage of which can be efficient at existing and forecasted indicators of power installations for existing and forecasted loads (energy demands).

To determine commercial potential of renewable and alternative energy sources, it is necessary to carry out a series of feasibility studies in definite points of their possible location (project technical and economic feasibility). Based on economic evaluations, technologies of renewable and alternative energy sources were ranked in their efficiency. The quantitative evaluations of potential generation of renewable and alternative energy sources competing with diesel power stations were given for the areas of decentralized power supply system where energy was generated by diesel power stations.

Such an analysis allows outlining the approaches to the development program of renewable and alternative energy sources in Tomsk Oblast including pilot projects for which design of technical and economic feasibility is reasonable.

Technologies of renewable and alternative energy sources can be ranked in descending order in terms of their efficiency in the following way:

1. Wood gasification power stations are efficient in the areas of decentralized power supply. In case of competition with diesel power stations they generate essentially (several times) cheaper electrical energy and are more preferable energy sources for isolated consumers.

Tomsk Oblast is rich in forest resources. Within the period of intensive development of forest industrial activity the industrial wastes amounted 260 000 tons. Even at the moment of maximum timber harvest, forest industry used only about quarter of forest resources subjected to cutting down that resulted in accumulation of mature and overmature planting.

Taking into account the fact that with the efficiency coefficient of gasification power stations 20%, the potential of electrical energy generation for Tomsk Oblast amounts in general approximately 160 - 640 mln kWh (i.e. 3 – 12 % of total energy generation).

2. Wind energy units are competitive in comparison with diesel power stations in the areas of decentralized power supply at average wind velocity more than 3,4 m/sec (such a velocity is typical for large number of residential areas). Their use allows reducing energy generation at diesel power stations and saving diesel fuel.

3. Small diversion hydrostations in some cases appear to be efficient for power supply of isolated consumers; however, conditions for building such hydro power stations are not favorable in Tomsk

IOP Conf. Series: Earth and Environmental Science **27** (2015) 012068 doi:10.1088/1755-1315/27/1/012068

Oblast due to plain relief. Potential scales of small hydropower industry development are not great – several plants with unit capacity in the order of hundreds of kilowatts.

4. Large peat reserves permit (in case of buildup of its production and manufacture of fuel briquettes) significant reduction of Tomsk Oblast dependence on foreign coal consumed by boilers. No limitations for peat production were stated.

5. Geothermal heat supply is competitive with boilers over the whole territory of the Oblast. Areas where thermal deposits are discovered have demand for thermal energy, hence, it is reasonable to build thermal power systems in: Asino, Teguldet, Zyryanskoye, Pervomaiskoye, Beluy Yar, Bakchar, Podgornoye, Stepanovka, Nazino, Kolpashevo, Kataiga, Narym, Parabel', Kargasok and other settlements.

6. Thermal pumps using low-temperature heat in the areas of centralized power supply (where the energy is cheaper) are nearly equally-economic in comparison with boilers. Their implementation can be reasonable due uneconomic (ecological and social) reasons.

7. Geothermal power stations could theoretically be built in Tomsk Oblast. However, low temperature of thermal waters requires special equipment, which is not available at the moment.

8. Solar energy reaching our planet is known to exceed human energy demand 20000 times. Of its total amount approximately one-fourth is spent for water evaporation and, in fact, continuously more or less evenly accumulates in the atmosphere over any point of the world [4]. However, utilizing solar energy for electrical- or thermal supply is economically inefficient and application of its technologies is not reasonable on a large scale.

An integral part of attracting investments for development of alternative and renewable energy sources is arrangement of high energy efficiency demonstration zones.

Demonstration zone of high energy efficiency is a system of complex energy supply, where there are conditions for efficient use of fuel-power resources, solution of management, engineering, economic, and legal problems in the priority power supply fields, concentration of production and engineering-research foreign and domestic experience for further development in economy and social services.

Such zones may be arranged in Kolpashevo, Asino, and other settlements of Tomsk Oblast [5].

In the structure of such high energy efficiency zones there should be energy service companies engaged in production, transport, and distribution of fuel-power resources, ongoing engineering centers should be arranged to train and retrain specialists in the sphere of energy saving, to perform advisory service, hold seminars and other activities for information support in rational use of fuel-power resources [6].

When arranging demonstration zones of high energy efficiency, it is necessary to use the following primary principles:

- application of new promising technologies and facilities to produce, transport, and distribute heat energy;
- development of energy saving activity;
- development of commercial appeal and cost-efficiency of the demonstration projects.

3. Conclusions

In conclusion it should be noted that Tomsk Oblast possesses great potential of alternative and renewable energy sources, among which the most efficient is implementation of gas-generating power stations in order to save fuel consumed by diesel power stations as well as build wind power plants.

Alternative and renewable energy sources are most efficient in the areas of decentralized power supply where they are competitive with diesel power stations.

To develop the utilization of clean alternative renewable energy sources one needs to take actions at the national level.

1. To assume obligations in development of renewable sources at the legislation level.
2. To design market techniques for renewable energy development at the legislation level.
3. To provide consistent return on investments placed in the projects of "renewable energy".

4. To reform electrical energy markets forming the guaranties:
   • Elimination of barriers on the way to renewable energy source development;
   • Elimination of unfavorable conditions in the market;
   • Consideration of social and ecological costs resulted from environmental pollution as a part of energy cost.

References

[1] Magomedov A M 1996 *Alternative renewable energy sources* (Makhachkala: Yupiter) p 245
[2] Golitsyn M V, Golitsyn A M and Pronina N V 2004 *Alternative energy source* (Moscow: Nauka) p 159
[3] Veselovskiy O N and Shteiberg Ya A 1993 *Essays on the history of power engineering* (Moscow: Izdatel'stvo MEI) p 252
[4] Styrikovich M A and Shpil'rain E E 1981 *Power engineering. Problems and perspectives* (Moscow: Energiya) p 192
[5] The Order of State Duma of Tomsk Oblast № 1008 "On Energetic strategy of Tomsk Oblast for the period up to 2020" 2008 *Ofitsial'nuye vedomosti Gosudarstvennoy Dumy Tomskoy Oblasti* 13(135) **I** 28
[6] Rozhkova D S, Khadkevich I A, Glyzina T S and Matyugina E G 2014 Industrial management of peat sorbent production in Tomsk region *IOP Conf. Ser.: Earth Environ. Sci.* **21(1)** 1–4

Petroleum staff reluctance and adjustment to innovative changes

Yu S Makasheva[1], N P Makasheva[1], V V Remnyakov[2], B S Burykhin[2] and I V Shenderova[1]

[1] Institute of Natural Resources, National Research Tomsk Polytechnic University, 30 Lenin Ave., Tomsk, 634050, Russia
[2] National Research Tomsk State University, 36 Lenin Ave., Tomsk, 634050, Russia

E-mail: yuliam@pptsu.ru

Abstract. The modern economy is developing in the direction of innovations implementation. Innovations are becoming the basic prerequisite for the competitiveness of the enterprises. The Russian oil and gas sector innovation issue is very crucial. Low innovation activity of companies could result in a serious threat due to the strong global competition, increased uncertainty and risks. The need for innovative changes often meets reluctance. The reasons of it vary and require serious research. Managers should give special attention to the development of adjustment ability of the staff, to introduce modern methods for improving the adjustment potential of the enterprise staff.

1. Introduction
Innovative component is becoming increasingly valuable in the dynamic development of the production. It has become a key factor of economic growth. Innovations, introducing something new, means launching to the market of new products with a high scientific and technical potential and advanced consumer satisfying qualities. Innovative way of economic development involves a continuous search and implementation of innovative ideas, which allow increasing the efficiency of social production.

2. Materials and Methods
The innovative development of the economy is a complex dynamic process, which combines challenges related to - the necessity of society's intellectual potential building and development; scientific research development; studying of new technological breakthrough discoveries and inventions; advanced technology and equipment implementation; new types of raw materials research; choosing of the best contemporary forms of production organization and management [1].

These issues are extremely relevant for the leading sector of modern Russian economy – the petroleum industry. The oil and gas sector plays a strategic role in the country's economy, providing almost all economic sectors with fuel and energy resources and a wide range of petroleum products. Oil and gas resources estimate for over 70% of the primary energy balance of the country; fuel and energy complex enterprises provide more than 30% of industrial production volume in Russia [2]. The share of oil and gas revenues in the federal budget is about 50% at the contribution to country's GDP of about 1/3 and to the export of almost 2/3. A large concentration of the world's oil and gas resources on the territory of our country, at a large demand for these resources in the world economy, makes the industry highly required in the world market.

But the modern world market is characterized by the intensifying of global competition on the background of growing uncertainty and risks. In recent years there has been a significant transformation of the global energy markets. New technologies in oil production significantly increased the volume of available mineral reserves and revealed new leaders of the production growth (the USA and Iraq). These technologies also led to a steady increase of the oil recovery index. The rapid development of technologies of liquefied gas transportation, led to the serious changes in global gas market.

In order to keep the position in the market, to provide competitive advantages and dynamic growth, a company needs to pay close attention to the innovations.

According to the research data of the International PwC Practice on strategy development in the field of innovations, covering the representatives of petroleum companies from 18 countries, a company, leading in the field of innovative solutions development, gains the significant competitive advantage. The growth rate of 20% of the surveyed companies, leading in the field of innovations developments, turned to be on 16% higher than the growth rate of the least innovation oriented companies. In the next five years the increase of the world's average growth rate is forecasted to double. [3]

The innovative activity of Russian companies is quite low. The experts noted a significant technological gap in the oil and gas industry [4]. The innovative products account in exports of about 1.5%, which is ten times lower than in the world in average; 60.3% expenses on technological innovation falls on the purchase of machinery and equipment, while the share of self development does not exceed 12%. The share of R & D expenditures of Russian companies does not exceed 0.02% from the revenue. The unit costs per 1 ton of oil equivalent are less than 0.2 US dollars. Similar costs of ExxonMobil and Total are in average 1$, Chinese and Brazilian companies – from 2, 35 to 3.22 US dollars. The lag in technological development is reflected, in particular, in the low level of productivity of the Russian oil and gas sector, which estimates about 35% from the US production level [5].

Production, which does not involve innovative developments, becomes weak and uncompetitive. Therefore, the policy of oil and gas complex senior management is to improve the competitiveness of the companies to the international level on the basis of the latest achievements in science and technology. Innovations are the background for the successful development of the company.

But innovative production - is not only the introduction, development and offer of a new product. Innovation - is, above all, the ambition of company's staff to the continuous development of creative solutions for tasks solving. However, innovations are always associated with changes. A human due to its nature is inclined to be apprehensive to changes, which bring the uncertainty and breake the usual order of things. Isaac Newton defined inertia as an inherent power of substance having the ability to reluctance [6]. The phenomenon of reluctance occurs to some extent at any changes. Unambiguous introduction of changes is the exception rather than the rule. Studies have shown that up to 80% of innovations fail, and in more than 50% of cases, the reasons are intercompany factors [7].

With the development of innovative processes, the issue of reluctance to changes becomes more and more relevant. Science and practice are currently focused on the management of changes with the aim to explore the possibility of change management and the ways of overcoming negative attitudes towards them.

Reasons for reluctance to the innovative changes are numerous and different. They can be caused by both external and internal factors, to be of objective and subjective character.

The well-known classification of J. Cotter and J. Schlesinger highlights the reasons of reluctance to changes at different levels:

- the individual level - self interest, misunderstanding and lack of trust, low tolerance, deviation in the situation evaluation;

- the group level (collective level) - peer pressure, fatigue caused by changes, previous bad experience [8].

V.M. Konovalov in the "Innovatsionnaya saga" argues that, ultimately, all the reasons could be combined in two groups: wrong understanding of the situation and the fear that if the innovations are introduced it will lead to the impairment of personal interests. These reasons are often interconnected [9].

Reluctance to innovation could be of any types. It could emerge actively and openly or to be of passive and hidden character. Passive reluctance reflects in the form of absenteeism, delay in assignments fulfilling, indifference and reducing working ability; active reluctance reflects in strong changes neglecting: abandonment of innovations implementation, open opposition, sabotage.

The modern theory and practice has developed a number of approaches to overcoming the issue of reluctance. I.Ansoff highlights four groups of methods which allow predicting, reducing and controlling the reluctance to changes: forced method; method of adjustment changes; crisis management and reluctance management. J. Cotter and L. Schlesinger developed six ways to overcome the reluctance, which could be divided into "mild" and "strong". Strong ways are focused on the forcing. In this case, the correct actions are achieved via penalty. In the innovation activity, where the creative component prevails, such methods are ineffective. Their application could lead to the success only in the exceptional cases of urgent necessity and only for the short term period.

Mild ways to overcome reluctance to innovation imply informing, discussion, conviction, help and support. There are a great number of such methods.

Not only managers but also sociologists, psychologists are involved in the research of methods, techniques and forms of overcoming reluctance. The practical application of these methods increases the efficiency of innovation activity at the enterprises and allows successful solving of the problem staff adjustment to innovative changes.

Currently the adjustment ability is one of the key challenges of personnel management. This concept emerged in the management with the development of systematic approach as the characteristic patterns of functioning and development of different systems at the interaction of internal and external conditions of their existence. In general sense, the adjustment is the adaptation of the system to the character of the individual influence or to the condition changes in general. The process of adjustment occurs when in the system “body-environment” the significant changes emerge. This provides the generation of a new state, adjusted to the changing conditions. Due to the fact that the body and the environment are not static, but are in a dynamic equilibrium, their correlation constantly changes. Therefore, the process of adjustment is a continuous process.

The adjustment in staff management is considered from the point of a new employee adjustment to the new working conditions and place. Such interpretation of adjustment is called "probation period", "career guidance", which are understood as the process preceding the direct entry of a new employee into organization and getting of new position. These processes are closely connected with training, coaching and consulting of the employee.

But this approach to the adjustment is not sufficient. The adjustment of staff in the organization should not only be limited to the orientation in the professional sphere. It is a multifaceted phenomenon, and it could be considered much deeper with the respect to a variety of situations.

Currently, two models of adjustment have been developed in management practices: adjustment at employment and adjustment to the changing conditions in the production process. The first model prevails. The second model is also known as working position adjustment related not only to new employees, but also to the whole company's staff, who has to adjust to changes under the usual working conditions. The changes may be related to the need to develop additional job functions, the increased working load on the staff, the gap between actual and desired qualification of the employee, etc. Many employers experience stress due to the need to be in a constant state of readiness to changes in the future. Therefore, the issue of adjustment is becoming increasingly important not only in the initial stages of adjusting to the new team, but throughout the working life. This issue acquires a special attention under the conditions of innovative changes. Proper adjustment management allows mitigating the negative aspects, foreseeing and overcoming of the reluctance to changes, encouraging the creative attitude to work and increasing the innovation activity of workers.

It should be noted that traditionally the management understands the adjustment as a process which influences greatly the factors, predetermining its progress, time and reduction of the negative consequences. But the adjustment is also a self-regulating property of any live system. In the process of adjustment the person stands as active subject of this process. Therefore, the understanding of adjustment should not be limited only to the understanding of it as a process. It should also be considered from the point of personal property.

The notion of adjustment was introduced into economic science from biology. Regarding the socio-economic systems, the adjustment is the property, which characterizes the ability of a system to normal function under the environmental parameters changing and its adjustment to these changes. Adjustment could be characterized by sensitivity of the system to internal and external changes.

The employees' adjustment ability is a certain individual characteristics that allow quick and adequate challenging situation overcoming, normalizing the interaction with the environment. Man as an active subject of the adjustment process, who adjusts to the environment considering his needs, interests and ambitions.

Adjustment is the result of interaction between employee's abilities and regulatory requirements of his professional activity. These abilities could be congenital - the instincts, temperament, emotions, appearance, physical condition; and acquired - social status, education, competence.

Adjustment ability of the employees may increase or decrease under the influence of the social environment, working conditions, training, lifestyle. Thus, according to the research, people develop greater ability to adjust in the case of serious changes by demonstrating great stress resistance.

Individual reactions to the changes are different, as well as peoples' ability to adjust. These abilities should be developed and are necessary to be developed. Under the rapid and often unpredictable changes in innovations it is particularly important. High adjustment ability allows employees to cope effectively with the changes. According to the leading Western employers, adjustment is among the top ten skills which are required from the employees. The reports on personnel management from 2014 states that HR-directors are more inclined to hire the employees taking into account their ability to cope with changes and uncertainty, and in every second case they note that the ability of the staff to cope with emergency issues is a key attribute for future success in business.

Employee with high ability to adjust is more open to innovations, easily accepts new ideas and does not afraid changes. Experts identify some important personality traits that contribute to the development of adjustment ability in humans. These are - openness and tolerance, psychological stability, dominance, self-perception and self-estimation. At innovation-oriented organizations, these qualities should be paid a special attention. Cultivating such qualities in employees, allows influencing the depth and the extent of their adjustment ability.

3. Conclusion

Human resource management of the petroleum industry starts to be more and more oriented to the tendencies connected with the global strategic objectives as well as the innovations and investment development of the industry. According to a number of experts, a principally new approach for providing the country's petroleum sector by human resources of new type – employees with high adjustment ability to innovations. This requires the intense application of modern techniques to stimulate the growth of adjustment ability and innovation activity of workers. For the successful implementation of innovative development, it is necessary to create favorable conditions in the working team, to introduce modern technologies of working with the staff which enable the employees to realize their potential.

4. References

[1]. Makasheva N P 2013 The state support and financing of innovative activity in Russia and the

world countries *Tomsk state university journal* **3** 161–73
[2]. Pavlovskaya A V 2013 *Effektivnost innovatsionnogo razvitiya neftegazovogo kompleksa na evropeyskom Severe* (UGTU: Uhta) p 266
[3]. Gateway to growth: innovation in the oil and gas industry [web-site] URL: http://www.pwc.com/
[4]. Sharf I, Malanina V and Kamynina L 2014 Features of the marketing strategy of oil and gas companies in exploration drilling *IOP Conf. Ser.: Earth Environ. Sci.* **21** 012047 URL:http://iopscience.iop.org/1755-1315/21/1/012047
[5]. Makov V M 2010 Faktornyiy analiz innovatsionnoy deyatelnosti neftegazovogo sektora Rossii *Audit i finansovyiy analiz* **2** 194–98
[6]. Nyuton I 1989 *Matematicheskie nachala naturalnoy filosofii* (Moscow: Nauka) p 688
[7]. Rossinskiy E O 2002 Organizatsionnyiy mehanizm preodoleniya soprotivleniya izmeneniyam, PhD thesis 08.00.05, Tver, p 189
[8]. Kotter J P and Schlesinger C A 1979 Choosing Strategies for Cnange *Harvard Business Review* **57** 106–14
[9]. Konovalov V M 2005 *Innovatsionnaya saga* (M.-SPb-Kiev: Vilyams) p 224

Compensation for oil pollution damage

E G Matugina[1], T S Glyzina[1], Yu V Kolbysheva[1], A S Klyuchnikov[1] and O V Vusovich[2]

[1] National Research Tomsk Polytechnic University, 30 Lenin Ave., Tomsk, 634050, Russia
[2] National Research Tomsk State University, 36 Lenin Ave., Tomsk, 634050, Russia

E-mail: inosine@mail.ru

Abstract. The commitment of national industries to traditional energy sources, as well as constantly growing energy demand combined with adverse environmental impact of petroleum production and transportation urge to establish and maintain an appropriate legal and administrative framework for oil pollution damage compensation. The article considers management strategies for petroleum companies that embrace not only production benefits but also environmental issues.

1. Introduction

Dependence of many economies on the consumption of traditional energy sources, as well as the limited use of alternative energy sources, makes it necessary to improve hydrocarbon production efficiency both in terms of volume of extraction and environmental impact reduction. The fulfilment of these vital conditions is complicated by fast-developing technologies applied in petroleum industries, which in its turn leads to the increase in natural resource production and technogenic environmental impact. For example, the number of accidental oil spills and leaks annually increases not even in proportion to the production but significantly faster. According to NPO "Centre of Ecology FEC", petroleum product losses make up 3.5-4.5% during production and transportation stages in Russia; according to the current level of production it is nearly 18-23 million tons of oil losses or 4.2-17.2 billion dollars per year [1]. However, this data is not accurate as it does not take into account environmental damage costs. Thus, it is necessary to develop and establish an appropriate legal and administrative framework for oil pollution damage compensation, suitable not only for petroleum companies but also corresponding environmental organizations.

The necessity of taking into account the ecological effects significantly influences production engineering. It becomes absolutely apparent in a case of oil production and transportation in order to secure stable supply of natural resources as the basis of national economy. In this regard, it is necessary to emphasize a significant negative impact produced by petroleum industry on the atmosphere, water and forest resources (production companies use approximately 2000 million cubic meters of fresh water including 701.5 million cubic meters in crude oil and gas production), flora and fauna (oil fields are one of the main sources of forest fires resulting in the death of 20-40% of trees in tundra forests) [2] and etc. The diversity of the environmental impact effects makes it necessary to differentiate the activities and responsibilities of petroleum companies. The task is complicated as various business entities having different rights and interests in the field of the environmental management are usually involved in legal and administrative framework development.

The development of legal and administrative framework for oil pollution damage compensation involves regulation of allocation procedure and use of economic resources with regard to environmental purposes and industry structure changes, as well as national economy as a whole. Thus, this brings up an additional structural segment – environmental. The above-mentioned tendencies are rather complex and interconnected. They are referred not only to the industry in question but to all sectors of the economy defining such important socio-economic indicators as income levels of people, their health and work capacity, life expectancy, and etc. It is supposed that the compensation for the environmental damage is carried out both by oil production and transportation companies and corresponding environmental organizations.

Therefore, the aim of the study is to elaborate the approaches to develop legal and administrative framework for oil pollution damage compensation, taking into account environmental segment of the national economy.

2. Theoretical Bases

The introduction of the environmental factor into petroleum industry has a harmful effect as it is required to enhance the corresponding production and transport technologies (core activity) and maintain environment (imputed; and in the absence of effective stimuli – secondary activity). The importance and the exclusive role of each component in production processes indicate the formation of the integrated system "production-environment" (with possible prevalence of the second component). It is facilitated by the development of appropriate legal framework, which defines the restrictions and is directly related both to the regulative sphere (for example, the federal laws "Subsoil Law", "Environmental Protection Act" and others) and standard of living as a whole (e.g., Civil, Tax, Land Codes, and etc.). Such a combination of these spheres is aimed to ensure continuous and effective production engineering in terms of needs satisfaction and environmental impact reduction.

The current administrative limitations encourage enterprises to adapt to new conditions determining the methods of compliance with introduced regulations and rules.

The development of corresponding legal and administrative framework can be initiated by the following entities (depending on the scale and significance of the problem), allowing the integration of their efforts:

- the state (for example, within the framework of the federal target program "Measures for reduction of risks and mitigation of consequences of the emergency situations of natural and technogenic character in Russia", it is planned to construct landfill for safe recycling of oil toxic wastes in the site of the Samotlor field (Yugra) [3];
- the administrative region (in Tomsk Oblast the use of peat sorbent in emergency oil spill response is discussed[4]);
- the company itself (for example, in 2012 JSC "Tatneft" spent 6 billion rubles on the implementation of measures to improve the environmental safety of production facilities [5].

However, the list of entities can be continued. Joining efforts of environmental organizations to find the ways and develop the methods in order to ensure the compliance with environmental requirements results in business isolation and at the same time indicates the formation of ecological segment. Its peculiar feature is the combination of the state priorities and companies' interests (both the segment itself and oil producing companies). This partnership has following advantages. Producer-enterprises are able to exclude the non-core sector using the services of environmental segment which provides standard and custom services. Environmental organizations consider the ways to minimize the anthropogenic pressure. As they are not burdened by other production activity, they are interested in getting breakthrough innovations. The result is the development of new economic relations, generation of new ideas and possibility of their rapid dissemination, improvement of customer-enterprises efficiency. For example, the company "Composite" which produces universal technical materials and means for the oil spill response, is a partner of JSC "Lukoil"; JSC "Transneft"; Sakhalin - I, Sakhalin - II; JSC "NK Rosneft"; "TNK-BP Holding"; sea and river ports, and steamship lines; Russian Emergency Situations Ministrydivisions [6].

The significance of the environmental segment for oil production is confirmed by the following data: the world oil recovery equipment market totalled 300 million Euros, more than half of it falls on three firms – Finnish "Lamor", Danish "Roclean-Desmi" and British "Vicoma". The sorbent market in Russia is estimated as 1.5 billion dollars [7]. These figures do not take into account all the variety of activities. In this regard, it is appropriate to structure the environmental segment by different features:

1. depending on the profile of companies' activity:
   - materials production,
   - equipment/ technologies,
   - services;
2. depending on partners:
   - with the state participation,
   - with the administrative region participation,
   - with the special trading companies participation;
3. depending on the level of products/ services compliance to existing requirements:
   - relevant,
   - forward-looking;
4. depending on the administrative geography:
   - local,
   - regional,
   - national;
5. depending on temporal aims:
   - related to the elimination of accumulated damage,
   - related to the current activity,
   - related to preventive activities.

The development of legal and administrative framework for oil pollution damage compensation is related to the distribution and redistribution of entity resources, including oil production companies (for example, in the above-mentioned companies, expenditures on the environmental protection are planned to be 90 billion rubles in 2014-2016: more than 40 billion rubles are planned on the construction of environmental infrastructure and about 50 billion rubles – other environmental protection measures including the elimination of accumulated environmental damage [8]. Information exchange among the industries and the integration of efforts contribute to solving environmental problems. Thus, CJSC Scientific-Technical Company "ModulNefteGasComplect" unites the design institute "Research and Design Institute ONGM", the Oilfield equipment manufacture and assembly plant.

The criteria on the environment damage mitigation are definitely set by the state that is authorized with the unique (compared to other entities) powers and the motivation tools. However, one can hardly exclude the initiative activity of the companies that develop new and unique approaches. It is these actions that indicate a progressive approach to the environmental management and awareness of the importance of environmental issues. The activity towards the best compliance to the requirements or "knowledge of the future" creates the conditions to disseminate the best practices and, as a consequence, change petroleum company regulation, including the state level. For example, since 2010 JSC "NK Rosneft" has been executing the corporate program to eliminate the accumulated environmental damage, which has no analogues in Russia (1.5 thousand hectares of polluted and degraded lands were recultivated, more than 2 million cubic meters of drilling and oil sludge waste were recycled). The company has sent a proposal to the Russian Ministry of Nature Management to include this program (for the period 2014-2020) in to the draft federal target program "The liquidation of accumulated environmental damage in 2014-2025" [9]. It indicates the development of legal and administrative framework from the micro-level (enterprise) to the macro-level (state).There is also room for possible cooperation between the state and the special companies at the stage of developing new approaches to minimize the negative production impacts. Thus, OJSC "Poliglin" has developed a polymer-clay sorbents for water purification and disinfection based on the results of the RF Ministry Grants Program "Development of Scientific Potential of Higher Education", Ministry of Education program of the Russian Federation and the American Foundation for civil Research and Development "Fundamental studies and higher education", the Foundation for Assistance to Small Innovative enterprises in Science and technology [9-12].

3. Results and discussion

The important role of petroleum industry in maintaining the relevant dynamics of the national economy is out of question, however, the significant environmental damage requires a change in the existing approaches to the production management. The case in point is the introduction of the environmental component in to petroleum company management. The first step is the adaption and consolidation of standards and rules initiated by the state, which would regulate petroleum company management in terms of combination of economic and environmental components. Therefore, activity priorities are defined, the implementation of which is based either on the introduction of environmental component in the company structure, or cooperation with the companies providing services to eliminate the negative environmental effects (environmental segment).

A leading role in environmental segment development belongs to the state which is responsible for initiating corresponding administrative and legal framework aimed at performance control and entity motivations. Moreover, the work of corresponding environmental organizations is more defined, since the development of techniques to minimize environmental impact is their core business. They have guaranteed the market for their services (in case of providing services of high quality). The development of legal and administrative framework for oil pollution damage compensation can be initiated by local authorities and oil production/transport companies with the possibility to apply the same regulations at the macro-level in terms of not only best practice exchange but also laws, regulations and targeted programs.

Due to a great variety of adverse environmental effects caused by oil production and transport and the need for an integrated approach to environmental issues involving entities at all levels of management, the environmental segment can be structured according to a wide range of criteria.

References

[1] Russian oil tears: the pipeline accident gives rise a cancer [web-site] URL: http://top.rbc.ru/economics/10/04/2012/645532.shtml
[2] Golubchikov S Oil production ecology [web-site] URL: http://www.ng.ru/energy/2006-12-12/11_eco.html
[3] Drilling slurries [web-site] URL: http://www.rg.ru/2012/07/26/reg-urfo/poligon.html
[4] Akasaki I, Sota S, Sakai H, Tanaka T, Koike M and Amano H 1996 *Electron. Lett.* **32** 1105
[5] Rozhkova D S 2014 Industrial management of peat sorbent production in Tomsk region *XVIII International Scientific Symposium in Honour of Academician M.A. Usov: Problems of Geology and Subsurface Development (7–11 April 2014)* (Tomsk: TPU Press) **7** 1–4
[6] Sultanova G 2013 Ecology and oil production *Oil and Gas Vertical* **17** 84–86
[7] Oil skimmers [web-site] URL: http://m.expert.ru/countries/2005/01/01co-oil_56136/
[8] Rosneft: Health and safety environment [web-site] URL:http://www.rosneft.ru/docs/report/2013/care/ecology.html
[9] OJSC "Poliglin" [web-site] URL: http://www.rvf.ru/upload/files/2011/exposition/polyglin-2011-rus.pdf
[10] Tsibulnikova M R, Pogharnitskaya O V and Strelnikova A B 2015 Designing economic and legal mechanism of land management in oil and gas companies *IOP Conf. Ser.: Earth Environ. Sci.* **24** 012032
[11] Sharf I V, Malanina V A and Kamynina L A 2014 Features of the marketing strategy of oiland gas companies in exploration drilling *IOP Conf. Ser.: Earth Environ. Sci.* **21** 1–6

[12] Sizov A V, Tretjyakov K N, Boyarko G Y and Shenderova I V 2015 Liability of the supervisor under petroleum drilling contract *IOP Conf. Ser.: Earth Environ. Sci.* **24** 012029

IOP Conf. Series: Earth and Environmental Science **27** (2015) 012071 doi:10.1088/1755-1315/27/1/012071

Intensive use of energy resources regarding sustainable cities

E G Matugina[1,2], O V Pogharnitskaya[1,3], K O Petrova[1] and A B Strelnikova[1]

[1] National Research Tomsk Polytechnic University, 30 Lenin Ave., Tomsk, 634050, Russia

e-mail: [2] emk512542@mail.ru, [3] pov@tpu.ru

Abstract. Intensive development of cities being a factor of fuel-power complex (FPC) development increases the load on its capacity. This has the negative impact on human health and capacity to work. Therefore, it is important to integrate environmental and recreation components into the urban area through designing the system "production–ecology–recreation" and thus ensuring sustainable development of the city.

1. Introduction

An urban area prospects critically depend on its fuel-power complex (FPC), with the latter being the major factor source of pollution, too. As the urban economic system develops, the load on the FPC increases, which negatively affects the environment. Being an attributive factor of modern society development, urban sprawl boosts social production. This is both the cause and the effect for demand to evolve: currently, about 75% of energy carriers are consumed by the population of the cities, and about a half of this energy is spent for public transport. [1] It is also noteworthy that industrial production is concentrated in the cities, which creates a lot of prospects for self-realization and provides higher income. As a result, the city becomes an attractive place for living. The scale and the significance of the urban sprawl is proved by the following facts: currently, about 50% of world population live in urban areas and it is predicted that by 2050 70% will be city dwellers. The number of people living in cities and towns increases by 73 million every year. [2]

2. Materials and Methods

The advantages of urban sprawl are qualified staff, developed transport, educational, and cultural infrastructures, etc. All together these aspects provide intensive social and economic development. According to the rating by the Institute of Spatial Planning, the main industrial centres are located in the cities of the central part of Russia (27%), the Volga region (24%), and in Ural (16%). [3] On the other hand, the development of urban system has an inevitable negative environmental impact: man-made landscapes, reduction of nature component, negative impacts on health and capacity to work caused by near-by location of industrial and residential areas. The latter negatively affects both the environment and human health: chronic fatigue syndrome, steady-state stress, emotional overload, etc. The following data prove the statement above. More than 75% of pollution is caused by urban sprawl and the ecological situation in the region is determined by the absence or presence of the big cities. According to the data by Greenpeace, the plume of pollution and heat stretches up to 50 km and cover the area of 800–1000 km^2. The demand in energy resources increased dramatically at the end of the XXth – the beginning of the XXI century, which caused an increase in oil and gas recovery and, as a result, great amount of pollutants released. The share of pollutants produced by FPC has increased by

10% over the past years and currently is about 56%, with coal power plant being the major source of pollution. [4, 5, 6]

This negative environmental impact makes it necessary to take into account ecological and recreation components of industrial production, which is the first step towards "production–ecology–recreation" system. This system will lay the foundation for sustainable development of the city. The aim of the research is to outline the approaches which will provide sustainable development of the urban areas and protect the environment in the course of the urban sprawl.

Firstly, it is necessary to determine the notion of sustainable development of the city in the perspective of FPC development. In our point of view, FPC should ensure production continuity and be aimed at the increase in social and economic level of the region, including continuous energy supply and maintaining / improving the quality of human resources. To hit the goal, two different trends in the development of FPC as the source of energy should be taken into account:

- the benefits which determine the level and the mode of consumption;
- the negative impact on the quality of human resources. This impact is double: on the one hand, it is the damage caused to the environment; on the other hand, it is the negative impact on human health and human capacity to work. These impacts may be neutralized either independently (by means of purification plants, de-dusters, etc.) or in combination (creating green zones). Let us turn to implementing the second option, when the territory is used for ecological recovery and recreation. The green zone is created to compensate for the impacts on environment (such as air pollution, noise, etc.) and human health (fatigability, performance decrement etc.).

Integrating the recreation zone into the urban area, one should take into account the following aspects:

- Planning. To create / maintain the consistent architectural composition and to save the traditions, the recreation zones should be incorporated into the existing urban landscape. For example, in Copenhagen the 7000 m^2 garden is on the roof of the hospital while in London on the roof of the 23-storey building they plant about 100 different types of flowers and shrubs (the unique experience of planting at such height) [7].
- Ecology. It is necessary to identify the particular negative environmental impacts and to create the recreation zone to minimize them. For instance, sulfur dioxide produced by heat power plants and boiler-houses is well neutralized by hawthorn, pear tree, walnut tree, poplar, honeysuckle, white willow, ash-leaved maple, etc. [8]
- Nature and climate. It is important to choose the plants which will ensure continuous compensative effect of the recreation zone. For example, willow, asp, birch, and pine-tree are tolerant of wet soil conditions while ornamental shrubs will not survive in the swampy area. [8]
- Arrangement. The zones should be accessible to visitors and not overloaded. For example, the limit recreational load for forest parks is 50 persons per hectare and for park – 300 persons per hectare.[9]

Recreation can be described in terms of structure and infrastructure. Since recreation as a process is incoherent and diversified, there are different groups of characteristic features to specify it. Here are some of the bases on which varied types of recreation can be identified:

- origin – natural or artificial recreational objects (or their cohabitation);
- type of use – continuous, repetitive, or occasional recreation;
- accessibility – walk-up or remote recreation zone;
- absence / presence of buildings – integrated or newly created recreation zone.

Due to varied classifications, there are different approaches to monitoring recreation and integrating it into the urban environment. The report "Planning Sustainable Cities: Global report on Human Settlements 2009" assesses the effectiveness of urban planning as a tool for dealing with the unprecedented challenges facing 21st-century cities and for enhancing sustainable urbanization. The

Global Report identifies a number of promising trends for bridging the green and brown agendas, including:

- the development of sustainable energy in order to reduce cities' dependence on non-renewable energy sources;
- the improvement of eco-efficiency in order to enable the use of waste products to satisfy urban energy and material needs;
- the development of sustainable transport in order to reduce the adverse environmental impacts on the cities dependent on fossil fuel-driven cars;
- the development of 'cities without slums' so as to address the pressing challenges of poor access to safe drinking water and sanitation as well as vulnerability to natural hazards. [10]

United Nations Conference on Sustainable Development (Rio de Janeiro, 2012) focused on the green economy in the context of sustainable development and the institutional framework for sustainable development. The concept of sustainable cities is supposed to provide:

- reducing use of non-renewable resources;
- sustainable use of renewable resources;
- assimilation of patterns of consumption at both regional and global levels.

As a result, one will face a dilemma in the attempt to reconcile green (environment protection) and brown (human wellbeing) agendas. The brown agenda is very important as it recognizes that all urban dwellers have needs for healthy and safe living and working environments, and places its emphasis on creating economic opportunities which have always been a development factor throughout the history of cities. Cities gradually substitute natural habitat and consume natural resources producing wastes. Today, there is a measure of the impact humans have on the environment, which is called ecological footprint. The ecosystems are characterized by the limits for the safe use of their resources. The principle aim of spatial planning is bridging green and brown agendas, i.e. to save the environment and to enhance the human habitat.

Today, there are new trends (though partially overlapping ones) aimed at achieving the synergetic effect of both green and brown agendas, which is quite important for the current stage of economic development. These trends are detailed below.

Development of renewable energy sources

The use of renewable energy sources results in healthy living and working environments in the city, reduces the consumption of natural fuels and minimizes the negative environmental impact (sun, wind, hydro- and geothermal energy). For example, 0.0125% of sun energy is enough to supply the global demand in energy while 0.5% could totally supply this demand in future. The sun energy which reaches the Earth within a minute is enough to supply the global demand in energy over a year. The developed countries intensively use this source of energy, for instance, in Germany, there are about 15–20 GW of solar photovoltaic plants while their total global capacity is about 40 GW. [11]

Establishing cities with zero balance of carbon dioxide emission

Saving green planting and growing new trees reduce the amount of carbon dioxide emissions and cool buildings and premises, which decreases the amount of electricity used for air conditioners.

The distribution of water and electricity supplies

The water- and energy distribution systems in big cities tend to turn from big centralized networks to smaller district ones. This will reduce the ecological footprint as the city is effectively supplied with energy and water thanks to the advantages of electronic management systems, as well as the efficiently constructed system of water pipelines.

Expanding the surface area for photosynthesis within environmentally friendly infrastructure

It is supposed to create ecological zones within the city for both environment protection and recreation.

What is the situation in Russia in terms of implementing green economy and ensuring sustainable development? The adjusted net savings index which is calculated with regard to macroeconomic indicators adjusted to natural capital depreciation and the negative environmental impact, is negative while for the majority of high human development countries this index is positive. The ecological footprint of Russia (4.4 ha) is less than that of above mentioned countries, which proves that individual load on the biosphere caused by a citizen of Russia is relatively less.

Russia is the global leader in forestry and, as a result, in corresponding ecosystem services. However, the share of fossil fuel in the total amount of energy consumed in greater, as well as the amount of carbon dioxide emission per capita, the amount of city pollutants, the level of natural resources depletion. The social indicator of people's satisfaction with environment protection measures is lower than that in the majority of other countries.

It is worth noting that over the period of 1990–2010 the negative environmental effect of FPC decreased, which was caused by the reduction of oil and gas recovery. Over the two decades the amount of waste water discharge reduced by 40%, as well as air pollutant emissions did. The withdrawal of water from water bodies decreased by 32%. However, the tendency to reduce the use of natural resources and environment pollution changed in 1999 when the economic recovery started. The level of air pollution caused from stationary sources increased and the recovery of natural resources (particularly oil) intensified. The oil recovery increased 1.6 times in comparison with that in 2000. [12]

3. Conclusion

To understand the key issues of the environment and human habitat, one might turn to the effect of decoupling, which is the indicator of the green economy. "Decoupling is the basis for the strategy towards the sustainable and environmentally sound economy. It implies decoupling natural resource use and environmental impacts from economic growth and human welfare". [13] Retention / decrease in natural resources use and pollution accompanied by significant GDP growth in 2000s resulted in reduction of natural resources intensity in the country, i.e. the national economy was decoupled.

References

[1] *Mir gorodov* 2009 **2** p 4
[2] Urbanizatsiya mira trebuet soglasovannoy politiki po razvitiyu naselennikh punktov URL: http://www.un.org/russian/news/story.asp?NewsID=22297#.VDShJVfzuDk
[3] 250 krupneyshikh promishlennikh tsentrov Rossii (Top–250) 2010 URL: http://urbanica.spb.ru/?p=1321
[4] Urovni i tempy, etapy i perspectivy urbanizatsii URL: ttp://www.sociocity.ru/scitys-949-1.html
[5] Chronic fatigue syndrome URL: http://www.neboleem.net/sindrom-hronicheskoi-ustalosti.php
[6] Revich B A 2010 *Studies on Russian economic development* **4** 87–99
[7] Blagoustroistvo i ozelenenie za rubezhom kak eto vyglyadit? URL: http://baurchi.ru/uslugi/blagoustroistvo-i-ozelenenie-za-rubezhom-kak-eto-vyglyadit
[8] Erilin G N 2012 *Ekologiya na predpriyatii* **6** 25–32 (in Russian)
[9] Moscow city government department for environmental management and protection URL: http://www.dpioos.ru/eco/ru/qa/o_10529
[10] Planning Sustainable Cities: Global report on Human Settlements 2009 (London, Sterling, Virginia) p 120
[11] Shutkin O I 2012 *Ekologicheskie sistemy* **1** 47–58 (in Russian)
[12] Navstrechu "Zelyonoy" ekonomike Rossii URL: http://www.sustainabledevelopment.ru/upload/File/Reports/ISD_UNEP_GE_Rus.pdf
[13] Bobylev S N and Zakharov 2012 Green economy and modernization *Economic and environmental outlines of Sustainable Development* **60** p 90

Corporate social policy – problems of institutionalization and experience of Russian oil and gas companies

E Nekhoda[1,2,3], Yu Kolbysheva[2,4] and V Makoveeva[1]

[1] National Research Tomsk State University, 36 Lenin Ave., Tomsk, 634050, Russia
[2] National Research Tomsk Polytechnic University, 30 Lenin Ave., Tomsk, 634050, Russia

E-mail: [3] sheyna@sibmail.com, [4] safia78@mail.ru

Abstract. The article examines a range of problems related to the process of institutionalization in the corporate social policy, characterizing the social responsibility of business and representing a part of the general strategy of corporate social responsibility. The experience of the social policy implementation in oil and gas companies is analyzed.

1. Introduction

Recently both in Russia and abroad a number of researches, based on the prediction of social existence including those of futuristic character, and basic social and economic indicators of the global community have appeared. The following researchers such as F. Fukuyama, P. Drucker, J. Stiglitz, R. Ackoff, E. Gontmakher, E. Balatsky, A. Buzgalin, D. Petrosyan and others study this problem.

One of the scientific discussions concerns the change of views on the role of business, especially the large one, in the life of modern society. It requires revealing contradictions in the use of indicators of economic growth, which is the main goal and means to resolve national and global challenges of our time. It is necessary to determine the increasing significance of the social environment in the human life, company and national economy. Therefore, there is a need to supplement the economic efficiency parameters with social development indicators and subjective indicators of human perception of the environment (working life and a company, family values, the natural environment).

The active business participation in resolving various social issues and problems is determined by a variety of objective reasons among which are: 1) the search for a new paradigm of social development and the evaluation criteria of social progress; 2) the crisis of the welfare state models; 3) the increasing significance of social problems and risks of our time; 4) the ongoing debates about the social role and functions of business. So far there is a viewpoint among western and Russian scientists that the development of the concept of the corporate social responsibility and corporate social policy is no more than an "attempt of the state to empower business to resolve social problems". However, the contemporary realities of today's world force business and management of companies to take into account the interests of all internal and external social partners. They understand that a favorable business environment greatly depends on it.

The issues related to the implementation of corporate social policy are a part of the overall strategy of corporate social responsibility (CSR) in most foreign companies. This social policy includes a wide range of issues from the regulation of social and labor relations and staff training to the local community development programs. They are also involved in protecting the environment and solving social problems at the level of society through social investment.

2. Materials and methods

Currently, the researchers study the theoretical and methodological imperfection of corporate social policy content. This is rather a weak argument for its implementation due to the lack of developed goals, stages, methods and tools for its formation [2, 4, 5, 14].

According to Olga Kanaeva the special theoretical and practical significance lies in "the development of the standard (reference) model of the development and implementation of social policy of the company" [5, p. 41]. The author understands the model of corporate social policy as a "generalizing scheme, which gives an idea of the main stages, processes, methods and tools of the development and implementation of social policy. This algorithm presents a certain sequence of actions and management decisions that enable to pass from the declaration of CSR principles to social activities. The social activity, integrated into the company's strategy, has a systematic character and particular results – to step away from the implementation of certain social programs and projects to the formalized systematic social policy [5, p. 41].

Thus, there is an urgent need to institutionalize such a formalized sustainable corporate social policy, which will be a part of the overall company's strategy with its own goal, structure, principles, formation algorithm, methods, and structured social programs. The evaluation of social policy in the creation of shared values, as well as its efficiency, is very significant.

3. Results and discussion

In this article the corporate social policy is understood as a complex of measures aimed at the human control and human development to create necessary conditions for the labor activity and form a particular social development of business and management activities. The main directions of the corporate social policy are: 1) development of human resources; 2) health, safety and environment; 3) social support of employees; 4) environmental protection; 5) contribution to the development of the regions of operation; 6) contribution to the development of society as a whole through the implementation of significant social programs. Within the framework of corporate social policy, companies can develop a variety of social programs: non-state pension fund, social and medical insurance (mostly optional), training and retraining of workers. It can also be a program of incentive tourism and leisure at the expense of the enterprise, programs to improve working conditions and health, and others.

Until now there is no universal algorithm of CSR strategy development. In our opinion, the issues, related to CSR and corporate social policy, are very subjective. They depend on such factors as the size and scope of company activities, the labor market, the need to keep and develop staff, the economic necessity and advantage of the maintenance and development of social infrastructure. It also includes the conditions of business development and expansion, interests and pressure from employees, individual values of top management.

Meanwhile, western researchers have already proposed several guidelines for the implementation of CSR in the company. The generalization of the existing approaches allows offering the following, quite universal, integrative structure of CSR [4, 8, 9].

The starting point in the development of CSR strategy consists in the identification of the subjective value of CSR and the development of a definition of corporate social policy taking into account organizational values and norms (Step 1). From the beginning it specifies the direction for the organization activities of CSR and corporate social policy. The interests of stakeholders are taken into account on the next step (Step 2). These interests are of great significance to set objectives and values of the company formation. The careful selection of key stakeholders will enable the company to concentrate its time, resources and energy in the most efficient ways. The following step is to develop an integration plan for CSR, which consists of three phases: to determine the causes of CSR implementation (step 3.1), to control the internal and external conditions (step 3.2), to plan the relevant activities of CSR and corporate social policy (step 3.3). Step 4 is directly connected with Step 2. It identifies the strategy and communication channels for different groups of stakeholders according to their influence on the company and the related expectations. The two final steps are considered to be rather standardized. They are the last in all strategic approaches.

However, each proposed step should be considered in the context of a particular company, the available information on its activities and other relevant data. The steps which are efficient for one company will not be appropriate for another.

At the moment a number of large Russian companies form the policy of socially responsible business (in this case it is necessary to realize that this process is long and progressive, not giving immediate benefits and results). Companies make non-financial reports[5] on a voluntary basis, in particular social ones; place the main results of social activities and social investments on their own sites, for instance, the site of the Russian Union of Industrialists and Entrepreneurs (RUIE). Among the Russian companies, first of all, major oil and gas companies, iron and steel companies, large transport building companies, energy companies and the banking sector provide the policy of socially responsible business. Some Russian companies have a great experience in implementing the principles of socially responsible business and corporate social policy. These companies are "Aeroflot - Russian Airlines", "Gazprom", "Lukoil", "Rosneft", "Mosenergo", "Rusal", "Norilsk Nickel", "SberBank" and many others. As it can be seen from table 1 the oil and gas industry takes the second place in the number of published reports and the third place among companies providing non-financial reports. JSC "NK Rosneft", JSC "Lukoil", JSC "Tatneft", JSC "TNK-BP Holding" are among the Russian oil and gas companies, which are the leaders in the field of social policy and providing social initiatives.

This article presents the results of applied research conducted by the authors on the basis of the analysis of non-financial reports of Russian oil companies, available in the public domain[6] [10]. The company's official websites were also used. The purpose of this analysis is to determine the ranking of companies in the field of social responsibility. It can serve as a basis for further researches, in particular - to calculate the correlation between the ranking and the socio-economic efficiency of companies, to evaluate the social welfare of workers and the efficiency of implemented social programs. To determine the rankings of oil companies in social responsibility the method of N.A. Krichevsky, S.F. Goncharov "Corporate Social Responsibility was applied. Methods of determining the ranking of social responsibility" was used [6].

A number of important conclusions were drawn during the study.

Companies prepare non-financial reports in accordance with the Sustainability Reporting Guidelines of the Global Reporting Initiative (guidelines of GRI). Companies are also guided by the basic efficiency indicators developed by the Russian Union of Industrialists and Entrepreneurs (RUIE) [3, 12, 14] and guideline for voluntary reporting of sustainable development in the oil and gas industry (Oil and Gas Industry Guidance on Voluntary Sustainability Reporting, IPIECA/ API, 2010). The reports reflect the achieved progress by oil companies in implementing the principles of the UN Global Compact. All of them were approved by the RUIE Council on non-financial reporting.

After analyzing the reports of the companies, the main indicators of economic, social and environmental performance on sustainable development were chosen. The data presented in table 1, were selected from the reports published by companies in the public domain. They reflect the main financial, social and environmental performance of companies in 2013 [7]. The indicators are presented both in quantitative and qualitative terms. The indicators of social and environmental

[5] The non-financial reports refer to officially published documents of companies and organizations in which they report all or some key aspects of their activities. As a rule these reports include the CSR strategy in relation with the development strategy of the company, compliance with ethical business principles, quality of corporate management, contribution to the environmental protection, human rights, human resource management, quality of products and services, support for local communities, the company's impact on society in all key areas of its activity [11].

[6] Non-financial reports of Russian oil companies were taken as the object of applied research due to several reasons, primarily, for reasons of information availability in the public domain and the role they now play in the Russian economy.

performance are divided into three groups. It is done in order to continue calculations of the coefficients according to the method of N.A. Krichevsky and S.F. Goncharova [6].

Table 1. Main performance indicators on sustainable development.

Indicators	Companies			
	JSC "NK Rosneft"	JSC "Lukoil"	JSC «Gazprom Neft»	JSC «Bashneft»
General economic indicators, mln. rubles				
Sales proceeds	3078000	4314301	1230266	532502
Taxes paid	1641000	1216440	311000	83100
Dividends	215500	197284	34433	5457
Net profit	342000	370915	176296	56570
Salaries	94200	57302	32803	35564
Staff and social programs expenses	16191	7980	1869	983
Training expenses	No data	No data	249,3	188
Labour safety expenses	3500	3300	н/д	648,5
Industrial safety expeses	4400	9310	н/д	2255
Quantitative indicators				
Group 1				
Average monthly salary, rub.	50667	51156	59828	39564
Staff turnover rate,%	12	12	15,4	10,81
Total number of personnel, pers.	166100	112014	54829	57329
Number of young specialists, pers.	2868	894	368	352
Number of trained staff, pers.	71500	60300	16990	30957
Number of accidents	34	30	47	36
Social insurance, ths. rub.	16191	7980	1868,7	983
Group 2				
Charity expenses, 2012, ths. rub.	6223	4960	2399	1298,6
Charity expenses, 2011, ths. rub.	6092	5704	1440	1448,9
Group 3				
Environmental protection expenses, 2012, ths. rub.	22100	23374	3380,6	2405,1
Environmental protection expenses, 2011, ths. rub.	17500	22200	3656,7	2147,5
Qualitative indicators				
Labour contract	+	+	+	+
organized structure on social policy (the department)	+	+	+	+
Annual social report	+	+	+	+
Sociological researches	+	+	+	+
Good business practice	+	+	+	+

The study of non-financial reports of the largest oil companies in Russia such as JSC "NK RosNeft", JSC "LUKOIL", JSC "Gazprom Neft", JSC "Bashneft" allowed identifying the main directions of corporate social policy: 1) development of human resources; 2) industrial safety and labor protection; 3) social support of employees; 4) environmental protection; 5) input to the development of regions of operation.

To determine the total ranking of the company, N.A. Krichevsky and S.F. Goncharov offer the following algorithm: the average quantitative indicator is calculated on the basis of quantitative indicators, and then it is adjusted according to the value of quality index. The ranking of corporate social responsibility of JSC "NK Rosneft", JSC "LUKOIL", JSC "Gazprom Neft" and JSC "Bashneft" is calculated. The results of coefficients calculations and final ranking are presented in table 2.

Table 2. Ranking of oil companies social responsibility.

Indicators	JSC "NK Rosneft"	JSC "Lukoil"	JSC «Gazprom Neft»	JSC «Bashneft»
Group 1 - social responsibility indicators				
Salary growth rate	116,00	106,10	100,00	77,55
Staff turnover rate	12,00	12,00	15,40	10,81
Coefficient of young specialists employment	1,73	0,80	0,67	0,61
Coefficient of Education growth	43,05	53,83	30,99	54,00
Coefficient of industrial injuries	73,33	170,67	88,59	44,21
Coefficients of staff social insurance expenses				
Coefficient per one employee	115,53	113,49	95,33	42,67
Coefficient of social expenses to net profit	4,73	2,15	1,06	1,74
Group 2 - social responsibility indicators to the citizens				
Coefficient of expenses on social support of the community	102,15	86,96	166,60	89,63
Coefficient of expenses on social support of the community to net profit	1,82	1,34	1,36	2,30
Group 3 - environmental responsibility indicators				
Coefficient of environmental protection expenses to net profit	6,46	6,30	1,92	4,25
Integral quality index	1	1	1	1
Final rating	**476,80**	**553,64**	**501,91**	**327,77**

According to the results of the study the most socially responsible company in 2013 was the company JSC "LUKOIL".

4. Conclusion

Finally, it should be noted that since the mid-2000's, the concept of integrated reporting began to be developed in the world. The integrated reporting brings together financial and non-financial information and reflects the ability of a company to create and maintain its value in the short, medium and long term [13]. The concept of integrated reporting is supported by over 80 global companies and 25 institutional investors included in the Pilot program of the International Integrated Reporting Council (IIRC). Among the participants are Danone, Deloitte, HSBC, KPMG, Microsoft, Coca-Cola, Unilever, Volvo and others. Only two Russian companies are involved in this program - State Corporation "Rosatom" and JSC "NK Rosneft". The participation of JSC "NK Rosneft" confirms the leadership of the Russian oil industry in the development and implementation of social initiatives and management innovations.

References

[1] Andriof J and Waddock S 2002 *Unfolding stakeholder engagement* (Greenleaf, UK) p 168

[2] Buzgalin A V 2011 Budushhee «skandinavskoj modeli» (novyj proekt sotsial'no orientirovannogo razvitiya) *Al'ternativy. Teoreticheskij i obschestvenno-politicheskij zhurnal* **1** 4–16

[3] Prokopov F, Feoktistova E 2008 *Bazovye pokazateli rezul'tativnosti. Rekomendatsii po ispol'zovaniyu v praktike upravleniya i korporativnoj nefinansovoj otchetnosti* (Moskva: RSPP) p 68

[4] Hohnen P 2007 Corporate social responsibility: An implementation guide for business p 104 (http://www.iisd.org/pdf/2007/csr_guide.pdf)

[5] Kanaeva O A 2011 Corporate Social Policy: Basic Premises of Analysis *Vestnik St. Petersburg University* **5(4)** 41–52

[6] Krichevskij N A, Goncharov S F 2007 *Korporativnaya sotsial'naya otvetstvennost'* (Moskva: Dashkov i Ko) p 216

[7] *Annual Reports Competition* (http://konkurs.micex.rts.ru)

[8] McElhaney K 2008 *Just Good Business: The Strategic Guide to Aligning Corporate Responsibility and Brand* p 194

[9] Maignan I, Ferrell O C, Ferrell L I 2005 A Stakeholder Model for Implementing Social Responsibility in Marketing *European Journal of Marketing* pp 956-977 (http://www.kantakji.com/media/3461/z117.pdf)

[10] *Russian Union of Industrialists and Entrepreneurs* (http://xn--o1aabe.xn--p1ai/simplepage/157)

[11] *Povyshenie informatsionnoj otkrytosti biznesa cherez razvitie korporativnoj sotsial'noj nefinansovoj ot-chetnosti. Analiticheskij obzor korporativnykh nefinansovykh otchetov za 2008-2011* 2012 (Moskva: RSPP) p 104

[12] *JSC "Lukoil"* (http://www.lukoil.ru)

[13] *Integrated reporting* (http://ir.org.ru)

[14] Chirikova A 2006 Business as the subject of social policy in modern Russia *Society and Economics* **9** 100–19

Regulatory framework of pricing and estimate standards in the petroleum sector

V Romanyuk[1,2], U Baitova[1,3], E Safronova[1], G Pozdeeva[1,4]

[1] Institute of Natural Resources, National Research Tomsk Polytechnic University, 30 Lenin Ave., Tomsk, 634050, Russia

E-mail: [2] romanyukvb@rambler.ru, [3] baitova-juliya@rambler.ru, [4] galyunya-tom@yandex.ru

Abstract. The article is devoted to the improvement of the regulatory framework of pricing and estimate standards in the petroleum sector. Regulatory framework reform in the construction sector is necessary at the state, territorial and branch levels. Taking into account the provision and opportunity for potential relations between sellers and buyers, peculiarities of industrial sectors and investment projects, sources of financing, technology pre-investment studies, pre-project, project, tender and production documentation.

1. Introduction

Nowadays, in the formation of the construction products cost, a main element of the market value definition of market is the estimate and regulatory framework of pricing system existing in Russia. It is based on the system of construction base prices. Annually, a large number of year-to-year itemized average cost estimate handbooks for construction products for are published.

Estimate standards are a common name of complex rules, pricing and prices are published in separate estimate handbooks. Regulations and rules determining the construction estimate cost embrace essential requirements.

In accordance to the estimate norm, an established resource base for construction, assembly and other works is considered.

2. Materials and methods

According to CG (Construction Guidelines) 81-35.2004 the following types of estimate standards are defined (figure 1).

Branch estimate standards (BES) for construction estimate documents within the relevant branch and, as a rule, in specific areas. State estimate standards should not contradict or duplicate above-mentioned ones.

Territorial estimate standards include documents introduced for the construction within the territory of the Russian Federation and approved by the Regional administration of the Russian Federation constituent unity.

Firm estimate standards and house-in regulatory framework include estimate documents in compliance to the actual conditions of a particular company (construction supervisor). As a rule, the structure is based on SES, BES, and TES in terms of aspects and human resources.

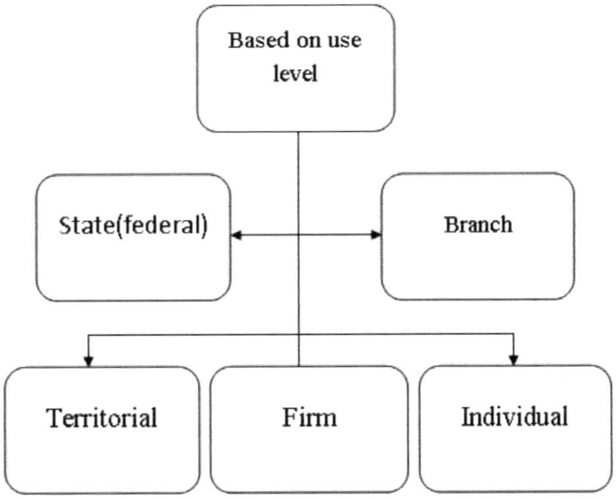

Figure 1. Types of estimate standards [1].

IES unit and specifications are developed in the cases when not one of the existing estimate and rates handbooks for this or that technology project standard is envisaged in the work and specific conditions of production including all complicating factors. They are approved by the investor (the customer) as part of the construction design. When applying individual standards and rates, the increasing coefficients of production condition performance is not included.

Branch, territorial and firm regulatory frameworks are based on the federal regulations and /or standards. Above-mentioned frameworks include a specific branch or territory, as well as the contractor's (firm) focused orientation.

State, branch territorial and individual standards are evaluated to the forms of estimate normative base of "pricing and assessment in the construction industry". This system does not include free (market) and regulated prices and tariffs for the manufacturing of technical products and services.

Based on the structure and intensive enlarged estimate standards are divided into eight groups (figure 2).

According to intensive enlarged estimate standards are divided into two groups:

- elemental estimate standards (ESN) and unit prices (UP);
- enlarged estimate standards (EES) and indicators (EEI).

EES and EEI determine price of construction products in the initial investment process stage.

EES and EEI and used to compile investor estimates, develop a project construction feasibility study, draft business plans, determine tender initial cost and cost estimate in selecting the most effective project and etc. [2].

EES and EEI determine a simple and affordable way of project cost and activities, based on the structural and other parameters of buildings and facilities, as well as integrated units or work volume.

3. Results and Discussions

Existing system of the construction pricing is regulated by state, territorial and branch regulatory framework and methodological guidelines. Based on the regulatory framework documents, estimate standards and norms handbooks, the cost construction rules and procedures are defined at state, territorial and branch levels. There are strong differences and contradictions between branches, state and territorial documents. These differences are mainly associated with different pricing methodologies in the construction industry.

The 1984 regulatory framework being used in domestic practice of construction cost computation does not meet the modern requirements in the scope of work and quantitative physical indicators of labor costs, machinery and material consumption of elemental and enlarged estimated standards.

1991 estimate regulatory base (ERB-91) was developed in terms of eliminating many shortcomings of the 1984 estimate regulatory base, which was incomplete in content. The principle of formation of elemental cost estimate standards, overhead costs and other expenses for time scheduled technology of construction and assembly work was not implemented. This reduces the calculation accuracy and results in significant contradictions between the customers and the contractors involved in capital construction for determining the negotiated prices.

The shortcomings of published handbooks - 2001 SICAS (state itemized cost estimate standards) 1991 ERB - are connected with the inadequate listed modern domestic and foreign construction technology, new means of mechanization and transport, materials and equipment, as well as the application of required standard production costs, including direct and overhead costs, profit estimates and other costs. Similar shortcomings are inherent for territorial estimate standards an regulations. Based on 2001 SICAS Federal Unit Rates (FUR) and price tags were developed for Russian Federation territorial areas. In some Russian regions with extensive experience in pricing development, Territorial Unit Rates (TUR) and price tags based on the territorial element estimated standards were developed. However, in terms of inflation, an index for determining the construction projects value of current prices is introduced into these rates. As domestic and foreign experience demonstrates, precision and calculation accuracy of value determination based on the index approach is insufficient.

In addition, this approach leads to a certain contradictions between the customers and the contractors in capital construction concerning negotiated costs and settlement payments pricing for work performed. New approaches should be introduced based on resource and resource-ranked calculation methods including market prices monitoring for basic resources

Detailed or enlarged data involved in the basic design, investment justification for construction, business plans, project and (or) working design is the determination basis for the project cost. In addition to regulatory framework for assessment of basic work types, it is necessary to develop price-lists of norms and standards, requirements for all resources and their current cost in rubles (foreign currency) for the project. Although creating a data file is time-consuming, it is objectively necessary as it includes specific index of labour requirements, material output ratio, as well as the cost of a project per a unit power of construction products. One of the main principles evaluating construction project cost is the improvement of the regulatory framework itself to provide investor estimates formation, as well as to improve tender documentation, business plans, settlement payments documentation for work performed. Capital construction contractors need a new regulatory framework in contracting, settlement payments for work performed, preparing bids, planning and operational management of the construction project progress and estimated prime cost of contract works.

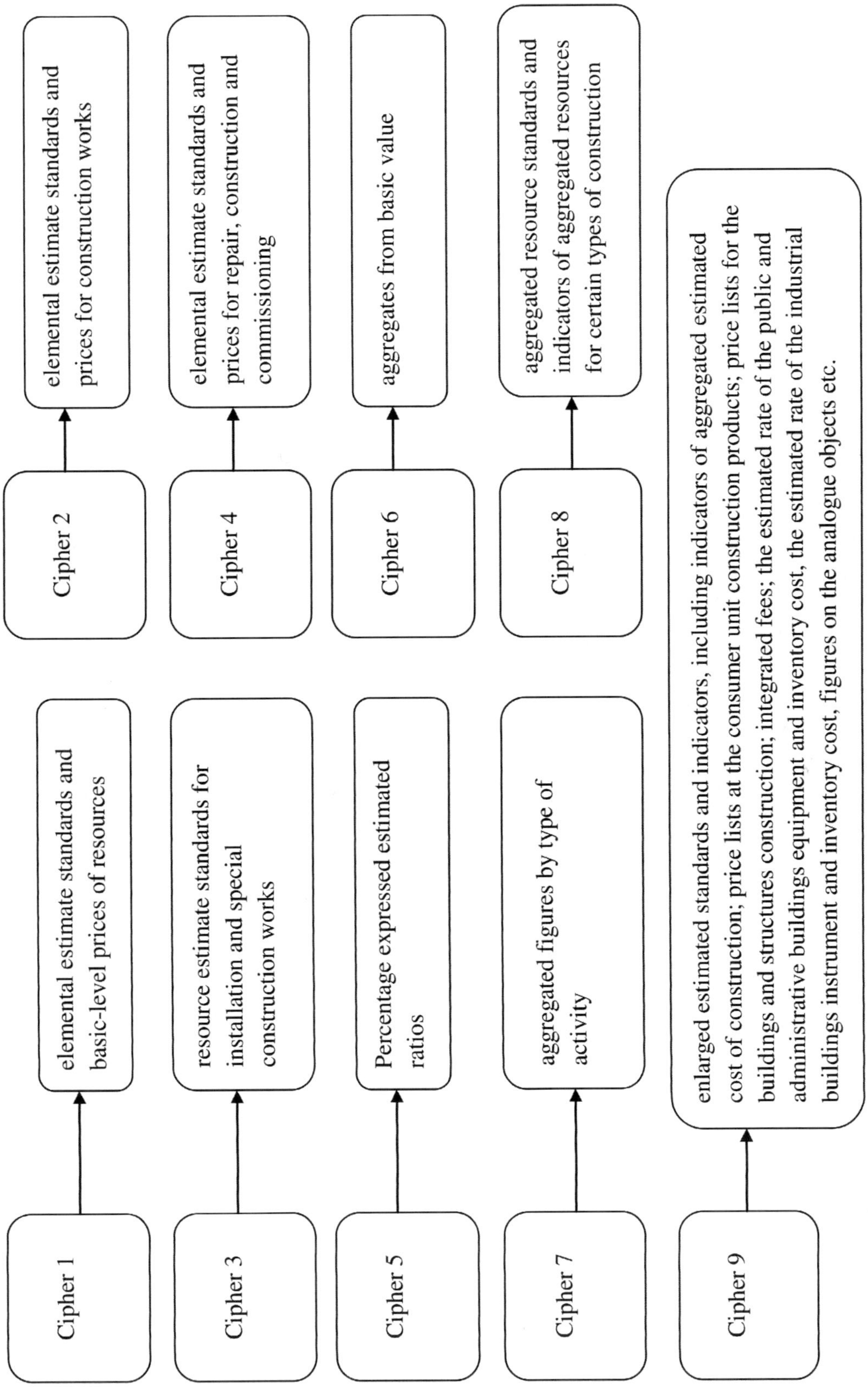

Figure 2. Structure types of estimate standards.

To determine capital investment volume (for each investment project), being the main tool, methodologies and regulatory framework bases for monitoring evaluation different project types exist. For instance, petroleum construction includes such projects as field facilities, surface infrastructure, linear gas pipelines, and subsurface pipeline facilities.

4. Conclusion

Improvement of standardized construction in the petroleum sector is to increase efficient investment, reliability and durability of buildings and structures, saving material, energy, labor and financial resources, sustainable management of land and other natural resources, environmental protection, and to establish necessary requirements for healthy and safe working and living conditions on the project enterprises, buildings and structures.

References

[1] Official site of the State standards and regulations library [Electronic resource]. - Access: http://ohranatruda.ru/, free
[2] Official site of the Federal Center of Pricing [Electronic resource]. - Access: http://www.faufccs.ru/, free
[3] Official site of the Ministry of Energy of the Russian Federation. [Electronic resource]. - Access: http://minenergo.gov.ru/, free.
[4] Badaguev B T 2013 *Organizatsiya stroitel'nogo proizvodstva. Proizvodstvennaya i tekhnicheskaya dokumentatsiya* (Moscow: Al'fa-Press) p 456
[5] Chaya V T, Pankratova VV 2009 Normativno-pravovaya baza formirovaniya otchetnosti neftegazovykh predpriyatij Rossii. *Audit i finansovyj analiz* **6** 99–109

The influence of mechanochemical modification on prevention of toxic ability of humic acids towards phenanthrene in aquatic environment

N S Shekhovtsova[1], E V Maltseva[2], T S Glyzina[1,3] and I S Ovchinnikova[1]

[1] National Research Tomsk Polytechnic University, 30 Lenin Ave., Tomsk, 634050, Russia
[2] Institute of Petroleum Chemistry SB RAS, Akademichesky Ave., 4, Tomsk, 634021, Russia

E-mail: [3] inosine@mail.ru

Abstract. The aim of the research work is to quantify interaction between phenanthrene with modified humic acids in aquatic environment. The changes in the structure and properties of humic acids after modifications were studied with 1H NMR spectroscopy and potentiometric titration methods. Our research demonstrates that the application of thiourea as a modified agent increases the binding capacity of humic acids towards phenanthrene.

1. Introduction

Polycyclic aromatic hydrocarbons (PAH) – are organic compounds, consisting of three or more condensed aromatic rings and have particular interests regarding their toxic and cancerogenic properties [1]. Typical representative of PAH is phenanthrene, anthropogenic sources among which are coal tar, oil and products of its processing, cigarettes smoke, etc. Extended exposure to phenanthrene leads to the decrease of the amounts of hemoglobin in the blood and causes leucocytosis [2]. Up-to-date methods to purify water from phenanthrene and it various oxidation compounds are used (water oxidation with ozone, hydrogen peroxide and UV exposure). These thod are not environmentally friendly and costly [3]. Therefore, the effort is going towards finding new methods, which are based on organic compounds, like humic acids (HA).

Humic acids are natural compounds having a detoxifying ability towards organic pollutants. Authors of the research paper [4] for the first time proved the relationship between aromaticity of humic acids and their detoxifying ability towards PAH. Prior research showed that mechanoactivation modifies the content and properties of humic acids, increasing their detoxifying abilities [5, 6].

Therefore, the main purpose of our work is the quantification of interaction of phenanthrene with modified humic acids in aquatic environments.

2. Materials and methods

The research was focused on humic acids, which were extracted from initial transitional peat of Tomsk region (HA1), mechanoactivated without any reagents (HA2) and mechanochemically modified in the presence of thiourea (HA3). Peat was treated in a planetary activator mill rotating at 1820 revolution per minute (RPM).

The influence of the modification on composition and acid-base properties of humic acids was determined in Tomsk Regional Center for Collective Use SB RAS using the methods of 1H NMR spectroscopy.

The amount of basic oxygen-containing functional groups was estimate using the method of potentiometric titration. The experiment was carried out on Multitest IPL - 103 [7-9]. The degree of dissociation and the dissociation constants of each type of functional groups were calculated according to the Henderson-Hasselbalch equation [5].

3. Results and discussion

Table 1 presents the data concerning changes of fragments compositions of humic acids after mechanoochemical modifications. According to our results, the above-mentioned humic acids consist of high amounts of aromatic and carbohydrate portions. After the process of modification the slight decline of the Hα and Hβ is occured, which could be associated with breaks within carbons bonds. At the same time, the increase of aromatic fragments in HA3 structure up to 8% was observed (HAr).

Table 1. The fragment structure of modified humic acids according to 1H NMR spectroscopy.

Sample	Hydrogen content in structural fragments of HAs, %				
	H_{Ar}	H_{α}-C-Ar, H_{α}-C=C-	H_{β}-C-C=C-	H_{γ} -C-R	$H_{carbohyd.}$
HA1	25.20	18.70	15.10	6.10	34.90
HA2	23.93	16.05	14.75	6.88	38.38
HA3	33.55	15.15	12.98	8.36	29.96

As it shown in table 1, applying thiourea as a modifying agent leads to the increase of acid groups amount, which is caused by air oxidation of HA during the mechanochemical modification of peat. In addition, the table shows that the modification of objects structures leads to the increase of the dissociation constants of HA fragments functional groups, thus weakening their acidic properties and increasing their hydrophobicity in aqueous solutions.

Figure 1 shows the dependence of pH from dissociation of functional groups of HA fragments, exposed to different conditions of modification. The data shows, that modification increases the dissociation of functional groups at the aromatic ring for ArOH and decreases for ArCOOH and CnCOOH comparing to the initial HAs. Thus, aptitude of functional groups of these modified HA is reduced compared to the initial sample.

For the quantity analysis of interaction of phenanthrene with modified humic acids in aquatic environment the methods of molecular spectrophotometry were used. The experiment was conducted to compare the absorbance of the standard and sample solution.

For this purpose, a series of standard solutions of hydro-alcoholic phenanthrene within the concentration range - C0(fen) from 0.5×10^{-4} up to 4×10^{-4} g/L were prepared. All measurements were performed on a spectrophotometer Agilent Cary Win. The absorbance of these solutions was measured in the wavelength range from 190 to 400 nm and a thickness of cuvette 1 mm. The results of absorbance dependency of standard solutions of phenanthrene represented from the wavelength are presented in figure 2.

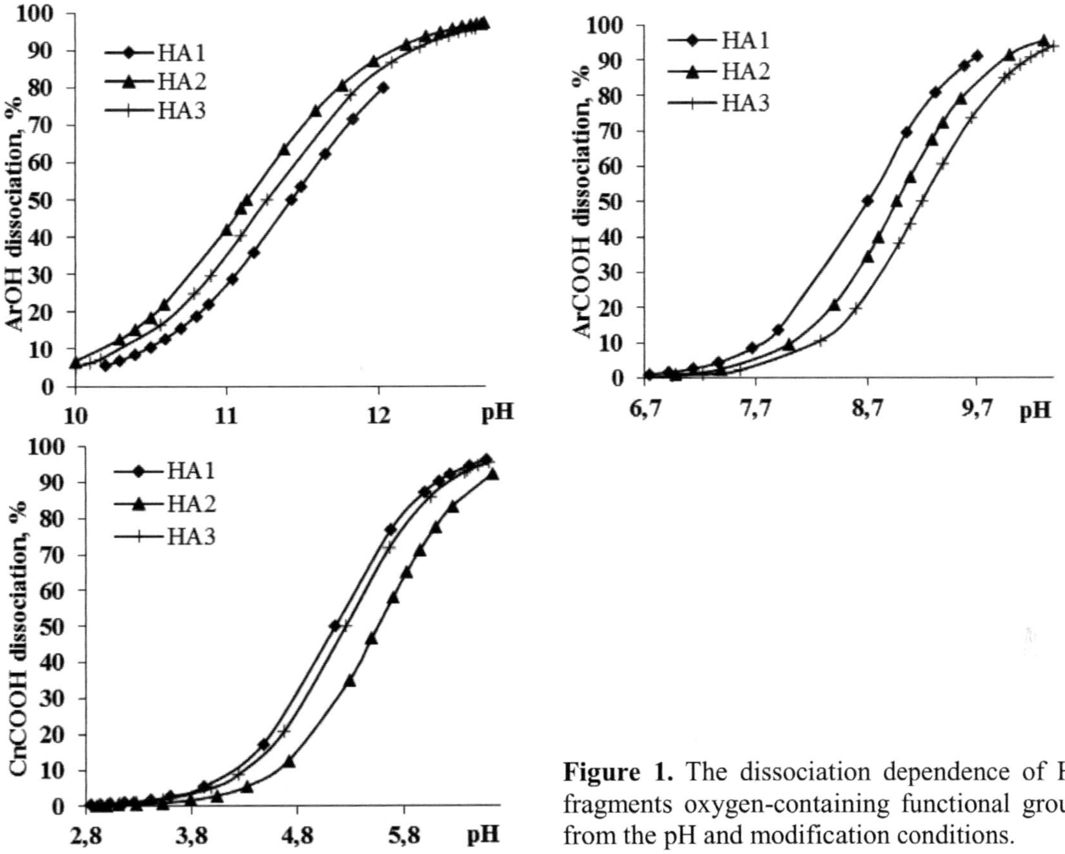

Figure 1. The dissociation dependence of HA fragments oxygen-containing functional groups from the pH and modification conditions.

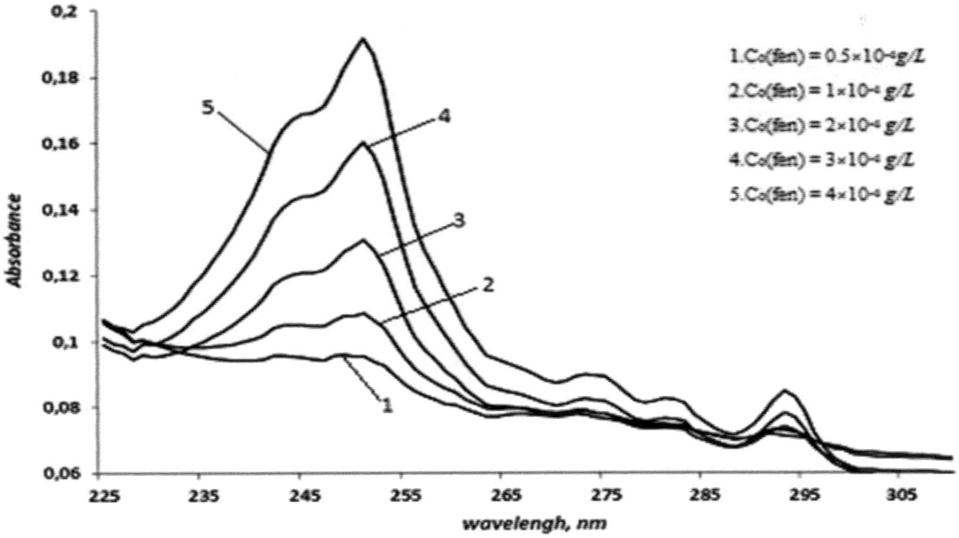

Figure 2. The dependence of the absorbance of the standard solutions of phenanthrene from the wavelength.

To the test solution with a known concentration of phenanthrene the different samples of humic acids with concentration of 0.01 g/L were added, and the absorbance was measured at the same wavelength. Mathematical treatment of the obtained spectra, comprising the steps of smoothing and consideration of baseline were carried out with the program Assayer.

Table 2 represents the concentration of phenanthrene, absorbed by the molecules. Thereafter, the concentration of phenanthrene, absorbed by the molecules of the modified humic acids is determined by the following equitation:

$$C_{abs}(\text{fen}) = 100 \cdot \left[1 - \frac{A_x}{A_0}\right],$$

where A_0 – the optical density of the standard solution of hydro-alcoholic phenanthrene at a wavelength 250 nm, A_x – the optical density of the standard solution of hydro-alcoholic phenanthrene by adding of HA at a wavelength 250 nm.

Table 2. The results of calculation of the absorbed concentration of phenanthrene by humic acids.

C_0(fen) $\times 10^4$, g/L	C_{abs}(fen), %		
	HA1	**HA2**	**HA3**
0.5	42.85±5.23*	66.65±4.62	86.67±5.53
1	40.91±5.15	57.89±3.12	61.04±4.05
2	27.50±4.23	30.00±2.82	39.24±3.87
3	21.21±3.70	28.12±1.90	38.28±3.57
4	17.99±1.53	20.69±0.43	26.70±2.35
5	11.76±1.23	19.10±1.52	18.40±2.25
10	9.92±1.45	10.64±0.40	13.85±1.74

* – confidence interval, m = 3, P = 0.95.

The data in the table shows that the application of mechanochemical modification increases the humic acids binding ability towards phenanthrene in the row HA3>HA2>HA1. The application of thiourea as a modifying agent provides the greatest absorption of phenanthrene in C0 = 0.5×10^{-4} g/L with sample HA3 up to 86.66% compared with HA1 (42.85%). With the increase of phenanthrene concentration the residua ratio of toxic substance in the solution increases, which points to the limited absorption capacity of humic acids.

4. Conclusion

The modification of the structure of peat humic acids in the presence of thiourea and mechanochemical action provides an increase in the hydrophobicity of macromolecules structure, increasing the number and availability of aromatic fragments involved in interaction with various PAHs. Phenanthrene as an example we modeled the system and quantified the interaction of the initial and modified HAs with PAH. It was established that modified HA3 bind up to 86.66% of phenanthrene. At the same time, the limited absorbed concentration of toxic PAH by macromolecules HA3 increased up to 40% compared with the initial sample HA1.

References

[1] Neff J M 1979 *Polycyclic Aromatic Hydrocarbon in the Aquatic* Environment (London: Applied Sci. Pub. LTD) p 262
[2] Klar E 1971 *Polycyclic Hydrocarbons* (Moscow: Chemistry)
[3] Shiyan L N 2004 *The chemistry of water. The water conditioning* (Tomsk: TPU Press)
[4] Perminova I, Grechishcheva N and Petrosyan V 1999 Relationships between structure and binding affinity of humic substances for polycyclic aromatic hydrocarbons: relevance of molecular descriptors *Environ. Sci. Technol* **33** 3781–87

[5] Maltseva E V, Filatov D A, Yudina N V and Chaikovskaya O N 2011 Role of humic acids of peat in detoxication of tebuconazole *Solid fuel chemistry* **1** 65–69
[6] Maltseva E V, Yudina N V, Chaikovskaya O N et al. 2011 Association constant of modified humic acids with triazoles-cyproconazole and tebuconazole series of biocides *Journal of Physical Chemistry* **85** 1676–79
[7] Mirzobekzoda M P, Maltseva E V and Shekhovtsova N S 2015 Analysis of Structural Features of Humic Acids after Modification *Advanced Materials Research* **1085** 3–6
[8] Rozhkova D S, Khadkevich I A, Glyzina T S and Matyugina E G 2014 Industrial management of peat sorbent production in Tomsk *region IOP Conf. Ser.: Earth Environ. Sci.* **21** 1–4
[9] Maltseva E V, Shekhovtsova N S and Glyzina T S 2014 Acid-base Properties of Modified Peat Fulvic Acid *Procedia Chemistry* **10** 500–3

Supervising development in petroleum industry of the Russian Federation

A Sizov[1,2], K Tretyakov[1,3], G Boyarko[1,4], I Shenderova[1,5] and A Shakirov[1,6]
[1] National Research Tomsk Polytechnic University, 30 Lenin Ave., Tomsk, 634050, Russia

E-mail: [2] Sizov_aleksey@rambler.ru, [3] Magg_1@mail.ru, [4] Gub@tpu.ru, [5] inna-shenderova@yandex.ru, [6] shakiroff_as93@vtomske.ru

Abstract. Supervising has strongly consolidated in the domestic petroleum services market of the Russian Federation. But despite the rapid growth of the supervising services market, there is a definite need for its further development. In the article, the developments of supervising in petroleum industry of the Russian Federation, as well as possible ways of its improvement are considered.

1. Introduction

Supervising in petroleum industry of the Russian Federation presents a combination of existing types of quality control. The services rendered by the supervisor could be extended and supplemented on a contractual basis considering the Customer's needs. The lack of legal norms and regulation standards in the Russian Federation Legislation results in a diversity of contractual terms and conditions. The failure to comply with these conditions often leads to the complexity and uncertainty of supervising services in the Russian Federation, as well as misunderstanding of the supervisor's role and his main functions in the production process. This creates the necessity to develop supervising activity and increase its efficiency. The purpose of this article is to assess the current issues related with supervising in petroleum industry of the Russian Federation, as well as to offer further ways of its development, taking into account the international experience.

2. Materials and methods

Historical and logical methods. The study is based on the chronological order of supervising development in the Russian Federation since the period of its origin and gradual integration into petroleum industry together with logical generalizations and consequences.

Method of scientific abstractions. The study is based on the abstraction from the amount of demands in the petroleum supervising services market of the Russian Federation. The authors consider such services as marketable.

3. Results and discussion

The modern concept of organizing oil and gas production in terms of market relations in the Russian Federation is based on the outsourcing, i.e., the involvement of third-party production resources. The main functions of the subsurface user are in the acquisition of the license, economic and organizational management. Oil and gas field services in this case are rendered by specialized service organizations.

In international practice such activity has been applied for a long period of time. The mechanism of these relations is constantly being improved developing from traditional contractual relations, providing a particular service which is a part of the whole project, to the alliance of integrated project management.

Historically, petroleum industry has undergone restructuring during the market economy development in the Russian Federation. First, it became independent from the state. Then, the following diversification of organized companies led to the emerging of the detached independent business units. Therefore, the diversification resulted in the establishment of separate organizations carrying out the single activity: drilling, repairing of equipment, transportation, etc. One of the main reasons for such restructuring was to improve the operating efficiency of specialized organizations. On the one hand, if these organizations were a part of the company, the latter should provide them with necessary equipment and materials, staff maintenance, and what is the most important - the scope of work. In case if the company fails to provide the scope of work, it will bear the substantial losses. On the other hand, these organizations were unable to search for extra work orders independently, thus being forbidden to gain extra profits from their activities [1].

Nowadays the process of maintaining oil and gas production involves a large diversity of technologically complex operations for the implementation of which expensive equipment (for example, the telemetry equipment to study wells) and highly qualified personnel are required. Self maintenance of such operations is economically unprofitable for the company. Thus, drilling, exploration and other services markets (provided by the Contractors, Subcontractors and the Suppliers) were formed. Drilling contractors have their own drilling rigs and qualified personnel, and can directly carry out the construction of oil and gas wells. In most cases the Customer organizes a tender on choosing of the Contractor. Under conditions of developed competition in the market the Customer tends to get the better and reasonable priced services. There are several ways to organize production processes: "Design-Build" and "Turnkey". For example, one of the "Turnkey" projects is field development. For the "Turnkey" field development project, not only the wells, but also the additional facilities for product transportation, accommodation, and etc. are constructed. In this case the Contractor can hire the Subcontractor who will perform certain construction works, such as well construction, under the conditions of the signed contract. At this, the General Contractor, personally, performs well construction, enters the Contracts with the Subcontractors and the Suppliers. Considering "Design-Build" project, the oil and gas company organizes itself the field development process by entering into contracts with drilling contractors, equipment suppliers, installers and other service organizations [2].

The process of oil and gas wells construction and exploration work is characterized by the complexity of production operations. The quality of the obtained results at the first stages of production could be easily lost during further production operation, due to the effect of geological, technical and technological factors, especially in the sector of oil and gas wells construction and workover. Thus the compliance with the relevant production work technology and the quality of its performance is of great importance. Therefore, in the case of "Turnkey project" or "Design-Build project" the Customer and the General Contractor have to control the quality of works performed by the Contractor and the Subcontractor carrying out supervising – technological control over the production process.

Supervising in petroleum industry of the Russian Federation is an adopted element of international experience. It is a kind of service for construction quality control, rendered to the Customer by an independent engineer (engineering organization), through the Contractor, hired for field facilities construction. Business between the Supervisor and the Customer is based on Civil Law relations. However, there are no special norms to regulate the relations between the Customer and the Supervisor in the Civil Legislation. Consequently, the Contract is designed in accordance with the Civil Code of the Russian Federation: Chapter 39 "Compensated rendering of services" - general form of Contract and its main conditions; Chapter 749 "Share of the engineer (engineering organization) in execution of rights and liabilities of the customer" - rights and liabilities of parties, functions and relations between the Supervisor and the Customer [3].

Quality of oil and gas well construction in the Russian Federation is based on designing instruments to control the construction process of different kinds. Basic instruments for oil and gas well construction supervision in the Russian Federation are shown in Table 1.

Table 1. Basic instruments for oil and gas well construction supervision in the Russian Federation.

Legislative regulation	Parties	Subject of control
Design supervision		
1)Urban Planning Code of the Russian Federation 2)SP* 11-110-99 «Building works design supervision» __________ * set of rules	The Customer signs a Design Works Contract with the Contractor	Correspondence of the design works performance to the project documentation.
Production supervision		
1) Urban Planning Code of the Russian Federation. 2)SNiP* 12-01-2004 «Construction management» __________ *construction rules and regulations	Realized by the Contractor	1) incoming control of Customer's project documentation, materials and tools applied during construction works; 2) production supervision during construction works performance and completion; 3) mid-term evaluation of the construction process.
Process supervision		
1) Urban Planning Code of the Russian Federation. 2)SNiP 12-01-2004 «Construction management»	Realized by the Contractor	1) correspondence of order and structure of technological process to the regulatory standards requirements; 2) correspondence of construction works quality indexes to the requirements of project, technological and regulatory standards.
Customer supervision		
1) Urban Planning Code of the Russian Federation. 2)SNiP 12-01-2004 «Construction management»	Realized by the Customer	1) checking of Customer's license possession on materials, equipment and tools applied during construction works, as well as control over the storage and keeping conditions; 2) checking of Customer's operational control execution; 3) control over project documentation correctness and over State Inspection procedure and Local Self-Governance regulations; 4) control over construction work volume, evaluation of working conditions and correspondence to the working schedule.
State construction supervision		
1) Urban Planning Code of the Russian Federation. 2)SNiP 12-01-2004 «Construction management»	State supervision bodies	Final evaluation of the object under construction correspondence to the requirements of technical regulations, project and standard regulations, law regulations, as well as safety regulations.
Administrative supervision		
1) Administrative Violations	Law enforcement	Construction supervisory control for limiting the

Legislative regulation	Parties	Subject of control
Code of the Russian Federation 2) Law of Local Self-Governance in the Russian Federation 3) SNiP 12-01-2004 «Construction management»	authorities; Public administration; Self-governance authorities	unfavorable influence on people and territory of the construction site.
Supervising (engineering supervision, construction supervision)		
Civil Code of the Russian Federation, Article 749. No direct control.	Competent engineer or engineering organization signs Service Contract with the Customer	Providing of technical inspection over the object under construction. Combination of different types of existing controls. Supervising services could be extended on the Contract base considering Customers needs.

It becomes clear from Table 1 that there is a variety of instruments to control the quality of oil and gas well construction. The State construction supervision and the Administrative control for construction works are obligatory in accordance with the Legislation of the Russian Federation [4]. These types of supervision imply the single inspection control, which highlights the work performance for a definite period of time, thus making such control inefficient during supervising of hazardous production facility construction – oil and gas well construction. According to Table 1, from the point of self quality control, the production and operation supervision of the Contractor's work performance is inefficient. For instance, under such control the Contractor can conceal the deviations and irregularities in construction technology foreseen in work performance regulations, as well as failures which were not detected in time, due to the penalties stipulated by the Contract. In order to eliminate such precedents, the Customer undertakes technical supervision over the Contractor (see Table 1) by establishing the separate infrastructural department which is responsible to control work performance. However, Customer's internal Supervisor acts under the Heads order. Consequently, this type of supervising is also an inefficient one as the company Heads are key indicator performance oriented, thus the facts of faults and irregularities in construction process could be concealed as well. In addition to the above described instruments of oil and gas well construction quality control, the Customer may also realize design supervision. Legislation of the Russian Federation restricts the procedure of design supervision due to the fact that it could be carried out only by project developer, the representative of project design organization. By definition, design supervision is carried out to ensure compliance with the decisions contained in the project documentation of construction work performance on the site [5]. Design supervision is not compulsory and relates mainly to the Customer competence. The only instrument of oil and gas wells construction quality control, which is not enshrined in the law, is supervising [6]. It is targeted to control the Contractor's work technology performance, thus the Supervisor is constantly on-site, realizing control over the process. Supervising could comprise both individual and all duties of the Customer to control the construction process, established by the Contract from the perspective of an independent party. Involvement of the Supervisor in the production process is mandatory for oil and gas well construction industry in accordance with the Legislation of the Russian Federation [7].

In the European Union, the economic relations of construction work performance and other relations connected with construction works are governed by the Civil Law, and general forms of contracts are designed by FIDIC – International Federation of Consulting Engineers. Suggested forms of FIDIC contracts cover almost all the types of construction works relations.

Forms of FIDIC Contracts which could be applied to oil well construction depending on type of production organization:

1. Conditions of Contract for Design-Build and Turnkey, (Orange Book); the Orange Book contains all the rest forms of FIDIC Contracts, used during integrated construction works. In accordance with

this Contract, the Contractor should hand the objects of construction to the Customer after the completion of works. The objects of construction, in order to meet the Contract requirements, conditions and target purposes, should be subjected to work, performance of which is necessary for quality correspondence provision. Objects, which are outside the Contract, but considered to be necessary for the provision of quality, construction completion, further safe and efficient exploitation, are also included in the Contract terms and conditions.

The Contractor is not permitted to enter into the Subcontract for the construction of the whole complex of objects if:

(a) the agreement on buying of the materials or entering into the Subcontract, is not required from the Contractor;

(b) there is no preiminary agreement from the Customer representative on entering into the Subcontract with the other Subcontractors;

(c) no later than 28 days before work performance by the Subcontractor, the Customer representative did not receive notification from the Contractor about work initiation;

(d) the Customer did not submitt fair and sensible arguments for entering into the Subcontract.

The Contractor is responsible for the compliance of all Contract terms and liabilities by the Subcontractors. The Contractor has an obligation to control work performance of the Subcontractors, their representatives and employees, as well as to detect failures to comply with their Contract liabilities.

2. Integrated Contract relations comprise the following forms of FIDIC Contracts:

- Conditions of Contract for Electrical and Mechanical Works, (Yellow Book)
- Conditions of Contract for Construction, (New Red Book)
- Conditions of Contract for Plant and Design and Build, (New Yellow Book)

3. In case of the Contract for Design-Build and Turnkey or the Integrated Contract, the Customer or the Contractor (when inviting the Subcontractor to realize the control over the work performance) is permitted to hire building consultants. Relations are regulated in accordance with the Client/Consultant Model Services Contract, (White Book), which is the analogue of Russian supervising.

4. Conclusion

Supervising as a form of legal relations is considered to be one of the priorities for the developments of economic relations of the Russian Federation. The international supervising experience is borrowed and adapted with regard to the existing reality of the relations in the Russian Federation for the production control and creation of special rules governing these relations. Thus, supervising will be mainly implemented in the development of model regulations, industry standards of supervisors' activities on the basis of foreign experience. The regulation of relations in the European countries and the Russian Federation is different. In the European countries contracts can be developed by international organizations which are the part of the legal system. In the Russian Federation contracts are protected by laws. The contract forms developed by various international organizations are rarely applied in the Russian Federation. In the Russian Federation non-standard forms of the contracts, such as supervising service contract, are often developed by the Parties on the basis of the relevant laws, regulations, normative and technical documents.

References

[1] Sizov A V, Boyarko G Y and Shenderova I V 2014 Development of well construction and workover supervising in Russian Federation [Electronic resources] *IOP Conf. Ser.: Earth Environ. Sci.* **21** 012048 Mode of access: http://iopscience.iop.org/1755-1315/21/1

[2] Shirenkova E A 2010 Legal aspects of supervising services *Drilling and Oil* **9** 52–55

[3] Sizov A V, Tretyakov K N, Boyarko G Y and Shenderova I V Liability of the supervisor under petroleum drilling contract [Electronic resource] *IOP Conf. Ser.: Earth Environ. Sci.* **24** 012029 Mode of access: http://dx.doi.org/10.1088/1755-1315/24/1/012029

[4] SNiP (Construction norms and regulations) 12-01-2004 "Construction management" approved by Gosstroy (Russian State Committee for Construction), 19th April 2004 70
[5] SNiP (Construction norms and regulations) "Construction management" approved by Gosstroy (Russian State Committee for Construction) 10th June 1999 44
[6] Nesterova T N 2009 Domestic drilling supervising – challenges and experience Part 1 *Drilling and oil* **5** 52–54
[7] Federal norms and rules of industrial safety "Safety rules in petroleum industry" (Resolution №101, 12th March 2013 Federal serves of ecological, technological and nuclear control) p 387

Economic assessment of environmental impact in the course of oil field development and production

M R Tsibulnikova[1,3], O S Kupriyanova[2,4] and A B Strelnikova[1]

[1] National Research Tomsk Polytechnic University, 30 Lenin Ave., Tomsk, 634050, Russia
[2] National Research Tomsk State University, 36 Lenin Ave., Tomsk, 634050, Russia

E-mail: [3] tsibulnikova2011@yandex.ru, [4] kupriyanovaos@sibmail.com

Abstract. The article considers the variety of impacts that oil exploration and production operations have on the environment at different stages of the process. To provide accurate economic assessment, an oil field development project was designed, with various development options. These options being analyzed, the strategy with the minimal environmental impact was identified. This has allowed preparation of a guideline on how to prevent deterioration of the environment and to reduce the negative environmental impact

1. Introduction

Economic growth and success depend on both natural resources management and the intention to reduce the negative impacts caused by implementing development projects [10]. This can be achieved by careful project selection and improvement of project management, i.e. planning, design, and implementation [8]. Environmental impact assessment (EIA) includes two stages. The first stage is determining and measuring the environmental impact, the second one is calculating monetary value, which is necessary for further economic analysis [4]. Intensive industrial development over the past years exacerbates ecological problems, namely pollutant emissions, smoke, accumulation of industrial wastes. This causes ecological imbalance in the territory of the region, which has a negative impact on human health [5]. Economic assessment of the environmental impact is based on studying and measuring biophysical changes caused implementation of a certain project.

The main objectives of environmental impact assessment used for oil field development and production are determination of economic, ecological, and social impacts caused by different field development options; setting guidelines to minimize the negative environmental impact and to prevent environmental degradation [2]. Potential environmental impacts of the project activities while field development and production are as follows: air pollution, destruction of habitat, surface and ground water contamination, soil contamination, soil and ground disturbance.

To prevent pollution, environment protection measures should be taken in the area of oil field development and production.

2. Materials and Methods

In our research, environmental impact assessment (EIA) was made for the oil field located in Kargasok region, Tomsk Oblast. The Upper-Jurassic formations of the field are at the depth of 2625 m.

The calculation was made for each component of the environment with the application of the relevant calculation methods [6].

3. Results and Discussion

The oil field development will have negative environmental impact on the atmosphere, flora and fauna, lithosphere and terrain, hydrosphere, and natural resources. The negative impacts are caused by constructing well pads and access roads; well drilling, cementing, completion; testing; well performance; welding, painting, and coating; leakages and accidents (pipeline failures, oil, gas, and water showings); well abandonment.

Mechanical impact is a challenge of major concern because of heavy trucks and construction machines which cause soil, plants, and water contamination. The operations performed by the machines and vehicles result in mechanical disturbance of soil and subsoil, destroying soil structure, mixing of soil layers [7], and negative impact on soil fertility. In the result, biological characteristics of the soil change to the worse and the biological toxicity increases [1]. Chemical contamination by oil products, drilling muds, and cementing slurries is an inevitable consequence of bund failures and poor pit tightness, as well as mud pit overflows. Pollution potential of the drilling mud is determined by the oil product content, surface active substances (surfactants), heavy metals etc. Oil spills have the most dangerous and long-term effects. The rate of contamination is determined by the depth of oil penetration, the amount of oil, and mechanical properties of subsoil.

In the process of oil field development, hydrogeological properties of the formation change significantly, as well as those of shallow aquifers of fresh groundwater do. The changes within the formation are caused by the recovery of reserves within production zone. This decreases the formation pressure and causes water coning and fingering. The latter is the reason for the chemical composition of the water to change.

Taking into account all factors, types, and sources of pollution mentioned above, three alteration projects of oil field development and production have been considered. Each of the projects involves drilling 66 wells.

Project 1 details: one drilling crew, which drills 12 wells per year over the period of 5.5 years; constructing a modular cluster pump station (MCPS); drilling a well for water supply up to Cenomanian strata; constructing an oil treatment unit, a compressor and a gas dehydration unit. Gas flaring does not take place. The gathering pipeline stretches from the modular cluster pumping station to the oil treatment crossing the river and is constructed using the trenchless horizontal directional drilling (HDD) technique.

Project 2 details: two drilling crews, which drill 24 wells per year over the period of 2.8 years; constructing two MCPSs; constructing an oil treatment unit, a compressor and a gas dehydration unit. As in the case of Project 1, the gathering pipeline between the MCPSs and the oil treatment unit crosses the river and is constructed using HDD.

Project 3 details: one drilling crew, which drills 12 wells per year over the period of 5.5 years; drilling a well for water supply up to Cenomanian strata; constructing a MCPS; constructing an oil treatment unit, a compressor, and a gas dehydration unit; constructing flares as 5% of gas is flared. The gathering pipeline stretches from the MCPS to the oil treatment unit crossing the river and is constructed using conventional trenching technique.

Project 1 is considered the most appropriate because of the following advantages. The rate of the environmental impact caused by Project 1 implementation is less than that of Project 2. The amount of water recovered per year from the Cenomanian aquifer is less, as well as the rate of air pollution because the amount of emissions caused by two MCPSs is twice as much as that caused by one. Moreover, the more water is pumped into the formation, the more significant the changes are in the chemical composition of the formation fluid. When the stage of drilling is completed, the second MCPS is out of use, therefore Project 1 is more preferable than Project 2 in terms of economy as well.

Project 1 is also more appropriate than Project 3. Gas flaring will have negative atmospheric impact as more pollutants are emitted into the air. When the pipes are lowered into the trenches, the banks of the river will be destroyed and a lot of river creatures die. If there are any pipeline failures, petroliun products will get into the river, which poses a deadly threat for both river creatures and the animals whose habitat includes the banks of the river [3]. In terms of economy, trenches are cheaper than HDD technique, but one should take into account the additional costs for flares construction and the payment for gas flaring. As a result, if one considers the whole life cycle of the field, Project 1 is the most effective.

Economic assessment of the environmental impact includes the payments for the negative atmospheric impact over the period of the field development and production calculated for three designed facilities: a diesel power station, welding and painting units, and a steel vertical stock tank. The total is 2674852.203 rubles.

All wastes produced over the project implementation can be split into industrial wastes (cuttings) and municipal solid wastes (MSW). The industrial wastes are supposed to be accumulated within a special place in the territory of the field and then transported to the industrial waste disposal site once every three months. While the wastes are in the territory of the filed, they have the negative impacts on the atmosphere (as they contain volatile substances), surface and ground waters, soil.

The payment for accumulation and transferring the wastes to the disposal site over the period of drilling is 212256157.4 rubles. MSW are packed in special containers, which are transported to the MSW recycling site and the payment over the period of the field development is 1555533.92 rubles.

The Project implies pumping water into the formation to flood oil towards production well and maintaining the pressure. For these purposes, the water supply well will be drilled up to Cenomanian strata. The water will be supplied from this source over the whole period of the field production, which is 39 years. The Project also implies construction of oil treatment unit, where the water is treated and desilted before being pumped back into the formation. This provides water circulation and the amount of Cenomanian water recovered decreases every year. The implementation of circulation system allows reducing the negative impact on underground waters of Cenomanian strata and cutting back on water tax. The artesian water for the domestic use will be supplied from another well also drilled in the territory of the field.

The total amount of water tax adjusted for the inflation over the period of the oil field development and production is 6077487 rubles.

From the oil treatment unit, the main pipeline stretches to the nearest oil field. Within the Project, the most probable pipeline accident was designed, i.e. pipeline break because of the intensive corrosion. If the amount of oil at the pipeline inlet were 400 tons per day, the repair services took an hour, the amount of oil leaked would be 16.7 tons. The accident would extend throughout the forests and marshes, the oil would get into bog lakes avoiding the rivers. In this case, the maximum probable payment for the negative environmental impact including that on the forests is 4033594.87 rubles.

The total amount of payments for the negative environmental impact, water tax, and transferring wastes to the disposal site over the period of the field development and production is represented in table 1.

Table 1. Negative environmental impact over the period of the field development and production.

Sphere	Environmental Impact	Amount of Payment, rubles
Atmosphere	Air pollutant emissions	2674852.203
Hydrosphere	Water tax adjusted for inflation	6077487
Lithosphere	Accumulation and transferring of industrial and municipal solid wastes to the disposal and recycling sites	213811691.3
Accident	Negative environmental impact	4033594.87
Total, rubles		226597625.4

The payment for the negative impact on the lithosphere makes up about 94%, which results from the number of wells and the periods of drilling and production.

Alongside with the impacts mentioned above, the consequences of the project implementation will be reduction of natural habitat, destruction of mineral metabolism, destruction of plants, deforestation, decrease in the population of animals, noise and vibration impacts. However, within the Project, these impacts are not assessed and neither are ecosystem services [9]. EIA project is the calculation of costs for oil field development and exploration and is not regarded as a financial instrument to compensate for the negative environmental impact, stimulate the companies to implement efficient resources management and minimize the project effects on the environment.

4. Conclusion

Reflecting on the facts represented above, we arrive at the following conclusions:

– the atmospheric impact might be evaluated as over-limit because no research has been conducted to set the limits (for more accurate analyses, maximum allowable concentration should be calculated);

– the impact on lithosphere, hydrosphere, and fauna is evaluated as acceptable;

– the impact on flora over the periods of field development and exploration is evaluated as moderate; when the well is abandoned, vegetation is recovered;

– social and economic impacts are positive for the development of the local economy, however, the major part of ecosystem services is not considered.

References

[1] Bezrodniy Yu G 2002 *Zaschita okruzhayuschey sredy v neftegazovom komplekse* **2** 15–20

[2] Carpenter S R, Mooney H A, Agard J, Capistrano D, Defries R S, Diaz S, Dietz T and Whyte A 2009 *Proc. Natl Acad. Sci. U.S.A.* **106 (5)** 1305–1312

[3] Dashevskiy A V 2002 *Spravochnik ingenera po dobyche nefti i gaza* (Ufa: State oil technical University Journal) p 278 (in Russian)

[4] Dixon J A et al. 2000 Economic Analysis and Environmental Assessment (Moscow: Vita-Press) p 272

[5] Knoche S and Lupi F 2007 *Ecological Economics* **64(2)** 313–320

[6] Metodika opredeleniya uscherba okruzhatuschey prirodnoy srede pri avariyakh na magistral'nykh nefteprovodakh 1995 (in Russian)

[7] Khaustov A P and Redina M M 2006 Okhrana orkuzhayusxhey sredy pri dobyche nefti (Moscow: Delo) p 552 (in Russian)

[8] Sharf I V, Malanina V A and Kamynina L A 2014 *IOP Conf. Ser.: Earth Environ. Sci.* **21** pp 1–6

[9] Tsibulnikova M R 2012 *Tomsk State University Journal* **360** 193–199

[10] Tsibulnikova M R, Pogharnitskaya O V and Strelnikova A B 2015 *IOP Conf. Ser.: Earth Environ. Sci.* **24** pp 1–5

The comparative analysis of payments for negative environmental impact in Russia and Kazakhstan

M R Tsibulnikova[1,2,3], A T Ospanov[1], D V Salata[1], A B Strelnikova[1]

[1] National Research Tomsk Polytechnic University, 30 Lenin Ave., Tomsk, 634050, Russia
[2] National Research Tomsk State University, 36 Lenin Ave., Tomsk, 634050, Russia

e-mail: [3] tsibulnikova2011@yandex.ru

Abstract. The article represents the calculation of the payment for negative environmental impact caused by the development of the uranium ores deposits in the Republic of Kazakhstan. To compare the deposits in Kazakhstan and the Russian Federation, the event is simulated as if the object were located in the territory of the Russian Federation. The comparison of the results serves as an evidence to substantiate the experts' claim that the financial mechanism of land management in Russia should be reformed.

1. Introduction

Charges for the negative environmental impact is a way to compensate the damage to the environment [1] caused by the companies and other enterprises through contamination and waste disposal [2]. The scientists state that current charges perform neither compensation [3, 4] nor regulatory functions [5]. Even with the penal sanctions being accounted, the payment of companies make up hundreds and tenths of a percent within the expenses and income respectively [6]. It is still more profitable for Russian companies to pay the charges for the damage to the environment than to carry out environmental actions [7]. The absence of economic incentives to motivate companies to perform environmental actions and insignificance of charges increase the level of environmental pollution [8, 9, 10].

2. Materials and Methods

The authors of the article analyze the payments for the negative environmental impact made by an exploration company in Kazakhstan in 2015. The company is Volkovgeology, which explores Irkol deposit in Kazakhstan. The calculation was made for all possible sources of contamination when constructing a well for in situ leaching of uranium, which makes a considerable negative impact on the environment. Special attention has been paid to the process waste management system as the wastes of this particular type are produced in great amounts. The same calculation has been made in compliance with the laws and regulations of the Russian Federation.

3. Results and Discussion

The major sources of negative environmental impact are the mobile drill rig BPU-1200 M equipped with the drill ZIF-1200 MR, the bulldozer T-165-2, trailers, the mobile power plant AKSA AJD-200 with the capacity of 160 kW, the excavator EK-18, the water tank truck KRAZ-6322, the maintenance crew vehicle GAZ-66, the repair vehicle on the basis of ZIL-131.

All moving machines (plants and vehicles) firm and shave the soil, which causes dust. The vehicles and the power plant emit exhaust gases. Constructing pits, clarifiers, and sumps add even more dust [11]. Potential air pollutants, wastewater releases, and wastes caused by the construction works are represented in table 1.

Table 1. Air pollutants, wastewater releases, and wastes caused by the construction works.

Activities	Air pollutants	Waste water releases	Wastes
Pre-drilling activities (construction of access roads, site, three sumps)	Dust caused by moving vehicles and excavating works, exhaust gases	No	Barren rocks at the site, domestic garbage
Wellbore enlargement with the implementation of clay-based drilling mud. The access to the orebody.	Dust caused by moving vehicles, exhaust gases	No	Cuttings (resulted from drilling both barren rocks and orebody), domestic garbage
Well completion: implementation of water-based mud, 3-phase airlift pumping until complete clarification of the mud	Dust caused by moving vehicles, exhaust gases	No	Cuttings
Geological, hydrogeological, and radioecological surveys, including topographical survey	Dust caused by moving vehicles, exhaust gases	No	Domestic garbage
Mining	Dust caused by moving vehicles and excavating works, exhaust gases	No	Domestic garbage

Well construction undoubtedly has a negative environmental impact on:

- air (due to the vehicles and special machines);
- soil and cover crop (due to the pits, clarifiers, and annular seal failure);
- surface and ground waters (mud losses, fuel and lubricant storage, formation fluid seeps);
- subsurface resources;
- flora, fauna, and human beings.

The implemented waste management system makes it possible to minimize the negative environmental impact. Non-radioactive process wastes are split into two groups: one group is further recovered and recycled while the other one is buried. Radioactive and highly radioactive substances are sent to be buried at the special burial site for radioactive wastes. Broken rock debris and cuttings are put into wide ditches and covered with the humus layers, which were put aside and generated while digging the ditches. There is radiation safety and ecological control service which is in charge of environmental monitoring, i.e. accounting, keeping, transferring, and transporting all wastes including radioactive ones [12, 5].

In the process of works, about 22.2 ha of soil will be disturbed. Therefore, during the performance period mine reclamation will be provided: backfilling, further covering of sumps, pits, and clarifiers with the soil layer, afterward furrow firming and irrigation. The experience of similar activities performed in the bordering region shows that the vegetation on such saline soils is restored within three years. As a result, by the end of the performance period a certain part of the lands will be reclaimed [8].

Moreover, actual mining will take place in the territory when drilling is completed. Therefore, the question on compensation for agricultural losses caused by different land-disturbing activities should be solved only after actual mining is over, reclamation activities are performed, and the land is transferred to the government.

If the rate of pollution exceeds the regulated limits, the charge for emissions is calculated in compliance with the Code of the Republic of Kazakhstan for Taxes and Other Obligatory Payments to the State Budget. The calculation of the charge for harmful emissions given below is done in accordance with tariffs approved by the Solution of the Maslikhat of Kyzylorda region № 121, dated 26.12.2008 (tables 2, 3, 6).

Table 2. The calculation of payment for air pollution in the Republic of Kazakhstan.

№	Pollutants	Tariff per 1 ton (minimum calculation index) 2015 – 1982 tenge	Tariff, tenge (2015)	Emissions, ton per year	Amount of tax, tenge
1	Nitrous oxides	20	39 640	3.554	140 881
2	Sulphur oxides	20	39 640	1.115	44 199
3	Carbon oxides	0.32	634	3.043	1 930
4	Hydrogen sulphide	124	245 768	0.000003	1
5	Hydrocarbons	0.32	634	0.856	543
6	Formaldehyde	332	658 024	0.036	23 689
7	Inorganic dust	10	19 820	3.052	60 491
8	Carbon soot	24	47 568	0.142	6 755
9	Iron oxides	30	59 460	0.0123	731
10	Benzo(a)pyrene	996.6 per 1 kg	1408195.8	0.0000004	1
	Total per year				**279 219**

Table 3. The calculation of payment for pollution caused by vehicles and mobile machines within the regulated limits.

№	Fuel	Tariff per 1 ton (minimum calculation index) 2015 – 1982 tenge	Tariff, tenge (2015)	Fuel consumption, ton per year	Amount of tax, tenge
1	Petrol	0.66	1308.12	92.64	121184.24
2	Diesel fuel	0.9	1783.8	259.63	463127.99
	Total				**584312.23**

The comparative analysis of payments for the negative environmental impact on Irkol deposit was made on the basis of current tariff rates in the RF. The calculation was made as per rate of exchange 1 ruble = 0.28 tenge (dated 10.04.2015). Current tariff rates in the RF are approved by the Government Resolution № 344 (dated 12.06.2003) and N 410 (dated 01.07.2005). Both inflation and ecological situation factors are taken into account. The results are represented in tables 4, 5.

Table 4. The comparative calculation of payments for pollution caused by stationary sources in compliance with the laws and regulations of the RF.

№	Pollutants	Tariff in Kazakhstan, rubles per ton (2015)	Tariff in Russia, rubles per ton (2015), with all factors taken into account	Amount of pollution, ton per year	Amount of tax in Kazakhstan, rubles	Amount of tax in compliance with the RF legislation, rubles
1	Nitrous oxides	11218.24	205.21	3.554	39869.75	729.32
2	Sulphur oxides	11218.24	123.48	1.115	12508.45	137.68
3	Carbon oxides	179.42	3.528	3.043	546.2	10.74
4	Hydrogen sulphide	69553.8	123.48	0.000003	0.28	0.0004
5	Hydrocarbons	179.42	29.4	0.856	153.67	25.17
6	Formaldehyde	186222.77	4016.04	0.036	6755	144.58
7	Inorganic dust	5609.12	241.08	3.052	17119.3	735.78
8	Carbon soot	13461.89	470.4	0.142	1911.69	66.8
9	Iron oxides	16827	305.76	0.0123	206.88	3.76
10	Benzo(a)pyrene	394294.824	12052829.88	0.0000004	0.28	4.8
	Total per year				**79019.81**	**1858.63**

Table 5. A comparative calculation of payments for air pollution emissions caused by mobile sources in the Republic of Kazakhstan and the Russian Federation.

№	Fuel	Tariff in Kazakhstan, rubles (2015)	Tariff in Russia, rubles (2015)	Fuel consumption, ton per year	Amount of tax in Kazakhstan, rubles	Amount of tax in Russia, rubles
1	Petrol	300	7.6	92.64	121184.24	704.064
2	Diesel fuel	504.82	14.7	259.63	463127.99	3816.561
	Total				**584312.23**	**4520.62**

If the same deposit were located in the territory of the Russian Federation, the total amount of payment for air pollution emissions in the Republic of Kazakhstan would be 42.5 times as much as that in the Russian Federation. While the tariff for benzo(a)pyrene emissions in Russia is 30 times as much as that in Kazakhstan, the tariffs for all other pollutants are significantly lower. For example, the tariff for hydrogen sulphide emission in Kazakhstan is 563 times as much as that in Russia. The results of the comparative analysis of payments for air pollution emissions caused by mobile sources are even more impressive: the total payment in Kazakhstan is 129 times as much as that in Russia.

However, the payments for waste disposal in Kazakhstan and Russia are commensurable: the payment in Russia is 38% more than that in Kazakhstan. The total payment for the negative environmental impact in Kazakhstan over 2015 will be 263571.73 rubles. If the deposit were located in the territory of Russia, the payment would be 32108.55 rubles. The total amount of payment in Russia would be an 8th of that in Kazakhstan (table 6).

Table 6. A comparative calculation of payments for the negative environmental impact caused by wastes disposal.

№	Waste	Tariff per 1 ton (minimum calculation index) 2015 1982 tenge	Tariff, tenge, 2015	Amount of wastes, ton per year	Amount of tax, tenge	Amount of tax in Kazakhstan, rubles	Amount of payment in Russia, rubles
1	Ferrous and non-ferrous metal junk	2	3964	4	15856	4487.3	5842.4
2	MSW (municipal solid wastes)	0.19	376.58	3.6	1355.688	383.66	5258.16
3	Cuttings from barren rocks	0.004	7.928	5785	45863.48	12979.51	13606.32
5	Oily rags	2	3964	0.7	2774.8	785.28	1022.42
	Total				**65849.968**	**18635.74**	**25729.3**

The comparison of payments for the negative environmental impact has allowed us to identify two different approaches to economic regulation of environmental protection. In the Republic of Kazakhstan, the regional authorities are in charge of approving the tariffs, with the minimum calculation indexes being specified in the tax laws of the republic. In Russia, the tariffs are approved by the Government of the RF, with the regional differences in the amount of the tax being determined by the ecological situation factor. In Russia, the payments for the air pollution emissions are lower than those in Kazakhstan, while the payments for wastes disposal in general are higher. The wastes in Russia are differentiated in accordance with the classes of hazard.

4. Conclusion

The analysis of the systems of payments proves that the charges in Russia do not encourage the companies to reduce the negative environmental impact, as the charges are low enough and incommensurable with the costs for environmental constructions.

Currently, it is necessary to reform the system of charges for the negative environmental impact, with the average fifty-fold increase in the rate of the tariff for air pollution emissions.

While reforming the system of charges, the experience of the Republic of Kazakhstan should be taken into account and the subjects of the RF might be vested with the power to set the tariffs for the negative environmental impact.

References

[1] Bengtsson N 2015 *Journal of Development Economics* **115** 85–98
[2] Kerr J M, Vardhan M and Jindal R 2014 *International Journal of the Commons* **8** 595–616
[3] Yesina E I 2009 *Finansovye issledovania* **1** 45–52
[4] Khalishkhova L Z 2014 *Aktualnye Voprosy Sovremennoy Ekonomiki* **4** 744–50
[5] Metodika rascheta vybrosov zagryaznyayuschikh veschestv v atmosferu ot statsionarnykh dizelnykh ustanovok, RND 211.2.02.04-2004
[6] Kopytova A I 2012 *Tomsk State University Journal* **12 (127)** 155–60
[7] Yankovich E, Osipova N, Yazikov E and Talovskaya A 2009 *3rd International Symposium on Trace Elements in the Food Chain Deficiency or Excess of Trace Elements in the Environment as a Risk of Health (TEFC2009)* (*Budapest: Hungarian Academy of Sciences (HAS))* pp 56–60
[8] Metodicheskiye rekomendatsii po raschyotu vybrosov ot neorganizovannykh istochnikov, utverzhdyonnye prikazom Ministra okhrany okruzhayuschey sredy, RK № 100-p, dated 18.04.2008/URL: http://online.zakon.kz/Document/?doc_id=30203232
[9] Tagaeva T O 2011 *Studies on Russian Economic Development* **3** 143–153
[10] Ponomaryov M A 2011 *Terra Economicus* **9** 62–65
[11] Berikbolov B R, Kayukov P G, Dubchinin P P, Yefremov G F et al. 2002 *Otsenka vozdeystviya dobychi urana metodom podzemnogo vyshchelachivaniya na okruzhayushchuyu sredu na otrabatyvaemykh mestorozhdeniyakh Severniy Karamurun, Kanzhugan, Uvanas i Mynkuduk* (Moscow: Geoiformark) p 834
[12] Metodika rascheta kontsentratsiy v atmosfernom vozdukhe vrednykh veschestv, soderzhaschikhsya v vybrosakh predpriyatiy, RND 211.2.01.01-97 URL:http://tengrinews.kz/zakon/docs?ngr=V1200007664#z4

Associated petroleum gas utilization in Tomsk Oblast: energy efficiency and tax advantages

A Vazim[1,2], V Romanyuk[1], K Ahmadeev[1,3] and I Matveenko[1,4]
[1] Institute of Natural Resources, National Research Tomsk Polytechnic University, 30 Lenin Ave., Tomsk, 634050, Russia

E-mail: [2] vazim@tpu.ru, [3] ahmad1994@mail.ru, [4] mia2046@gmail.com

Abstract. This article deals with oil production companies activities in increasing the utilization volume of associated petroleum gas (APG) in Tomsk Oblast. Cost-effectiveness analysis of associated petroleum gas utilization was carried out using the example of gas engine power station AGP-350 implementation at Yuzhno-Cheremshanskoye field, Tomsk Oblast. Authors calculated the effectiveness taking into account the tax advantages of 2012. The implementation of this facility shows high profitability, the payback period being less than 2 years.

1. Introduction

Russia's support for the sustainable development idea defined the trends for improving economy competitiveness through reduction of resource intensity, GDP energy intensity, as well as harmful environmental effects. In 2007, since the volume of cost-effective associated petroleum gas (APG) usage was about 50%, it was decided to increase the volume of APG up to 95% with the help of tax penalties and advantages. The first resolution of the RF Government determined a target value of APG flaring no more than 5% [1]. According to the preliminary results of this resolution implementation in 2012, as well as considering some counteractions from several oil companies [9], the Government decided to strengthen reprisal measures and increase tax penalties for APG flaring from 1 January, 2013. Multiplying factors were established in the payment calculation formula from 4.5 in 2012, to 12 in 2013 and to 25 in 2014. Without metering devices, the factor is 120, regardless of flaring volume (6 in 2012). As a result, double growth of annual APG utilization investments is expected as well as refining capacity construction will increase up to 50 billion rub. This allows increasing APG utilization level from 76.2 in 2012, up to 78.8 in 2013, up to 80 in 2014, up to 95 % in 2015. [2]

Success of the associated gas emission reduction program has been already remarkable - Russian oil companies will have reduced APG flaring by 29% from 15.8 billion cum in 2013 to 11.2 billion cum in 2014. APG production in 2014 will decrease only by 3.9% to 71.5 billion cum [3]. The leaders in utilization, i.e. vertically-integrated oil companies, which recycled APG in 2013, are "Surgutneftegas" (99.3%) and "Tatft" (93.6%). Other companies utilized fewer APG: "Lukoil" (87.8%), "Gazprom oil" (79.5%), "Rusneft" (77.1%), "Slavneft" (76.6%), "Bashneft" (72.7%), "Rosneft" (63.6%).

We should mention some disadvantages of associated gas emission reduction program. According to the results of 2012, the total volume of oil companies' tax penalties for APG flaring reached 16.5 billion rub. Total tax penalties sum for the period from January to September of 2013 was 181.34 billion rub, while payment sum for APG flaring and/or dessipation was 1.122 billion rub.

Taking those facts into account, oil companies were offered to balance unit investments to reach 5% flaring - about 4800 rub for 1000 cum, and the unit payment for 1000 cum of flared APG is 20 rub. Oil companies supported this offer. For example, LUKOIL invested 8.25 billion rub into APG utilization over 9 months of 2013, being the second after «Rosneft» 17.48 billion rub. [4].

Now let us analyze another aspect of the politics, i.e. stimulating. It is commonly known that combination of both measures is optimal, as stimulating actions define realization of some politics and restrictive ones – its usage [13].

The direct support programs, including concessionary financing and tax incentives, are an example of stimulating actions. Such measures are thought to be important financial mechanisms which the Government uses for increasing investments into energy efficiency. Program of direct support and credit accommodation, with guaranties from the Government to pay the interest rate, contribute to technological investments and gaining valuable experience, which could result in further technological development, knowledge acquiring, and potential market expansion. However, such programs and credit accommodation are rather costly and should be used to assist in implementation new energy efficiency technologies, i.e. they should provide consumers' awareness and develop the market [11].

The legislation provides the following measures for business companies aimed at energy saving and energy efficiency increasing: government loan guaranties; tariff regulation; tax advantages; investment tax credit. RF Internal Revenue Code provides the following tax advantages and tariff incentives for the companies: firstly, exemption from business property tax (p.21, 3 art. 381); secondly, increasing factor of depreciation value for income tax (p.1, art. 259.3); thirdly, investment tax credit (art. 67 IRC).

Property tax relief has been provided for the companies investing in energy efficient technologies, fixed asset since1 January (p.21, art.381 IRC RF):

1) with regard to newly commissioned projects of high energy effectiveness in accordance with the list of such projects provided by the Government;
2) with regard to newly commissioned projects of high energy effectiveness class, if determination of energy effectiveness classes is provided by RF legislation.

Within 3 years from the day of such property registration, it is not included into the base of property tax [5].

2. The Tomsk Oblast APG utilization features

The main peculiarity of oil production in the Tomsk Oblast is medium and small size fields with low APG resources. Their location in lightly settled land with the absence of large fuel buyers results in low amounts of APG utilization. Gas-to-oil ration in extracted oil for various fields can vary significantly within 30-80 cub. m/t.

Having analyzed the table one can see that over 3 billion cum was extracted in Tomsk Oblast. The largest license holders of the region extract 2.9 billion cum: LtD "Tomskneft VNK", LLC "Gazpromneft-vostok", LtD "Tomskgazprom". LtD "Tomskneft VNK" holds the first place (1.76 billion cum) exceeding LLC "Gazpromneft-vostok" 2. 27 times, and LtD "Tomskgazprom" 5.2 times at the APG production level. However, smaller companies are leaders at the level of APG utilization: LtD "VTK" (99 %), LLC "Sredne-Vasyuganskoye" (97 %). Among bigger companies leaders are LtD "Tomskneft VNK" – 85 %, LtD "Tomskgazprom"- 69 %, LLC "Gazpromneft-vostok" is left behind significantly – 19 %. Nevertheless, it should be mentioned that no company reached required 95 % of APG utilization.

Industry experts distinguished three main factors which prevent regional companies from reaching 95% level of APG usage.

Table 1. APG production profiles and usage by the Tomsk Oblast license holder for 2012, thous. cum. [6].

License holder	Crude-oil production, thous. T.	APG production, mln cub. m	APG usage, APG	APG usage level, %
LtD "Tomskneft" VNK	6969.8	1762.9	1491.7	85
LLC "Gazpromneft-vostok"	1191.5	338.7	62.9	19
LtD "Tomskgazprom"	1059.3	775.1	534.1	69
NC "Rusneft"	1045.2	89.7	6.5	7
Imperial Energy	621.1	78.2	23.9	31
LtD "VTK"	424	21.3	21	99
LLC "Matyushkinskaya vertical"	206,2	12,0	8,4	69
LLC "Stimul-T"	104.8	5.9	4.7	80
LLC "South-Ohteurskoye"	74.8	5.8	2.2	37
LtD "Tomskgeoneftegaz"	29.4	1.0	0.3	29
LLC "Sredne-Vasyuganskoye"	17	1.0	1.0	97
LLC "Giant"	1.5	0.3	0	14

Firstly, Tomsk resources are mainly oil-based: out of 11.7 mil tons of produced oil in 2012, there were 4.6 billion cum of APG. Technological, logistic, and economic peculiarities are mainly based on oil. Secondly, Tomsk oil and gas industry is distinguished by geographic dispersion of oil fields and scarcity of their reserves which prevents from considering APG usage as a cost-effective project. Thirdly, main gas pipeline "Nizhnevartovsk – Parabel - Kuzbas" runs far away from main oil-producing areas of the region. Therefore, transport access to the facilities is rather a costly task.

At Luginetskiy oil, gas and condensate field, which provides over 70% of "Tomskneft" APG amount, gas treatment and gas-compressor station (GCS) is in operation. In late 2011 it transported jubilee 10 billion cubic meter to the gas pipeline system. In 2012 the gas volume transported to the main gas pipeline exceeded 1.2 billion cum.

In 2012 at LtD "Tomskneft" oilfields the following construction was carried out as a part of the gas program: main gas pipeline "West-Moiseevskoye oilfield – Dvurechenskaya GTPP"; vacuum GCS Free Water Knock Out (FWKO) unit at Igolsko-Talovoye oilfield; vacuum compressor station FWKO unit–9 of Sovetskoye oilfield; vacuum compressor station FWKO unit–5 of Vakhskoye oilfield and flare facilities upgrade; gas engine power station of 5 MW at Yuzhno-Cheremshanskom oilfield; GTPP at Dvurechenskaya oilfield.

In 2012 GTPP of 12 MW power was commissioned at Igolsko-Talovoye oilfield, in addition to 24 MW station, which has been operated since 2004 [7].

Application of modular gas engine power unit for APG utilization at oilfield (for example, Yuzhno-Cheremshanskoye oilfield). Yuzhno-Cheremshanskoye oilfield is located in the Karagasok district of Tomsk Oblast, the Russian Federation, 40 km North from Noviy Vasugan settlement. Oil and gas production unit №7 (OGPU – 7), which provides service to Yuzhno-Cheremshanskoye oilfield, controls extraction operations at Vasugan district as well, where there are 7 oilfields - Olenye, Severo-Olenye, Lomovoe, Yuzhno-Cheremshanskoe, Kolotushnoe, Katylginskoe, and Ontonigajskoe. Most of them have been developed over the past 30 years. Oilfield production is 1500 t oil per day, water cut amounts 80%.

Daily, at Yuzhno-Cheremshanskoe oilfield high-pressure flare (HPF) 37 thous cum of APG are flared, 26 t of them accounts for methane. Since 01.01.2012, according to RF Government Regulation d.d. 8

January, 2009 №7 "On Measures to stimulate decrease of atmospheric air pollution with APG flaring combustion gases" payments for flaring over 5% of APG has been increased in 4.5 times [8].

Implementation of gas engine power station will decrease air pollution and large energy costs.

In this particular paper, we analyze gas engine power station AGP-350, produced by CJSC "PFK "Rybinskkompleks". Gas engine power station is a system developed on the basis of conventional engine working on nutural gas or any other gas fuel. Production of two energy types – caloric and electricity – is possible.

Cost-effectiveness analysis of Yuzhno-Cheremshanskoe oilfield was carried out by the authors. In 2012, over 28 thous. tons of oil and over 1.5 mil cum of APG were produced, in this case the utilization level was a vanishingly small value. Therefore, it was chosen as a pilot oilfield for estimation of implementation efficiency of the most common way of APG utilization - gas engine power station.

AGP-350 gas engine power system cost is 6 011.27 thous. rub. The approximate cost of testing and commissioning is 10% of equipment cost; maintenance service cost is 10% of equipment cost. AGP-350 power is 332.50 kWh. According to OPNG UPNG, gas cost, including VAT, is 631.3 rub/thous cum. Electricity cost according to Electric Board data, is 2.54 rub kW. Energy consumption demand is 3 200.00 kWh. According to log sheet of 15 February, 2012 the APG flaring volume is 36 497.00 cub. m/day

The energy cost for facility operations is 71 201.28 thousand rub per year. Flaring emission cost at the high pressure flare (simplified calculation) for methane is 12 571.93 thousand rub per year. In total, the cost before implementation is 83 773.21 thousand rub per year.

Cost calculation after implementation of 10 AGP-350. The total annual cost after implementation is 81 492.78 thousand rub per year. The total cost advantages - 2 280.43 thousand rub per year. Calculation of cost after two year implementation: the total annual cost after implementation is 15 244.11 thous. rub per year; the total cost advantages - 68 529.10 thousand rub per year.

Thus, using AGP-350 will save 68 529.10 thous. rub paid for electricity per year. It will lead to emission reduction and, consequently, decreasing penalties.

3 Conclusions

Resolution of the RF Government to reach 95% of APG utilization can be achieved in 2015. Tax penalties and tax advantages are mechanisms for reaching this factor. Tax penalties prevails tax advantages. Multiplying coefficient can reach 25 (without metering devices – 120). Whereas, tax advantages are determined by income tax related to accelerated cost recovery, though such a measure is applied to facilities for APG utilization as a special case, tax advantages for corporate property tax related to the equipment installed in 2013. Besides, a special case is tax advantages for APG utilization facilities.

Therefore, in APG utilizing gas engine power stations are a priority. This particular article analyzes an example of AGP-350 gas engine power station implementation at Yuzhno-Cheremshanskoe oilfield. Economical indicator calculation has revealed the high effectiveness of this APG utilization method: cost effectiveness is over 75%, payback time is less than 2 years. Moreover, annual energy saving is 68 529.10 thous. rub per year.

References

[1] RF Government Regulation 2009 d.d. 8 January *On Measures to stimulate decrease of atmospheric air pollution with APG flaring combustion gases* **7** URL: http://www.consultant.ru/document/cons_doc_LAW_137666/

[2] Fomchenko T S Profit out of the way *Russkaya gazeta* **6403 (131)** URL: http://www.rg.ru/201/06/16/utilizaciya.html

[3] Shapovalova A and Yagova O 2011 Oil flare will raise *Commersant* **85**(8)

URL: http://www.kommersant.ru/doc/1640995
[4] *LUKOIL offers to reduce Tax penalties volume for APG flaring* URL: http://itar-tass.com/ekonomika/859313
[5] Dzhaarbekov S M *Tax advantages for energy efficient objects* URL: http://www.abok.ru/for_spec/articles.php?nid=5136
[6] Tsibulnikova M R and Sharf I V 2014 *Contemporary problems of science and education* **2** URL: www.science-education.ru/116-12817
[7] Andreev A Associated achievements. URL: http://www.gazprom-neft.ru/sibneft-online/arhive/276/1094464/
[8] Balandina AS 2012 *Creating investment-oriented and resources-saving tax system for gas producing enterprises* (Tomsk) 150–152
[9] Røland T H 2010 Associated Petroleum Gas in Russia: Reasons for Non-utilization FNI Report 13.2010. Lysaker, FNI [Electronic resource] *Fridtjof Nansen Institute*. Recent FNI Publications [site]. URL: www.fni.no/doc&pdf/FNI-R1310.pdf
[10] Sharf I V, Chukhareva N V and Kuznetsova L P 2014 Financial and tax risks at implementation of "ChayandaLensk" section of "Sila Sibiri" gas transportation system construction project [Electronic resource] *IOP Conf. Ser.: Earth Environ. Sci.* **21**
[11] Chukhareva N V 2012 Associated petroleum gas efficient utilization system in pipelines operating *International scientific research magazine* **4(4) 5**
[12] *Energy Strategy of Tomsk for the period till 2020*. URL: [http://www.prime-tass.ru/news/show]
[13] Tsibulnikova M R, Pogharnitskaya O V and Strelnikova A B 2015 Designing economic and legal mechanism of land management in oil and gas companies *IOP Conf. Ser.: Earth Environ. Sci.* **24** 1–5

IOP Conf. Series: Earth and Environmental Science **27** (2015) 012079 doi:10.1088/1755-1315/27/1/012079

Tax incentives as the tool for stimulating hard to recover oil reserves development

I V Sharf[1,3], D N Borzenkova[1,4] and L S Grinkevich[2,5]

[1] Institute of Natural Resources, National Research Tomsk Polytechnic University, 30 Lenin Ave., Tomsk, 634050, Russia
[2] National Research Tomsk State University, 36 Lenin Ave., Tomsk, 634050, Russia

E-mail: [3] irina_sharf@mail.ru, [4] borzenkova.darya@mail.ru, [5] nasty_saturn@mail.ru

Abstract. The share of hard-to-recover oil reserves, principally from unconventional hydrocarbon sources, has significantly increased in the world petroleum market. Russian policy of subsurface management is directed to stimulate the development, survey and involvement into production of hard-to-recover oil reserves by tax-financial and economic-organizational tools among which tax incentives is the most effective one. The article highlights different categories of hard-to-recover oil reserves as a basis for generating tax incentives. Also the aspects of tax influence on petroleum business (involved in production of had to recover reserves) in Tomsk region are revealed, both positive and negative.

Introduction

The issue concerning hard-to-recover oil reserves is becoming now extremely relevant in rapidly changing macroeconomic conditions. This could be determined by the following factors:

1) change of quantitative and qualitative characteristics of hydrocarbon reserves as well as the growth of total share of hard-to-recover oil reserves. For the year 1960, 50-60% of oil was produced from newly discovered deposits; in 1990 – 20-25%; at present time – 12-15% and according to the geological forecast, the number will be 7-10%. It should be considered that Russian oil industry raw material base could be characterized by high concentration of hydrocarbons. State Register of Mineral Reserves estimates 2885 deposits with oil reserves; more than a half of recovered reserves are buried in 11 unique deposits (with the reserves of more than 300 mln.t.) and in 85 large deposits (60-300 mln.t.); among them 9 unique and 56 large deposits are situated in West Siberian petroleum basin. Besides, most of newly developed deposits in Russia refer to the category of small with the average reserves amount of 4 mln.t. [1, P. 54].

2) change of energy producing and energy consuming structure due to the increase in alternative energy sources significance and production of hydrocarbons (HC) from unconventional reserves [2,3]. This fact influences strongly the strategy of financial trading organizations. Long-term and risky investments for production and development of HC from conventional reserves are becoming less attractive. This reflects on the financial possibilities of petroleum enterprises and principally on Russian enterprises. Another factor is the growing interest of international petroleum enterprises to shale hydrocarbon reserves. For instance, «ExxonMobil» quit the Shtokman deposit development project in September 2012 and got the full control over the shale oil production on Bakken shale formation in the north of the USA.

Thus, fulfilling the task "the effective use of natural energy resources and potential for sustained economic growth and increasing the life quality of people" [4], as well as saving natural-resource potential for further generations is impossible without resource-innovative operation of petroleum enterprises involved in the production of hard-to-recover hydrocarbon reserves.

Consequently, state tax policy is directed to exploration, production and development of hard-to-recover hydrocarbon reserves, especially oil.

The article aims to analyze current tax policy – positive and negative aspects as well as its influence on production operation of petroleum enterprises on the example of Tomsk region petroleum sector.

Hard to recover oil reserves: general characteristics

Currently, there is no precise definition for hard-to-recover hydrocarbon reserves in the Russian legislation. In accordance with the order of Ministry of Natural Resources and Environment of Russian Federation from 1st November, 2013, to the recoverable reserves refer "the part of geological reserves, which could be extracted from the formation (reservoir) during the whole period of developing within optimal project decision with the appliance of available technology, considering environmental and subsurface using requirements". Taking into account this definition, the reserves of the developing formations could be referred to the recoverable reserves, and the reserves from the exploration deposits – to the hard-to-recover reserves (ranging of deposits according to the degree of commercial development [5]).

For petroleum industry the basic criteria is the quality of oil, consequently, considering abnormal physical and chemical properties, the following types of oil could be defined: hard, viscous, sulfurous, paraffin oil, tarry oil, with high (more than 500 m^3 /t) or low (less than 200 m^3/t) gas saturation, with the presence of more than 5% of aggressive components (hydrogen sulfide, carbon dioxide) in non-associated and (or) associated gas [6, C. 32]. The data, obtained by Institute of Petroleum Chemistry, Siberian Division of the Russian Academy of Sciences, shows that the present types of oil are common for many deposits in the world.

According to the instruction of oil reserves, resources and flammable gas classification, oil (considering the content and physical properties) could be subdivided depending on its properties, hydrocarbon-type content, fractional composition, sulfur and other non-hydrocarbon components, asphaltenes and tars [5].

Geology considers reservoir properties of surrounding formation which influence physical and chemical properties of raw hydrocarbons. The key property of the reservoir is porosity – when water and gas can flow freely through pores at pressure changing [7].

Depending on the degree of porosity, the productive formations could be subdivided into: low permeable (from 0 to 100 mD); medium permeable (from 100 mD to 500 mD); high permeable (more than 500 mD). Reservoirs could be classified into 5 types (μm^2): extra permeable (>1); high permeable (0.1 - 1); medium permeable (0.01 – 0.1); low permeable (0.001 – 0.01); extra low permeable (<0,001).

According to the decree of the Russian Federation № 700-r from 3rd May 2012, four categories of hard to recover oil extraction projects, based on reservoir permeability and oil viscosity, are defined:

1) projects to extract oil from low permeable reservoirs - from 1,5 to 2 mD inclusive (from $1{,}5 \cdot 10^{-3}$ μm^2 to $2 \cdot 10^{-3}$ μm^2 inclusive);
2) projects to extract oil from ultra low permeable reservoirs - from 1 to 1,5 mD inclusive (from $1 \cdot 10^{-3}$ μm^2 to $1{,}5 \cdot 10^{-3}$ μm^2 inclusive);
3) projects to extract oil from extra low permeable reservoirs – up to 1 mD inclusive (up to $1 \cdot 10^{-3}$ μm^2 inclusive);
4) projects to extract ultra-viscous oil under formation conditions of more than 10 000 mPa·s.

The other characteristics of formations are – low porosity, deep bedding or permafrost, interbedding temperature (100° C > t <20° C), high water content in the extracted oil [6, C.33].

Economic efficiency of hydrocarbon reserves production is the basis for oil producing companies' business strategy. If under current economic conditions and tax system, at application of hydrocarbon production technique and technology, which provide the compliance with the requirements of rational subsurface use and environmental protection, the development of the reserves is not profitable, then these reserves will be considered as hard-to-recover.

It should be taken into consideration that the category of hard-to-recover is constantly changing due to the development and implementation of innovative technologies. For example, in 80-90's of the last century in Western Siberia, Achimovskaya and Bazhenov suites, as well as Paleozoic deposits were not included in the category.

Hard to recover oil reserves of Tomsk region

At present, hard-to-recover reserves account for 65% (18.7 bln. t.) from 28.9 bln. t. of oil extracted in Russia from the reservoirs of ABC1+C2 category. The initial reserves amount:

a) The Bazhenov suite in Western Siberia - 330 mln.t.

b) Domanic formations of Timano-Pechorskoj and Volgo-Uralskoj provinces - 300 mln.t.

Besides, Low Silurian formations of Kaliningrad region are very perspective for unconventional hydrocarbon resources development (the initial gas reserves amount - 530 bln. m^3), as well as low permeable reservoirs of Vendian Cambrian formations (Talakanskoe, Verhnechonskoe, Danilovskoe deposits) in Western Siberia.

"Russia is not among the leading countries in production of shale hydrocarbons. There are only 35 wells (from which 700 thousand tons of shale oil was extracted last year) while in the USA there are 113 wells. This fact shows that the USA is the leading country in shale hydrocarbons production (40% of the world's production) [8, p. 24]."

The potential of Tomsk region in developing of the Bazhenov and Tjumen suites, as well as Low Jurassic deposits, is very high (table 1).

Table 1. Potential of hard to recover oil reserves development in Tomsk region.

Stratigraphic complex	Reserves, mln.t.		% of hard to recover reserves from approved reserves
	Total	Hard to recover	
Paleozoic	90.1	14.5	16.1
Low Middle-Jurassic	279.7	193.7	69.3
Upper Jurassic	908.7	103.6	11.4
Bazhenov	37.6	37.6	100.0
TOTAL	1278.5	349.4	26.5

Thus, for oil production companies of Tomsk region the developing hard-to-recover reserves is one of the promising areas. Tomsk region, due to its geological history, has a lot of small and middle size deposits. This is the reason why petroleum business of Tomsk region is marked only by one leading company OAO "Tomskneft" VNK and some small producing companies, which develop the sites where the amount of hydrocarbon reserves is less than 5 mln.t.

Hard to recover reserves and tax legislation

One of the most effective tools to stimulate hard-to-recover reserves is tax incentives.

Tax legislation has the following incentives for the production of oil [9]:

a) in the following Russian regions:

- Republics of Bashkortostan and Tatarstan (article 343.2);
- Republic of Sakha (Yakutia), Irkutsk region, Krasnoyarsk region (subparagraph 2 paragraph 4 article 342.5);

– Nenets Autonomous region, Yamal peninsula in Yamal-Nenets Autonomous region (subparagraph 5 paragraph 4 article 342.5);

b) from offshore deposits of Azov, Baltic, Pechora, White, Japan, Okhotsk, Caspian, Black, Barents, Kara, Laptev, West-Siberian, Chukotsk, Bering seas (paragraph 5 article 338);

c) from the formations placed north of the Arctic Circle, fully or partially on the borders of inland sea waters and territorial waters of Russian Federation continental shelf;

d) from the formations placed within the borders of inland sea waters, territorial waters of Russian Federation continental shelf or in the Russian sector of Caspian sea floor;

e) from the types of geological formations – deposits (subparagraph 21 paragraph 1 article 342), referred to Bazhenov, Abalaksk, Hadumsk or Domanikov productive formations [9];

f) from the formations containing oil with the viscosity of 10 000 mPa x s and more (in reservoir conditions). Formations containing oil with the viscosity of more than 200 mPa x s (in reservoir conditions) are having zero tax rate. Thus, raising marginal threshold points to the dual character of tax incentive, firstly introduced in 2006, which stimulated business to apply new technologies in order to reduce the tax burden. In case when oil viscosity varies in the range of more than 200 mPa·s and less than 10 000 mPa·s (in reservoir conditions), the coefficient K_{rp} (coefficient characterizing region of oil production and oil properties) will be equal to 0 in the equation of calculating Severance Tax (ST) (equation 1).

$$\text{Total ST} = Q \times (K_{ST} \times K_p - D_s), \text{ where} \tag{1}$$

D_s – index, characterizing the specificity of oil production (Equation 2)

$$D_s = K_{ST} \times K_p \times (1 - K_{fd} \times K_r \times K_c \times K_{rd} \times K_{rp}), \tag{2}$$

where

K_{rp} – coefficient, characterizing region of oil production and oil properties

K_p – coefficient, characterizing dynamics of world oil prices

K_{fd} – coefficient, characterizing degree of definite oil field depletion

K_c – coefficient, characterizing the degree of oil production complexity

K_{rd} – coefficient, characterizing degree of definite oil reservoir depletion

Decreased value of K_c coefficient at calculating the amount of ST is applied to oil, which is extracted from the definite hydrocarbon deposit depending on permeability and thickness of the formation (subparagraph 2,3, paragraph 1, article 342.2 Tax Code of the Russian Federation):

0.2 – at permeability not more than $2 \cdot 10^{-3}$ μm^2 and effective oil height not more than 10 meters;

0.4 – at permeability not more than $2 \cdot 10^{-3}$ μm^2 and effective oil height not more than 10 meters.

At the extraction of oil from definite formation of Tyumen suite - $K_c = 0.8$.

For the republics of Bashkortostan and Tatarstan, tax deductions are applied to the calculated sum of ST concerning oil which was extracted from deposits with initial reserves (on 1st January, 2011) of 2 500 mln.t., 200 mln.t. and more. Calculation of tax deductions depends on the amount of export tax.

For oil extracted from the formations, which are situated fully or partially on the territorial entity of Russian Federation (Sakha republic (Yakutia), Irkutsk region, Krasnoyarsk region) - $K_{rp} = 0$.

At the development of new offshore hydrocarbon deposits, special order of tax calculation is applied as well as special ST rate of 15% is added to the taxation base. Taxation base is defined as the cost of crude hydrocarbons. Cost of crude hydrocarbons is the product of extracted hydrocarbon amount and minimal marginal value of crude hydrocarbon unit. Minimal marginal value of crude hydrocarbons (here oil) is defined as the product of average oil price in the world market per barrel in US dollars during the last tax period and the average rate of US dollar to Russian ruble (during that tax period) established by Central Bank of Russia.

Production of oil from the Bazhenov suite is beneficial for oil companies. For a long period of time, the Bazhenov suite (square -1 mln. km^2, thickness - from 5 to 40 m) was considered to be a good impermeable environment for oil and gas traps. Current scientific research proved the presence of large amount of high quality light industrial reserves of oil. The Bazhenov suite differs from traditional reservoirs in microcavity, flaggyness, foliation and bedding. This fact points the necessity of using special technology, and consequently applying qualitative approach to choosing of service company and staff [10].

Thus 313.3 mln.t. of oil in Tomsk region deposited in the Bazhenov suite, 927.5 mln.t. of oil from Tyumen suit, not less than 50% from 1150 mln.t. of oil from Low Jurassic formations could get under preferential taxation.

Conclusion

1. Variety of approaches to understand the notion, hard-to-recover reserves, explains the necessity of applying different stimulating tools for the development of such reserves.
2. State tax policy is characterized by the increase in tax burden on oil and gas production enterprises due to the constant raise of specific rate at calculating of Severance Tax (ST) on oil from 2008 till 2014 and changing of calculating algorithm in 2015. This significantly reduces financial and investment ability of the companies. Comparative analysis showed that change of ST calculating algorithm leads to the increase in tax amount by 17% per one ton of oil.
3. Variety of tax incentives for different types of hard-to-recover reserves (zero ST rate, reduced coefficients in the equation of calculating ST, special order of tax base calculation for a number of deposits), complicate the calculation of ST and negatively influence the management of tax system.
4. Large amount of hard-to-recover hydrocarbon reserves in Russia and their variety, requires considerable financial and investment resources as well as applying innovations in production-technological process and attracting highly qualified specialists, thus there is a necessity in rational state tax-financial policy.
5. Incentives are more tangible for large companies involved in developing large deposits as they allow multiplying financial and investment resources for implementing new technologies. Small petroleum companies in Tomsk region, which are involved in production of hydrocarbons from small deposits, do not have significant financial benefits from the reduction of tax burden on the development of hard-to-recover reserves. High cost of special technology and equipment, need of highly qualified staff, obtaining investment funds – all these is a challenging task for small petroleum companies.
6. The effective method to support small petroleum business is, according to the authors' opinion, application of additional income tax instead of severance tax during 5 years. Shortfalls in taxes into budgetary system will be partially compensated by income tax.

References

[1] Shpurov I 2014 Novaya klassifikaciya zapasov uglevodorodov – sredstvo regulirovaniya innovacionnogo processa v TEK *Neftegazovaya vertical* **16** 46–56
[2] Bajus M 2014 Shale gas and tight oil, unconventional fossil fuels *Petroleum & Coal* **56** 206–21
[3] Bryden K, Federspie M, Habib E Th and Schiller Jr R 2014 Processing Tight Oils in FCC: Issues, Opportunities and Flexible Catalytic Solutions *Grace Catalysts Technologies Catalagram.* **114** 1–23
[4] Energy strategy of Russia for the period till 2030 [Electronic resource] // Ministry of Energy of the Russian Federation [web-site]. URL: http://minenergo.gov.ru/aboutminen/energostrategy/ (25th April, 2015).
[5] Oil and combustible gas classification manual//State Reserves Committee [web-site]. URL: http://www.gkz-rf.ru/index.php?option=com_content&view=article&id=189:2014-04-30-12-17-36&catid=53:docsuvs&Itemid=70 (20th March, 2015).

[6] Yashchenko I G 2014 Difficult-to-recover oils: physical and chemical properties and environmental impact of production *Exposition Oil & Gas* **1** 30–35
[7] Technical library / neftegaz.ru [web-site]. URL: http://neftegaz.ru/tech_library/view/4601 (20th March, 2015).
[8] Fesenko V 2014 Neftegazovij kompleks – oborudovanie, tekhnologii i videnie razvitiya *TEK. Strategiya razvitiya. Infirmacionno-analiticheskij zhurnal* **3** 22–25
[9] Tax Code [Electronic source] // Information reference bureau "ConsultantPlus".
[10] Sharf I V, Malanina V and Kamynina L 2015 Features of the marketing strategy of oil and gas companies in exploration drilling *IOP Conf. Ser.: Earth Environ. Sci.* **21** 012047

Production operation of small petroleum enterprises in Tomsk region

I Sharf [1,2], V Filjushin[1,3], I Shenderova[1,4] and O Kochetkova[1,5]

[1] Institute of Natural Resources, National Research Tomsk Polytechnic University, 30 Lenin Ave., Tomsk, 634050, Russia

E-mail: [2] irina_sharf@mail.ru, [3] filushinvv@mail.ru, [4] Inna-shenderova@yandex.ru, [5] kochetkova@tpu.ru

Abstract. Implementing resource-innovative strategy to develop Russian fuel-and-energy sector implies the promotion of small oil production enterprises developing fields with the reserves of less than 5 mln. tons. However, the efficiency of such enterprises, investigated on the example of Tomsk region (considering the indexes of oil extraction, growth of reserves, amount of capital expenditures and geological surveys), signifies the presence of definite tendencies. Consequently, management decisions concerning economic, organizational, financial and fiscal character should be taken in order to eliminate detrimental factors.

1. Introduction

The objective set by Russian Federation Government to create favorable conditions to stimulate the activity of small and middle-size enterprises in the sphere of small and challenging field development, is supported by small petroleum enterprises in all oil producing regions of the Russian Federation, as well as in Tomsk region. However, the realization of these objectives does not always give positive results.

At present, the majority of fields, discovered in the Russian Federation, refer to the category of small ones, with the average reserves amount of about 4 mln.t. [1]. The large number of such fields is located in Uralo-Povolzhje and West Siberia.

Usually, small and middle-size fields are not very attractive for large petroleum enterprises. This predetermines the substantial contribution of small enterprises to the development of petroleum sector. For instance, there are more than 7000 of such enterprises in the USA and the total share of these enterprises in the whole petroleum production is more than 46%.

Considering the fact that in Russia, as well as in the whole world, small and medium-size enterprises develop the fields with the reserves of less than 5 mln. tons and that in the unallocated reserve fund there are only fields where it is unlikely to develop large deposits by implementing modern scientific knowledge and technologies, the social and economic significance of small and medium-size enterprises will be growing in future.

The present article aims to analyze features, tendencies and challenges of small petroleum enterprises in Tomsk region, which are the subsidiaries of large corporations or refer to the independent oil producing companies.

2. Small petroleum enterprises. Characteristics

Information field analysis has revealed the essential criteria-based differences when referring the company to the category of individual oil producer (IOP).

At present, according to the Ministry of Energy of the Russian Federation data on January 1st, 2014, among 294 of petroleum enterprises 111 are within 10 vertically integrated companies (VIC), the share of which is 87.4 %; 3 enterprises operate in accordance with the Federal Law "On Production Sharing Agreement"; 180 independent producing companies are outside VIC.

IOPs are the enterprises that produce oil with the amount of liquid hydrocarbons being refined of less than 1 mln. t., regardless the amount of extraction [2]. The same criterion could be applied when referring the company to the sphere of small subsurface users.

The Skolkovo Innovation Center and noncommercial association "Assoneft" suggest the following criteria of relevance to IOP (table 1) [3].

Table 1. IOP relevancy criteria.

	Criteria	Definition
1	Oil refining limit	Outside the Group of Persons; total amount of liquid hydrocarbons being refined within the last year is more than 1 mln. t.
2	State freedom	Outside the Group of Persons, which includes legal persons, with the share of the Russian Federation in the authorized fund of more than 50%.
3	Retail trade limit	Outside the Group of Persons; total amount of hydrocarbon retail trade within the last year is more than 50 million rubles (excluding VAT and excise duties).
4	Group of Persons	In accordance with the Federal Law "On Protection of Competition" (Article 9) or in accordance with The Tax Code of Russian Federation (Article 20).
5	Scale of production operation	*Small enterprises:* outside the Group of Persons; total amount of hydrocarbon production within the last year is not more than 300 thousand tons. *Middle-size enterprises:* outside the Group of Persons; total amount of hydrocarbon production within the last year is not more than 5 mln tons.

This approach focuses on the legal background of business operations and annual oil output. It should be noted that contrary to VIC, the production operation of small enterprises does not include all the technological processes such as exploration, transporting, refining and marketing. Basic operation areas are development and production of hydrocarbons.

According to the data, collected by "Skolkovo", the share of IOP in exploration drilling for the year 2012 was 17%, in production drilling – 3.8 % [3]. Besides, 38 % of the companies were involved only in geological surveys while 52 % of the companies were producing oil. The rest 10 % of the companies were at the initial stage of their development.

The criteria-based differences, mentioned above, are clearly shown by oil production data. Considering "Skolkovo" data, the share of IOP in total oil production has reduced for the last 10 years to 3%. In 2012 it was 2.8 %, about 14.4 mln t. with more than 60 % of IOP producing less than 50 thousand tons of oil per year. However, according to the Ministry of Energy of the Russian

Federation, the share of IOP in total oil production is increasing. From 2005 to 2013 the growth was 5 %. In 2013 – 2.3 % (65.9 mln t). This underlines the social and economic significance of IOP both in national and regional scale of production operation.

3. Production Operation of Small Enterprises in Tomsk Region

Tomsk region, due to its geological history, has a great number of small and middle-size deposits. According to "Tomsknedr", 131 of hydrocarbon deposits were discovered in Tomsk region – 102 oil deposits, 21 oil and gas condensate, 8 gas condensate. Among 131 deposits, 112 refer to small deposits and 6 to large. 34 enterprises are operating on the territory of Tomsk region. Only 15 are engaged in hydrocarbon production.

Oil production analysis shows, that OAO "Tomskneft" has been the leader of oil production for many years. The other two groups are made by companies producing less than 1000 thousand tons and companies with oil production volumes less than 100 thousand tons. However, the role of small enterprises constantly increases. This allows maintaining the sustained level of oil production in the region – about 11 mln. t. (figure 1). Oil production targets were achieved on 96%. It should be noted that only three enterprises managed to increase oil production as compared with the year 2012 (OOO "Gazpromneft-Vostok", OAO "Tomskgazprom", OOO "Tomskgeoneftegaz").

Reasons that lead to such situation:

- The majority of small enterprises obtained exploration license during the last 10 years. Due to this, they are only at the initial stage of development.
- The majority of companies own from 1 to 3 licensed fields with small deposits. The leading enterprise OAO "Tomskneft" VNK has 51 deposits, among which 29 are at the stage of development. This points to the conservation of fields as deferred assets.

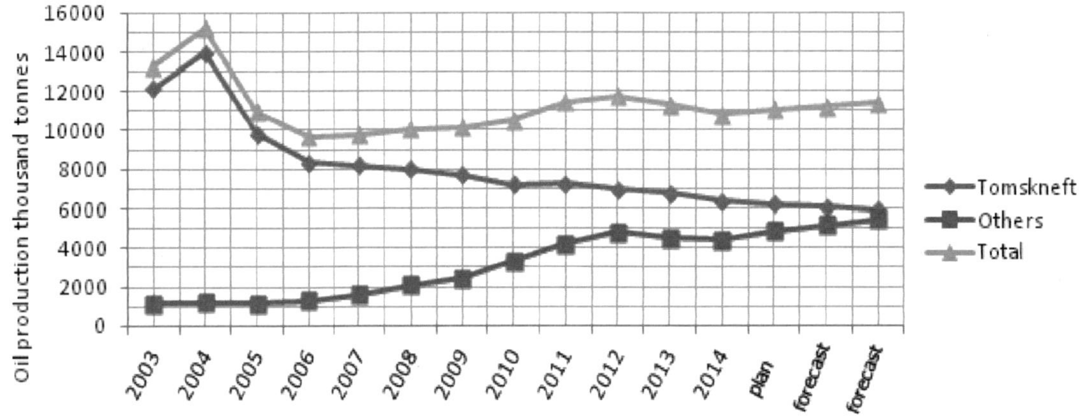

Figure 1. Tomsk region production profile.

- Reducing exploration drilling volume and the number of production wells.
- Geological and operational arrangements of enterprises turned to be insufficient to compensate the natural decrease in oil production due to reservoir depletion and water content.

Besides, long term prospective of enterprise production operation is defined by constant growth of hydrocarbon reserves and consequently by geological surveys and drilling.

At present, subsurface users try to rehabilitate the mineral reserves base. But because of the lack of own financial-investment resources and difficulties in obtaining credit sources due to high interest rate, the geological survey programs could not be implemented into scale. Furthermore, the dynamics of finance provision, considering macroeconomic factors, is volatile despite the total positive trend during the last ten years (figure 2, table 2). Innovative scientific methods play the basic role in the growth of reserves.

Table 2. Plans and execution of geological surveys in Tomsk region.

Year	Type of production operation					
	Drilling (thousand line meters)		2D seismic survey (line kilometers)		3D seismic survey (km3)	
	Amount	Compared to previous year	Amount	Compared to previous year	Amount	Compared to previous year
2009	71.2		4205		745	
2010	104.7	147.05%	1454	34.58%	535	71.81%
2011	92	87.87%	2514	172.90%	1350	252.34%
2012	52	56.53%	1996	79.40%	1140	84.44%
2013	38	73.08%	600	30.06%	1468	128.77%
2014	50.2	132.11%	673	121.17%	1335.8	90.99%
Plan 2015	64.6	128.69%	3750	557.21%	1885	141.11%

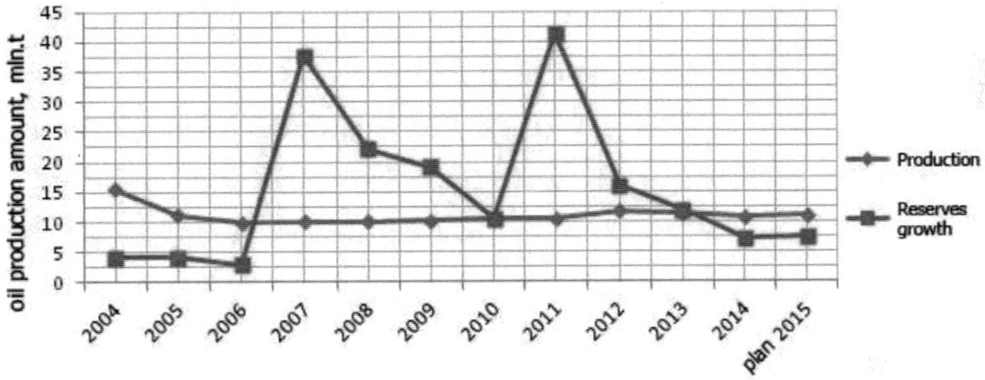

Figure 2. Production dynamics and reserves growth.

In general, enterprises investments into the basic stock are distributed as follows:

- field facilities construction;
- exploration drilling;
- oil and gas pipeline construction;
- equipment;
- well workover;
- design an survey works.

It should be mentioned that there is a growth in expenditures on field facilities construction (44% from capital expenditures in 2012, 49% – in 2014) and exploration drilling. As a result, further growth in oil production could be observed among small petroleum enterprises [4].

4. Fiscal stimulus of small petroleum enterprises production operation

In order to stimulate small petroleum enterprises production operation, the coefficient, characterizing the amount of reserves of definite field (K_r) (equation 1), was introduced on January 1st, 2012 to calculate the Severance Tax (*ST*). This coefficient allows decreasing tax burden as it depends from the amount of extracted oil reserves (V_r) and the degree of reserve depletion (C_{rd}) of definite field [5].

$$K_r = 0{,}125 \times V_r + 0{,}375 \quad (1)$$

Conditions of given calculation algorithm application:

- Amount of initial extracted oil reserves V_r should be less than 5 mln t.
- Degree of reserve depletion C_{rd} should be less or equal to 0.05.
- Date of issuing the license for the right of subsoil resource use (January 1st, 2012), which defines the period of fiscal preferences.
- The given algorithm could not be applied to the zero-rated oil and to the oil, which was produced in the amounts not exceeding the initial oil reserves. In this case K_r is equal to 1.
- Each well should be equipped with oil metering units.

Definite positive influence on oil production enterprises should be considered [6]. However, despite tax preferences, tax burden remained quite high. Calculations, made according to the data obtained from small oil fields in Tomsk region, showed that the share of *ST* in production costs was 50%. The share of *ST* in tax payment – 70%.

From January 1st, 2015 *ST* calculation formula has changed (equation 2). In accordance with the Article 342.5 of The Tax Code of the Russian Federation, the specific rate K_{st} changes as follows: 766 rub/t in 2015, 857 rub/t in 2016, 919 rub/t in 2017. It should be mentioned that earlier the planned rate for the year 2015 was 530 rub/t, for the year 2016 – 559 rub/t.

$$TotalST = Q \times (K_{ST} \times K_P - D_S) \quad (2)$$

where: D_s – index, characterizing oil production (equation 3)

$$D = K_{ST} \times K_P \times (1 - K_{fd} \times R_r \times R_c \times R_{rd} \times R_{rp}) \quad (3)$$

where: K_{rp} – coefficient, characterizing region of oil production and oil properties;
K_p – coefficient, characterizing dynamics of world oil prices;
K_{fd} – coefficient, characterizing degree of definite oil field depletion;
K_c – coefficient, characterizing the degree of oil production complexity;
K_{rd} – coefficient, characterizing degree of definite oil reservoir depletion.

The essence lies in transferring of tax burden to the production sector and cutting of export tax in order to stimulate oil refining industry [7, 8], as well as applying of tax incentive to stimulate the production of hard to develop reserves. It should be also mentioned, that in the world petroleum market unconventional oil reserves that are hard to develop will be more attractive in the long run [9,10]. Consequently, small oil production enterprises of Tomsk region have a large potential as the region is rich in unconventional oil reserves, for example Bazhenov suite. However, small enterprises, as it was mentioned above, are not engaged in oil refinery and foreign trade. The analysis showed that the amount of ST, per one ton of oil paid in December 2014 and January 2015, increased on 17%. This could exacerbate the unstable financial position of small enterprises.

Additional factors of economic and organization character, negatively affecting the production operation of small enterprises, could be defined.

- Small enterprises could not establish the oil sale price independently. Prices are established by:
  1. VIC when transporting oil to the oil refinery plant (ORP);
  2. Higher corporation or the Headquarters;
- Prices in the domestic market are lower than in the foreign market.
- Dominating petroleum corporations create additional barriers in order to block the access to oil pipeline system and trading markets.
- Higher specific expenditures, as compared to VIC, namely on preparation of commercial oil before transporting it to the main pipeline system.

IOP Conf. Series: Earth and Environmental Science **27** (2015) 012080 doi:10.1088/1755-1315/27/1/012080

- Limitation of access to financial and investment resource of banking sector and stock exchange due to relatively low income.

These facts significantly reduce the possibility of business development.

5. Conclusion

- Petroleum industry of Tomsk region is attractive in terms of investment potential due to the demand for small and middle-size deposits in the world market, as well as due to the presence of large number of unconventional oil reserves.
- Further development of financial and tax tools for stimulating small petroleum enterprises production operation is necessary. It is also more efficient to replace the Severance Tax with Tax on financial result for reserves which are hard to develop, as current financial benefits are more reasonable for financially stable large corporations, not for small oil production enterprises.
- It is necessary to adopt a special law to eliminate industrial factors restricting small oil production enterprises development.

References

[1] Shpurov I 2014 Novaya klassifikaciya zapasov uglevodorodov – sredstvo regulirovaniya innovacionnogo processa v TEK *Neftegazovaya vertical* **16** 46–56

[2] About ministry. Ministry of Energy of the Russian Federation [web-site]. – URL: http://minenergo.gov.ru/activity/oilgas/

[3] Est' li buduschee u sektora rossijskih nezavisimih neftyanih kompanij. Assoneft [web-site]. – URL: http://www.assoneft.ru/analytics/dictionary/

[4] Sharf I V 2012 Analiz rezultativnosti deyatelnosti neftedobivajuschih predpriyatij na primere Tomskoj oblasti *Ekonomika regiona* **3** 189–96

[5] Tax code [Electronic source] Information reference bureau "ConsultantPlus".

[6] Mazurina E V 2014 Hydrocarbons production – a source of incomes of the state and mineral developers [Electronic source] *Neftegazovaâ geologiâ. Teoriâ i practika. Electronic edition* **9** 3 URL: http://www.ngtp.ru/rub/3/33_2014.pdf

[7] Saenko V V June 5th 2014 Prospects for development of oil sector. Roundtable "Prospects for development of oil sector up to 2020" *Analytical Centre for the Government of Russian Federation*. Moscow.

[8] Grinkevich L, Sharf I and Gerasimovich G 2014 Analysis of Russian Federation budget revenues from the implementation of export duties "60-66-90" scheme [Electronic resource] *IOP Conference Series: Earth and Environmental Science* **21** 012050

[9] Bajus M 2014 Shale gas and tight oil, unconventional fossil fuels *Petroleum & Coal* **56** 206–21

[10] Bryden K, Federspie M, Habib E Th and Schiller Jr R 2014 Processing Tight Oils in FCC: Issues, Opportunities and Flexible Catalytic Solutions *Grace Catalysts Technologies Catalagram.* **114** 1–23